Statistical Decision Theory and Related Topics IV

Volume 2

Shanti S. Gupta James O. Berger
Editors

Statistical Decision Theory and Related Topics IV

Volume 2

With 24 Illustrations

Springer-Verlag
New York Berlin Heidelberg
London Paris Tokyo

Shanti S. Gupta
James O. Berger
Department of Statistics
Purdue University
West Lafayette, IN 47907
USA

AMS Classification: 62C05, 62-06

Library of Congress Cataloging-in-Publication Data
Statistical decision theory and related topics IV.
Papers from the Fourth Purdue Symposium on Statistical Decision Theory and Related Topics, held at Purdue University, June 15–20, 1986.
1. Statistical decision—Congresses. I. Gupta, Shanti Swarup. II. Berger, James O. III. Purdue Symposium on Statistical Decision Theory and Related Topics (4th : 1986 : Purdue University)
QA279.4.S744 1988 519.5′42 87-27499

Softcover reprint of the hardcover 1st edition 1988

ISBN-13: 978-1-4612-8365-2 e-ISBN-13: 978-1-4612-3818-8
DOI: 10.1007/978-1-4612-3818-8

Camera-ready copy prepared by the editors using TEX.

9 8 7 6 5 4 3 2 1

CONTENTS

CONTRIBUTORS

Numbers in parentheses indicate the pages on which the authors' contributions begin.

Hussein Al-Rawwash (225), Department of Statistics, Yarmouk University, Jordan

A. P. Basu (239), Department of Statistics, University of Missouri-Columbia, Columbia, Missouri 65211

Robert E. Bechhofer (3), School of Operations Research and Industrial Engineering, Cornell University, Ithaca, New York 14853-7501

Roger L. Berger (253), Department of Statistics, North Carolina State University, Raleigh, North Carolina 27695-8203

Peter J. Bickel (265), Department of Statistics, University of California, Berkeley, California 94720

Raymond J. Carroll (273), Department of Statistics, The University of North Carolina, Chapel Hill, North Carolina 27514

Pinyuen Chen (23), Department of Mathematics, Syracuse University, Syracuse, New York 13210

Arthur Cohen (33), Department of Statistics, Rutgers University, Rutgers, New Jersey 08903

Noel Cressie (107), Department of Statistics, Iowa State University, Ames, Iowa 50010

A. W. Davis (119), CSIRO, Division of Mathematics and Statistics, Glen Osmond, South Australia 5064

J. K. Ghosh (239), Stat./Math. Division, Indian Statistical Institute, Calcutta, India

Prem K. Goel (335), Department of Statistics, The Ohio State University, Columbus, Ohio 43210-1247

David M. Goldsman (3), Department of Industrial and Systems Engineering, Georgia Institute of Technology, Atlanta, Georgia 30332

Shanti S. Gupta (37), Department of Statistics, Purdue University, West Lafayette, Indiana 47907-2005

Jason C. Hsu (89), Department of Statistics, The Ohio State University, Columbus, Ohio 43210-1247

Wen-Tao Huang (51), Institute of Statistics, Academia Sinica, Taipei, Taiwan, Republic of China

S. N. Joshi (239), Department of Statistics, Purdue University, West Lafayette, Indiana 47907-2005

Woo-Chul Kim (89), Seoul National University, Seoul, South Korea

C. A. J. Klaassen (281), Department of Mathematics, University of Leiden, Leiden, Netherlands

Tze Leung Lai (131), Department of Statistics, Columbia University, New York, New York 10027

S. P. Lalley (145), Department of Statistics, Purdue University, West Lafayette, Indiana 47907-2005

Kinley Larntz (289), School of Statistics, University of Minnesota, St. Paul, Minnesota 55108

Lucien Le Cam (163), Department of Statistics, University of California, Berkeley, California 94720

Lii-Yuh Leu (65), Graduate Institute of Statistics, National Central University, Chung-Li, Taiwan, Republic of China

TaChen Liang (79), Department of Statistics, Purdue University, West Lafayette, Indiana 47907-2005

Yong B. Lim (351), Department of Experimental Statistics, Louisiana State University, Baton Rouge, Louisiana 70803

John I. Marden (225), Department of Statistics, University of Illinois, Urbana, Illinois 61820

Gary C. McDonald (363), Mathematics Department, General Motors Research Laboratories, Warren, Michigan 48090

H. Meister (299), Fachbereich Mathematik und Informatik, Lützowstr. 125, D-5800, Hagen 1, West Germany

Klaus J. Miescke (37), Department of Mathematics, Statistics, and Computer Science, University of Illinois at Chicago, P. O. Box 4348, Chicago, Illinois 60680

O. Moeschlin (299), Fachbereich Mathematik und Informatik, Lützowstr. 125, D-5800 Hagen 1, West Germany

Peter B. Morgan (107), Department of Economics, University of Western Ontario, London, Ontario, Canada

I. Olkin (373), Department of Statistics, Stanford University, Stanford, California 94305

S. Panchapakesan (79), Department of Mathematics, Southern Illinois University, Carbondale, Illinois 62901

Michael D. Perlman (289), Department of Statistics, University of Washington, Seattle, Washington 98195

D. Plachky (307), Inst. f. Math. Statistik, Einsteinstr. 62, 4400 Münster, West Germany

C. Radhakrishna Rao (319), Center for Multivariate Analysis, University of Pittsburgh, Pittsburgh, Pennsylvania 15260

H. E. Robbins (145), Department of Statistics, Columbia University, New York, New York 10027

Harold B. Sackrowitz (33), Department of Statistics, Rutgers University, Rutgers, New Jersey 08903

Jerome Sacks (385), Department of Statistics, University of Illinois, 1409 West Green St., Urbana, Illinois 61801

Ester Samuel-Cahn (177), Department of Statistics, Hebrew University, Jerusalem, Israel

Susannah Schiller (385), Department of Statistics, University of Illinois, 1409 West Green St., Urbana, Illinois 61801

Thomas Sellke (183), Department of Statistics, Purdue University, West Lafayette, Indiana 47907-2005

Pranab Kumar Sen (195), Biostatistics Department, School of Public Health, University of North Carolina, Chapel Hill, North Carolina 27514

Marc Sobel (23), Department of Statistics, University of Georgia, Athens, Georgia 30602

Milton Sobel (23), Department of Mathematics, University of California at Santa Barbara, Santa Barbara, California 93106

T. P. Speed (119), CSIRO Division of Mathematics and Statistics, Canberra, ACT 2601, Australia

Gunnar Stefansson (89), Marine Research Institute, Reykjavik, Iceland

W. J. Studden (351), Department of Statistics, Purdue University, West Lafayette, Indiana 47907-2005

Y. L. Tong (373), School of Mathematics, Georgia Institute of Technology, Atlanta, Georgia 30332

A. W. van der Vaart (281), Department of Mathematics, University of Leiden, Netherlands

W. R. van Zwet (281), Department of Mathematics, University of Leiden, Netherlands

Michael Woodroofe (209), Department of Statistics, University of Michigan, Ann Arbor, Michigan 48109

H. P. Wynn (351), University Statistical Laboratory, The City University, Northampton Square, London, England

Joseph Yahav (265), Department of Statistics, The Hebrew University, Jerusalem, Israel

Grace L. Yang (163), Department of Mathematics, University of Maryland, College Park, Maryland 20742

PREFACE

The Fourth Purdue Symposium on Statistical Decision Theory and Related Topics was held at Purdue University during the period June 15–20, 1986. The symposium brought together many prominent leaders and younger researchers in statistical decision theory and related areas. The 65 invited papers and discussions presented at the symposium are collected in this two-volume work.

The papers are grouped into a total of seven parts. Volume I has three parts: *Part 1 - Conditioning and Likelihood*; *Part 2 - Bayes and Empirical Bayes Analysis*; and *Part 3 - Decision Theoretic Estimation.* Part 1 contains the proceedings of a Workshop on Conditioning, which was held during the symposium. Most of the articles in Volume I involve either conditioning or Bayesian ideas, resulting in a volume of considerable interest to conditionalists and Bayesians as well as to decision-theorists.

Volume II has four parts: *Part 1 - Selection, Ranking, and Multiple Comparisons*; *Part 2 - Asymptotic and Sequential Analysis*; *Part 3 - Estimation and Testing*; and *Part 4 - Design and Comparison of Experiments and Distributions.* These articles encompass the leading edge of much current research in mathematical statistics, with decision theory, of course, receiving special emphasis. It should be noted that the papers in these two volumes are by no means all theoretical; many are applied in nature or are creative review papers.

We are extremely grateful to the administration of Purdue University, and to the School of Science, for encouragement and financial support of the symposium. The symposium was also supported by the National Science Foundation under Grant Number DMS-8514749, by the Office of Naval Research under Grant Number N00014-86-C-0050, by the Air Force Office of Scientific Research, and by the U. S. Army Research Office. We sincerely thank these agencies for their assistance, and in particular wish to thank Drs. Alvin Thaler, Peter Purdue, and Bruce Trumbo of the National Science Foundation, Dr. Major Brian Woodruff of the Air Force Office of Scientific Research, and Dr. Robert Launer of the U. S. Army Research office.

Many individuals contributed to the success of the symposium. The program for the symposium was developed under the guidance of an advisory committee composed of Lawrence D. Brown, Cornell University, Morris H. DeGroot, Carnegie-Mellon University, Persi Diaconis, Stanford University, Lucien Le Cam, University of California at Berkeley, Michael Perlman, University of Washington and Henry Wynn, The City University, London. We two acted as the co-chairmen of this committee. The excellence of the program was due in large part to the efforts of these colleagues.

Numerous faculty, staff, and students were instrumental in seeing that local arrangements and activities went smoothly. Especially invaluable in this regard were Norma Lucas and Teena Chase.

Colleagues at Purdue and elsewhere also graciously served as referees for the papers appearing in the volumes. We thank them for their efforts, which in many cases resulted in a strengthening of the papers. Finally, the superb appearance of the volumes is due to the extensive efforts of Norma Lucas, Teena Chase, and Deana Jolley, who not only did a masterful job of typing the papers, but also performed the difficult task of achieving a uniform format and appearance; Mark Senn was also invaluable in this regard, in his construction of the word processing format. To these individuals we are extremely grateful.

Shanti S. Gupta
James O. Berger

CONTENTS OF VOLUME 1

PART 1 - Selection, Ranking, and Multiple Comparisons

SEQUENTIAL SELECTION PROCEDURES FOR MULTI-FACTOR EXPERIMENTS INVOLVING KOOPMAN-DARMOIS POPULATIONS WITH ADDITIVITY

ROBERT E. BECHHOFER

School of Operations Research and Industrial Engineering
Cornell University
Ithaca, New York 14853-7501, U.S.A.

DAVID M. GOLDSMAN

Department of Industrial and Systems Engineering
Georgia Institute of Technology
Atlanta, Georgia 30332, U.S.A.

1. INTRODUCTION

The benefit achieved by factorial experimentation for ranking and selection problems in which there is *no interaction* between the (qualitative) factor-levels was pointed out by Bechhofer (1954), Section 4. That section considered a *single-stage* procedure for ranking *normal* means when the factor-level effects on the normal means are *additive*, and the common variance of the populations is assumed to be *known*. This problem was studied further by Bawa (1972) who indicated the magnitude of savings in total sample size that could be achieved in such settings by employing one multi-factor experiment in place of the corresponding independent single-factor experiments. Bawa's result was asymptotic for the situation in which the specified minimum probability of a correct selection approaches unity.

The purpose of the present article is to generalize these ideas and results in two directions: First, to point out how the *additive* model for normal means can be extended in a natural and meaningful way to other members of the Koopman-Darmois (K-D), i.e., exponential family. To that end we shall employ the notation and terminology used in Table 4.1.5 of Bechhofer, Kiefer and Sobel (1968) (referred to hereinafter as B-K-S). (In that table the normal means problem is referred to as Case A.) Second, to show how the basic sequential ranking procedure (Procedure $\mathcal{P}_B^*$) described in Section 12.3 of B-K-S, can be adapted to accommodate selection and ordering (ranking) goals for multi-factor experiments involving K-D and related frequency functions (fr.f.'s) when the *additive* model holds. We rely here heavily on the results in B-K-S, no new theoretical developments being required.

Although the procedures that we describe are applicable to p-factor experiments ($p \geqq 2$) we shall illustrate the methods only with 2-factor experiments (the generalization to multi-factor experiments being straightforward). Also, we shall focus our attention only on one goal for selection problems, and consider it for normal means, Bernoulli single-trial "success" probabilities, Poisson means and multinomial single-trial "cell" probabilities. Selection procedures for single-factor experiments involving these fr.f.'s are given in Sections 12.6.1.1, 12.6.1.4, 12.6.1.5 and 5.3.3, respectively, of B-K-S.

In Section 2 we set forth our K-D model for 2-factor experiments, and introduce our notation. We also state for arbitrary K-D fr.f.'s with "additivity" our selection goal and indifference-zone probability requirement. In Section 3 we describe our generalized sequential selection procedure for 2-factor experiments involving arbitrary K-D populations with "additivity." These procedures are then specialized in Section 4 to specific K-D and related fr.f.'s. Section 5 discusses two methods of sequential experimentation, namely one-factor-at-a-time vs. factorial experimentation, and demonstrates, via Monte Carlo (MC) sampling, the savings in expected total sample size that can result from using factorial experiments. Some concluding remarks are made in Section 6.

2. TWO-FACTOR EXPERIMENTS FOR KOOPMAN-DARMOIS FR.F.'S

For 2-*factor* experiments involving qualitative variables we assume that we have $a \cdot b$ fr.f.'s which are of the same univariate K-D or related family, i.e., for each (i, j) $(1 \leqq i \leqq a, \quad 1 \leqq j \leqq b)$ the random variable has a fr.f. of the form

$$f(x_{ij}, \theta_{ij}) = \exp\{P(x_{ij})Q(\theta_{ij}) + R(x_{ij}) + S(\theta_{ij})\} \tag{2.1}$$

for θ_{ij} in an open interval Θ; these fr.f.'s are assumed to differ at most from each other only in the values of the θ_{ij}. Here $P(x_{ij})$ and $R(x_{ij})$ do not depend on θ_{ij}, while $Q(\theta_{ij})$ and $S(\theta_{ij})$ do not depend on x_{ij}. We assume that Q is a strictly increasing function on Θ, and that $Q(\theta_{ij})$ can be expressed as

$$Q(\theta_{ij}) = \tau + \xi_i + \eta_j \qquad (1 \leqq i \leqq a, \quad 1 \leqq j \leqq b) \tag{2.2}$$

for some $\{\xi_i\}$ and $\{\eta_j\}$ satisfying $\sum_{i=1}^{a} \xi_i = \sum_{j=1}^{b} \eta_j = 0$. If (2.2) holds we shall say that the factor-level "effects" on $Q(\theta_{ij})$ are *additive*. In this article we are concerned only with the additive model. We refer to $\xi_i(\eta_j)$ as the "effect" of the ith (j^{th}) level of the first (second) factor on $Q(\theta_{ij})$. The ordered values of the ξ_i and η_j are denoted by

$$\xi_{[1]} \leqq \cdots \leqq \xi_{[a]}, \quad \eta_{[1]} \leqq \cdots \leqq \eta_{[b]}. \tag{2.3}$$

It is assumed that the values of the $\xi_i, \xi_{[i]}$ (all i) and $\eta_j, \eta_{[j]}$ (all j) are completely unknown, and that the pairing of the $\xi_{[i]}$ and $\eta_{[j]}$ with the $a \cdot b$ fr.f.'s

(2.1) is also completely unknown.

To illustrate the formulation of the problem we focus on the important goal of selecting the factor-level combination associated with $\xi_{[a]}$ and $\eta_{[b]}$. However, our methods apply equally well to the more general goals (and others) posed as Goals I and II for $t \geqq 1$ in Section 3.6.2 of B-K-S. We now formally state the objective of our experiment, and give the associated probability requirement.

Goal: To select that one of the $a \cdot b$ fr.f.'s (2.1) which is associated with $\xi_{[a]}$ and $\eta_{[b]}$. (2.4)

We limit consideration to procedures that guarantee the following indifference-zone requirement on the probability of a correct selection (CS):

Probability Requirement:

$$P\{CS\} \geqq P^* \text{ whenever } \begin{cases} \xi_{[a]} - \xi_{[a-1]} \geqq \delta_\xi^* \\ \eta_{[b]} - \eta_{[b-1]} \geqq \delta_\eta^*. \end{cases} \tag{2.5}$$

The constants $\{\delta_\xi^*, \delta_\eta^*, P^*\}$ with $(0 < \delta_\xi^*, \delta_\eta^* < \infty,\ 1/ab < P^* < 1)$ are specified prior to the start of experimentation.

The basic sequential selection procedure (Procedure $\mathcal{P}_B^*$) for K-D and related fr.f.'s and Goals I and II, which is given for *single-factor* experiments in Chapter 5 of B-K-S, takes observations a *vector-at-a-time*. Our corresponding generalized procedure (Procedure $\mathcal{P}_{B(2,+)}^*$) for *2-factor* experiments with *additivity* takes observations a *matrix-at-a-time*, i.e., $\{X_{ijs}(1 \leqq i \leqq a,\ \ 1 \leqq j \leqq b)\}$, $s = 1, 2, \ldots$. Except in Section 4.4 which deals with multinomial populations, the elements X_{ijs} $(1 \leqq i \leqq a,\ \ 1 \leqq j \leqq b)$ in each $a \cdot b$ matrix are independent for each $s = 1, 2, \ldots$, and the matrices are i.i.d.

At the mth stage of experimentation $(m = 1, 2, \ldots)$ the x_{ijs} enter our procedure only through the statistics

$$A_{im} = \sum_{j=1}^{b} \sum_{s=1}^{m} P(X_{ijs}) \quad (1 \leqq i \leqq a)$$

and (2.6)

$$B_{jm} = \sum_{i=1}^{a} \sum_{s=1}^{m} P(X_{ijs}) \quad (1 \leqq j \leqq b).$$

We denote the ordered values of the A_{im} and B_{jm} by

$$A_{[1]m} \leqq \cdots \leqq A_{[a]m}, \qquad B_{[1]m} \leqq \cdots \leqq B_{[b]m}. \tag{2.7}$$

Equal A_{im}-values and equal B_{jm}-values can occur with positive probabilities for discrete r.v.'s; when this happens they should be arranged in any definite order.

In the next section we describe our generalization of Procedure $\mathcal{P}_B^*$ for 2-factor experiments involving K-D populations with additivity.

3. SEQUENTIAL SELECTION PROCEDURE (PROCEDURE $\mathcal{P}_{B(2,+)}^*$) FOR 2-FACTOR K-D FR.F.'S WITH ADDITIVITY

BASIC SELECTION PROCEDURE (PROCEDURE $\mathcal{P}_{B(2,+)}^*$)

The Sampling Rule: At the mth stage of experimentation ($m = 1, 2, \ldots$) observe the random matrix $\{X_{ijm}(1 \leqq i \leqq a, \quad 1 \leqq j \leqq b)\}$. (3.1a)

The Stopping Rule: After the mth stage of experimentation, compute the statistic

$$Z_m = U_m + V_m + U_m \cdot V_m$$

where (3.1b)

$$\begin{cases} U_m = \sum_{i=1}^{a-1} \exp\{-\delta_\xi^*(A_{[a]m} - A_{[i]m})\} \\ V_m = \sum_{j=1}^{b-1} \exp\{-\delta_\eta^*(B_{[b]m} - B_{[j]m})\}. \end{cases}$$

Stop at the first value of m (call it n) for which $Z_n \leqq (1 - P^*)/P^*$.

The Terminal Decision Rule: Select the population associated with $A_{[a]n}$ and $B_{[b]n}$. If, because of ties between the A_{in} and/or B_{jn}, two or more populations are associated with $A_{[a]n}$ and/or $B_{[b]n}$, then select one such population at random. (3.1c)

REMARK 3.1. The stopping rule (3.1b) and the terminal decision rule (3.1c) can be derived by paralleling the development of Section 5.2.1 of B-K-S as follows: Consider the associated identification problem when

$$\xi_{[1]}^0 = \cdots = \xi_{[a-1]}^0 = \xi_{[a]}^0 - \delta_\xi^*$$

and (3.2)

$$\eta_{[1]}^0 = \cdots = \eta_{[b-1]}^0 = \eta_{[b]}^0 - \delta_\eta^*.$$

After cancelling out the common factors involving R and S, we obtain as our stopping rule: "Stop at the first value of m for which

$$\max_{\substack{1\leqq i\leqq a\\1\leqq j\leqq b}}\left[\frac{\exp\{\xi^0_{[a]}A_{im}+\eta^0_{[b]}B_{jm}+\xi^0_{[a-1]}\sum_{r\neq i}A_{rm}+\eta^0_{[b-1]}\sum_{s\neq j}B_{sm}\}}{\sum_{p=1}^{a}\sum_{q=1}^{b}\exp\{\xi^0_{[a]}A_{pm}+\eta^0_{[b]}B_{qm}+\xi^0_{[a-1]}\sum_{r\neq p}A_{rm}+\eta^0_{[b-1]}\sum_{s\neq q}B_{sm}\}}\right]$$

$$\geqq P^*,\text{"} \tag{3.3}$$

and as our terminal decision rule: "Select the population associated with the cell (i^*, j^*) which yields the maximum in the ℓ.h.s. of (3.3)." It is readily verified that the maximum of the ℓ.h.s. of (3.3) is given by the population yielding $A_{[a]n}$ and $B_{[b]n}$. Furthermore, after dividing the denominator of the ℓ.h.s. of (3.3) by the numerator, replacing $\xi^0_{[a]}-\xi^0_{[a-1]}$ by δ^*_ξ, and $\eta^0_{[b]}-\eta^0_{[b-1]}$ by δ^*_η, and then inverting both sides of equation (3.3) we obtain

$$1+\sum_{i=1}^{a-1}\exp\{-\delta^*_\xi(A_{[a]n}-A_{[i]n})\}+\sum_{j=1}^{b-1}\exp\{-\delta^*_\eta(B_{[b]n}-B_{[j]n})\} \tag{3.4}$$
$$+\sum_{i=1}^{a-1}\exp\{-\delta^*_\xi(A_{[a]n}-A_{[i]n})\}\times\sum_{j=1}^{b-1}\exp\{-\delta^*_\eta(B_{[b]n}-B_{[j]n})\}\leqq 1/P^*$$

which is equivalent to (3.1b).

REMARK 3.2. If (2.2) does *not* hold, i.e., if

$$Q(\theta_{ij})=\tau+\xi_i+\eta_j+\gamma_{ij}\qquad(1\leqq i\leqq a,\ 1\leqq j\leqq b) \tag{3.5}$$

where $\sum_{i=1}^{a}\xi_i=\sum_{j=1}^{b}\eta_j=\sum_{i=1}^{a}\gamma_{ij}=\sum_{j=1}^{b}\gamma_{ij}=0$ but $\gamma_{ij}\not\equiv 0$ (all i,j), then the factor-level "effects" on $Q(\theta_{ij})$ are *not* additive. An alternative objective which can be considered in such a case is:

Goal: To select that one of the $a\cdot b$ fr.f.'s which is associated with $\max_{\substack{1\leqq i\leqq a\\1\leqq j\leqq b}} Q(\theta_{ij})$. (3.6)

The problem is thus reduced to one involving a single factor with $a\cdot b$ levels, and Procedure $\mathcal{P}^*_B$ described in Section 12.3 of B-K-S can be used.

4. APPLICATION OF PROCEDURE $\mathcal{P}^*_{B(2,+)}$ TO SPECIFIC K-D AND RELATED FR.F.'S WITH ADDITIVITY

As pointed out in B-K-S, univariate fr.f.'s of the K-D form are widely used in applied studies. A list of six of these fr.f.'s is given in Table 4.1.5 of B-K-S, these six being referred to as Cases A–F; additional univariate fr.f.'s for which Procedure $\mathcal{P}_B^*$ can be used to solve ranking goals are given in B-K-S, Table 5.1.1. Three k-variate ($k \geqq 2$) fr.f.'s for which Procedure $\mathcal{P}_B^*$ can be used to solve ranking goals are given in Table 5.1.2 of B-K-S. The sequential selection procedure of our Section 3 is applicable to all of these fr.f.'s.

To illustrate our procedure we consider in some detail three univariate cases of Table 4.1.5, i.e., normal density functions with unknown means and a *common known* variance (Case A), Bernoulli fr.f.'s with unknown single-trial "success" probabilities (Case D), and Poisson fr.f.'s with unknown means (Case E), respectively. We also consider one multivariate case of Table 5.1.2, i.e., a single multinomial fr.f. with unknown single-trial "event" probabilities which we refer to here as Case M. To indicate that the *additivity* model is being assumed for these *2-factor* experiments, we use the notation Case A_2^+, ..., Case M_2^+.

4.1 THE NORMAL MEANS CASE (CASE A_2^+)

Let $\{X_{ijs}(1 \leqq i \leqq a,\ 1 \leqq j \leqq b)\}$ $(s = 1, 2, \ldots)$ denote independent matrix-observations from $a \cdot b$ independent normal populations Π_{ij} with unknown means θ_{ij} $(1 \leqq i \leqq a,\ 1 \leqq j \leqq b)$ and a *common known* variance σ^2 which without loss of generality we take to be unity. The fr.f. of X_{ij} is

$$f(x_{ij}, \theta_{ij}) = \frac{1}{\sqrt{2\pi}} \exp\{-\frac{1}{2}(x_{ij} - \theta_{ij})^2\}.$$

Here $P(x_{ij}) = x_{ij}, Q(\theta_{ij}) = \theta_{ij}, R(x_{ij}) = -\frac{1}{2}(\log 2\pi + x_{ij}^2)$ and $S(\theta_{ij}) = -\theta_{ij}^2/2$. Let $\mu = \tau$, $\alpha_i = \xi_i$ $(1 \leqq i \leqq a)$ and $\beta_j = \eta_j (1 \leqq j \leqq b)$. Then (2.2) becomes

$$\theta_{ij} = \mu + \alpha_i + \beta_j \tag{4.2}$$

which is the standard *additive (no interaction)* fixed-effects ANOVA model (qualitative variables). Since $\sigma^2 = 1$, the probability requirement can be stated as

$$P\{CS\} \geqq P^* \text{ whenever } \begin{cases} \alpha_{[a]} - \alpha_{[a-1]} \geqq \delta_\alpha^* \\ \beta_{[b]} - \beta_{[b-1]} \geqq \delta_\beta^* \end{cases} \tag{4.3}$$

where $\delta_\alpha^* = \delta_\xi^*, \delta_\beta^* = \delta_\eta^*$ and $\alpha_{[1]} \leqq \cdots \leqq \alpha_{[a]}, \beta_{[1]} \leqq \cdots \leqq \beta_{[b]}$ are the ordered values of the "effects" α_i $(1 \leqq i \leqq a)$ and $\beta_j (1 \leqq j \leqq b)$, respectively. The constants $\{\delta_\alpha^*, \delta_\beta^*; P^*\}$ with $(0 < \delta_\alpha^*, \delta_\beta^* < \infty, 1/ab < P^* < 1)$ are specified prior to the start of experimentation.

If observations are taken a matrix-at-a-time, then $A_{im} = \sum_{j=1}^{b} \sum_{s=1}^{m} X_{ijs}$, $B_{jm} = \sum_{i=1}^{a} \sum_{s=1}^{m} X_{ijs}$, and $A_{[i]m}(1 \leqq i \leqq a)$, $B_{[j]m}(1 \leqq j \leqq b)$ are defined by (2.7). The stopping rule is then given by (3.1b) and the terminal decision rule by (3.1c).

REMARK 4.1. For Case A_2^+ the common variance σ^2 is assumed to be *known*.

If the additive model (4.2) is valid but σ^2 is *unknown* then Procedure $\mathcal{P}^*_{B(2,+)}$ *will not* guarantee (4.3). A *two-stage* procedure that will guarantee (4.3) when (4.2) is valid and σ^2 is *unknown* has been proposed by Bechhofer and Dunnett (1986).

4.2 The Bernoulli Case (Case D_2^+)

Let $\{X_{ijs}(1 \leqq i \leqq a, 1 \leqq j \leqq b)\}(s = 1, 2, \ldots)$ denote independent matrix-observations from $a \cdot b$ independent Bernoulli populations Π_{ij} with unknown single-trial "success" probabilities θ_{ij} $(1 \leqq i \leqq a, 1 \leqq j \leqq b)$. When observations are taken a matrix-at-a-time, then $X_{ijs} = 1$ or 0 according as Π_{ij} yields a success or a failure on the sth matrix observation. The fr.f. of X_{ij} is

$$f(x_{ij}, \theta_{ij}) = \theta_{ij}^{x_{ij}}(1-\theta_{ij})^{1-x_{ij}} = \exp\{x_{ij}\log[\theta_{ij}/(1-\theta_{ij})] + \log(1-\theta_{ij})\}. \quad (4.4)$$

Here $P(x_{ij}) = x_{ij}, Q(\theta_{ij}) = \log[\theta_{ij}/(1-\theta_{ij})], R(x_{ij}) = 0$ and $S(\theta_{ij}) = \log(1-\theta_{ij})$.

If we assume the log linear (linear logistic) model

$$\log[\theta_{ij}/(1-\theta_{ij})] = \tau + \xi_i + \eta_j \qquad (\text{all } i, j) \qquad (4.5)$$

where $\sum_{i=1}^{a} \xi_i = \sum_{j=1}^{b} \eta_j = 0$, then the factor-level "effects" are *additive*. Define $\underset{\sim}{\alpha} = (\alpha_1, \ldots, \alpha_a)$ and $\underset{\sim}{\beta} = (\beta_1, \ldots, \beta_b)$ by $\xi_i = \log[\alpha_i/(1-\alpha_i)]$ and $\eta_j = \log[\beta_j/(1-\beta_j)]$, respectively. The model (4.5) has been employed by among others, Cox (1970), Sections 2.3-2.7, 6.1–6.5, Feinberg (1977), Sections 6.2-6.5, Haberman (1978), Chapter 5, and Breslow and Day (1980), Sections 6.1–6.4.

If observations are taken a matrix-at-a-time, then $A_{im} = \sum_{j=1}^{b}\sum_{s=1}^{m} X_{ijs}$, $B_{jm} = \sum_{i=1}^{a}\sum_{s=1}^{m} X_{ijs}$, and $A_{[i]m}(1 \leqq i \leqq a)$, $B_{[j]m}(1 \leqq j \leqq b)$ are defined by (2.7), and $\xi_{[i]}(1 \leqq i \leqq a)$ and $\eta_{[j]}(1 \leqq j \leqq b)$ are defined by (2.3).

For this model the distance function can be expressed as

$$\begin{aligned} \xi_{[a]} - \xi_{[a-1]} &= \log\left\{\frac{\alpha_{[a]}}{1-\alpha_{[a]}}\right\} - \log\left\{\frac{\alpha_{[a-1]}}{1-\alpha_{[a-1]}}\right\} \\ \eta_{[b]} - \eta_{[b-1]} &= \log\left\{\frac{\beta_{[b]}}{1-\beta_{[b]}}\right\} - \log\left\{\frac{\beta_{[b-1]}}{1-\beta_{[b-1]}}\right\}. \end{aligned} \qquad (4.6)$$

Since Q is a strictly increasing function, the goal (2.4), probability requirement (2.5), and procedure (3.1) can be stated equivalently as

Goal: To select that one of the $a \cdot b$ fr.f.'s. which is associated with $\alpha_{[a]}$ and $\beta_{[b]}$. (4.7)

Probability Requirement:

$$P\{CS\} \geqq P^* \text{ whenever } \begin{cases} \frac{\alpha_{[a]}}{1-\alpha_{[a]}} \Big/ \frac{\alpha_{[a-1]}}{1-\alpha_{[a-1]}} \geqq \theta^*_\alpha \\ \frac{\beta_{[b]}}{1-\beta_{[b]}} \Big/ \frac{\beta_{[b-1]}}{1-\beta_{[b-1]}} \geqq \theta^*_\beta \end{cases} \tag{4.8}$$

where $\theta^*_\alpha = e^{\delta^*_\xi}$ and $\theta^*_\beta = e^{\delta^*_\eta}$. The constants $\{\theta^*_\alpha, \theta^*_\beta; P^*\}$ with $(1 < \theta^*_\alpha, \theta^*_\beta < \infty$, $1/ab < P^* < 1)$ are specified prior to the start of experimentation. Note that the probability requirement is now stated in terms of the *"odds ratios"* of the α_i's and of the β_j's. The stopping rule is then given by (3.1b) with

$$U_m = \sum_{i=1}^{a-1} \left(\frac{1}{\theta^*_\alpha}\right)^{A_{[a]m} - A_{[i]m}}, \qquad V_m = \sum_{j=1}^{b-1} \left(\frac{1}{\theta^*_\beta}\right)^{B_{[b]m} - B_{[j]m}}. \tag{4.9}$$

4.3 THE POISSON CASE (CASE E_2^+)

Let $\{X_{ijs}(1 \leqq i \leqq a, 1 \leqq j \leqq b)\}$ $s = 1, 2, \ldots$ denote independent matrix-observations from $a \cdot b$ independent Poisson populations Π_{ij} with unknown means θ_{ij} $(1 \leqq i \leqq a, 1 \leqq j \leqq b)$. The fr.f. of X_{ij} is

$$f(x_{ij}, \theta_{ij}) = \theta_{ij}^{x_{ij}} \exp\{-\theta_{ij}\}/x_{ij}! = \exp\{x_{ij} \log \theta_{ij} - \log(x_{ij}!) - \theta_{ij}\}. \tag{4.10}$$

Here $P(x_{ij}) = x_{ij}, Q(\theta_{ij}) = \log \theta_{ij}$, $R(x_{ij}) = -\log(x_{ij}!)$ and $S(\theta_{ij}) = -\theta_{ij}$.

If we assume the log linear model

$$\log\ \theta_{ij} = \tau + \xi_i + \eta_j \qquad \text{all } (i, j) \tag{4.11}$$

where $\sum_{i=1}^{a} \xi_i = \sum_{j=1}^{b} \eta_j = 0$, then the factor-level effects are *additive*. Define $\underset{\sim}{\alpha}$ and $\underset{\sim}{\beta}$ by $\xi_i = \log \alpha_i$ and $\eta_j = \log \beta_j$, respectively. The model (4.11) was employed by, among others, Rasch (1960), pp. 13–23 and Haberman (1978), Section 2.5.

If observations are taken a matrix-at-a-time, then A_{im}, $A_{[i]m}$ $(1 \leqq i \leqq a)$, B_{jm}, $B_{[j]m}$ $(1 \leqq j \leqq b)$ are defined as for Case D_2^+, and $\xi_{[i]}$ $(1 \leqq i \leqq a)$, $\eta_{[j]}$ $(1 \leqq j \leqq b)$ are defined as in (2.3). The goal is given by (4.7), and the probability requirement (2.5) can be stated equivalently as

Probability Requirement:

$$P\{CS\} \geqq P^* \text{ whenever } \begin{cases} \frac{\alpha_{[a]}}{\alpha_{[a-1]}} \geqq \theta^*_\alpha \\ \frac{\beta_{[b]}}{\beta_{[b-1]}} \geqq \theta^*_\beta \end{cases} \tag{4.12}$$

where θ^*_α and θ^*_β are defined as below (4.8). The stopping rule is then given by (3.1b) with U_m and V_m given by (4.9).

4.4 The Multinomial Case (Case M_2^+)

This example does not have the structure (2.1). Nevertheless Procedure $\mathcal{P}^*_{B(2,+)}$ of Section 3 is applicable. Let $\{X_{ijs}(1 \leqq i \leqq a, 1 \leqq j \leqq b)\}$ $s = 1, 2, \ldots$ denote independent matrix-observations from a single $(a \cdot b)$-variate multinomial population having an unknown probability matrix $\{\theta_{ij}\} = \{\theta_{ij}(1 \leqq i \leqq a, 1 \leqq j \leqq b)\}$. Here $\theta_{ij}(0 \leqq \theta_{ij} \leqq 1, \sum_{i=1}^{a}\sum_{j=1}^{b} \theta_{ij} = 1)$ denotes the probability that $X_{ijs} = 1$ $(s = 1, 2, \ldots)$ on the sth matrix observation and 0 otherwise with $\sum_{i=1}^{a}\sum_{j=1}^{b} X_{ijs} = 1$ for all s.

The $a \cdot b$ variate fr.f. which is bivariate K-D is

$$f(\{x_{ij}\}, \{\theta_{ij}\}) = \prod_{i=1}^{a}\prod_{j=1}^{b} \theta_{ij}^{x_{ij}} = \prod_{i=1}^{a}\prod_{j=1}^{b} e^{x_{ij}\log\theta_{ij}}. \tag{4.13}$$

Here $P(x_{ij}) = x_{ij}$, $Q(\theta_{ij}) = \log\theta_{ij}$ and $R(x_{ij}) = S(\theta_{ij}) = 0$. Define $\alpha_i = \sum_{j=1}^{b} \theta_{ij} \geqq 0$ $(1 \leqq i \leqq a)$ and $\beta_j = \sum_{i=1}^{a} \theta_{ij} \geqq 0$ $(1 \leqq j \leqq b)$; then $\sum_{i=1}^{a} \alpha_i = \sum_{j=1}^{b} \beta_j = 1$. If we assume the log linear model

$$\log\theta_{ij} = \xi_i + \eta_j \qquad \text{(all } i, j) \tag{4.14}$$

where $\xi_i = \log\alpha_i$, $\eta_j = \log\beta_j$, then $\{\sum_{j=1}^{b} X_{ijs}(1 \leqq i \leqq a)\}$ $(s = 1, 2, \ldots)$ can be regarded as independent vector-observations from a single a-variate $(a \geqq 2)$ multinomial population with an unknown probability vector $\underset{\sim}{\alpha}$. A similar interpretation can be given to $\{\sum_{i=1}^{a} X_{ijs}(1 \leqq j \leqq b)\}$ $(s = 1, 2, \ldots)$ and $\underset{\sim}{\beta}$. Then (4.14) is equivalent to assuming the *multiplicative* model

$$\theta_{ij} = \alpha_i \cdot \beta_j \qquad \text{(all } i, j) \tag{4.15}$$

wherein the factor-levels for each of the two factors are independent. This model is being studied by Bechhofer and Goldsman (1986b) in the context of *single-stage* multinomial selection procedures for multi-factor experiments. It had been studied earlier by, among others, Feinberg (1977), Chapter 2, Upton (1978), Sections 3.1–3.7, and Plackett (1974), Sections 5.1–5.3, mostly from the viewpoint of estimation and hypothesis testing.

Proceeding as in Sections 4.2 and 4.3, we see that the goal (2.4) and probability requirement (2.5) can be stated equivalently as (4.7) and (4.12), respectively, where the α_i $(1 \leqq i \leqq a)$ and $\beta_j(1 \leqq j \leqq b)$ are defined as above, and the $\alpha_{[i]}$ and $\beta_{[j]}$ have the usual interpretation with θ^*_α and θ^*_β defined as below (4.8). The U_m and V_m of (3.1b) are defined as in (4.9).

5. SAVINGS IN EXPECTED TOTAL SAMPLE SIZE RESULTING FROM FACTORIAL EXPERIMENTATION

The analysis in this section wherein the emphasis is on expected total *sample size* is relevant for Cases A_p^+, D_p^+ and E_p^+ but *not* for Case M_p^+; for the latter the corresponding emphasis would be on expected total number of *stages*.

5.1 ONE-FACTOR-AT-A-TIME VS. FACTORIAL EXPERIMENTATION

In this section we compare the expected total sample sizes resulting from two possible methods of experimentation both of which guarantee the *same* probability requirement (2.5).

1) In the first method of experimentation two *independent* single-factor experiments are conducted, one of which guarantees the probability requirement

$$P\{CS\} \geqq P_1^* \text{ whenever } \xi_{[a]} - \xi_{[a-1]} \geqq \delta_\xi^*, \tag{5.1a}$$

while the second guarantees

$$P\{CS\} \geqq P_2^* \text{ whenever } \eta_{[b]} - \eta_{[b-1]} \geqq \delta_\eta^*. \tag{5.1b}$$

The constants $\{\delta_\xi^*, P_1^*\}$, $\{\delta_\eta^*, P_2^*\}$ with $(0 < \delta_\xi^* < \infty, 1/a < P_1^* < 1)$, $(0 < \delta_\eta^* < \infty, 1/b < P_2^* < 1)$ and $P_1^* \cdot P_2^* = P^*$ with $(1/ab < P^* < 1)$ are specified prior to the start of experimentation.

REMARK 5.1. This method of experimentation was suggested by the results reported for Case A_2^+ in Section 4 of Bechhofer (1954). There the $P\{CS\}$ of our (2.5) factored into the product of the $P\{CS\}$ of our (5.1a) and the $P\{CS\}$ of our (5.1b). Of course this factorization is a consequence of the independence present in Case A_2^+.

For each of the two single-factor experiments we employ Procedure $\mathcal{P}_B^*$ as described in Section 12.3 of B-K-S using $\{\delta_\xi^*, P_1^*\}$ for the first experiment and $\{\delta_\eta^*, P_2^*\}$ for the second. Since the experiments are independent, it is clear that the two, regarded as a combined experiment, will guarantee (2.5). Let $n_1'(a, \delta_\xi^*, P_1^*)$ and $n_1''(b, \delta_\eta^*, P_2^*)$ denote the random number of *stages* to terminate sampling for the first and second experiment, respectively. These depend on the $\xi_{[a]} - \xi_{[i]}$ $(1 \leqq i \leqq a-1)$ and the $\eta_{[b]} - \eta_{[j]}$ $(1 \leqq j \leqq b-1)$; in our present study we shall restrict consideration to the so-called *least-favorable* (LF) configurations, i.e., $\xi_{[a]} - \xi_{[i]} = \delta_\xi^*$ $(1 \leqq i \leqq a-1)$ and $\eta_{[b]} - \eta_{[j]} = \delta_\eta^*$ $(1 \leqq j \leqq b-1)$. For these configurations we will be interested in $a \cdot E\{n_1'(a, \delta_\xi^*, P_1^*)|LF\}$ $+ \, b \cdot E\{n_1''(b, \delta_\eta^*, P_2^*)|LF\}$, the expected total number of *observations* to terminate sampling using Procedure $\mathcal{P}_B^*$ with the two independent single-factor experiments. The optimal choice of (P_1^*, P_2^*) to minimize this quantity depends on the form of the fr.f., and on $(a, \delta_\xi^*; b, \delta_\eta^*)$. For the *completely symmetric* setup in which $a = b = k$ (say) and $\delta_\xi^* = \delta_\eta^* = \delta^*$ (say) it is clear that one should set $P_1^* = P_2^* = \sqrt{P^*}$. Then the expected total number of *observations* reduces to

$2kE\{n'_1(k,\delta^*,\sqrt{P^*})|LF\} = 2kE\{n_1|LF\}$ (say).

2) In the second method of experimentation one 2-factor experiment is conducted as described in Section 3. Let $n_2(a,b,\delta^*_\xi,\delta^*_\eta;P^*|LF)$ denote the random number of *stages* to terminate sampling for this 2-factor experiment when $\xi_{[a]}-\xi_{[i]} = \delta^*_\xi$ $(1 \leqq i \leqq a-1)$ and $\eta_{[b]}-\eta_{[j]} = \delta^*_\eta$ $(1 \leqq j \leqq b-1)$. To obtain direct comparability with the setup described in 1), above, we again consider the situation in which $a = b = k$ and $\delta^*_\xi = \delta^*_\eta = \delta^*$. Here we will be interested in the expected total number of *observations*, $k^2E\{n_2(k,k,\delta^*,\delta^*;P^*)|LF\} = k^2E\{n_2|LF\}$ (say). Then

$$RE(k,\delta^*,P^*|LF) = \frac{k^2E\{n_2|LF\}}{2kE\{n_1|LF\}} \tag{5.2}$$

can be regarded as a relative efficiency (RE) measure associated with the two methods of experimentation.

REMARK 5.2. Bawa (1972) showed that if the *single-stage* procedures of Bechhofer (1954) for Case A_2^+ are employed in the completely symmetric setup, then asymptotically $(P^* \to 1)$, the RE (5.2) is equal to 1/2, *uniformly* in (k,δ^*), i.e., a 2-factor experiment which guarantees (4.3) requires one-half as many total observations on-the-average as do two independent single-factor experiments when the combined single-factor experiments also guarantee the probability requirement (4.3). More generally, in the completely symmetric setup for Case A_p^+, the asymptotic RE is equal to $1/p$ where p is the number of factors.

5.2 MONTE CARLO SAMPLING RESULTS

We used Monte Carlo (MC) sampling to obtain estimates of RE$(k,\delta^*, P^*|LF)$ for Case A_2^+ of Section 4.1 for selected (k,P^*) with $\delta^* = 0.2$. These estimates which are given in Table 5.3 are based on the corresponding estimates of $E\{n_1|LF\}$ and $E\{n_2|LF\}$ in Tables 5.1 and 5.2, respectively. Both of these latter tables also contain estimates of the *achieved* probability of a correct selection, $P\{CS_1|LF\}$ and $P\{CS_2|LF\}$, for the corresponding procedures. As in Tables 18.4.1–18.4.6 of B-K-S, two estimates of these probabilities are given, the first being simply the observed proportion of correct selections while the second is the average value of $W_{[k]n_1}$ (or $W_{[k^2]n_2}$) at termination (defined by (12.1.10) of B-K-S for single-factor experiments); both are unbiased estimates of the achieved probability of a correct selection, the estimate based on $\bar{W}$ having much smaller variance than the one based on the observed proportion. All of the entries in a given row of Table 5.1 (Table 5.2) are based on the same 4000 independent replications of Procedure $\mathcal{P}_B^*$ (Procedure $\mathcal{P}^*_{B(2,+)}$) for the single-factor (two-factor) experiment. The number in parentheses below each entry is the estimated standard error of the estimate above it. The entries in Table 5.1 for k =2, 3, 5 and 10 are directly comparable to the corresponding entries for

$P^* = 0.75$, 0.90, 0.95 and 0.99 given in Tables 18.4.1, 18.4.2, 18.4.4 and 18.4.5, respectively, of B-K-S; these provided a partial check on the correctness of our present computer simulation program. The B-K-S entries are less precise being based on only 800 replications. The entries in Table 5.3 are derived from those in Tables 5.1 and 5.2.

We note from Tables 5.1 and 5.2 that the achieved $P\{CS|LF\}$ is estimated with very high precision by $\bar{W}_{[k]n_1}$ and $\bar{W}_{[k^2]n_2}$; for both methods of experimentation the $\bar{W}$ estimate always exceeds P^*, and the magnitude of the "excess" (which is quite large for $P^* = 0.75$ and $(0.75)^2$) decreases with increasing P^*. From Table 5.3 we see that for every (k, P^*) combination we have $\bar{W}_{[k^2]n_2} > (\bar{W}_{[k]n_1})^2$, i.e., the 2-factor experiment *achieves* a higher $P\{CS|LF\}$ than does the combined independent single-factor experiment. This superiority in achieved $P\{CS|LF\}$ is purchased at the cost of an increase in expected total sample size for Procedure $\mathcal{P}^*_{B(2,+)}$.

We also point out that for fixed P^* =0.9025, 0.9801 or 0.998001, the expected number of *stages* to termination $(E\{n_2|LF)\}$ decreases rapidly with increasing k for Procedure $\mathcal{P}^*_{B(2,+)}$ (Table 5.2). This consequence of matrix-at-a-time sampling could have practical advantages in certain situations.

From Table 5.3 we see that for each k the MC estimate of $RE(k, \delta^*, P^*|LF)$ tends to decrease (very slowly) with increasing P^*; it can be shown analytically (by using *mean path* approximations as in Section 14.2.2 of B-K-S) that $\lim_{P^* \to 1} RE(k, \delta^*, P^*|LF) = 1/2$, the result reported by Bawa (1972) for the Case A_2^+ *single-stage* procedures in the symmetric setup.

REMARK 5.3. We have conducted additional MC samplings (the detailed results of which are not reported herein) in order to study the rate of approach of $RE(k, \delta^*, P^*|LF)$ to 1/2 for P^* approaching unity. These were carried out for $k = 2$ and can be regarded as extensions of the results given in Table 5.3. For $P^* = (1 - 10^{-5})^2$, $(1 - 10^{-7})^2$ and $(1 - 10^{-9})^2$ we obtained 0.587, 0.574 and 0.564, respectively.

If the *achieved* $P\{CS|LF\}$ for the two methods of experimentation were equal (which could be accomplished by adjusting downward the P^* used on the r.h.s. of the equation for Z_n in (3.1b)), then this would decrease $E\{n_2|LF\}$ and hence make the RE measure (5.2) somewhat closer to 1/2. However, even without any adjustment, we see from Table 5.3 that for Case A_2^+ over the range of (k, P^*) studied, the expected total sample size using Procedure $\mathcal{P}^*_{B(2,+)}$ is only approximately 60 to 65 percent of that which would be required for two independent single-factor experiments each of which used Procedure $\mathcal{P}^*_B$. We would anticipate analogous results for Cases D_2^+ and E_2^+ although the rate of approach to 1/2 with increasing P^* can be anticipated to depend on the fr.f..

Table 5.1

Monte Carlo Estimates of $P\{CS_1|LF\}$ and $E\{n_1|LF\}$ obtained with Procedure $\mathcal{P}_B^*$ for single-factor Case A_1^+ experiments with selected $k = a = b$ and P^* when $\delta^* = 0.2$

k	P^*	Estimate of $P\{CS_1\|LF\}$: Observ. Proportion of Correct Selections	Estimate of $P\{CS_1\|LF\}$: $\bar{W}_{[k]n_1}$	Estimate of $E\{n_1\|LF\}$: $\bar{n}_1(LF)$
2	0.75	0.7762 (0.0066)	0.7794 (0.0004)	17.6 (0.2)
	0.90	0.9100 (0.0045)	0.9138 (0.0002)	48.8 (0.6)
	0.95	0.9538 (0.0033)	0.9572 (0.0001)	70.6 (0.8)
	0.99	0.9925 (0.0014)	0.9915 (0.00002)	117.0 (1.2)
	0.999	0.9988 (0.0005)	0.99915 (<0.00001)	178.4 (1.5)
3	0.75	0.7650 (0.0067)	0.7766 (0.0003)	38.9 (0.4)
	0.90	0.9020 (0.0047)	0.9129 (0.0002)	77.6 (0.8)
	0.95	0.9568 (0.0032)	0.9569 (0.0001)	102.6 (1.0)
	0.99	0.9898 (0.0016)	0.9914 (0.00002)	153.9 (1.3)
	0.999	0.9988 (0.0005)	0.99915 (<0.00001)	218.8 (1.6)

Table 5.1 (continued)

Monte Carlo Estimates of $P\{CS_1|LF\}$ and $E\{n_1|LF\}$ obtained with Procedure $\mathcal{P}_B^*$ for single-factor Case A_1^+ experiments with selected $k = a = b$ and P^* when $\delta^* = 0.2$

k	P^*	Estimate of $P\{CS_1\|LF\}$: Observ. Proportion of Correct Selections	Estimate of $P\{CS_1\|LF\}$: $\bar{W}_{[k]n_1}$	Estimate of $E\{n_1\|LF\}$: $\bar{n}_1(LF)$
5	0.75	0.7762 (0.0066)	0.7759 (0.0003)	64.3 (0.6)
	0.90	0.9135 (0.0044)	0.9124 (0.0002)	109.5 (1.0)
	0.95	0.9580 (0.0032)	0.9566 (0.0001)	136.5 (1.2)
	0.99	0.9898 (0.0016)	0.9914 (0.00002)	196.0 (1.5)
	0.999	0.9995 (0.0004)	0.99914 (<0.00001)	262.1 (1.7)
10	0.75	0.7665 (0.0067)	0.7742 (0.0003)	97.2 (0.8)
	0.90	0.9075 (0.0046)	0.9117 (0.0002)	149.8 (1.2)
	0.95	0.9585 (0.0032)	0.9564 (0.0001)	178.0 (1.3)
	0.99	0.9913 (0.0015)	0.9913 (0.00002)	238.4 (1.7)
	0.999	0.9990 (0.0005)	0.99913 (<0.00001)	313.8 (1.9)

All estimates in this table are based on 4,000 replications. The numbers in parentheses are the estimated standard errors of the estimates above them.

Table 5.2

Monte Carlo Estimates of $P\{CS_2|LF\}$ and $E\{n_2|LF\}$ obtained with Procedure $\mathcal{P}^*_{B(2,+)}$ for 2-factor Case A_2^+ experiments with selected $k = a = b$ and P^* when $\delta^* = 0.2$

k	P^*	Estimate of $P\{CS_2|LF\}$		Estimate of $E\{n_2|LF\}$
		Observ. Proportion of Correct Selections	$\bar{W}_{[k^2]n_2}$	$\bar{n}_2(LF)$
2	0.5625 $= (0.75)^2$	0.6028 (0.0077)	0.6102 (0.0006)	11.1 (0.1)
	0.8100 $= (0.90)^2$	0.8388 (0.0058)	0.8383 (0.0004)	31.5 (0.3)
	0.9025 $= (0.95)^2$	0.9185 (0.0043)	0.9188 (0.0002)	45.0 (0.4)
	0.9801 $= (0.99)^2$	0.9825 (0.0021)	0.9838 (0.00004)	73.5 (0.6)
	0.998001 $= (0.999)^2$	0.9990 (0.0005)	0.99838 (<0.00001)	107.9 (0.7)
3	0.5625 $= (0.75)^2$	0.6200 (0.0077)	0.6155 (0.0007)	16.3 (0.1)
	0.8100 $= (0.90)^2$	0.8390 (0.0058)	0.8424 (0.0004)	32.4 (0.3)
	0.9025 $= (0.95)^2$	0.9270 (0.0041)	0.9213 (0.0002)	43.2 (0.3)
	0.9801 $= (0.99)^2$	0.9815 (0.0021)	0.9843 (0.00005)	63.6 (0.4)
	0.998001 $= (0.999)^2$	0.9983 (0.0007)	0.99844 (<0.00001)	88.8 (0.5)

Table 5.2 (continued)

Monte Carlo Estimates of $P\{CS_2|LF\}$ and $E\{n_2|LF\}$ obtained with Procedure $\mathcal{P}^*_{B(2,+)}$ for 2-factor Case A_2^+ experiments with selected $k = a = b$ and P^* when $\delta^* = 0.2$

k	P^*	Estimate of $P\{CS_2\|LF\}$: Observ. Proportion of Correct Selections	Estimate of $P\{CS_2\|LF\}$: $\bar{W}_{[k^2]n_2}$	Estimate of $E\{n_2\|LF\}$: $\bar{n}_2(LF)$
5	0.5625 $= (0.75)^2$	0.6328 (0.0076)	0.6291 (0.0009)	16.4 (0.1)
	0.8100 $= (0.90)^2$	0.8525 (0.0056)	0.8511 (0.0005)	27.6 (0.2)
	0.9025 $= (0.95)^2$	0.9245 (0.0042)	0.9252 (0.0003)	34.1 (0.2)
	0.9801 $= (0.99)^2$	0.9830 (0.0020)	0.9853 (0.0001)	47.5 (0.3)
	0.998001 $= (0.999)^2$	0.9973 (0.0008)	0.99853 (<0.00001)	63.0 (0.3)
10	0.5625 $= (0.75)^2$	0.6553 (0.0075)	0.6540 (0.0011)	12.7 (0.1)
	0.8100 $= (0.90)^2$	0.8520 (0.0056)	0.8640 (0.0006)	19.0 (0.1)
	0.9025 $= (0.95)^2$	0.9363 (0.0039)	0.9332 (0.0003)	22.6 (0.1)
	0.9801 $= (0.99)^2$	0.9880 (0.0017)	0.9867 (0.0001)	29.4 (0.2)
	0.998001 $= (0.999)^2$	0.9973 (0.0008)	0.99870 (<0.00001)	37.8 (0.2)

All estimates in this table are based on 4,000 replications. The numbers in parentheses are the estimated standard errors of the estimates above them.

Table 5.3

Monte Carlo Estimates of $P\{CS|(k,\delta^*,P^*|LF)\}$ and $RE(k,\delta^*,P^*|LF)$ for case A_2^+ with selected $k=a=b$ and P^* when $\delta^*=0.2$

k	P^*	Estimates of $P\{CS\|(k,\delta^*,P^*\|LF)\}$		Estimate of $RE(k,\delta^*,P^*\|LF)$
		$(\bar{W}_{[k]n_1})^2$	$\bar{W}_{[k^2]n_2}$	$k^2\bar{n}_2(LF)/2k\bar{n}_1(LF)$
2	0.5625	0.6075	0.6102	0.631
	0.8100	0.8350	0.8383	0.645
	0.9025	0.9162	0.9188	0.637
	0.9801	0.9831	0.9838	0.628
	0.998001	0.9983	0.9984	0.605
3	0.5625	0.6031	0.6155	0.629
	0.8100	0.8334	0.8424	0.626
	0.9025	0.9157	0.9213	0.632
	0.9801	0.9829	0.9843	0.620
	0.998001	0.9983	0.9984	0.609
5	0.5625	0.6020	0.6291	0.638
	0.8100	0.8325	0.8511	0.630
	0.9025	0.9151	0.9252	0.625
	0.9801	0.9829	0.9853	0.606
	0.998001	0.9982	0.9985	0.601
10	0.5625	0.5994	0.6540	0.653
	0.8100	0.8312	0.8640	0.634
	0.9025	0.9147	0.9332	0.635
	0.9801	0.9827	0.9867	0.617
	0.998001	0.9982	0.9987	0.602

6. CONCLUDING REMARKS

We have proposed a generalized version of Procedure $\mathcal{P}_B^*$ of B-K-S which can be used for selection and ordering (ranking) problems arising from multi-factor experiments involving K-D (and related) fr.f.'s which possess a certain additivity property. This procedure can be used for p-factor ($p \geqq 2$) experiments having complete additivity, and in particular for normal means with a common known variance (Case A_p^+) and many of the standard log linear models arising in categorical data analysis, e.g., the Bernoulli (Case D_p^+), Poisson (Case E_p^+) and multinomial (Case M_p^+).

We have demonstrated conclusively using MC sampling experiments that this generalized procedure, Procedure $\mathcal{P}_{B(p,+)}^*$, for p-factor experiments with additivity when applied to Case A_p^+ for $p = 2$, requires a substantially smaller expected total number of observations than that which is required by two independent single-factor experiments when both methods of experimentation guarantee the same requirement on the probability of a correct selection. In addition, the 2-factor experiment achieves a higher probability of a correct selection than do the combined two independent single-factor experiments.

Although our MC studies were only for Case A_2^+ in symmetric settings we anticipate savings (although not as great) in non-symmetric settings, with the same qualitative results for $p > 2$ and other K-D fr.f.'s with additivity.

Procedure $\mathcal{P}_B^*$ and Procedure $\mathcal{P}_{B(p,+)}^*$ are open procedures. The values of $E\{n_1\}$ and $E\{n_2\}$ as well as $\text{Var}\{n_1\}$ and $\text{Var}\{n_2\}$ can be decreased by *truncation* of the procedures for Case A_p^+ ($p \geqq 1$). Truncation of Procedure $\mathcal{P}_B^*$ for single-factor experiments involving the multinomial has been carried out by Bechhofer and Goldsman (1985), (1986a) with the objective of decreasing $E\{\text{Stages}\}$ and $\text{Var}\{\text{Stages}\}$. Such truncation studies can also be carried out for Case M_p^+ ($p > 1$). These problems are currently under investigation by the authors.

ACKNOWLEDGMENTS

The authors are greatly indebted to Professors Thomas Santner and Ajit Tamhane who read earlier versions of this article and made many constructive comments. This research was partially supported by the U.S. Army Research Office through the Mathematical Sciences Institute of Cornell University and by U.S. Army Research Office Contract DAAL03-86-K-0046 at Cornell University.

BIBLIOGRAPHY

Bawa, V. S. (1972). Asymptotic efficiency of one R-factor experiment relative to R one-factor experiments for selecting the best normal population. *Journal of the American Statistical Association*, **67**, 660–661.

Bechhofer, R. E. (1954). A single-sample multiple decision procedure for ranking means of normal populations with known variances. *Annals of Mathematical Statistics*, **25**, 16–39.

Bechhofer, R. E. and Dunnett, C. W. (1986). Two-stage selection of the best factor-level combination in multi-factor experiments: common unknown variance. *Statistical Design – Theory and Practice*, Proceedings in honor of Professor Walter T. Federer, Biometrics Unit, Cornell University.

Bechhofer, R. E. and Goldsman, D. M. (1985). Truncation of the Bechhofer-Kiefer-Sobel sequential procedure for selecting the multinomial event which has the largest probability. *Communications in Statistics — Simulation and Computation*, **14**, 283–315.

Bechhofer, R. E. and Goldsman, D. M. (1986a). Truncation of the Bechhofer-Kiefer-Sobel sequential procedure for selecting the multinomial event which has the largest probability (II): Extended tables and an improved procedure. *Communications in Statistics — Simulation and Computation*, **15**, 829–851.

Bechhofer, R. E. and Goldsman, D. M. (1986b). Single-stage selection procedures for multi-factor multinomial experiments. In preparation.

Bechhofer, R. E., Kiefer, J. and Sobel, M. (1968). *Sequential Identification and Ranking Procedures.* The University of Chicago Press, Chicago.

Breslow, N. and Day, N. (1980). *Statistical Methods in Cancer Research*, Vol 1. *The Analysis of Case Control Studies*, IARC Scientific Publications No. 32, International Agency for Research on Cancer, Lyon, France.

Cox, D. R. (1970). *The Analysis of Binary Data*, Methuen, London.

Feinberg, S. E. (1977). *The Analysis of Cross-Classified Categorical Data*, MIT Press, Cambridge.

Haberman, S. J. (1978). *Analysis of Qualitative Data*, Academic Press, New York.

Plackett, R. L. (1974). *The Analysis of Categorical Data*, Griffin Statistical Monograph No. 35, London.

Rasch, G. (1960). *Probabilistic Models for Some Intelligence and Attainment Tests*, Nielsen and Lydiche, Copenhagen.

Upton, G. J. G. (1978). *The Analysis of Cross-Tabulated Data*, John Wiley & Sons, New York.

SELECTION PROBLEM FOR A MODIFIED MULTINOMIAL (VOTING) MODEL

PINYUEN CHEN

Department of Mathematics
Syracuse University
Syracuse, New York 13210, U.S.A.

MARC SOBEL

Department of Statistics
University of Georgia
Athens, Georgia 30602, U.S.A.

MILTON SOBEL

Department of Mathematics
University of California at Santa Barbara
Santa Barbara, California 93106, U.S.A.

1. INTRODUCTION AND FORMULATION OF THE PROBLEM

Suppose there is given a modified multinomial model with four candidates running for office denoted by 1, 2, 3 and 4. Voters chosen at random from a population Ω are asked to vote (independently) for at most two candidates. Let p_i $(i = 1,2,3,4)$ denote the proportion of voters in the population Ω that will vote for candidate i alone and let $p_{ij}(= p_{ij})$ denote the proportion that will vote for candidates i and j jointly. Clearly then we have

$$\sum_{i=1}^{4} p_i + \sum_{1\le i<j\le 4} p_{ij} = 1 \tag{1.1}$$

One question that arises almost immediately concerns the weight to be given to single votes versus that for a double-vote. Suppose we give weight a_S to single votes and weight a_D to double votes. Then we wish to compare the candidates on the basis of their (theoretical) parameter ratings defined by

$$p_{i\cdot} = a_S p_i + a_D \sum_{\substack{j=1 \\ j\neq i}}^{4} p_{ij} \qquad (i = 1,2,3,4) \tag{1.2}$$

and select the candidate with the highest rating, i.e., if the ordered values of these four ratings are denoted by $p_{[1]\cdot} \le p_{[2]\cdot} \le p_{[3]\cdot} \le p_{[4]\cdot}$, then, for $p_{[3]\cdot} < p_{[4]\cdot}$, we want to select the candidate associated with $p_{[4]\cdot}$. Thus the weights a_S and a_D reflect what is perceived to be the relative importance of single versus double votes. In this paper we consider a single vote to be two votes for the same candidate and hence take $a_S = 2$ and $a_D = 1$ but the same methods can be used for other values of a_S and a_D.

We are considering a fixed-sample size voting scheme based on n votes. To avoid inefficiency and unfairness, we will observe and make use of all n votes as cast (i.e., without discarding any and without altering any of the votes). Let N_i denote the number of single votes for candidate i and let N_{ij} denote the number of votes for the pair of candidates (i, j). In analogy with the parameter ratings $p_{i\cdot}$ in (1.2), we define the scores

$$N_{i\cdot} = a_S N_i + a_D \sum_{\substack{j=1 \\ j \neq i}}^{4} N_{ij} \qquad (i = 1, 2, 3, 4). \tag{1.3}$$

The four candidates will be ranked according to their scores, the notation using being $N_{[1]\cdot} \leq N_{[2]\cdot} \leq N_{[3]\cdot} \leq N_{[4]\cdot}$ for the ordered values of the four scores. We also use the notation $N_{(i)\cdot}$ for the score of the candidate with parameter rating $p_{[i]\cdot}$.

Our main interest is in the problem of selecting the best candidates, i.e., the one with parameter rating $p_{[4]\cdot}$. The proposed procedure R is based on the observed scores $N_{i\cdot}$ (or equivalently on $N_{[i]\cdot}$). We select the candidate with the largest score, i.e., with score $N_{[4]\cdot}$. Ties are possible but the sample size will usually be large enough to allow us to disregard ties and make use of large sample theory.

Clearly a correct selection is equivalent to the event that $N_{(4)\cdot} = \max_{i \leq j \leq 4} N_{(j)\cdot}$ and the requirement we wish to satisfy can therefore be written in the form

$$P(CS) = P\{N_{(4)\cdot} = \max_{1 \leq j \leq 4} N_{(j)\cdot} | PZ\} \geq P^* \tag{1.4}$$

where P^* is preassigned for all parameters in a set called the preference zone (PZ) which we define presently.

For convenience, in writing p_i and p_{ij}, we assume from this point on that the subscript 4 (without parentheses) refers to the best candidate. We introduce a constant $\Theta \geq 1$ and assume that there are more double votes than single votes: in particular that, for every pair i, j $(1 \leq i,\ j \leq 3)$,

$$\min_{i \leq i,j \leq 3} p_{ij} \geq \Theta \max_{1 \leq i \leq 3} p_i. \tag{1.5}$$

We also introduce constants δ_1, δ_2 (both at least 1) such that

$$p_4 \geq \delta_1 \max_{1 \leq i \leq 3} p_i,$$

$$\min_{1 \leq i \leq 3} p_{4i} \geq \delta_2 \max_{1 \leq i,j \leq 3} p_{ij}. \tag{1.6}$$

We wish to use the three inequalities above together with the inequality,

$$p_{[4]\cdot} \geq \delta^* p_{[3]\cdot} \tag{1.7}$$

to define a PZ for this problem.

The idea of introducing the extra inequalities in (1.5) and (1.6) is to prevent the problem from degenerating into one where only single votes predominate. In addition we wish to define a candidate as best only if he is best with respect to double votes alone and also with respect to single votes alone.

We claim (as will be shown later) that, under slightly stronger restrictions than those presented above, there will be a least favorable configuration (LF) with the properties that

$$p_{12} = p_{13} = p_{23};\ p_{14} = p_{24} = p_{34};\ p_1 = p_2 = p_3;\ p_{4\cdot} = \delta^* p_{j\cdot} \tag{1.8}$$

for $j = 1$,2, and 3. Furthermore, in this LF configuration,

$$p_{12} = \Theta p_1;\ p_4 = \delta_1 p_1;\ p_{14} = \delta_2 p_{12}. \tag{1.9}$$

Since the set of all p's add to 1, it easily follows from (1.9) that

$$p_1 = \frac{1}{3\Theta + 3 + \delta_1 + 3\Theta\delta_2},\ p_4 = \delta_1 p_1;\ p_{12} = \Theta p_1;\ p_{14} = \Theta\delta_2 p_1, \tag{1.10}$$

and we denote the quantity $3\Theta + 3 + \delta_1 + 3\Theta\delta_2$ by s.

The PZ will now be defined in such a way that, in approaching the LF configuration, all similar parameters (e.g., p_{12}, p_{13}, p_{23} etc.) will be moving in the same direction, namely, upward or downward. In place of (1.7) we now write

$$p_{[4]\cdot} \geq \frac{2\delta^*}{3+\delta^*} > \frac{2}{3+\delta^*} \geq p_{[3]\cdot}. \tag{1.11}$$

Note that since $a_S = 2$ and $a_D = 1$, it follows that $p_{[1]\cdot} + p_{[12]\cdot} + p_{[3]\cdot} + p_{[4]\cdot} = 2$ and hence the LF configuration we have $p_{[4]\cdot} = \delta^* p_{[1]\cdot}$ and $p_{[1]\cdot} = p_{[2]\cdot} = p_{[3]\cdot} = 2/(3+\delta^*)$. Hence the two inner quantities in (1.11) represent the LF values for $p_{[4]\cdot}$ and $p_{[1]\cdot}$ respectively. In place of (1.6) we write

$$p_4 \geq \delta_1/s \geq 1/s \geq \max_{1\leq i\leq 3} p_i, \tag{1.12}$$

$$\min_{1\leq i\leq 3} p_{4i} \geq \Theta\delta_2/s \geq \Theta/s \geq \max_{1\leq i,j\leq 3} p_{ij}. \tag{1.13}$$

In place of (1.5) we now write

$$\min_{1\leq i,j\leq 3} p_{ij} \geq \Theta/s \geq 1/s \geq \max_{1\leq i\leq 3} p_i. \tag{1.14}$$

The remark after (1.11) about the 2 inner values also applies to (1.12), (1.13) and (1.14) and this explains the origin of the inner values. The region where all the inequalities in (1.11), (1.12), (1.13) and (1.14) hold will constitute our preference zone (PZ).

We have assumed that $\Theta \geq 1$. The case of Θ large (i.e., $\Theta \approx \infty$) corresponds to having only double votes or in other words the single votes become negligible. In the case Θ close to 0 (which we have omitted) the double votes become negligible and the problem then reduces to an ordinary multinomial selection problem which has already been treated in the literature [1] and for which tables are available. The general topic of multinomial selection problems has a huge literature and the reader can find reference to some of this literature in books on ranking and selection (cf. [2] and [3]).

2. SOME PROOFS

Let $\underset{\sim}{p} = \{p_i(1 \leq i \leq 4), p_{ij}(1 \leq i \leq j \leq 4)\}$ denote the true parameter configuration. For notational convenience in this section we use $N_{4\cdot}$ without parentheses to denote the score of the best candidates, i.e., $N_{4\cdot} = N_{(4)\cdot}$. The probability generating function for the vector of scores $\underset{\sim}{N}$ with sample size n is easily seen to be

$$\phi_n(\underset{\sim}{t}|\underset{\sim}{p}) = \left(\sum_{i=1}^{4} p_i t_i^2 + \sum_{1 \leq i < j \leq 4} t_i t_j p_{ij}\right)^n, \tag{2.1}$$

where the subscript $j = 4$ now refers to the best candidate. Hence the probability that $N_{i\cdot} = j_i$ $(i = 1,2,3,4)$ is given by

$$D_4^{j_4} \cdots D_1^{j_1} \phi_n(\underset{\sim}{t}|\underset{\sim}{p}) / \prod_{i=1}^{4} j_i!\Big]_{\underset{\sim}{t}=\underset{\sim}{0}} = \prod_{i=1}^{4} D_i^{j_i} \phi_n(\underset{\sim}{0}|\underset{\sim}{p}) / \prod_{i=1}^{4} j_i!, \tag{2.2}$$

where $D_4^{j_4}$, for example denotes a derivative with respect to t_4 taken j_4 times and the result is evaluated at $\underset{\sim}{t} = 0$.

Hence the probability of a correct selection $P_n(CS|\underset{\sim}{p})$ under the configuration $\underset{\sim}{p}$ based on n observations is

$$P_n(CS|\underset{\sim}{p}) = \sum_{j \,\varepsilon\, S_n} \prod_{i=1}^{4} D_i^{j_i} \phi_n(\underset{\sim}{0}|\underset{\sim}{p}) / \prod_{i=1}^{4} j_i!, \tag{2.3}$$

where the summation is over the set S_n of vectors (j_i, j_2, j_3, j_4) such that $j_4 = \max_{1 \leq \alpha \leq 4} j_\alpha$ and $\sum_{\alpha=1}^{4} j_\alpha = 2n$. This is a small sample result and includes ties as correct selection. If we include ties as correct selections, then this is equivalent to writing

$$P_n(CS|\underset{\sim}{p}) = P\{N_{4\cdot} \geq N_{i\cdot} \qquad (i = 1,2,3)\}. \tag{2.4}$$

Asymptotically $(n \to \infty)$ the case of ties is negligible and hence we can write for large values of n

$$P_n(CS|\underset{\sim}{p}) \approx P\{N_{4\cdot} > N_{i\cdot} \qquad (i = 1,2,3)\}. \tag{2.5}$$

Our main result needs five similar lemmas and, because of their similarity, we prove only one and state the other four. Our lemmas compare the $P_n(CS|\underset{\sim}{p})$ for two configurations, $\underset{\sim}{p}$ and $\underset{\sim}{p}^{(\Delta)}$, where $\Delta > 0$ is small, $\underset{\sim}{p}^{(\Delta)}$ is defined by

$$p_{14}^{(\Delta)} = p_{14} - \Delta, \qquad p_{13}^{(\Delta)} = p_{13} + \Delta, \tag{2.6}$$

and the other components of $\underset{\sim}{p}$ are unchanged.

Lemma 1. *For n large and Δ small we have*

$$P_n(CS|\underset{\sim}{p}) \geq P_n(CS|\underset{\sim}{p}^{(\Delta)}). \tag{2.7}$$

PROOF. Using (2.3) we wish to differentiate the right side of (2.7) with respect to Δ and interchange the order of differentiation so that D_Δ is the first operator. We then wish to show that the result is non-positive, i.e., that

$$\sum_{\underset{\sim}{j}\,\varepsilon\, S_n} \frac{\left(\sum_{\alpha=1}^{4} D_{\alpha}^{j_\alpha}\right)\left(D_\Delta \phi_n(\underset{\sim}{0}|\underset{\sim}{p}^{(\Delta)})\right)}{\prod_{i=1}^{4} j_\alpha!} \leq 0. \tag{2.8}$$

The left side (LS) of (2.8), after operating with D_Δ and using (2.6) is

$$LS = \sum_{\underset{\sim}{j}\,\varepsilon\, S_n} \left(\prod_{\alpha=1}^{4} D_{\alpha}^{j_\alpha}\right) \frac{(t_1t_3 - t_1t_4)}{\prod_{i=1}^{4} j_i!} \phi_{n-1}\left(\underset{\sim}{t}|\underset{\sim}{p}^{(\Delta)}\right)\Big]_{\underset{\sim}{t}=0} \tag{2.9}$$

and a factor n has been deleted in (2.9), since it does not affect our proof. Using Leibniz's rule for differentiating a product in (2.9), we now obtain after evaluating at $\underset{\sim}{t} = 0$,

$$\begin{aligned} LS = {} & \sum_{\underset{\sim}{j}\,\varepsilon\, S_n} j_1 j_3 \frac{D_1^{j_1-1} D_2^{j_2} D_3^{j_3-1} D_4^{j_4}}{\prod_{i=1}^{4} j_i!} \phi_{n-1}\left(\underset{\sim}{0}|\underset{\sim}{p}^{(\Delta)}\right) \\ & - \sum_{\underset{\sim}{j}\,\varepsilon\, S_n} j_1 j_4 \frac{D_1^{j_1-1} D_2^{j_2} D_3^{j_3} D_4^{j_4-1}}{\prod_{i=1}^{4} j_i!} \phi_{n-1}\left(\underset{\sim}{0}|\underset{\sim}{p}^{(\Delta)}\right). \end{aligned} \tag{2.10}$$

If we make the obvious change of variable in the first summation of (2.10), $j_1' = j_1$, $j_2' = j_2$, $j_3' = j_3 - 1$, $j_4' = j_4 + 1$, and drop the primes, then all the corresponding terms vanish except for those in which $j_3 = 0$. However these terms have a negative sign in (2.10) and this proves the desired result in (2.8), namely that the $P_n(CS|\underset{\sim}{p}^{(\Delta)})$ is decreasing in Δ and hence Lemma 1 holds. □

Lemma 2. *Using a new* $\underset{\sim}{p}^{(\Delta)}$ *defined by*

$$p_4^{(\Delta)} = p_4 - \Delta, \; p_1^{(\Delta)} = p_1 + \Delta \quad \textit{(other components unchanged)} \tag{2.11}$$

we have for n *large and* Δ *small,*

$$P_n(CS|\underset{\sim}{p}) \geq P_n(CS|\underset{\sim}{p}^{(\Delta)}). \tag{2.12}$$

Lemma 3. *Using a new* $\underset{\sim}{p}^{(\Delta)}$ *defined by*

$$p_{14}^{(\Delta)} = p_{14} - \Delta, \quad p_1^{(\Delta)} = p_1 + \Delta \quad \textit{(other components unchanged)}, \tag{2.13}$$

we have, for n *large and* Δ *small,*

$$P_n(CS|\underset{\sim}{p}) \geq P_n(CS|\underset{\sim}{p}^{(\Delta)}). \tag{2.14}$$

Lemma 4. *Using a new* $\underset{\sim}{p}^{(\Delta)}$ *defined (for* $1 \leq i < j < 4$*) by*

$$p_{i4}^{(\Delta)} = p_{i4} - \Delta, \quad p_{j4}^{(\Delta)} = p_{j4} + \Delta \quad \textit{(other components unchanged)},$$

we have for n *large and* Δ *small,*

$$P_n(CS|\underset{\sim}{p}) \geq P_n(CS|p^{(\Delta)}). \tag{2.16}$$

Lemma 5. *Using a new* $\underset{\sim}{p}^{(\Delta)}$ *defined (for* $1 \leq i,j < k < 4; i \neq j$*) by*

$$p_{ij}^{(\Delta)} = p_{ij} - \Delta, \quad p_{ik}^{(\Delta)} = p_{ik} + \Delta \quad \textit{(other components unchanged)}, \tag{2.17}$$

we have, for n *large and* Δ *small,*

$$P_n(CS|\underset{\sim}{p}) \geq P_n(CS|\underset{\sim}{p}^{(\Delta)}), \tag{2.18}$$

From these lemmas, we use the following heuristic argument to prove our main

Theorem. *For* n *large,*

$$\inf P_n(CS|\underset{\sim}{p}) \geq P_n(CS|\underset{\sim}{p}^{(LF)}), \tag{2.19}$$

where the infimum is taken over all $\underset{\sim}{p}$ *in the PZ (cf.* (1.5), (1.6) *and* (1.7)*) and* $\underset{\sim}{p}^{(LF)}$ *denotes the LF configuration (cf.* (1.8), (1.9) *and* (1.10)*).*

PROOF. The proof is accomplished by using the above lemmas to push the parameters $\underset{\sim}{p}$ (assumed to lie originally in the PZ) in the direction of the LF configuration. Let us use the notation p_4' for any of the parameters p_{14}, p_{24}, p_{34} and p_4; similarly we use p_{ij}' for any of the remaining parameters. We will say that a pair (p_{ij}', p_4') are associable if they can be simultaneously moved (p_{ij}' upwards and p_4' downwards) without increasing the $P(CS)$. The strategy we use for moving the parameters is to take (at each stage) the smallest p'_{ij} and the largest p'_4 that are associable and move them simultaneously until there are no more associable parameters to move. At this stage we use lemmas 4 and 5 to put some p'_4 and/or some p'_{ij} into a position where there are again pairs of associable parameters to move. Either this strategy leads to the LF configuration or some further movement will always be possible. Since all movements are in directions of the LF configuration, we have the desired result. □

APPENDIX

We have obtained some results concerning the sample size $n = n(\delta^*, \theta, \delta_1)$ needed in order to get a PCS of both $P^* = .9$ and $P^* = .95$ under the LF configuration given by (1.8), (1.9), and (1.10) (when $k = 4$). The following table was obtained by an asymptotic normal approximation to the probability of correct selection (in the LF configuration) by using tables of the ranking and selection normal integral adopted by Milton (1963).

The LF configuration is determined through assigning values to the parameters θ, δ^* and δ_1 and δ_2 (with $\theta \geq 1$, $1 \leq \delta^* \leq 3$, and $\delta_1 > 1$). The table gives $n = n(\delta^*, \theta, \delta_1)$ when $\theta = 1, 3, 1000$; $\delta^* = 1.2$, 1.6, 2.0; and δ_1, δ_2 are given as one of the following 5 points on the line

$$\delta_2 = \frac{2\delta^*(1+\theta) - 2\delta_1}{\theta(3-\delta^*)}; \tag{A.1}$$

corresponding to $\delta_1 = 1$, δ^*,

$$\left[\frac{2\delta^*(1+\theta)}{2+\theta(3-\delta^*)}\right], \left[\frac{2\delta^*(1+\theta) - \theta(3-\delta^*)\delta^*}{2}\right], \text{ or } \left[\frac{2\delta^*(1+\theta) - \theta(3-\delta^*)}{2}\right].$$

We will say that we are in cases $J = 1$, 2, 3, 4, or 5 according as δ_1 takes on the first, second, etc. (respectively) of its possible values listed above. The δ_2 value in (A.1) can be obtained by formulas (1.8), (1.9) and (1.10) together with an extra relation from (1.1) and (1.2) and $a_S = 2$ and $a_D = 1$:

$$\sum_{i=1}^{4} p_{i\cdot} = a_S \sum_{i=1}^{4} p_i + a_D \sum_{i=1}^{4} \sum_{\substack{j=1 \\ i \neq j}}^{4} p_{ij} = 2. \tag{A.2}$$

Table 1. Sample Size Needed for Desired PCS

J = Case of (configuration) Number	θ	δ^*	$n = n(\delta^*, \theta, J)$ (.90)	(.95)
1	1	1.2	370	253
1	3	1.2	287	406
1	1000	1.2	219	309
1	1	1.6	41	58
1	3	1.6	33	47
1	1000	1.6	27	38
1	1	2.0	14	20
1	3	2.0	12	16
1	1000	2.0	10	14
2	1	1.2	392	556
2	3	1.2	295	461
2	1000	1.2	219	309
2	1	1.2	48	68
2	3	1.6	36	56
2	1000	1.6	27	38
2	1	2.0	18	25
2	2	2.0	13	20
2	1000	2.0	10	14
3	1	1.2	399	567
3	3	1.2	300	425
3	1000	1.2	219	309
3	1	1.6	52	74
3	3	1.6	38	54
3	1000	1.6	27	38
3	1	2.0	21	30
3	3	2.0	15	21
3	1000	2.0	10	14
4	1	1.2	406	577
4	3	1.2	311	442
4	1000	1.2	234	333
4	1	1.6	55	79
4	3	1.6	43	61
4	1000	1.6	34	48
4	1	2.0	23	33
4	3	2.0	18	27
4	1000	2.0	15	21
5	1	1.2	477	692
5	3	1.2	380	550
5	1000	1.2	288	417
5	1	1.6	66	96
5	3	1.6	56	82
5	1000	1.6	47	69
5	1	2.0	28	41
5	3	2.0	24	36
5	1000	2.0	21	31

BIBLIOGRAPHY

Bechhofer, R. E., Elmaghraby, S. and Morse, N. (1959). A single-sample multiple-decision procedure for selecting the multinomial event which has the highest probability. *Ann. Math. Statist.* **30**, 102–119.

Gibbons, J. D., Olkin, I. and Sobel, M. (1977). *Selecting and Ordering Populations: A New Statistical Methodology.* John Wiley and Sons, New York.

Gupta, S. S. and Panchapakesan, S. (1979). *Multiple Decision Procedures*, John Wiley and Sons, New York.

Milton, R. C. (1963). Tables of equally correlated multivariate normal probability integral. Tech. Report No. 27, Department of Statistics, University of Minnesota, Minneapolis.

A DECISION THEORY FORMULATION FOR POPULATION SELECTION FOLLOWED BY ESTIMATING THE MEAN OF THE SELECTED POPULATION

ARTHUR COHEN AND HAROLD B. SACKROWITZ

Department of Statistics
Rutgers University
New Brunswick, New Jersey 08903, U.S.A.

1. INTRODUCTION AND SUMMARY

The problem of estimating the selected mean of several normal populations has been studied by Dahiya (1974), Cohen and Sackrowitz (1982), and others. Estimating the selected mean of exponential populations has been studied by Sackrowitz and Samuel-Cahn (1984) who also considered other types of densities (1986). In the above studies dealing with normal distributions the selected population is always the one corresponding to the largest sample mean. In all other cases the selection rule is also predetermined.

Cohen and Sackrowitz (1986) gave a decision theory formulation to inference following model selection. Their formulation can be adapted to the problem of deciding on which population to select and then estimating the mean of the selected population.

In this note Cohen and Sackrowitz' (1986) formulation is adapted. It is seen that for the normal model the usual selection rule, "choose the population with the largest sample mean", is clearly justified. It corresponds to the generalized Bayes procedure with respect to Lebesgue measure and would also result as a Bayes procedure with respect to any multivariate conjugate prior distribution. On the other hand it is demonstrated that such a selection rule need not be optimal for other location parameter families. In fact a sensible example is given where it is optimal (generalized Bayes with respect to Lebesgue measure), to sometimes (depending on the sample point) to select the population with the *smaller* sample mean.

2. MAIN RESULTS

Let $X = (X_1, X_2, \ldots, X_m)$ be a random sample from a population whose density is $f(x - \mu_1)$ and let $\underset{\sim}{Y} = (Y_1, \ldots, Y_n)$ be an independent random sample from a population whose density is $f(y - \mu_2)$. Assume μ_1 and μ_2 represent the means of populations 1 and 2 respectively. The problem is to select a population and then to estimate the mean of the selected population where we implicitly are assuming that a larger mean is desirable. For ease of presentation only we let $m = n$ and we study only the case of two populations.

The loss function is $L((a,t),\underset{\sim}{\mu}) =$

$$(t-\mu_1)^2 I(a,1) + (t-\mu_2)^2 I(a,2) + M(1-I(a,\lambda)) \tag{2.1}$$

where $\underset{\sim}{\mu} = (\mu_1, \mu_2)$, t represents the estimate of the mean of the selected population; $\lambda = 1$, represents the case where $\mu_1 - \mu_2 \geq 0$, $\lambda = 2$, represents the case where $\mu_1 - \mu_2 < 0$; $a = 1$ represents the selection of population 1, $a = 2$ represents the selection of population 2; $I(a,\lambda) = 1$ if $a = \lambda$, $I(a,\lambda) = 0$ if $a \neq \lambda$; M is a constant.

For loss function (2.1) only non-randomized estimators t need be considered and so a decision function $\delta(\underset{\sim}{z})$ may be written as the pair $(\phi(\underset{\sim}{z}),\ t(\underset{\sim}{z}))$ with $\underset{\sim}{z} = (\underset{\sim}{x}, \underset{\sim}{y})$, $\underset{\sim}{x}$ is a realization of $\underset{\sim}{X}$, $\underset{\sim}{y}$ is a realization of $\underset{\sim}{Y}$, $\underset{\sim}{Z} = (\underset{\sim}{X}, \underset{\sim}{Y})$ and $\phi(\underset{\sim}{z})$ is the probability of choosing $a = 2$. A generalized prior distribution is denoted by $\xi(\underset{\sim}{\mu})$. For given $\xi(\underset{\sim}{\mu})$ and $i = 1, 2$ let

$$r(i,\underset{\sim}{z}) = E_\xi\{(t-\mu_i)^2 | \underset{\sim}{Z} = \underset{\sim}{z}\} \tag{2.2}$$

and let $t_\xi^{(i)}(\underset{\sim}{z})$ minimize (2.2). Denote the minimal value of (2.2) by $r^*(i,\underset{\sim}{z})$, where we have suppressed the dependence on ξ.

Theorem 2.1. *For each fixed* $\underset{\sim}{Z} = \underset{\sim}{z}$ *a generalized Bayes procedure is* $(a, t_\xi^{(a)}(z))$ *where* $a = 2$ *whenever*

$$P\{\lambda = 2 | \underset{\sim}{Z} = \underset{\sim}{z}\} - P\{\lambda = 1 | \underset{\sim}{Z} = \underset{\sim}{z}\} \geq [r^*(2;\underset{\sim}{z}) - r^*(1,\underset{\sim}{z})]/M, \tag{2.3}$$

where $r^*(i,\underset{\sim}{z})$ *is obtained from* (2.2) *replacing* $t(\underset{\sim}{z})$ *with* $t^{(i)}(\underset{\sim}{z})$.

PROOF. The a posteriori risk given $\underset{\sim}{Z} = \underset{\sim}{z}$ is

$$E\{(1-\phi(\underset{\sim}{z}))[(t-\mu_1)^2 + M(1-I(1,\lambda))] \tag{2.4}$$

$$+ \phi(\underset{\sim}{z})[(t-\mu_2)^2 + M(1-I(2,\lambda))] | \underset{\sim}{Z} = \underset{\sim}{z}\}$$

$$= (1-\phi(\underset{\sim}{z}))r(1;\underset{\sim}{z}) + (1-\phi(\underset{\sim}{z}))MP\{\lambda = 2 | \underset{\sim}{Z} = \underset{\sim}{z}\}$$

$$+ \phi(\underset{\sim}{z})r(2;\underset{\sim}{z}) + \phi(\underset{\sim}{z})MP\{\lambda = 1 | \underset{\sim}{Z} = \underset{\sim}{z}\}.$$

From (2.4) we see that whenever $a = i$, $t^{(i)}(\underset{\sim}{z})$ should be chosen to minimize $r(i;\underset{\sim}{z})$ which is part of the statement of the lemma. Replacing r^* for r in (2.4), collecting terms and minimizing for fixed $\underset{\sim}{z}$ we find $\phi(\underset{\sim}{z}) = 1$ whenever (2.3) holds. □

Suppose for the moment that the problem was just to decide on which population to select. This amounts to testing the hypothesis H_0: $\mu_1 - \mu_2 \geq 0$ vs H_1: $\mu_1 - \mu_2 < 0$. For any continuous permutation invariant (in (μ_1, μ_2))

generalized prior distribution we note that the generalized Bayes test amounts to rejecting H_0 if and only if

$$P\{\lambda = 2|\underset{\sim}{Z} = \underset{\sim}{z}\} - P\{\lambda = 1|\underset{\sim}{Z} = \underset{\sim}{z}\} > 0. \tag{2.5}$$

Now we return to the problem of selecting followed by estimating the mean of the selected population.

EXAMPLE 2.2. Suppose $f(x)$ and $f(y)$ are normal densities with unknown means μ_1 and μ_2 respectively and with known and equal variances. Let $\xi(\underset{\sim}{\mu})$ be Lebesgue measure or any conjugate permutation invariant prior distribution. Then the generalized Bayes selection rule is to select $a = 2$ if and only if $\overline{Y} > \overline{X}$.

PROOF. Note that $r^*(i; \underset{\sim}{z})$ is the posterior variance of μ_1. For the $\xi(\underset{\sim}{\mu})$ considered $r^*(i; \underset{\sim}{z})$ is independent of $\underset{\sim}{z}$ and the permutation invariance of $\xi(\underset{\sim}{\mu})$ implies $r^*(1; \underset{\sim}{z}) = r^*(2; \underset{\sim}{z})$. From (2.3) then we see that the selection rule is identical to the generalized Bayes test of H_0. See (2.5). The posterior distribution of $(\mu_2 - \mu_1)$ is normal with mean $\overline{Y} - \overline{X}$ and constant variance (independent of $\underset{\sim}{Z}$). Hence the Bayes test rejects if and only if $\overline{Y} - \overline{X} > 0$. □

Now suppose we assume $f(x)$ and $f(y)$ are uniform densities on $\mu_1 \pm 1/2$ and $\mu_2 \pm 1/2$ respectively. Let $m = n = 2$ and let $\xi(\underset{\sim}{\mu})$ be Lebesgue measure.

Lemma 2.3. *Let X and Y be uniformly distributed as above. Then the generalized Bayes test of H_0 vs H_1 is to reject H_0 if and only if $\overline{Y} > \overline{X}$.*

PROOF. Consider the rectangle formed by the following four points in the (μ_1, μ_2) plane: $(X_{(1)} + \frac{1}{2}, Y_{(1)} + \frac{1}{2})$, $(X_{(1)} + \frac{1}{2}, Y_{(2)} - \frac{1}{2})$, $(X_{(2)} - \frac{1}{2}, Y_{(1)} + \frac{1}{2})$, $(X_{(2)} - \frac{1}{2}, Y_{(2)} - \frac{1}{2})$, where $X_{(1)} \le X_{(2)}$; $Y_{(1)} \le Y_{(2)}$. The posterior density of (μ_1, μ_2) is constant on this rectangle. Therefore the location of the vertices of the rectangle related to the equiangular line can help determine when $P\{\lambda = 2|\underset{\sim}{z}\} > P\{\lambda = 1|\underset{\sim}{z}\}$. In fact this will be true in the following cases:

(a) $Y_{(2)} > X_{(2)}$ and $Y_{(1)} > X_{(1)}$

(b) $Y_{(2)} > X_{(2)}, Y_{(1)} < X_{(1)}$, and $\overline{Y} > \overline{X}$

(c) $Y_{(1)} < X_{(2)}, Y_{(1)} > X_{(1)}$, and $\overline{Y} > \overline{X}$.

Condition {(a) or (b) or (c)} is equivalent to $\overline{Y} > \overline{X}$. □

Now we return to the problem of selection and estimating the mean of the selected population. We will observe that the selection rule is no longer select $\lambda = 2$, if $\overline{Y} > \overline{X}$ for all $(\underset{\sim}{x}, \underset{\sim}{y})$.

EXAMPLE 2.4. Let X and Y be uniformly distributed as above. Then the generalized Bayes procedure with respect to Lebesgue measure is as follows: Whenever $a = 1$, estimate μ_1 by $\overline{X}$. Whenever $a = 2$, estimate μ_2 by $\overline{Y}$.

Choose $\lambda = 2$ if and only if

$$P\{\lambda = 2|\underset{\sim}{z}\} - P\{\lambda = 1|\underset{\sim}{z}\} \geq (1/12M)[(1-(y_{(2)}-y_{(1)}))^2 - (1-(x_{(2)}-x_{(1)}))^2]. \tag{2.6}$$

PROOF. Use Lemma 2.1. We need to compute $r^*(1;\underset{\sim}{z})$. Using the fact that the posterior density of $\mu_1|\underset{\sim}{z}$ is uniform on $(x_{(2)} - \frac{1}{2}, x_{(1)} + \frac{1}{2})$ we derive the posterior variance as $(1-(x_{(2)}-x_{(1)}))^2/12$. Similarly for $r^*(2;\underset{\sim}{z})$. □

From (2.6) we see that the usual selection rule will not always follow. In particular if $\overline{Y} - \overline{X}$ is positive but small the left hand side of (2.6) will be positive but close to zero. Clearly we can find values of (y_1, y_2), (x_1, x_2) so that $\overline{Y} - \overline{X}$ is small and positive and yet the right hand side of (2.6) is positive and greater than the left hand side. The explanation in these cases is that we prefer selecting the population whose mean can be estimated with a posterior risk that is relatively small, even though the corresponding sample mean of that population is smaller than the sample mean of the other population.

Clearly the constant M controls when a reversal can occur. For $M \to \infty$, no reversals can occur whereas if M tends to zero the population chosen is the one with the smaller posterior variance.

REMARK 2.5. The loss (2.1) is felt to be justified in many practical circumstances. Other losses may sometimes be appropriate.

ACKNOWLEDGMENTS

This research was supported by NSF-Grant DMS-84-18416.

BIBLIOGRAPHY

Cohen, A. and Sackrowitz, H. B. (1982). Estimating the mean of the selected population. Third Purdue Symposium on Statistical Decision Theory and Related Topics, III. Gupta, S. S. and Berger, J. O. eds. Academic Press, New York.

Cohen, A. and Sackrowitz, H. B. (1986). An approach to inference following model selection with applications to transformation-based and adaptive inference. Submitted to *Journal of the American Statistical Association.*

Dahiya, R. C. (1974). Estimation of the mean of the selected population. *J. Amer. Statist. Assoc.* **69**, 226–230.

Sackrowitz, H. B. and Samuel-Cahn, E. (1984). Estimation of the mean of a selected negative exponential population. *JRSS. B.* **46**, 242–249.

Sackrowitz, H. B. and Samuel-Cahn, E. (1986). Evaluating the winner: A Bayes and minimax approach. IMS-Lecture Notes and Monograph Series: Adaptive Statistical Procedures and Related Topics.

ON THE PROBLEM OF FINDING THE LARGEST NORMAL MEAN UNDER HETEROSCEDASTICITY

SHANTI S. GUPTA

Department of Statistics
Purdue University
West Lafayette, Indiana 47907, U.S.A.

KLAUS J. MIESCKE

Department of Mathematics, Statistics, and Computer Science
University of Illinois at Chicago
Chicago, Illinois 60680, U.S.A.

1. INTRODUCTION

Let $\mathcal{P}_1, \ldots, \mathcal{P}_k$ be $k \geq 3$ given normal populations with unknown means $\theta_1, \ldots, \theta_k \in \mathrm{R}$, and a common known variance $\sigma^2 > 0$. Suppose we want to find the population with the largest mean, where k independent samples of sizes $n_1, \ldots, n_k$ from $\mathcal{P}_1, \ldots, \mathcal{P}_k$ are available with sample means $\overline{X}_1, \ldots, \overline{X}_k$, respectively.

The natural decision rule d^N, which selects that population which is associated with the largest sample mean, has been studied by many authors since it was introduced in the pioneering paper of Bechhofer (1954). It was found that it is the uniformly best permutation invariant procedure if the sample sizes $n_1, \ldots, n_k$ are all equal. The most general version of this so-called "Bahadur-Eaton-Goodman-Lehmann Theorem" is presented in Gupta and Miescke (1984), where the risk function of multi- stage selection rules with screening is studied under a permutation invariant loss structure.

The situation changes drastically when the assumption of equal sample sizes is dropped. Besides being asymptotically consistent when the sample sizes tend to infinity, no optimum property of the natural rule d^N is known so far. On the contrary, Lam and Chiu (1976), and more generally Tong and Wetzell (1979), have brought to light quite pathological behavior of the probability of a correct selection under $d^N, P(CS|d^N)$, say. If $\theta_1, \ldots, \theta_k$ are sufficiently close together and if $\theta_1, \ldots, \theta_{k-1} < \theta_k$, then its value is strictly decreasing in n_k.

It should be noted that technically there will be no great changes if we assume that $\mathcal{P}_1, \ldots, \mathcal{P}_k$ have different but known variances. However, we feel that the chosen model provides a better motivation for our considerations. Nevertheless, our analysis will be based on k independent random variables $X_i \sim N(\theta_i, p_i), i = 1, \ldots, k$, where $p_1, \ldots, p_k$ are known, and can be thus applied to the more general case, too.

Whenever comparisons with a control are incorporated into the problem, difficulties caused by heteroscedasticity can be overcome more easily. This has been done for example by Miescke (1981) and Gupta and Miescke (1985). However, the transition to the corresponding problem without a control, as it is described in Miescke (1979), cannot be made in the given situation.

Although some work has been done already to solve the given selection problem, no modification or substitute of d^N has been found so far which can be considered to be better in some reasonable sense. Some insight into the structure of the problem has been gained by Bechhofer and Tamhane (1986), who looked for the best allocations of observations, subject to $n_1 + \ldots + n_k$ being fixed, to maximize $P(CS|d^N)$ for the case of known but unequal variances.

The problem under concern, although being rarely mentioned in the literature, e. g. Berger (1983) and Miescke (1984), is well known to the statistical community. A recent simulation study by Zaher and Heiny (1984), where d^N is compared with two similar rules which are based on medians and rank-sums, respectively, under $n_1 = \ldots = n_k$ but different variances of $P_1, \ldots, P_k$, corroborates this fact. It should be pointed out that the problem of selecting a subset for unequal sample sizes (or unequal variances) has been studied by Gupta and Huang (1976).

In the next section, the minimax approach is used to detect weak points in the performance of d^N. However, no alternative decision rule can be found in this approach. Therefore, Bayes rules with respect to various priors are studied in the subsequent sections to find reasonable modifications of or alternatives to d^N. Similar techniques have been used previously by Ehrman, Krieger and Miescke (1986) in the related subset selection problem. Several promising candidates to be used as alternatives to d^N will be derived and proposed in this paper. Comparisons of the performance characteristics of all rules considered in a simulation study is planned to be made in the future.

2. MINIMAXITY

The problem, which will be considered throughout this paper, can be formulated in a concise form as follows. Given are independent random variables $X_i \sim N(\theta_i, p_i), i = 1, \ldots, k$, where $p_1, \ldots, p_k$ are fixed known positive numbers. To be found is the index i_0, say, with $\theta_{i_0} = max\{\theta_1, \ldots, \theta_k\}$, which we may assume to be unique for the sake of simplicity. Under the $0-1$ loss function, the probability of a correct selection and the risk function of a (possibly randomized) decision rule d at a parameter configuration $\underline{\theta} = (\theta_1, \ldots, \theta_k) \in \mathrm{R}^k$ are connected through

$$P_{\underline{\theta}}(CS|d) = 1 - R(\underline{\theta}, d). \tag{1}$$

Thus all decision theoretic formulations in terms of risk can be translated immediately into the "P(CS)-language" used in the area of ranking and selection.

We begin our study with minimax considerations since this will lead us directly to the weak points in the performance of the natural decision rule d^N, which selects in terms of the largest value among $X_1, \ldots, X_k$. We shall see that the performance of d^N becomes unsatisfactory whenever the parameters $\theta_1, \ldots, \theta_k$ are lying closely together. This complements the findings of Lam and Chiu (1976) and of Tong and Wetzell (1979) and indicates that d^N cannot be considered to be a universally acceptable decision rule.

Let φ and Φ denote the density and cumulative distribution function, respectively, of $N(0,1)$ in the sequel. The first result is a reformulation of the findings by Tong and Wetzell (1979), presented however in a form which is more suitable for our further considerations, and proved differently.

Lemma 1. *The function*

$$H(\gamma_1, \ldots, \gamma_{k-1}) = \int_{\mathsf{R}} \prod_{i=1}^{k-1} \Phi(\gamma_i z)\varphi(z)dz \tag{2}$$

it is strictly increasing in $\gamma_i \geq 0, i = 1, \ldots, k-1$.

PROOF. The partial derivative of H with respect to γ_1 is equal to

$$\int_{\mathsf{R}} \prod_{i=2}^{k-1} \Phi(\gamma_i z) z\varphi(\gamma_1 z)\varphi(z)dz. \tag{3}$$

After combining the two φ-functions, and then integrating by parts, it can be seen that

(3) equals

$$(2\pi)^{-\frac{1}{2}}(1+\gamma_1^2)^{-1} \int_{\mathsf{R}} M(w)\varphi(w)dw, \tag{4}$$

where

$$M(w) = \frac{\partial}{\partial w} \prod_{i=2}^{k-1} \Phi((1+\gamma_1^2)^{-\frac{1}{2}}\gamma_i w), w \in \mathsf{R}, \tag{5}$$

is clearly positive over the whole real line. □

As an immediate consequence, we can state the following.

Corollary 1. *The function*

$$G(\sigma_1,\ldots,\sigma_k) = \int_{\mathbb{R}} \prod_{i=1}^{k-1} \Phi(\sigma_i^{-1}z)\sigma_k^{-1}\varphi(\sigma_k^{-1}z)dz \tag{6}$$

is strictly decreasing in $\sigma_1,\ldots,\sigma_{k-1}$, *and strictly increasing in* σ_k.

Now we can state the main result of this section. The points of weakness of d^N, which we have mentioned before, will become visible in the course of the proof.

Theorem 1. *For the given problem, the natural decision rule* d^N *is minimax if and only if* $p_1 = p_2 = \ldots = p_k$. *Moreover, the minimax- value of the problem is* $1 - 1/k$.

PROOF. Consider the no-data rule d^0, which selects every $i \in \{1,\ldots,k\}$ with the same probability $1/k$. It has clearly the constant risk $1 - 1/k$.

The risk function of d^N can be represented in a convenient way by using the following notation. For any vector $\underline{a} \in \mathbb{R}^k$, let $a_{[1]} \leq \cdots \leq a_{[k]}$ denote the ordered coordinates. Moreover, whenever $\underline{\theta} = (\theta_1,\ldots,\theta_k)$ and $\underline{X} = (X_1,\ldots,X_k)$ are considered jointly in the sequel, let subscript $(j) = i$, if $\theta_i = \theta_{[j]}, i,j = 1,\ldots,k$. As mentioned before, we may assume that no ties occur among the θ_i's. This simplifies our considerations without losing generality. Introducing generic random variables $N_1,\ldots,N_k$, which are independent standard normals, we can represent the risk of d^N at $\underline{\theta} \in \mathbb{R}^k$ by

$$\begin{aligned} R(\underline{\theta}, d^N) &= 1 - P_{\underline{\theta}}\{X_{(k)} = X_{[k]}\} \\ &= 1 - P\{\theta_{[i]} + p_{(i)}^{\frac{1}{2}} N_i \leq \theta_{[k]} + p_{(k)}^{\frac{1}{2}} N_k, i < k\}. \end{aligned} \tag{7}$$

And since this is an increasing function of $\theta_{[i]}, i < k$, we conclude that

$$\begin{aligned} &\sup_{\underline{\theta}} R(\underline{\theta}, d^N) \\ &\quad = 1 - \inf_{\underline{\theta}} \int_{\mathbb{R}} \prod_{i=1}^{k-1} \Phi((p_{(k)}/p_{(i)})^{\frac{1}{2}} z)\varphi(z)dz \\ &\quad = 1 - \int_{\mathbb{R}} \prod_{i=2}^{k} \Phi((p_{[1]}/p_{[i]})^{\frac{1}{2}} z)\varphi(z)dz, \end{aligned} \tag{8}$$

where the second equation is a consequence of Lemma 1. Moreover, from Lemma 1 we see that

$$\sup_{\underline{\theta}} R(\underline{\theta}, d^N) \geq 1 - 1/k, \tag{9}$$

with equality holding if and only if $p_1 = p_2 = \ldots = p_k$.

Thus, to complete the proof, we have to show that the minimax value of the given problem is equal to $1 - 1/k$. Since the no-data rule d^0 has constant risk $1 - 1/k$, it suffices to find a sequence of priors such that the sequence of associated Bayes risks tends to this value. The following class of conjugate priors will be seen to contain such a sequence.

Let, apriori, $\Theta_1, \ldots, \Theta_k$ be independent random variables with $\Theta_i \sim N(\mu_i, r_i), \mu_i \in \mathsf{R}, r_i > 0, i = 1, \ldots, k$. Then, as it is well known, aposteriori, given $\underline{X} = \underline{x}, \Theta_1, \ldots, \Theta_k$ are independent normals with expectations $(p_i\mu_i + r_i x_i)/(p_i + r_i)$ and variances $r_i p_i/(p_i + r_i), i = 1, \ldots, k$, respectively. And marginally, $X_1, \ldots X_k$ are independent normals with expectations μ_i and variances $p_i + r_i, i = 1, \ldots, k$, respectively.

At $\underline{X} = \underline{x}$, the Bayes rule d^B, say, minimizes the posterior expected loss, and it yields the posterior risk

$$\begin{aligned} &\min_{i=1,\ldots,k} (1 - P\{\Theta_i = \Theta_{[k]} | \underline{X} = \underline{x}\}) \\ &= 1 - \max_{i=1,\ldots,k} \int_{\mathsf{R}} \prod_{j \neq i} \Phi((\alpha_j p_j)^{-\frac{1}{2}} (\alpha_i x_i + (1 - \alpha_i)\mu_i \\ &\quad - \alpha_j x_j - (1 - \alpha_j)\mu_j + (\alpha_i p_i)^{\frac{1}{2}} z)) \varphi(z) dz, \end{aligned} \tag{10}$$

where $\alpha_s = r_s/(p_s + r_s), s = 1, \ldots, k$.

For the special case of $r_s = 1/n$ and $\mu_s = 0, s = 1, \ldots, k$, we see that (10) tends to $1 - 1/k$, if n tends to infinity. And since the marginal densities of $X_1, \ldots, X_k$ are bounded by a constant, a routine application of Lebesgue's dominated convergence theorem shows that the sequence of Bayes risks tend in fact to $1 - 1/k$, if n tends to infinity. This completes the proof of the theorem. □

From (8) in the last proof, we can see now clearly what might go wrong in the performance of the natural rule d^N. If the parameters $\theta_1, \ldots, \theta_k$ are close together, and if the variance $p_{(k)}$ of $X_{(k)}$, which is associated with $\theta_{[k]}$, is relatively small in comparison with $p_{(i)}, i \neq k$, then the rule d^N performs "worse than at random".

One natural way out of this dilemma, and to possibly save the reputation of d^N, is to look at the average risk over all $k!$ permutations of a given parameter vector $\underline{\theta}$, rather than taking the risk function as a measure of performance. The average risk of a rule d at $\underline{\theta} \in \mathbb{R}^k$ would be

$$\tilde{R}(\underline{\theta}, d) = (1/k!) \sum_{\pi} R(\pi(\underline{\theta}), d), \tag{11}$$

where $\pi(\underline{\theta}) = (\theta_{\pi(1)}, \ldots, \theta_{\pi(k)})$, and the summation being taken over $k!$ permutations π of $(1, 2, \ldots, k)$. The average risk reflects perhaps better the prevailing attitude of researchers in the area of ranking and selection, which states that "the pairing between the θ_i's and the P_i's is completely unknown."

It can be shown that with respect to the average risk $\tilde{R}, d^N$ is in fact minimax. This result, however, is not of great support for d^N, since it shares this property with a large class of monotone decision rules, as we shall see in the next theorem.

Theorem 2. *Let d^h be the decision rule which selects in terms of the largest $h_i(X_i), i = 1, \ldots, k$, where $h_1, \ldots, h_k$ are strictly increasing functions. Then d^h is minimax with respect to the average risk $\tilde{R}$, and the minimax value of the problem is again $1 - 1/k$.*

PROOF. For every decision rule d, and for every permutation symmetric prior with density ρ w. r. t. the Lebesgue measure on $\mathbb{R}^k$, the Bayes risk satisfies

$$\begin{aligned} r(\rho, d) &= \int_{\mathbb{R}^k} R(\underline{\theta}, d)\rho(\underline{\theta})d\underline{\theta} \\ &= \int_{\{\underline{\theta}|\theta_1 < \ldots < \theta_k\}} \tilde{R}(\underline{\theta}, d)\rho(\underline{\theta})d\underline{\theta} \leq \sup_{\underline{\theta}} \tilde{R}(\underline{\theta}, d). \end{aligned} \tag{12}$$

Since the sequence of priors chosen below of (10) consists of such symmetric priors, it follows similarly as in the proof of Theorem 1 that the no-data rule d^0 is minimax w. r. t. $\tilde{R}$, and that the minimax value is again $1 - 1/k$. It remains thus to show that every rule of type d^h has supremal average risk $1 - 1/k$. Let d^h be any such rule, and let $\underline{\theta} \in \mathbb{R}^k$ be fixed, where we may assume without loss of generality that $\theta_1 < \theta_2 < \ldots < \theta_k$ holds. Then similarly as before in (7),

$$\begin{aligned} &(1/k!) \sum_{\pi} P_{\pi(\underline{\theta})}(CS|d^h) \\ &= (1/k!) \sum_{\pi} P_{\pi(\underline{\theta})}\{h_{\pi^{-1}(k)}(X_{\pi^{-1}(k)}) > h_{\pi^{-1}(j)}(X_{\pi^{-1}(j)}), j < k\} \\ &= (1/k!) \sum_{\pi} P\{h_{\pi^{-1}(k)}(\theta_k + \beta_{\pi^{-1}(k)}N_k) > h_{\pi^{-1}(j)}(\theta_j + \beta_{\pi^{-1}(j)}N_j), j < k\}, \end{aligned} \tag{13}$$

where $N_1, \ldots, N_k$ are independent standard normals, and $\beta_i = p_i^{\frac{1}{2}}, i = 1, \ldots, k$. A lower bound of (13) is attained if all $\theta_1, \ldots, \theta_{k-1}$ are put equal to θ_k, because of the monotonicity of $h_1, \ldots, h_k$. Doing so, and then splitting the sum into a suitable double sum, we see that the lower bound is

$$(1/k!)\sum_{i=1}^{k}\sum_{\pi,\pi(i)=k} P\{h_i(\theta_k+\beta_i N_k)$$
$$> h_{\pi^{-1}(j)}(\theta_k+\beta_{\pi^{-1}(j)}N_j), j<k\}$$
$$= (1/k)\sum_{i=1}^{k} P\{h_i(\theta_k+\beta_i N_i) > h_j(\theta_k+\beta_j N_j), j\neq i\} = 1/k. \quad (14)$$

Thus, in view of (1), the supremal average risk of d^h is equal to $1-1/k$, and the proof of the theorem is completed. □

Our conclusions of this section are (1) that the natural rule d^N cannot be accepted as a universally good decision rule, and (2) that the minimax principle does not lead to a convincing alternative to d^N. Therefore it seems to be reasonable to study the form of Bayes rules with respect to various priors in more detail, in the hope to learn more about how such good decision rules act in different situations. Our main interest thereby will focus on permutation symmetric (exchangeable) and on conjugate priors. This will be done in the subsequent sections.

3. BAYES RULES FOR EXCHANGEABLE PRIORS

Permutation invariant (exchangeable) priors appear to be the suitable priors to adopt if there is no initial knowledge available as to how the ordered parameters $\theta_{[1]},\ldots,\theta_{[k]}$ are associated with the populations $\mathcal{P}_1,\ldots,\mathcal{P}_k$. They reflect the prior opinion that each of the k populations may equally likely be the one which has the largest mean.

Since we are considering Bayes rules, we may restrict considerations to nonrandomized decision rules d, which can be represented simply by measurable functions $d\colon \mathbb{R}^k \to \{1,2,\ldots,k\}$, where $d(\underline{x}) = i$ means that at $\underline{X}=\underline{x}$, d selects population $\mathcal{P}_i, i=1,\ldots,k, \underline{x}\in\mathbb{R}^k$.

Now, for any prior τ, after $\underline{X}=\underline{x}$ has been observed, the Bayes rule selects that population which is associated with the smallest posterior expected loss. This decision process consists thus of pairwise comparisons of the k competing posterior risks. Ignoring a common factor, which depends on $\underline{x}$ and $\underline{p}$, the Bayes rule d^B can be written as

$$d^B(\underline{x}) = i \ \text{ if } \ \mathcal{G}(i|\underline{x}) = \max_{j=1,\ldots,k} \mathcal{G}(j|\underline{x}), \quad (15)$$

where

$$\mathcal{G}(s|\underline{x}) = \int_{\{\underline{\theta}|\theta_s=\theta_{[k]}\}} \prod_{j=1}^{k} \varphi((x_j - \theta_j)/p_j^{\frac{1}{2}})d\tau(\underline{\theta}), \quad i,s = 1,\ldots,k.$$

To find out under which conditions one population is preferred over another one if τ is symmetric, let us compare without loss of generality $\mathcal{G}(2|\underline{x})$ and $\mathcal{G}(1|\underline{x})$, say, to keep the notation simple. After exchanging the variables θ_1 and θ_2 in the integral representation of $\mathcal{G}(2|\underline{x})$, and some standard calculations, we see that

$$\mathcal{G}(2|\underline{x}) - \mathcal{G}(1|\underline{x})$$
$$= \int_{\{\underline{\theta}|\theta_1=\theta_{[k]}\}} [M_{2,1}(\underline{x},\underline{\theta}) - 1] \prod_{i=1}^{k} \varphi((x_i - \theta_i)/p_i^{\frac{1}{2}})d\tau(\underline{\theta}), \tag{16}$$

where

$$M_{2,1}(\underline{x},\underline{\theta})$$
$$= exp\{(\theta_1 - \theta_2)[(x_2 - (\theta_1 + \theta_2)/2)/p_2 - (x_1 - (\theta_1 + \theta_2)/2)/p_1]\}.$$

Although the Bayes rules may have in general very complicated forms, several conclusions can be drawn from (16). The first one is

Theorem 3. *Under a symmetric prior τ, suppose that for two populations P_a and P_b, say, the variances p_a and p_b are equal. Then the Bayes rule relatively ranks P_a and P_b in the same way as the natural rule d^N, namely according to the larger of the two values x_a and x_b, no matter what $x_i, i \neq a, b$, might actually be.*

Another finding is the following. Suppose we know a constant lower (upper) bound d to $\theta_1, \ldots, \theta_k$. Then if $p_a > (<) p_b$ and if $(x_a - d)/p_a > (x_b - d)/p_b$, every Bayes rule w.r.t. a symmetric prior prefers P_a to P_b.

If the prior knowledge asserts that the parameters $\theta_1, \ldots, \theta_k$ are in a slippage configuration $\theta_1 = \ldots = \theta_{i-1} = \theta_{i+1} = \ldots = \theta_k = \delta$, and $\theta_i = \delta + \Delta$, where $\delta \in \mathbb{R}$ and $\Delta > 0$ are known, and where apriori each $i \in \{1, \ldots, k\}$ may be, with the same probability $1/k$, the index of the slipped population, then from (16) it follows that the Bayes rule is given by

$$d^{\delta,\Delta}(\underline{x}) = i \;\; if \;\; (x_i - \delta - \Delta/2)/p_i = \max_{j=1,\ldots,k} (x_j - \delta - \Delta/2)/p_j. \tag{17}$$

It should be noted that it is quite different from the decision rule d^t, say, which selects in terms of the smallest p-value of the best 1-sample tests for $H_i: \;\; \theta_i = \delta$ versus $K_i: \;\; \theta_i = \delta + \Delta$, and thus selects in terms of the largest $(x_i - \delta)/p_i^{\frac{1}{2}}, i = 1, \ldots, k$.

Exchangeable normal priors give Bayes rules which are in general quite complicated in their structure. Although we know that the Bayes rule is determined by

$$d^B(\underline{x}) = i \text{ if } P\{\Theta_i = \Theta_{[k]}|\underline{X} = \underline{x}\}$$
$$= \max_{j=1,\ldots,k} P\{\Theta_j = \Theta_{[k]}|\underline{X} = \underline{x}\}, \tag{18}$$

and a prior $\underline{\Theta} \sim N(\underline{\mu}, A)$ with $\underline{X}|\underline{\Theta} = \underline{\theta} \sim N(\underline{\theta}, \Sigma)$ would result in $\Theta|\underline{X} = \underline{x} \sim N(\underline{x} - \Sigma(\Sigma + A)^{-1}(\underline{x} - \underline{\mu}), (\Sigma^{-1} + A^{-1})^{-1})$ where, marginally, $\underline{X} \sim N(\underline{\mu}, \Sigma + A)$, there is not much simplification to gain if we assume that $\underline{\mu} = \mu_0\underline{1}$, and $A = aI + b\underline{1}\,\underline{1}^T$, where $\underline{1} = (1, \ldots, 1)^T$, I is the $k \times k$ identity matrix, $\mu_0, b \in \mathrm{R}, a > 0$, and $a + kb > 0$ to have A positive definite, even if, as in the present setting, Σ is diagonal with diagonal elements $p_1, \ldots, p_k$.

One limiting case, however, the noninformative prior case, is of natural interest and leads in fact to an interesting decision rule. Suppose we are in the situation which led to (10), but now letting $r_1, \ldots, r_k$ tend to infinity. Then the generalized Bayes rule d^∞, say, can be seen to be based, formally, on $\Theta_i \sim N(x_i, p_i), i = 1, \ldots, k$, independent, at $\underline{X} = \underline{x}$, and to be given by

$$d^\infty(\underline{x}) = i \;\; if \;\; \mathcal{H}(i|\underline{x}) = \max_{j=1,\ldots,k} \mathcal{H}(j|\underline{x}), \tag{19}$$

where

$$\mathcal{H}(s|\underline{x}) = \int_{\mathrm{R}} \prod_{j \neq s} \Phi(p_j^{-\frac{1}{2}}(x_s - x_j + p_s^{\frac{1}{2}} z))\varphi(z)dz, \quad i, s = 1, \ldots, k.$$

One interesting feature of d^∞ is that it selects in terms of the largest variance among $p_1, \ldots, p_k$, whenever $x_1, \ldots, x_k$ are lying closely together. This is an immediate consequence of Lemma 1. We conclude this section by proposing two other type of decision rules which seem to be reasonable alternatives to d^N, worth to be studied in more detail in the future. The first is given by

$$d^\epsilon(\underline{x}) = d^N(\underline{x}), \;\text{ if }\; \underline{x} \notin B, \text{ and}$$
$$d^\epsilon(\underline{x}) = i, \;\text{ if }\; \underline{x} \in B \text{ and } p_i = \max_{j=1,\ldots,k} p_j, \tag{20}$$

where $B \subseteq \mathrm{R}^k$ is an area where the coordinates of the vectors are close to each other, e. g. where $\max_{i,j} |x_i - x_j| < \epsilon$ for some $\epsilon > 0$. The other type of decision rule is of the form (19), where $\mathcal{H}$ is replaced by $\tilde{\mathcal{H}}$, say, with

$$\tilde{\mathcal{H}}(s|\underline{x}) = (x_s - \tilde{x})/p_s, \quad s = 1, \ldots, k, \tag{21}$$

and where $\tilde{x}$ is an average of $x_1, \ldots, x_k$, e. g. the weighted average with weights $p_1^{-1}, \ldots, p_k^{-1}$.

4. BAYES RULES FOR POSTERIORS WITH (DT).

One of the basic facts which lead to the "Bahadur *et. al.* Theorem" mentioned in the introduction is the following. Suppose that at every $\underline{X} = \underline{x}$, the posterior depends on $\underline{x}$ through $\underline{g}(\underline{x}) = (g_1(\underline{x}), \ldots, g_k(\underline{x}))$, where $g_1, \ldots, g_k$ are given functions. Then if the posterior is (DT) in $(\underline{\theta}, \underline{g}(\underline{x}))$, and if the loss function is permutation invariant and favors selection of larger parameters, then the posterior risk acts like the loss function where $\underline{g}(\underline{x})$ plays the role of $\underline{\theta}$. For details see Gupta and Miescke (1984). Thus the Bayes rule selects here in terms of the largest $g_i(\underline{x}), i = 1, \ldots, k$.

Under a normal prior $\underline{\Theta} \sim N(\underline{\mu}, A)$, as considered after the statement of (18), the posterior is (DT) if and only if the covariance matrix associated with it is of the form

$$(\Sigma^{-1} + A^{-1})^{-1} = a^2[(1-\rho)I + \rho \underline{1}\,\underline{1}^T], \tag{22}$$

where $a \in \mathbb{R}$ and $-(k-1)^{-1} < \rho < 1$ are necessary and sufficient for this matrix to be positive definite.

If (22) holds true, the conditional expectation of $\underline{\Theta}$, given $\underline{X} = \underline{x}$, can be seen to be

$$\begin{aligned} E\{\underline{\Theta}|\underline{X} = \underline{x}\} &= \underline{\mu} + \gamma(\underline{x})\underline{1} + \\ &+ a^2(1-\rho)((x_1 - \mu_1)/p_1, \ldots, (x_k - \mu_k)/p_k), \end{aligned} \tag{23}$$

where $\gamma(\underline{x})$ is a certain function which is, as we shall see, of no relevance for the Bayes rule. Namely, if we set $g_i(\underline{x}) = E\{\Theta_i|\underline{X} = \underline{x}\}, i = 1, \ldots, k, \underline{x} \in \mathbb{R}^k$, then the posterior is (DT) in $(\underline{\theta}, \underline{g}(\underline{x}))$, and the Bayes rule is given by

$$d^B(\underline{x}) = i \ \text{ if } \ \mathcal{F}(i|\underline{x}) = \max_{j=1,\ldots,k} \mathcal{F}(j|\underline{x}), \tag{24}$$

where $\mathcal{F}(s|\underline{x}) = \mu_s + a^2(1-\rho)(x_s - \mu_s)/p_s, \quad i, s = 1, \ldots, k.$

Of special interest hereby is the case of $\mu_1 = \ldots = \mu_k = \mu$, where the Bayes rule d^μ, say, assumes the simple form

$$d^\mu(\underline{x}) = i \ \text{ if } \ (x_i - \mu)/p_i = \max_{j=1,\ldots,k} (x_j - \mu)/p_j, \tag{25}$$

$i = 1, \ldots, k$, which is almost the same as that one of the Bayes rule for the slippage situation, given by (17).

The interesting feature of the rule d^∞, discussed just after the statement of (19), has an analog in the rules given by (25) and (17). If there are (almost) tied x_i's, which are smaller (larger) than μ or $\delta + \Delta/2$, respectively, then the Bayes rules prefer the population with the larger (smaller) variance.

The choice of a prior, which results in a posterior with the (DT)- property and ultimately in a Bayes rule of simple structure, is made not only for convenience. It has also a statistical justification since it leads to a posterior situation where the information about the unknown parameters $\theta_1, \ldots, \theta_k$ is equally and thus fairly balanced. This can be seen perhaps most easily in the case where

$\rho = 0$ in (22). Then A is diagonal with diagonal elements $r_1, \ldots, r_k$, say, which brings us back to the situation considered at (10), where now we have

$$p_i^{-1} + r_i^{-1} = a^{-2}, \quad i = 1, \ldots, k. \tag{26}$$

Calling, as usual, the inverse of a variance "precision," the sum of the prior precision and the sampling precision is constant across the k populations, if (26) holds.

Returning to the original form of the problem, as it was presented in the introduction, we can state the following interesting fact. Suppose that the prior is known, as it should be, before the sampling is performed. Suppose further that the sample sizes from the populations $\mathcal{P}_1, \ldots, \mathcal{P}_k$ can be chosen in such a way that (22) holds, which means that in case of a diagonal A, the condition (26) is fulfilled. Then the information about the unknown parameters $\theta_1, \ldots, \theta_k$ is fairly balanced, and the Bayes decision rule assumes the simple form given by (24) or (25), respectively. It should be pointed out clearly, that in this case the Bayes rule is the same under every loss which is permutation invariant and favors selection of larger parameters.

5. CONCLUDING REMARKS

It can be seen easily that the natural rule d^N is an extended Bayes rule. Since if apriori, $\Theta_1, \ldots, \Theta_k$ are i. i. d. $N(0, n)$, then the Bayes risk of d^N with respect to this prior tends to 0 if n tends to infinity. This is not surprising, as we know that the performance of d^N is only unsatisfactory if the parameters $\theta_1, \ldots, \theta_k$ are close together. On the other hand, we saw that d^N cannot be the Bayes rule for any normal prior $\underline{\Theta} \sim N(\underline{\mu}, A)$.

We could not settle, however, the interesting question of whether or not d^N is admissible under the $0-1$ loss function on the parameter space $\Omega = \{\underline{\theta} \in \mathbb{R}^k | \theta_{[k]} \text{is unique}\}$. The restriction of parameters to Ω is made for simplicity, and does not cause any loss of generality. For other loss functions, however, this restriction may not simplify matters and may not be made, as e. g. in the example given below.

The $0-1$ loss function was adopted in our study because it connects the risk function in a natural way through (1) with the probability of a correct selection, which is the performance characteristic of decision rules considered primarily in the area of ranking and selection. With respect to other loss functions, however, the natural rule d^N may in fact be a proper Bayes rule and admissible on $\mathbb{R}^k$, as the following example demonstrates.

Assume that the loss for selecting population $\mathcal{P}_i$ at parameters $\theta_1, \ldots, \theta_k$ is of the form

$$L(\underline{\theta}, i) = \theta_{[k]} - \theta_i, \quad i = 1, \ldots, k, \quad \underline{\theta} \in \mathsf{R}^k. \tag{27}$$

Then the Bayes rule at $\underline{X} = \underline{x}$ is given by

$$d^*(\underline{x}) = i \text{ if } E\{\Theta_i | \underline{X} = \underline{x}\} = \max_{j=1,\ldots,k} E\{\Theta_j | \underline{X} = \underline{x}\}, \quad i = 1, \ldots, k. \tag{28}$$

Therefore, if apriori, $\Theta_i \sim N(\mu_i, r_i), i = 1, \ldots, k$, independent, the Bayes rule at $\underline{X} = \underline{x}$ turns out to be

$$d^b(\underline{x}) = i \;\; if \;\; \mathcal{M}(i|\underline{x}) = \max_{j=1,\ldots,k} \mathcal{M}(j|\underline{x}), \tag{29}$$

where $\mathcal{M}(s|\underline{x}) = (p_s\mu_s + r_s x_s)/(p_s + r_s), s = 1, \ldots, k$. And it can be seen now that d^b is the natural rule d^N if $\mu_i = \mu$ and $r_i = c\ p_i, i = 1, \ldots, k$, for some fixed $\mu \in \mathsf{R}$ and $c > 0$.

The admissibility of all Bayes rules considered in this paper, those under $0-1$ loss on Ω, as well as those under the loss function (27) on R^k, follows from the fact that the risk function of every selection rule in these problems is continuous in $\underline{\theta}$. This is an immediate consequence of the well known fact that the expectation of a bounded function under a multi-parameter exponential family is continuous in these parameters.

ACKNOWLEDGMENTS

The research of the first author was supported by the Office of Naval Research Contract N00014–84–C–0167 and NSF Grant DMS–860694 at Purdue University. The research of the second author was supported by the Air Force Office of Scientific Research Contract AFOSR–85–0347 at the University of Illinois at Chicago.

BIBLIOGRAPHY

Bahadur, R. (1950). On a problem in the theory of k populations. *Ann. Math. Statist.*, **21**, 362–375.

Bahadur, R. and Goodman, L. (1952). Impartial decision rules and sufficient statistics. *Ann. Math. Statist.* **23**, 553–562.

Bechhofer, R. E. (1954). A single-sample multiple decision procedure for ranking means of normal populations with known variances, *Ann. Math. Statist.* **25**, 16–39.

Bechhofer, R. E. , and Tamhane, A. C. (1986). Private Communication.

Berger, J. O. (1983). Comment on [16]. *J. Amer. Statist. Assoc.* **78**, 55–57.

Eaton, M. L. (1967). Some optimum properties of ranking procedures. *Ann.*

Math. Statist. **38**, 124–137.

Ehrman, C. M., Krieger, A., and Miescke, K. J. (1986). Subset selection towards optimizing the best performance at a second stage. 25 pp. in ms. *Journal of Business and Economic Statistics*. To appear.

Gupta, S. S. and Huang, D. Y. (1976). Subset selection procedures for the means and variances of normal populations: unequal sample size case. *Sankhyā*, Series B, **38**, 112–128.

Gupta, S. S. and Miescke, K. J. (1984). Sequential selection procedures: A decision theoretic approach. *Ann. Statist.* **12**, 336–350.

Gupta, S. S., and Miescke, K. J. (1985). Minimax multiple t- tests for comparing k normal populations with a control. *J. Statist. Plann. Inference* **12**, 161–169.

Lam, K., and Chiu, W. K. (1976). On the probability of correctly selecting the best of several normal populations. *Biometrika* **63**, 410–411.

Lehmann, E. L. (1966). On a theorem of Bahadur and Goodman. *Ann. Math. Statist.* **37**, 1–6.

Miescke, K. J. (1979). Identification and selection procedures based on tests. *Ann. Statist.* **7**, 207–219.

Miescke, K. J. (1981). Gamma-minimax selection procedures in simultaneous testing problems. *Ann. Statist.* **9**, 215–220.

Miescke, K. J. (1984). Two-stage selection procedures based on tests. *Design of Experiments: Ranking and Selection (Essays in honor of R. E. Bechhofer)*, T. J. Santner, A. C. Tamhane eds., Marcel Dekker, 1984, 165–178.

Morris, C. N. (1983). Parametric empirical Bayes inference: Theory and applications. *J. Amer. Statist. Assoc.* **78**, 47–65.

Tong, Y. L., and Wetzell, D. E. (1979). On the behavior of the probability function for selecting the best normal population. *Biometrika* **66**, 174–176.

Zaher, A. M., and Heiny, R. L. (1984). A comparison of selection procedures for selecting the best normal population under heterogeneity of variance. *Commun. Statist.-Simula. Computa.* **13**, 635–654.

ON LEAST FAVORABLE CONFIGURATIONS FOR SOME POISSON SELECTION RULES AND SOME CONDITIONAL TESTS

WEN-TAO HUANG

Institute of Statistics
Academia Sinica, Taipei
Taiwan, Republic of China

1. INTRODUCTION

Consider a non-homogeneous Poisson process $\{X(t),\ t \geq 0\}$ with unknown intensity function $\lambda(t)$. Suppose k samples are observed at some specified time $t_1, t_2, \ldots, t_k$. We are interested in determining which t_i has the largest value of $\lambda(t)$. This is an analogue of ranking and selection of Poisson populations. It is also related to the estimation of $\max\limits_{1\leq i\leq k} \lambda(t_i)$. For the selections of Poisson populations in subset selection formulations, some selection rules have been proposed by Gupta and Huang (1975) and Gupta and Wong (1976). For our convenience, we state these rules as follows.

Let $\pi_1, \pi_2, \ldots, \pi_k$ denote k Poisson populations such that the parameter space is given by $\Omega = \{(\lambda_1, \lambda_2, \ldots, \lambda_k)|\lambda_j > 0\}$, where λ_i is the parameter associated with π_i. Let X_i denote an observation from π_i. Let $\lambda_{[1]} \leq \lambda_{[2]} \leq \ldots \leq \lambda_{[k]}$ denote the ordered values of the λ_i and let $\pi_{(i)}$ denote the population associated with $\lambda_{[i]}$, $(i = 1, 2, \ldots, k)$. The population $\pi_{(k)}$ is considered as the best. Based on the observations of experiments $\underset{\sim}{X} = (X_1, X_2, \ldots, X_k)$, it is desired to select the best population. The event of a correct selection (CS) is defined to be the event that $\pi_{(k)}$ is selected. An unconditional rule R_1 of Gupta and Huang (1975) is defined as follows.

$$R_1: \quad \text{Select } \pi_i \text{ if and only if } X_i + 1 \geq c_1 \max_{1\leq j\leq k} X_j \text{ for some } 0 < c_1 \leq 1.$$

Conditioning on the total sum of observations, a conditional rule R_2 is defined by

$$R_2: \quad \text{Select } \pi_i, \text{ if and only if } X_i + 1 \geq c_2 \max_j X_j, \text{ given } \sum_1^k X_i = t$$

where $0 < c_2 = c_2(t) \leq 1$.

The monotonicity and strong monotonicity properties have been shown by Gupta and Wong (1976).

A criterion of P^*-condition is considered to determine c_1 and c_2, i.e. $\inf\limits_{\Omega} P_\lambda(CS|R_i) \geq P^*(i = 1, 2)$, where P^* is a preassigned value. To satisfy the P^*-condition, values of c_i are not unique. In order to keep the size of the subset selected as small as possible, it is desired to take the maximum c_i such that the P^*-condition holds. We call this maximum c_i the exact c_i-value for R_i. By a standard argument, it has been shown by Gupta and Huang (1975) that $\inf_{\Omega} P_{\underset{\sim}{\lambda}}(CS|R_1) = \inf_{\Omega_0} P_{\underset{\sim}{\lambda}}(CS|R_1)$ where $\Omega_0 = \{(\lambda, \lambda, \ldots, \lambda)|\lambda > 0\}$. However, the exact c_1-value for R_1 has not been obtained either in Gupta and Huang (1975) or in Gupta and Wong (1976). Instead, in Gupta and Huang (1975), a conservative solution c_1' ($c_1' \leq c_1$) is proposed which is derived by using two conservative inequalities. A normal approximation for the conservative value c_1' is also mentioned; however, the closeness of c_1' to the exact c_1-value has not been examined. For a slippage configuration, an upper bound for the size of the subset selected is given in Gupta and Huang (1975). This bound depends on the exact c_1-value and it is obtained by using pairwise comparison type of arguments. It is therefore important to consider evaluation of the exact c_1-value. In this paper, we study the least favorable configuration $\underset{\sim}{\lambda}_0$ (i.e. some $\underset{\sim}{\lambda}_0$ in Ω_0 at which $P(CS|R_1)$ attains its infimum) for R_1 and propose an algorithm using recursive techniques to evaluate the exact value of c_1. These, together with the exact expected subset size for a slippage configuration, are given in Section 1.

On the other hand, for the conditional rule R_2, it has been shown earlier that $\inf_{\Omega} P(CS|R_2) = \inf_{\Omega_1} P(CS|R_2)$, where $\Omega_1 = \{(\lambda, \lambda, \ldots, \lambda, \delta\lambda)|\delta \geq 1\}$, and the infimum of $P(CS|R_2)$ depends only on δ. When $k = 2$, it is known that the infimum occurs at $\delta = 1$; however, it is not true for $k \geq 3$ (see Gupta and Wong (1976)). The exact c_2-value has not been obtained. Some conservative solutions for c_2 are given in Gupta and Huang (1975) in which three inequalities are applied for its derivations. As a matter of fact, the information that $\Sigma X_i = t$ has not been sufficiently utilized, since it is based on the pairwise comparison arguments. Some analogous conservative evaluations are also obtained in Gupta and Wong (1976). In Section 2, we propose a method to find the least favorable configuration δ by solving a polynomial equation, and give an algorithm for evaluating the exact c_2.

In Section 3, we consider some tests of homogeneity against slippage alternatives. Two tests of different types are proposed and exact powers of the tests are derived. Relative efficiency of the two tests is considered with numerical results.

Three lemmas and a sketchy proof of Theorem 1 are given in the Appendix.

Finally, it needs to mention that the problem of exact evaluations of subset selection rules for binomial populations has been studied in Huang (1985).

2. EVALUATION OF EXACT C_1 OF R_1

Let $\Delta \subset R^k$. We use $S(\Delta)$ to denote the summation of $\prod_1^k g(x_i)$ over all $(x_1, x_2, \ldots, x_k)\epsilon\Delta$ where $g(\cdot)$ is some given function. When $g(x) = 1/x!$, $S_1(\Delta)$ is used instead of $S(\Delta)$. Further, [a] denotes the largest integer not exceeding a. Define $< a > \equiv -[-a]$. $a \vee b \equiv \max(a, b)$, $a \wedge b \equiv \min(a, b)$. When there is no confusion, $\{h(x_1, x_2, \ldots, x_k) = 0\}$ is used to denote the set $\{(x_1, x_2, \ldots, x_k)|h(x_1, x_2, \ldots, x_k) = 0\}$. Let $\Omega_0 = \{(\lambda, \lambda, \ldots, \lambda)|\lambda > 0\} \subset R^k$.

Now define

$$\Delta(k; t, a) \equiv \{\max_{1\le i\le k} x_i = a, \ \sum_1^k x_i = t\} \tag{1.1}$$

$$\xi(k; t, c) \equiv \{x_k + 1 \ge c \max_{1\le j\le k-1} x_j, \ \sum_1^k x_i = t\}. \tag{1.2}$$

Without loss of generality, let $\lambda_k = \lambda_{[k]}$. Now for $\underset{\sim}{\lambda}\epsilon\Omega_0$, the probability of selecting $\pi_{(k)}$, i.e. π_k, is given by

$$\begin{aligned} P_{\underset{\sim}{\lambda}}(CS|R_1) &= \sum_{t=0}^{\infty} P_{\underset{\sim}{\lambda}}(X_k + 1 \ge c \max_{1\le j\le k-1} x_j | \sum_1^k X_i = t) P_{\underset{\sim}{\lambda}}(\sum_1^k X_i = t) \\ &= \sum_{t=0}^{\infty}\{\sum_{\xi(k;t,c)} \frac{1}{x_1! x_2! \ldots x_k!}(\frac{t!}{k^t})\} P_{\underset{\sim}{\lambda}}(\sum_1^k X_i = t) \\ &= \sum_{t=0}^{\infty} a_t(c) b_t(\lambda) \end{aligned} \tag{1.3}$$

where

$$a_t(c) = S_1(\xi(k; t, c))t!/k^t. \tag{1.4}$$

Note that $a_t(c)$, which is the probability of selecting π_k given that the total observation is t, is independent of λ and is given by Lemma 1 in Appendix.

Let

$$b_t(\lambda) = e^{-k\lambda}(k\lambda)^t/t! \tag{1.5}$$

Hence, $P_{\underset{\sim}{\lambda}}(CS|R_1)$ is a convex combination of $\{a_t\}$ with associated weights $b_t(\lambda)$ of Poisson probability.

Let

$$f(\lambda, c) \equiv P_{\underset{\sim}{\lambda}}(CS|R_1) = \sum_{t=0}^{\infty} a_t(c) b_t(\lambda). \tag{1.6}$$

Since $|a_t(c)b_t(\lambda)| \le b_t(\lambda)$, the right hand side of (1.6) converges uniformly in λ

for any c. Setting $\frac{\partial f(\lambda,c)}{\partial\lambda}=0$, we have

$$\sum_{t=0}^{\infty} d_t(c)\lambda^t = 0 \tag{1.7}$$

where

$$d_t(c) = (a_{t+1}(c) - a_t(c))k^t/t! \tag{1.8}$$

Taking $\frac{\partial f(\lambda,c)}{\partial\lambda}=0$ and $\frac{\partial^2 f(\lambda,c)}{\partial\lambda^2}>0$, by careful compositions and using (1.7), we have

$$\sum_{t=0}^{\infty} e_t(c)\lambda^t > 0 \tag{1.9}$$

where

$$e_t(c) = (a_{t+2}(c) - a_{t+1}(c))k^{t+1}/t! \tag{1.10}$$

Theorem 1.

(i) *For given $c(0<c<1)$, if $\lambda=\lambda_0$ satisfies (1.7) and (1.9) with $d_t(c)$ and $e_t(c)$ given by (1.8), (1.10), (1.4) and Lemma 1 (take S_1 instead of S), then $\inf_{\Omega} P(CS|R_1) = f(\lambda_0,c)$ where $\lambda_0=\lambda_0(c)$.*

(ii) *For given $P^*(\frac{1}{k}<P^*<1)$, there exists a unique exact c_1-value such that $f(\lambda_0(c_1),c_1)\geq P^*$.*

The proof is straightforward and is thus omitted.

For practical evaluations of c_1-value, we may use, for example, bisection method. For each given c, we can find $\lambda_0(c)$. Start from $c=1$ and check if $f(\lambda_0(c),c)\geq P^*$ or $<P^*$. By using the usual bisection method, we can obtain a sequence, say, $\{c_n\}$ such that $c_n\to c_1$.

Let Ω_1 denote the slippage configuration defined by (2.1) and T_1 denote the random number $(1\leq T_1\leq k)$ of populations included in the subset selected using R_1. Then we have the following corollary immediately.

Corollary 1. *For $\lambda(\delta)\epsilon\Omega_1$, we have*

$$E_{\underset{\sim}{\lambda}(\delta)}T_1 = \sum_{t=0}^{\infty}(E_{\underset{\sim}{\lambda}(\delta)}T_2(t))\{((k-1+\delta)\lambda)^t exp(-(k-1+\delta)\lambda)/t!\}$$

where $E_{\underset{\sim}{\lambda}(\delta)}T_2$ the expected size of the subset selected using R_2, is given explicitly by Corollary 2.

3. EVALUATION OF EXACT VALUE OF C_2 OF R_2

Define

$$\Omega_1 = \{(\lambda,\lambda,\ldots,\lambda,\delta\lambda)|\lambda>0,\delta\geq 1\}\subset R^k. \tag{2.1}$$

For $\underset{\sim}{\lambda}(\delta) = (\lambda, \lambda, \ldots, \lambda, \delta\lambda) \epsilon \Omega_1$, define

$$g(k;t,c,\delta) = P_{\underset{\sim}{\lambda}(\delta)}(CS|R_2)$$

$$= \{ \sum_{\xi(k;t,c)} \frac{1}{x_1! x_2! \ldots x_k!} \delta^{x_k} \} / \{(k-1+\delta)^t / t!\} \tag{2.2}$$

where $\xi(k;t,c)$ is defined by (1.2). Note that the function g is independent of λ. Now, define the following polynomials in δ.

$$f_1(\delta; p, t, a_i, b_i)$$
$$= (k-1)a_1 b_1 + \{(k-1)a_2 b_2 + a_1 b_1 - t a_1\}\delta + \{(k-1)a_3 b_3 + a_2 b_2 - t a_2\}\delta^2$$
$$+ \ldots + \{(k-1)a_p b_p + a_{p-1} b_{p-1} - t a_{p-1}\}\delta^{p-1} + (a_p b_p - t a_p)\delta^p. \tag{2.3}$$

$$f_2(\delta; p, t, a_i, b_i)$$
$$= \sum_{i=1}^{p} \{(k-1)^2 a_i b_i (b_i - 1)\delta^{b_i - 2} + 2(k-1) a_i b_i (b_i - t - 1)\delta^{b_i - 1}$$
$$+ (t(t+1)a_i + a_i b_i (b_i - 1) - 2t a_i b_i)\delta^{b_i}\}. \tag{2.4}$$

We have then the following

Theorem 2. *For given $k, t (t \geq 1)$ and $c (0 < c \leq 1)$, if $\delta = \delta_0$ satisfies $f_1(\delta_0; p, t, a_i, b_i) = 0$ and $f_2(\delta_0; p, t, a_i, b_i) > 0$, then $\inf_{\Omega} P(CS|R_2) = g(k;t,c,\delta_0)$ $= t!(\sum_{i=1}^{p} a_i \delta_0^{b_i})/(k-1+\delta_0)^t$ where*

$$a_i = \sum_{a=\alpha_1}^{\alpha_2} S_1(\Delta(k-1; t-b_i, a)) \tag{2.5}$$

$$b_i = b_1 + i - 1, \; i = 1, 2, \ldots, p \tag{2.6}$$

$$p = t - \{< \frac{ct - k + 1}{k - 1 + c} > \vee 0\} + 1 \quad \text{if } ct \neq k - 1$$
$$= t \quad \text{if } ct = k - 1 \tag{2.7}$$

$$b_1 = < \frac{ct - k + 1}{k - 1 + c} > \vee 0 \quad \text{if } ct \neq k - 1$$
$$= 1 \quad \text{if } ct = k - 1 \tag{2.8}$$

$$\alpha_1 = < \frac{t - b_i}{k - 1} > \tag{2.9}$$

$$\alpha_2 = [\frac{b_i + 1}{c}] \wedge (t - b_i) \tag{2.10}$$

$\Delta(k;t,c)$ is defined by (1.1) and $S_1(\Delta(k;t,c))$ is given by Lemma 1 in the Appendix.

The proof is sketched in the Appendix.

For the evaluation of c_2-value, we start from $c = 1$ and decrease c until for the first time when $c = c_0$, $g(k; g, c_0, \delta_0) \geq P^*$, we take c_0 as c_2-value for R_2 corresponding to preassigned P^*. This process is always possible, since for any δ, $g(k; t, c, \delta)$ is decreasing and continuous in c for k, t fixed, and we have $g(k; g, 0, \delta) = 1$. Hence, there exists a largest c_2 such that $g(k; t, c_2, \delta_0) \geq P^*$. For practical computation of c_2, we may, for instance, follow the bisection method to obtain a bounded sequence $\{c_n\}$ which tends to exact c_2.

Let $T_2(t)$ denote the random size of subset selected using R_2 when total sum t is observed. For slippage configuration, we have

Corollary 2. *$\underset{\sim}{\lambda}(\delta) \epsilon \Omega_1$, we have*

$$E_{\underset{\sim}{\lambda}(\delta)} T_2(t) = \frac{(k-1)(t!)}{(k-1+\delta)^t} \{ \sum_{b=b_1}^{b_p} \frac{1}{b!} \sum_{r=0}^{t-b} \frac{\delta^r}{b! r!} A(k-2; t-x-b) \} + g(k; t, c_2, \delta)$$

where b_1, b_p and $g(k; t, c_2, \delta)$ are given in Theorem 2, and

$$A(k; t) = \sum_{i=0}^{t} \frac{1}{i!} A(k-1; t-i)$$

with

$$A(1; t) = 1/t!$$

PROOF. Since

$$E_{\underset{\sim}{\lambda}(\delta)} T_2(t) = \sum_{i=1}^{k} P_{\underset{\sim}{\lambda}(\delta)}(X_i + 1 \geq c_2 \max_j X_j | \sum_1^k X_i = t)$$

$$= (k-1) P_{\underset{\sim}{\lambda}(\delta)}(X_1 + 1 \geq c_2 \max_j X_j | \sum_1^k X_i = t) + g(k; t, c_2, \delta).$$

We note that

$$P_{\underset{\sim}{\lambda}(\delta)}(X_1 + 1 \geq c_2 \max X_j | \sum_1^k X_i = t)$$

$$= \{ \sum_{b=b_1}^{b_2} \sum_{\xi_1(k-1; t-b)} \frac{t! \delta^{x_k}}{b! x_2! \dots x_k!} \} / (k-1+\delta)^t$$

where $\xi_1(k; t) = \{\sum_1^k x_i = t\}$. The range of x_k is thus between 0 and $t - b$ with both ends included. The range of b is given by (2.6). Let $A(k; t) = S_1(\xi_1(k; t))$. Then the recursive relations for $A(k; t)$ is easy to obtain. The proof is thus complete. □

We give numerical results for comparisons between exact and approximate c_2-value and expected size of subset selected. For $k = 3,\ t = 6,\ P^* = 0.98$, we have exact $c_2 = \frac{2}{9} \fallingdotseq 0.222$. However, using the method proposed in Gupta and Huang (1975), we have the conservative value $c_2' = 0.166$. The least favorable configuration is given at $\delta = 4.833$. The exact expected size of subset selected is given by $E(T_2|t = 6) = 1.438$. However, through some numerical computations using the upper bound given in Gupta and Huang (1975), we obtain $E(T_2|t = 6) \leq 1.891$. It can be seen that the approximate c_2' is conservative.

4. TESTS OF HOMOGENEITY

Suppose $\pi_1, \pi_2, \ldots, \pi_k$ be k Poisson population such that π_i has density with parameter λ_i. Assume $\lambda_1 = \lambda_2 = \ldots = \lambda_{j-1} = \lambda_{j+1} = \ldots = \lambda_k = \delta\lambda_j$ for some j, unknown. Let X_i denote an observation from π_i. We want to test $H_0 :\ \ \delta = 1$ against $H_1 :\ \ \delta \neq 1$.

We propose the following two different types of non-randomized conditional tests which are conditioned on the total sum of observations.

$$\phi_1(k;t,c) = \begin{cases} 1 & \text{if } X_{max} - X_{min} > c \text{ given } \sum_1^k x_i = t \\ 0 & \text{otherwise.} \end{cases} \tag{3.1}$$

$$\phi_2(k;t,d) = \begin{cases} 1 & \text{if } X_{min} + 1 < dX_{max}, \text{ given } \sum_1^k x_i = t\ (0 < d < 1). \\ 0 & \text{otherwise.} \end{cases} \tag{3.2}$$

Define

$$\eta(k;t,\alpha) = \{\max x_i - \min x_j = \alpha, \Sigma x_i = t\}, \tag{3.3}$$

$$\tau_1(k;t,d) = \{\min x_i + 1 \geq d \max x_j,\ \Sigma x_i = t\}. \tag{3.4}$$

A. ϕ_1-test.

We have

$$E_{H_0}\phi_1(k;t,c) = \frac{t!}{k^t} \sum_{\alpha=c}^{t} S_1(\eta(k;t,\alpha)) \tag{3.5}$$

where $S_1(\eta(k;t,\alpha))$ is given by Lemma 3 in the Appendix. To compute the power of ϕ_1, we apply the method given in Section 1. We note that

$$\beta(\delta) = E_{H_1}\phi_1(k;t,c) = \sum_{\alpha=c}^{t}\{\sum_{\eta(k;t,\alpha)} \frac{1}{x_1!x_2!\ldots x_k!}\delta^{x_j}\}/\{(k-1+\delta)^t/t!\}$$

for some j $(1 \le j \le k)$. Without loss of generality, we may take $j = k$. Let

$$q_1(k;t,\alpha,\delta) = \sum_{\eta(k;t,\alpha)} \frac{1}{x_1!x_2!\dots x_k!}\delta^{x_k} = \alpha_1\delta^{\beta_1} + \alpha_2\delta^{\beta_2} + \dots + \alpha_r\delta^{\beta_r}. \quad (3.6)$$

Using arguments as in the previous section, we have

$$\beta_1 =< \frac{t-(k-1)\alpha}{k} > \text{ and } \beta_r = [\frac{t-\alpha}{k}] + \alpha \quad (3.7)$$

with $\beta_i = \beta_1 + i - 1$, $i = 1, 2, \dots, r$.

For given β_i, α_i is given by

$$\alpha_1 = \frac{1}{\beta_1!} S_1(\eta(k-1;t-\beta_1,\gamma)) \quad \text{if } [\frac{t-\beta_1-(\beta_1+\alpha)}{k-2}] \ge \beta_1 + 1$$

$$= \frac{1}{\beta_1!} S_1(\eta(k-1;t-\beta_1,\alpha)) \quad \text{otherwise.}$$

$$\alpha_i = S_1(\eta(k-1;t-\beta_i,\alpha)) \quad \text{for } 2 \le i \le r-1. \quad (3.8)$$

$$\alpha_r = \frac{1}{\beta_r - \alpha} S_1(\eta(k-1;t-(\beta_r-\alpha),\sigma)) \quad \text{if } [\frac{t-\beta_r-(\beta_r-\alpha)}{k-2}] \ge \beta_r - \alpha + 1$$

$$= \frac{1}{\beta_r - \alpha} S_1(\eta(k-1;t-(\beta_r-\alpha),\alpha) \quad \text{otherwise,} \quad (3.9)$$

where

$$\gamma = (\beta_1+\alpha) - [\frac{t-\beta_1-(\beta_1+\alpha)}{k-2}]. \quad (3.10)$$

$$\sigma = \beta_r - [\frac{t-\beta_r-(\beta_r-\alpha)}{k-2}]. \quad (3.11)$$

Hence, we have the following

Corollary 3.

(i) $E_{H_0}\phi_1(k;t,c) = \frac{t!}{k^t} \sum_{\alpha=c}^{t} S_1(\eta(k;t,\alpha))$ *where* $S_1(\eta(k;t,\alpha))$ *is given by Lemma 3.*

(ii) $E_{H_1}\phi_1(k;t,c) = t! \sum_{\alpha=c}^{t} \sum_{i=1}^{r} \alpha_i\delta^{\beta_i}/\{(k-1+\delta)^t\}$ *where the* β_i *are given by* (3.7), *the* α_i *by* (3.8) *and* (3.9), *and* $r = \beta_r - \beta_1 + 1$.

B. ϕ_2-test.

We note that

$$E_{H_0}\phi_2(k;t,d) = 1 - \{P_{H_0}(X_{min} + 1 \ge dX_{max}, \sum_1^k X_i = t)/P_{H_0}(\sum_1^k X_i = t)\}$$

$$= 1 - \{S_1(\tau_1(k;t,d))t!/k^t\} \quad (3.12)$$

where $\tau_1(k;t,d)$ is defined by (3.4).

By same method as before, we have

$$S_1(\tau_1(k;t,d)) = \sum_{x=x_1}^{x_2} S_1(\varsigma(k;t,[\frac{1+x}{d}])) \tag{3.13}$$

where

$$x_1 =< \frac{dt-k+1}{k-1+d} > \vee 0 \text{ and } x_2 = [\frac{dt-1}{1+dk-k}] \tag{3.14}$$

where $S_1(\varsigma(k;t,a))$ is given by Lemma 2 in Appendix.

The power of the $\phi_2 - test$ can be obtained by

$$E_{H_1}\phi_2(k;t,d) = 1 - \{ \sum_{\tau_1(k;t,d)} \frac{t!\delta^{x_k}}{x_1!x_2!\dots x_k!}/(k-1+\delta)^t\}. \tag{3.15}$$

Again, let

$$\sum_{\tau_1(k;t,d)} \frac{\delta^{x_k}}{x_1!x_2!\dots x_k!} = a_1\delta^{b_1} + a_2\delta^{b_2} + \dots + a_p\delta^{b_p}. \tag{3.16}$$

Then, analogously, we obtain

$$b_1 = x_1,\ b_p = [\frac{x_2+1}{d}],\ b_i = b_1 + i - 1, \tag{3.17}$$

where x_1 and x_2 are given by (3.14). For given b_i $(1 \le i < p)$,

$$a_i = S_1(\tau_1(k-1;t-b_i,d))\frac{1}{b_i!} \tag{3.18}$$

$$a_p = \sum_{y=y_1}^{y_2} S_1(\tau(k;b_p,y)) \tag{3.19}$$

where $y_1 = db_p - 1$ and $y_2 = b_p$ and $\tau(k;t,a,\alpha)$ is defined by (2) in the Appendix and $S_1(\tau(k;t,a,\alpha))$ is given by Lemma 3.

The relative efficiency of ϕ_2 with respect to ϕ_1, both of size α, is defined by

$$\text{Eff}_\alpha(\phi_2 : \phi_1) = E_{H_1}\phi_2(k;t,c_1)/E_{H_1}\phi_1(k;t,c_2)$$

where $c_1 = c_1(\alpha)$, $c_2 = c_2(\alpha)$.

Let a_i, b_i, α_i and β_i be given by (3.18), (3.19), (3.17), (3.8), (3.9) and (3.7), $p = b_p - b_1 + 1$, and $r = \beta_r - \beta_1 + 1$.

Corollary 4.

(i) $E_{H_0}\phi_2(k;t,d)$ *is given by* (3.15) *and* (3.16) *for* $\delta = 1$.

(ii) $EH_1\phi_2(k;t,d) = 1 - \{t! \sum_{i=1}^{p} a_i\delta^{b_i}/(k-1+\delta)^t\}$

(iii) $\mathit{Eff}_\alpha(\phi_2 : \phi_1) = \{(k-1+\delta)^t/t! - \sum_{i=1}^{p} a_i\delta^{b_i}\}/ \sum_{i=1}^{r} \alpha_i\delta^{\beta_i}$.

C. Numerical Comparisons between ϕ_1 and ϕ_2.

In Table 1, critical values of c and d are given for $k = 2(1)4$ and $t = 5(1)15$. Table 2 gives $\text{Eff}_\alpha(\phi_2 : \phi_1)$ for $k = 3,4$ and selected values of t, δ and α. For $k = 2$, ϕ_1 and ϕ_2 have the same power. However, for $k \geq 3$, numerical results indicate the superiority of ϕ_2 to ϕ_1.

Table 1. Critical values for conditional tests $\phi_1(t)$ and $\phi_2(t)$.

k		t	5	6	7	8	9	10	11	12	13	14	15
2	c	$(\alpha = 0.05)$	$*$	5	6	7	6	7	8	7	8	9	8
		$(\alpha = 01)$	4	5	6	5	6	7	6	7	6	7	8
	d	$(\alpha = 0.05)$	0.20	0.40	0.33	0.28	0.42	0.37	0.33	0.44	0.40	0.36	0.45
		$(\alpha = 0.1)$	0.50	0.40	0.33	0.50	0.42	0.37	0.50	0.44	0.55	0.50	0.45
3	c		5	6	6	6	7	7	7	7	8	8	8
			5	5	5	6	6	6	6	7	7	7	7
	d		0.25	0.20	0.20	0.20	0.20	0.25	0.25	0.25	0.25	0.25	0.28
			0.25	0.25	0.25	0.25	0.33	0.33	0.33	0.33	0.33	0.33	0.33
4	c		5	5	6	6	6	7	7	7	7	8	8
			4	5	5	5	6	6	6	6	7	7	7
	d		0.25	0.25	0.20	0.20	0.20	0.20	0.16	0.16	0.16	0.16	0.16
			0.33	0.25	0.25	0.25	0.20	0.20	0.20	0.20	0.25	0.28	0.28

Table 2. Relative efficiency, $\gamma = \text{Eff}_\alpha(\phi_2 : \phi_1)$.

	t	α		γ		γ		γ		γ
k=3	7	0.05	$(\delta = 0.1)$	1.000	$(\delta = 0.5)$	1.000	$(\delta = 0.9)$	1.000	$(\delta = 1.30)$	1.000
		0.1		1.000		1.000		1.000		1.000
	11	0.05		1.648		1.105		0.967		0.910
		0.1		1.000		1.000		1.000		1.000
	15	0.05		1.189		1.205		0.140		1.063
		0.1		1.067		1.125		1.114		1.100
k=4	7	0.1	$(\delta = 0.1)$	1.000			$(\delta = 0.9)$	1.000		
	11	0.1		0.961				0.828		
	15	0.1		1.584				1.274		

APPENDIX

Lemma 1.

(i) $S(\xi(k;t,c)) = \sum_{x=x_1}^{x_2} g(x) \sum_{a=a_1}^{a_2} S(\Delta(k-1;t-x,a))$

(ii) $$S(\Delta(k;t,a)) = \begin{cases} g^k(a) & \text{if } t = ka \\ \sum_{r=r_1}^{r_2} \{\binom{k}{r} g^r(a) \sum_{b=b_1}^{b_2} S(\Delta(k-r,t-ra,b))\}, & \text{otherwise} \end{cases}$$

$$S(\Delta(2;t,a)) = g^2(a) \text{ if } t = 2a$$
$$= 2g(a)g(t-a) \text{ otherwise },$$

$$S(\Delta(1;t,a)) = g(a) \text{ if } t = a$$
$$= 0 \text{ otherwise},$$

where

$$x_1 = \begin{cases} < \frac{ct-k+1}{k+c-1} > \vee 0 & \text{if } ct \neq k-1, \; x_2 = t, \; a_1 =< \frac{t-x}{k-1} > \\ 1 & \textit{if } ct = k-1, \end{cases}$$

$$a_2 = [\frac{x+1}{c}] \wedge (t-x), \; r_1 = \{t-k(a-1)\} \vee 0, r_2 = [\frac{t}{a}], \; b_1 =< \frac{t-ra}{k-r} >,$$
$$b_2 = (a-1) \wedge (t-ra).$$

PROOF. Since $(x_1, x_2, \ldots, x_k) \epsilon \xi(k;t,c)$ if and only if $(x_1, x_2, \ldots, x_k) \epsilon \bigcup_{a=a_1}^{a_2} \Delta(k-1;t-x_k,a)$ for some a_1 and a_2. We conclude $S(\xi(k;t,c)) = \sum_{x_k=\alpha}^{\beta} g(x_k) \sum_{a=a_1}^{a_2} S(\Delta(k-1;t-x_k,a))$ for some non-negative α and β.

By careful computations of lower and upper bounds of a and x_k, we can obtain a_1, a_2, α and β.

To find recursive relations for $S(\Delta(k;t,a))$, we assume that there are r x_i's which assume the maximum values a. Then there are $\binom{k}{r}$ vectors having exactly r components assuming a with remaining components having largest value a'. It is straightforward to obtain that r takes value from r_1 to r_2 and a' takes values from b_1 to b_2 which are given in the lemma.

For non-negative integer k $(k \geq 1)$, t, a, α and x_i we define the following sets

$$\varsigma(k;t,a) = \{\max_{1 \leq i \leq k} x_i \leq a, \sum_{1}^{k} x_i = t\}. \quad (1)$$

Then, we can compute $S(\varsigma(k;t,a))$ in terms of $S(\Delta(k;t,b))$. □

Lemma 2.

$$S(\varsigma(k;t,a)) = \sum_{b=b_1}^{b_2} S(\Delta(k;t,b))$$

where $b_1 = < \frac{t}{k} >$, $b_2 = a \wedge t$. *The proof is obvious and is omitted.*

Define

$$\tau(k;t,a,\alpha) = \{\max_i x_i = a, \ \min_j \ x_j = a - \alpha, \sum_1^k x_i = t\}. \qquad (2)$$

We have the following

Lemma 3.

(i) $S(\eta(k;t,\alpha)) = \sum_{a=a_1}^{a_2} S(\tau(k;t,a,\alpha))$,

(ii) $S(\tau(k;t,a,\alpha))$

$= \sum_{i=i_1}^{i_2} \binom{k}{i}(g(a))^i \sum_{j=j_1}^{j_2} S(\tau(k-i;t-ai,a-j,\alpha-j))$ *if* $j_2 \geq j_1$

$= 0$ *otherwise*

with

$$S(\tau(k;t,a,0)) = (g(a))^k \quad \text{if } t = ka$$
$$= 0 \quad \text{otherwise},$$
$$S(\tau(1;t,a,\alpha)) = 0 \quad \text{if } \alpha \neq 0$$

where

$a_1 = < (t+\alpha)/k >$, $a_2 = [(t+k\alpha-\alpha)/k]$, $i_1 = (t+k+\alpha-ka-1) \vee 1$,

$i_2 = [\frac{t-(a-\alpha)k}{\alpha}] \wedge k$, $j_1 = (ka-t-\alpha(k-i-1)) \wedge 1$,

$$j_2 = \begin{cases} [(ka-\alpha-t)/(k-i-1)] \wedge \alpha & \text{if } k-i-1 > 0 \\ \alpha & \text{if } k = i+1 \text{ and } ka = \alpha + t \\ 0 & \text{otherwise.} \end{cases}$$

PROOF. The derivations of $S(\eta(k;t,a))$ and of recursiveness for $S(\tau(k;t,a,\alpha))$ are analogous to the arguments given in Huang and Wang (1978) in which the functions H and T in Huang and Wang (1978) are respectively corresponding to $S(\tau(\cdot))$ and $S(\eta(\cdot))$. The proof is thus omitted. □

PROOF OF THEOREM 2. Let the admissible range for x_k be given by $b_1 \leq x_k \leq b_p$ such that

$$\sum_{\xi(k;t,c)} \frac{1}{x_1!x_2!\dots x_k!}\delta^{x_k} = a_1\delta^{b_1} + a_2\delta^{b_2} + \dots a_p\delta^{b_p}$$

where

$$a_i = \sum_{\epsilon(k-1;t-b_i,c,b_i)} \frac{1}{x_1!x_2!\ldots x_{k-1}!b_i!},\ b_{i+1} = b_i + 1,\ i = 1,2,\ldots,p$$

and

$$\epsilon(k;t,c,b) = \{b + 1 \geq c \max_{1\leq j\leq k-1} x_j,\ \sum_1^k x_j = t\}.$$

Then, by (2.2), we have

$$g(k;t,c,\delta) = t!(a_1\delta^{b_1} + \ldots + a_p\delta^{b_p})/(k-1+\delta)^t$$

which is rational in δ ($\delta > 0$). For given t ($t \geq 1$, $t = 0$ is a trivial case), set $\frac{\partial g(k;t,c,\delta)}{\partial \delta} = 0$. We have then $f_1(\delta;p,t,a_i,b_i) = 0$. Again, setting $\frac{\partial^2 g(k;t,c,\delta)}{\partial \delta^2} > 0$, we have $f_2(\delta;p,t,a_i,b_i) > 0$. Since $g(k;t,c,\delta)$ is bounded and continuous in δ, there exists at least one δ_0 which satisfies $f_1 = 0$ and $f_2 > 0$.

To determine b_1 and b_p, we note that, using analogous arguments in Lemma 1, when $ct \neq k-1$, $b_1 = < \frac{ct-k+1}{k-1+c} > \vee 0$. And, when $ct = k-1$, $b_1 = 1$ and $b_p = t$. Hence, when $ct \neq k-1$, $p = b_p - b_1 + 1 = t + 1 - \{< \frac{ct-k+1}{k-1+c} > \vee 0\}$, and $p = t$ when $ct = k-1$. To determine a_i, we note that $(x_1, x_2, \ldots, x_{k-1}, b_i) \epsilon\ \epsilon(k;t,c,b_i)$, if and only if $(x_1, x_2, \ldots, x_{k-1}) \epsilon \bigcup_{a=\alpha_1}^{\alpha_2} \Delta(k-1;t-b_i,a)$. Using analogous arguments, the lower and upper bounds of a are given by α_1 and α_2 which are given by (2.9) and (2.10). Hence, $a_i = \sum_{a=\alpha_1}^{\alpha_2} S_1(\Delta(k-1;t-b_i,a))$, $i = 1,2,\ldots,p$. Note that, in case there does not exist b_i ($b_1 \leq b_i \leq b_p$) for which $x_k = b_i$, then, we would have $a_i = 0$. The proof is complete. □

ACKNOWLEDGMENTS

The author would like to sincerely thank the referees and Professor Shanti Gupta, Professor S. Panchapakesan and Dr. TaChen Liang for their valuable comments, corrections and improvement of clarity of presentation.

BIBLIOGRAPHY

Gupta, S. S. and Huang, D. Y. (1975). On subset selection procedures for Poisson populations and some applications to the multinomial selection problems. *Applied Probability*, R. P. Gupta (ed.). North-Holland: Amsterdam, pp. 97–109.

Gupta, S. S. and Wong, W. Y. (1976). On subset selection procedures for

Poisson processes and some applications to the binomial and multinomial problems. In VIII. Oberwolfach-Tagung Über Operations Research, Verlag Anton Hain Meisenheim Ann Glan, H. P. Künzi (ed.), pp. 49–70.

Huang, W. T. (1985). Some optimization problems in selection procedures. *J. Statist. Comput. Simul.* **22**, pp. 83–91.

Huang, W. T. and Wang, K. C. (1978). On a sum of product of combinations that arises in a test of homogeneity of k binomial populations. *Tamkang Journal of Math.*, **9**, pp. 71–78.

SELECTION OF THE BEST NORMAL POPULATION BETTER THAN A CONTROL: DEPENDENCE CASE

LII-YUH LEU

Graduate Institute of Statistics
National Central University
Chung-Li, Taiwan, Republic of China

1. INTRODUCTION

Let $\pi_1, \pi_2, \ldots, \pi_k$ denote k treatments (populations) and let π_o denote a control population. It will be assumed that $\pi_o, \pi_1, \ldots, \pi_k$ have a multivariate normal distribution with means $\theta_o, \theta_1, \ldots, \theta_k$ and a specified type of covariance matrix. Let $\theta_{[1]} \leq \theta_{[2]} \leq \ldots \leq \theta_{[k]}$ be the ordered θ_i's. Our goal is to select the population associated with $\theta_{[k]}$ provided that $\theta_{[k]} > \theta_o$, otherwise, the population π_o is selected.

Analogous to the type I and type II errors in the testing hypothesis, we formulate the problem to satisfy the following two conditions:

(i) $P\{\pi_o \text{ is selected}\} \geq P_o^*$ whenever $\theta_{[k]} \leq \theta_o$,

(ii) $P\{\pi_{(k)} \text{ is selected}\} \geq P_1^*$ whenever $\theta_{[k]} \geq \max\{\theta_{[k-1]}, \theta_o\} + \Delta$, where P_o^*, P_1^* and $\Delta(> 0)$ are preassigned and $\pi_{(k)}$ denotes the population associated with $\theta_{[k]}$.

Selection procedures are considered for the cases when the control population is known or unknown. Also, two-stage procedures for unknown variance case are considered.

Paulson (1952) and Bechhofer and Turnbull (1978) considered this problem for the independent case. In this paper, we consider the dependent case. Existing tables which can be used to implement the procedures will be mentioned.

In the following discussion, we will let $\underline{X}_j = (X_{oj}, X_{1j}, \ldots, X_{kj})$ when the control population is unknown and $\underline{X}_j = (X_{1j}, X_{2j}, \ldots, X_{kj})$ when the control population is known. Similarly, we will let $\underline{\theta} = (\theta_o, \theta_1, \ldots, \theta_k)$ or $\underline{\theta} = (\theta_1, \ldots, \theta_k)$ without confusing.

Let $\Omega_o = \{\underline{\theta} \mid \theta_{[k]} \leq \theta_o\}$ and $\Omega_1 = \{\underline{\theta} \mid \theta_{[k]} \geq \max\{\theta_{[k-1]}, \theta_o\} + \Delta\}$, then property (i) is satisfied if $\inf_{\underline{\theta} \in \Omega_o} P_{\underline{\theta}}\{\pi_o \text{ is selected}\} \geq P_o^*$ and property (ii) is satisfied if $\inf_{\underline{\theta} \in \Omega_1} P_{\underline{\theta}}\{\pi_{(k)} \text{ is selected}\} \geq P_1^*$.

2. A SELECTION PROCEDURE WHEN THE CONTROL POPULATION IS UNKNOWN

Let $\underline{X}_j$, $j = 1, \ldots, n$ be a random sample from populations $\pi_o, \pi_1, \ldots, \pi_k$.

Suppose that $\underline{X}_j \sim N_{k+1}(\underline{\theta}, \Sigma)$, i.e. $\underline{X}_j$ is distributed as multivariate normal distribution with mean vector $\underline{\theta}$ and covariance matrix Σ. Further assume $\Sigma = (\sigma_{ij}), i,j = 0,1,\ldots,k$, is known and has the form $\sigma_{oo} = \sigma_0^2$, $\sigma_{11} = \ldots = \sigma_{kk} = \sigma^2$, $\sigma_{o1} = \ldots = \sigma_{ok} = \alpha$ and $\sigma_{ij} = \beta, i \neq j,\ i,j = 1,2,\ldots,k$. Thus all k treatments are assumed to have equal variances and covariances but the variance of the control, σ_o^2, may be different and the covariance between the control and a treatment, α, needs not be equal to the covariance between two treatments, β.

Let

$$\overline{X}_i = \frac{1}{n} \sum_{j=1}^{n} X_{ij},\ i = 0,1,\ldots,k$$

and

$$\underline{\overline{X}} = (\overline{X}_o, \overline{X}_1, \ldots, \overline{X}_k),$$

then

$$\underline{\overline{X}} \sim N_{k+1}(\underline{\theta}, \frac{1}{n}\Sigma)$$

and hence

$$(U_1, \ldots, U_k) \sim N_k(0, R)$$

where

$$\begin{aligned} R &= (r_{ij}) \\ r_{ii} = 1,\ r_{ij} &= (\sigma_o^2 + \beta - 2\alpha)/(\sigma_o^2 + \sigma^2 - 2\alpha),\ i \neq j \end{aligned} \tag{2.1}$$

and

$$U_i = [(\overline{X}_i - \overline{X}_o) - (\theta_i - \theta_o)]/[(\sigma_o^2 + \sigma^2 - 2\alpha)/n]^{1/2}. \tag{2.2}$$

Based on this property, we will define a selection procedure R_1 to satisfy conditions (i) and (ii).

Procedure R_1:

Select

$$\pi_o \text{ iff } \overline{X}_{[k]} < \overline{X}_o + d_1[(\sigma_o^2 + \sigma^2 - 2\alpha)/n]^{1/2},$$

or select the population associated with

$$\overline{X}_{[k]} \text{ iff } \overline{X}_{[k]} \geq \overline{X}_o + d_1[(\sigma_o^2 + \sigma^2 - 2\alpha)/n]^{1/2},$$

where $\overline{X}_{[k]} = \max_{1 \leq j \leq k} \overline{X}_j$ and d_1 is chosen to satisfy the following equation (2.3).

Theorem 1. *For a given P_o^*, if d_1 is chosen to satisfy*

$$P\{U_i \leq d_1,\ 1 \leq i \leq k\} = P_o^*, \tag{2.3}$$

where $(U_1, \ldots, U_k) \sim N_k(0, R)$ *are defined by* (2.1) *and* (2.2), *then the selection rule* R_1 *satisfies*

$$\inf_{\underline{\theta} \in \Omega_o} P_{\underline{\theta}}\{\pi_o \textit{ is selected}\} = P_o^*.$$

PROOF.

$$\begin{aligned} P_{\underline{\theta}}\{\pi_o \text{ is selected}\} &= P_{\underline{\theta}}\{\overline{X}_i < \overline{X}_o + d_1[(\sigma_o^2 + \sigma^2 - 2\alpha)/n]^{1/2},\ 1 \le i \le k\} \\ &= P_{\underline{\theta}}\{U_i \le (\theta_o - \theta_i)/[(\sigma_o^2 + \sigma^2 - 2\alpha)/n]^{1/2} + d_1,\ 1 \le i \le k\} \\ &\ge P_{\underline{\theta}}\{U_i \le d_1,\ 1 \le i \le k\}, \quad \text{if } \underline{\theta} \in \Omega_o. \end{aligned}$$

□

REMARK.

1. From Theorem 1, we can determine the values of d_1 using Table 1 of Gupta, Nagel and Panchapakesan (1973).
2. Let $\rho = (\sigma_o^2 + \beta - 2\alpha)/(\sigma_o^2 + \sigma^2 - 2\alpha)$. If ρ is non-negative, then

   $$P\{U_i \le d_1,\ 1 \le i \le k\} = \int_{-\infty}^{\infty} \Phi^k((\sqrt{\rho}x + d_1)/(1-\rho)^{1/2})d\Phi(x),$$

   where Φ is the c.d.f. of $N(0,1)$.
   Thus d_1 is chosen to satisfy

   $$\int_{-\infty}^{\infty} \Phi^k((\sqrt{\rho}x + d_1)/(1-\rho)^{1/2})d\Phi(x) = P_o^*. \tag{2.4}$$

In order to satisfy the condition (ii), we need to determine the values of n to satisfy

$$\inf_{\underline{\theta} \in \Omega_1} P_{\underline{\theta}}\{\pi_{(k)} \text{ is selected}\} \ge P_1^*,$$

for given P_1^* and $\Delta > 0$.

Theorem 2. *For given* P_1^* *and* $\Delta > 0$, *if* n *is chosen to satisfy*

$$\begin{aligned} P\{U_{ok} &\le \Delta/[(\sigma_o^2 + \sigma^2 - 2\alpha)/n]^{1/2} - d_1, \\ U_{jk} &\le \Delta/[(2\sigma^2 - 2\beta)/n]^{1/2}, 1 \le j \le k-1\} = P_1^*, \end{aligned} \tag{2.5}$$

where

$$(U_{ok}, U_{1k}, \ldots, U_{k-1,k}) \sim N_k(0, S)$$

$$S = (S_{ij}),\ S_{ii} = 1,\ S_{oj} = [(\sigma^2 - \beta)/2(\sigma_o^2 + \sigma^2 - 2\alpha)]^{1/2},\ S_{ij} = \frac{1}{2},\ i \ne j,$$

$$i, j = 1, 2, \ldots, k-1,$$

then

$$\inf_{\underline{\theta} \in \Omega_1} P_{\underline{\theta}}\{\pi_{(k)} \textit{ is selected}\} \ge P_1^*.$$

PROOF.

$P_{\underline{\theta}}\{\pi_{(k)}$ is selected$\}$

$$= P_{\underline{\theta}}\{\overline{X}_{(k)} \geq \overline{X}_o + d_1[(\sigma_o^2 + \sigma^2 - 2\alpha)/n]^{1/2},\ \overline{X}_{(k)} \geq \overline{X}_{(j)},\ 1 \leq j \leq k-1\}$$

$$= P_{\underline{\theta}}\{U_{ok} \leq (\theta_{[k]} - \theta_o)/[(\sigma_o^2 + \sigma^2 - 2\alpha)/n]^{1/2} - d_1,$$
$$U_{jk} \leq (\theta_{[k]} - \theta_{[j]})/[(2\sigma^2 - 2\beta)/n]^{1/2},\ 1 \leq j \leq k-1\}$$
$$\geq P_{\underline{\theta}}\{U_{ok} \leq \Delta/[(\sigma_o^2 + \sigma^2 - 2\alpha)/n]^{1/2} - d_1,$$
$$U_{jk} \leq \Delta/[(2\sigma^2 - 2\beta)/n]^{1/2}, 1 \leq j \leq k-1\}$$

where

$$U_{jk} = [(\overline{X}_{(j)} - \overline{X}_{(k)}) - (\theta_{[j]} - \theta_{[k]})]/[(2\sigma^2 - 2\beta)/n]^{1/2}, 1 \leq j \leq k-1$$

and

$$U_{ok} = [(\overline{X}_o - \overline{X}_{(k)}) - (\theta_o - \theta_{[k]})]/[(\sigma_o^2 + \sigma^2 - 2\alpha)/n]^{1/2}.$$

Here one can see that $(U_{ok}, U_{1k}, \ldots, U_{k-1,k}) \sim N_k(0, S)$, where S is defined as before. □

Corollary 3. *If $\sigma_o = \sigma$ and $\alpha = \beta$, then $S = (S_{ij})$, $S_{ii} = 1$, $S_{ij} = \frac{1}{2}$, $i \neq j$, and*

$$\inf_{\underline{\theta} \in \Omega_1} P_{\underline{\theta}}\{\pi_{(k)} \text{ is selected}\}$$
$$= \int_{-\infty}^{\infty} \Phi(x + \frac{\Delta}{[(\sigma^2 - \alpha)/n]^{1/2}} - \sqrt{2} d_1)$$
$$\times \Phi^{k-1}(x + \frac{\Delta}{[(\sigma^2 - \alpha)/n]^{1/2}}) d\Phi(x) \qquad (2.6)$$

If n is chosen such that (2.6) $\geq P_1^*$, then the condition (ii) is satisfied.

PROOF. If $\sigma_o = \sigma$ and $\alpha = \beta$, then

$$\inf_{\underline{\theta} \in \Omega_1} P_{\underline{\theta}}\{\pi_{(k)} \text{ is selected}\}$$

$$= P\{V_1 \leq \frac{\Delta}{[2(\sigma^2 - \alpha)/n]^{1/2}} - d_1,\ V_j \leq \frac{\Delta}{[2(\sigma^2 - \alpha)/n]^{1/2}},\ 2 \leq j \leq k\}$$

where

$$(V_1, \ldots, V_k) \sim N_k(0, S)$$
$$S = (S_{ij}),\ S_{ii} = 1,\ S_{ij} = \frac{1}{2},\ i \neq j.$$

Let $V_i = \frac{1}{\sqrt{2}} Z_i + \frac{1}{\sqrt{2}} Z_{k+1}$, $1 \leq i \leq k$, where $Z_1, \ldots, Z_{k+1}$ are i.i.d. $N(0,1)$.

Then

$$\inf_{\underline{\theta}\in\Omega_1} P_{\underline{\theta}}\{\pi_{(k)} \text{ is selected}\}$$

$$= P\{\frac{1}{\sqrt{2}}Z_1 + \frac{1}{\sqrt{2}}Z_{k+1} \leq \frac{\Delta}{[2(\sigma^2-\alpha)/n]^{1/2}} - d_1,$$

$$\frac{1}{\sqrt{2}}Z_j + \frac{1}{\sqrt{2}}Z_{k+1} \leq \frac{\Delta}{[2(\sigma^2-\alpha)/n]^{1/2}},\ 2 \leq j \leq k\}$$

$$= \int_{-\infty}^{\infty} \Phi(x + \frac{\Delta}{[(\sigma^2-\alpha)/n]^{1/2}} - \sqrt{2}d_1)\Phi^{k-1}(x + \frac{\Delta}{[(\sigma^2-\alpha)/n]^{1/2}})d\Phi(x).$$

□

REMARK.

1. By the result of Tong (1977), we have

$$\inf_{\underline{\theta}\in\Omega_1} P_{\underline{\theta}}\{\pi_{(k)} \text{ is selected}\}$$

$$\geq \Phi(\frac{\sqrt{n}\Delta}{[2(\sigma^2-\alpha)]^{1/2}} - d_1)\int_{-\infty}^{\infty} \Phi^{k-1}(x + \frac{\sqrt{n}\Delta}{(\sigma^2-\alpha)^{1/2}})d\Phi(x).$$

2. If $\sigma_o = \sigma$, $\alpha = \beta$, then d_1 is the solution of

$$\int_{-\infty}^{\infty} \Phi^k(x + \sqrt{2}d_1)d\Phi(x) = P_o^*, \tag{2.7}$$

and the procedure is reduced to

R_1 : Select π_o iff $\overline{X}_{[k]} < \overline{X}_o + d_1[2(\sigma^2-\alpha)/n]^{1/2}$, or

select the population associated with $\overline{X}_{[k]}$otherwise.

3. If $\sigma_o = \sigma$, $\alpha = \beta = 0$, then R_1 is defined by

R_1 : Select π_o iff $\overline{X}_{[k]} < \overline{X}_o + d_1\sigma(2/n)^{1/2}$, or

select the population associated with $\overline{X}_{[k]}$ otherwise .

This procedure was derived by Paulson (1952).

For this case, we have

$$\inf_{\underline{\theta}\in\Omega_o} P_{\underline{\theta}}\{\pi_o \text{ is selected}\} = \int_{-\infty}^{\infty} \Phi^k(x + \sqrt{2}d_1)d\Phi(x)$$

and d_1 is determined by the equation (2.7). Further,

$$\inf_{\underline{\theta}\in\Omega_1} P_{\underline{\theta}}\{\pi_{(k)} \text{ is selected}\}$$

$$\geq \Phi(\sqrt{\frac{n}{2}}\frac{\Delta}{\sigma} - d_1)\int_{-\infty}^{\infty} \Phi^{k-1}(x + \frac{\sqrt{n}\Delta}{\sigma})d\Phi(x). \tag{2.8}$$

Thus we can choose n to be the smallest integer such that

$$\Phi(\sqrt{\frac{n}{2}}\frac{\Delta}{\sigma} - d_1)\int_{-\infty}^{\infty}\Phi^{k-1}(x + \frac{\sqrt{n}\Delta}{\sigma})d\Phi(x) \geq P_1^*.$$

In fact, the result is easier than that of Paulson (1952).

3. A SELECTION PROCEDURE WHEN THE CONTROL IS KNOWN

In this section, we will assume that the control parameter is known. Let $\underline{X}_j = (X_{1j}, \ldots, X_{kj})$, $j = 1, \ldots, n$ be a random sample from populations $\pi_1, \pi_2, \ldots, \pi_k$. Suppose that $\underline{X}_j \sim N_k(\underline{\theta}, \sigma^2 R)$, where $R = (r_{ij}), r_{ii} = 1, r_{ij} = \rho$, $i \neq j$, σ^2 and ρ are known. We propose a procedure R_2 as follows:

Procedure R_2:

$$\text{Select } \pi_o \text{ iff } \overline{X}_{[k]} < \theta_o + d_2\sigma/\sqrt{n},$$

or select the population associated with $\overline{X}_{[k]}$ iff

$$\overline{X}_{[k]} \geq \theta_o + d_2\sigma/\sqrt{n},$$

where d_2 satisfies the equation (3.1).

Theorem 4. *For a given* P_o^*, *if* d_2 *is chosen to satisfy*

$$P\{U_i \leq d_2,\ 1 \leq i \leq k\} = P_o^* \tag{3.1}$$

where $(U_1, \ldots, U_k) \sim N_k(0, R)$, $R = (r_{ij})$, $r_{ii} = 1$, $r_{ij} = \rho$, $i \neq j$. *Then the proposed rule* R_2 *satisfies*

$$\inf_{\underline{\theta}\in\Omega_o} P_{\underline{\theta}}\ \{\pi_o \text{ is selected}\} = P_o^*.$$

PROOF.

$$\begin{aligned} P_{\underline{\theta}}\{\pi_o \text{ is selected}\} &= P_{\underline{\theta}}\{\overline{X}_i < \theta_o + d_2\sigma/\sqrt{n},\ 1 \leq i \leq k\} \\ &= P_{\underline{\theta}}\{U_i \leq \frac{\theta_o - \theta_i}{\sigma/\sqrt{n}} + d_2,\ 1 \leq i \leq k\} \\ &\geq P_{\underline{\theta}}\{U_i \leq d_2,\ 1 \leq i \leq k\} \text{ if } \underline{\theta} \in \Omega_o \end{aligned}$$

where $U_i = \frac{\overline{X}_i - \theta_i}{\sigma/\sqrt{n}}$ and $(U_1, \ldots, U_k)$ has the desired distribution. □

REMARK.

$$P\{U_i \leq d_2,\ 1 \leq i \leq k\} = \int_{-\infty}^{\infty}\Phi^k(\frac{\sqrt{\rho}x + d_2}{(1-\rho)^{1/2}})d\Phi(x).$$

Theorem 5. *For given P_1^* and $\Delta > 0$, if n is the smallest integer such that*

$$\int_{-\infty}^{\infty}\int_{\Psi}^{\infty} \Phi^{k-1}(y + \frac{\sqrt{n}\Delta}{\sigma(1-\rho)^{1/2}})d\Phi(y)d\Phi(x) \ \geq \ P_1^*,$$

where

$$\Psi = [d_2 - \frac{\sqrt{n}\Delta}{\sigma} - \sqrt{\rho}x](1-\rho)^{-1/2},$$

then

$$\inf_{\underline{\theta}\in\Omega_1} P_{\underline{\theta}}\{\pi_{(k)} \text{ is selected}\} \ \geq \ P_1^*.$$

PROOF.

$$\begin{aligned}
&P_{\underline{\theta}}\{\pi_{(k)} \text{ is selected}\}\\
&= P_{\underline{\theta}}\{\overline{X}_{(k)} \geq \theta_o + d_2\sigma/\sqrt{n},\ \overline{X}_{(k)} \ \geq \ \overline{X}_{(j)},\ 1 \leq j \leq k-1\}\\
&= P_{\underline{\theta}}\{U_k \ \geq \ \frac{\theta_o - \theta_{[k]}}{\sigma/\sqrt{n}} + d_2,\ U_k \ \geq \ U_j + \frac{\theta_{[j]} - \theta_{[k]}}{\sigma/\sqrt{n}},\ 1 \leq j \leq k-1\}\\
&\geq P\{U_k \geq d_2 - \sqrt{n}\Delta/\sigma,\ U_k \geq U_j - \sqrt{n}\Delta/\sigma,\ 1 \leq j \leq k-1\}
\end{aligned}$$

where $U_k = \sqrt{n}(\overline{X}_{(k)} - \theta_{[k]})/\sigma$.

Let $Z_1, \ldots, Z_{k+1}$ be i.i.d. $N(0,1)$, then

$$\begin{aligned}
&P_{\underline{\theta}}\{\pi_{(k)} \text{ is selected}\}\\
&\geq P\{\sqrt{1-\rho}Z_k + \sqrt{\rho}Z_{k+1} \geq d_2 - \sqrt{n}\Delta/\sigma, \sqrt{1-\rho}\ Z_k \ + \sqrt{\rho}\ Z_{k+1}\\
&\qquad \geq \sqrt{1-\rho}Z_j + \sqrt{\rho}Z_{k+1} - \sqrt{n}\Delta/\sigma, 1 \leq j \leq k-1\}\\
&= \int_{-\infty}^{\infty}\int_{\Psi}^{\infty} \Phi^{k-1}(y + \frac{\sqrt{n}\Delta}{\sigma(1-\rho)^{1/2}})d\Phi(y)\,d\Phi(x),
\end{aligned}$$

where

$$\Psi = [d_2 - \frac{\sqrt{n}\Delta}{\sigma} - \sqrt{\rho}x](1-\rho)^{-1/2}.$$

□

REMARK.

1. $P\{U_k \ \geq \ d_2 - \sqrt{n}\Delta/\sigma,\ U_k \ \geq \ U_j - \sqrt{n}\Delta/\sigma,\ 1 \ \leq \ j \ \leq \ k-1\}$ $\geq \Phi(\sqrt{n}\Delta/\sigma - d_2)\int_{-\infty}^{\infty}\Phi^{k-1}(x + \frac{\sqrt{n}\Delta}{\sigma(1-\rho)^{1/2}})d\Phi(x)$.
2. If $\rho = 0$, then

$$\inf_{\underline{\theta}\in\Omega_0} P_{\underline{\theta}}\{\pi_o \text{ is selected}\} = \Phi^k(d_2),$$

and

$$\inf_{\underline{\theta}\in\Omega_1} P_{\underline{\theta}}\{\pi_{(k)} \text{ is selected}\}$$

$$= \int_{-\infty}^{\infty}\int_{d_2-\frac{\sqrt{n}\Delta}{\sigma}}^{\infty} \Phi^{k-1}(y+\frac{\sqrt{n}\Delta}{\sigma})d\Phi(y)d\Phi(x)$$

$$= \int_{d_2-\frac{\sqrt{n}\Delta}{\sigma}}^{\infty} \Phi^{k-1}(y+\frac{\sqrt{n}\Delta}{\sigma})d\Phi(y)$$

which was derived by Bechhofer and Turnbull (1978).

4. KNOWN CONTROL, UNKNOWN VARIANCE CASE

Let $\underline{X}_j = (X_{1j},\ldots,X_{kj})$, $j = 1,\ldots,n$ be a random sample from $\pi_1,\ldots,\pi_k$ and $\underline{X}_j \sim N_k(\underline{\theta},\sigma^2 R)$ where σ^2 is unknown and R is a known correlation matrix. It should be pointed out that it is impossible in one-stage to find a procedure which achieves conditions (i) and (ii). A two-stage procedure due to Stein (1945) should be considered here. Let

$$S_{ij} = \sum_{e=1}^{n}(X_{ie}-\overline{X}_i)(X_{je}-\overline{X}_j)/(n-1), i,j = 1,\ldots,k, \tag{4.1}$$

and

$$S = (S_{ij}), i,, j = 1,\ldots,k.$$

Then the maximum likelihood estimator of σ^2 is

$$\hat{\sigma}^2 = (n-1)trR^{-1}S/nk.$$

We note that $nk\hat{\sigma}^2/\sigma^2 \sim \chi^2_{k(n-1)}$ and is independent of $\overline{\underline{X}}$. The unbiased estimator of σ^2 is $s^2 = trR^{-1}S/k$ and $\sqrt{n}(\overline{\underline{X}}-\underline{\theta})/s$ has a multivariate t-distribution with correlation matrix R and $k(n-1)$ d.f.. We propose a two-stage procedure R_3 as follows:

(1°) In the first stage, take $N_o > 1$ observations $\underline{X}_j$, $j = 1,\ldots,N_o$.

(2°) Calculate S_{ij} from (4.1) with $n = N_o$ and let $S_o^2 = trR^{-1}S/k$ where $S = (S_{ij})$.

(3°) Obtain a scalar d and a constant c which will be determined later.

(4°) In the second stage, take $N - N_o$ additional observations where $N = \max\{[(dS_o/c)^2]+1, N_o\}$ and $[x]$ denotes the largest integer less than x.

(5°) Compute the overall (first-stage plus second-stage) sample mean $\overline{X}_i = \frac{1}{N}\sum_{e=1}^{N} X_{ie}$, $1 \le i \le k$.

(6°) Select π_o iff $\overline{X}_{[k]} < \theta_o + c$, or select the population associated with $\overline{X}_{[k]}$ iff $\overline{X}_{[k]} \ge \theta_o + c$.

Theorem 6. *For a given P_o^*, if the scalar d in (3°) is the solution of the equation*

$$P\{T_i \le d, 1 \le i \le k\} = P_o^*, \tag{4.2}$$

where $(T_1, \ldots, T_k)$ has a multivariate t distribution with correlation matrix R and $k(N_o - 1)$ d.f., then

$$\inf_{\underline{\theta} \in \Omega_o} P_{\underline{\theta}}\{\pi_o \text{ is selected}\} \ge P_o^*.$$

PROOF.

$$\begin{aligned} P_{\underline{\theta}}\{\pi_o \text{ is selected}\} &= P_{\underline{\theta}}\{\overline{X}_i < \theta_o + c,\ 1 \le i \le k\} \\ &= P\{\frac{\sqrt{N}(\overline{X}_i - \theta_i)}{S_o} < \frac{\sqrt{N}(\theta_o + c - \theta_i)}{S_o},\ 1 \le i \le k\} \\ &\ge P\{T_i < \frac{\sqrt{N}c}{S_o},\ 1 \le i \le k\} \text{ if } \underline{\theta} \in \Omega_o \\ &\ge P\{T_i \le d,\ 1 \le i \le k\}, \text{ since } \frac{\sqrt{N}c}{S_o} \ge d. \end{aligned}$$

where $T_i = \frac{\sqrt{N}(\overline{X}_i - \theta i)}{S_o}$ and $(T_1, \ldots, T_k)$ has the desired distribution. □

REMARK. If $R = (r_{ij})$, $r_{ii} = 1$, $r_{ij} = \rho$, $i \ne j$ $(\rho \ge 0)$, then

$$S_o^2 = trR^{-1}S/k = \frac{[1 + (k-2)\rho]\sum_{i=1}^k S_{ii} - 2\rho \sum_{i>j}^k S_{ij}}{k(1-\rho)[1 + (k-1)\rho]}$$

and

$$S_o^2 = \sum_{i=1}^k S_{ii}/k \text{ if } \rho = 0.$$

Theorem 7. *Assume that R has the form $r_{ij} = \rho$, $i \ne j$. For given P_o^*, P_1^*, if d is the solution of equation (4.2) and c is the solution of*

$$P\{T_k \ge d(c - \Delta)/c,\ T_k \ge T_j - d\Delta/c,\ 1 \le j \le k-1\} = P_1^* \tag{4.3}$$

where $(T_1, \ldots, T_k)$ has the same distribution as that of Theorem 6. Then

$$\inf_{\underline{\theta} \in \Omega_1} P_{\underline{\theta}}\{\pi_{(k)} \text{ is selected}\} \ge P_1^*.$$

PROOF.

$$\begin{aligned} & P_{\underline{\theta}}\{\pi_{(k)} \text{ is selected}\} \\ &= P_{\underline{\theta}}\{\overline{X}_{(k)} \ge \theta_o + c,\ \overline{X}_{(k)} \ge \overline{X}_{(j)},\ 1 \le j \le k-1\} \\ &= P_{\underline{\theta}}\{\frac{\sqrt{N}(\overline{X}_{(k)} - \theta_{[k]})}{S_o} \ge \frac{\sqrt{N}(\theta_o + c - \theta_{[k]})}{S_o}, \frac{\sqrt{N}(\overline{X}_{(k)} - \theta_{[k]})}{S_o} \end{aligned}$$

$$\geq \frac{\sqrt{N}(\overline{X}_{(j)} - \theta_{[j]})}{S_o} + \frac{\sqrt{N}(\theta_{[j]} - \theta_{[k]})}{S_o},\ 1 \leq j \leq k-1\}$$

$$\geq P\{T_k \geq \frac{\sqrt{N}(c-\Delta)}{S_o},\ T_k \geq T_j - \frac{\sqrt{N}\Delta}{S_o},\ 1 \leq j \leq k-1\},\ \text{if } \underline{\theta} \in \Omega_1,$$

$$\geq P\{T_k \geq d(c-\Delta)/c,\ T_k \geq T_j - d\Delta/c,\ 1 \leq j \leq k-1\} \text{ if } c - \Delta < 0$$

where $T_i = \frac{\sqrt{N}(\overline{X}_{(i)} - \theta_{[i]})}{S_o}$ and $(T_1, \ldots, T_k)$ has the desired distribution. □

REMARK. If $\rho = 0$, then $S_o^2 = \sum_{i=1}^{k} \sum_{j=1}^{N_o} (X_{ij} - \overline{X}_i)^2 / k(N_o - 1)$, $\overline{X}_i = \frac{1}{N_o} \sum_{j=1}^{N_o} X_{ij}$ and d is the solution of

$$\int_o^\infty \Phi^k(dz) f_n(z) dz = P_o^*$$

and c is the solution of

$$\int_o^\infty \int_{(d-\frac{d\Delta}{c})z}^\infty \Phi^{k-1}(y + \frac{d\Delta}{c} z) d\Phi(y) f_n(z) dz = P_1^*$$

where $f_n(z)$ is the p.d.f. of $(\chi^2_{k(N_o-1)/k(N_o-1)})^{1/2}$. This result was derived by Bechhofer and Turnbull (1978).

5. UNKNOWN CONTROL, UNKNOWN VARIANCE CASE

Let $\underline{X}_j$, $j = 1, \ldots, n$ be a random sample from $\pi_o, \pi_1, \ldots, \pi_k$ and $\underline{X}_j \sim N_{k+1}(\underline{\theta}, \sigma^2 R)$ where σ^2 is unknown but $R = (r_{ij})$, $r_{ii} = 1$, $r_{01} = \ldots = r_{0k} = \alpha$ and $r_{ij} = \beta$, $i \neq j$, $i, j = 1, \ldots, k$ are known.
We propose a two-stage procedure R_4 as follows:

(1°) In the first stage, take $N_o > 1$ observations $\underline{X}_j$, $1 \leq j \leq N_o$.

(2°) Let $S = (S_{ij})$ $i, j = 0, 1, \ldots, k$,

$$S_{ij} = \sum_{e=1}^{N_o} (X_{ie} - \overline{X}_i)(X_{je} - \overline{X}_j)/(N_o - 1)$$

and

$$S_o^2 = trR^{-1}S/(k+1).$$

(3°) Obtain a scalar d and a constant c which will be determined later.

(4°) In the second stage, take $N - N_o$ additional observations where $N = \max\{[(ds_o/c)^2] + 1, N_o\}$ and $[x]$ denotes the largest integer less than x.

(5°) Compute the overall sample means $\overline{X}_i = \frac{1}{N} \sum_{e=1}^{N} X_{ie}$, $0 \leq i \leq k$.

(6°) Select population π_o iff $\overline{X}_{[k]} < \overline{X}_o + c$, or select the population associated with $\overline{X}_{[k]}$ iff $\overline{X}_{[k]} \geq \overline{X}_o + c$.

Theorem 8. *For a given P_o^*, if d is chosen so that*

$$P\{T_i \leq \frac{d}{(2(1-\alpha))^{1/2}},\ 1 \leq i \leq k\} = P_o^* \tag{5.1}$$

where $(T_1, \ldots, T_k)$ has a multivariate t-distribution with correlation matrix R^* and $(k+1)(N_o - 1)$ d.f., where

$$R^* = (r_{ij}^*),\ r_{ii}^* = 1,\ r_{ij}^* = \frac{1+\beta-2\alpha}{2(1-\alpha)},\ i,j = 1, \ldots, k,$$

then

$$\inf_{\underline{\theta} \in \Omega_o} P_{\underline{\theta}}\{\pi_o \text{ is selected}\} \geq P_o^*.$$

PROOF.

$$P_{\underline{\theta}}\{\pi_o \text{ is selected}\} = P_{\underline{\theta}}\{\overline{X}_i < \overline{X}_o + c,\ 1 \leq i \leq k\}$$

$$= P_{\underline{\theta}}\{\frac{\sqrt{N}[(\overline{X}_i - \overline{X}_o) - (\theta_i - \theta_o)]}{S_o(2(1-\alpha))^{1/2}} \leq \frac{\sqrt{N}(c + \theta_o - \theta_i)}{S_o(2(1-\alpha))^{1/2}},\ 1 \leq i \leq k\}$$

$$\geq P\{\frac{U_i}{S_o/\sigma} \leq \frac{\sqrt{N}c}{S_o(2(1-\alpha))^{1/2}},\ 1 \leq i \leq k\} \text{ if } \underline{\theta} \in \Omega_o$$

where

$$(U_1, \ldots, U_k) \sim N_k(0, R^*),\ (k+1)(N_o - 1)S_o^2/\sigma^2 \sim \chi^2_{(k+1)(N_o-1)}$$

is independent of $(U_1, \ldots, U_k)$.
If $T_i = \frac{U_i}{S_o/\sigma}$, then $(T_1, \ldots, T_k)$ has the desired distribution and

$$\inf_{\underline{\theta} \in \Omega_o} P_{\underline{\theta}}\{\pi_o \text{ is selected}\}$$

$$= P\{T_i \leq \frac{\sqrt{N}c}{S_o(2(1-\alpha))^{1/2}},\ 1 \leq i \leq k\}$$

$$\geq P\{T_i \leq \frac{d}{(2(1-\alpha))^{1/2}},\ 1 \leq i \leq k\}.$$

□

Theorem 9. *For given P_1^*, if c is chosen so that*

$$P\{T_o^* \leq \frac{d(\Delta - c)}{(2(1-\alpha))^{1/2}c},\ T_j^* \leq \frac{d\Delta}{(2(1-\beta))^{1/2}c},\ 1 \leq j \leq k-1\} = P_1^* \tag{5.2}$$

where $(T_o^*, T_1^*, \ldots, T_{k-1}^*)$ has a multivariate t-distribution with correlation matrix $R^{**} = (r_{ij}^{**})$, where

$$r_{ii}^{**} = 1,\ r_{oj}^{**} = ((1-\beta)/(1-\alpha))^{1/2}/2,\ r_{ij}^{**} = \frac{1}{2},\ i \neq j, i,j = 1, \ldots, k-1,$$

then

$$\inf_{\underline{\theta}\in\Omega_1} P_{\underline{\theta}}\{\pi_{(k)} \text{ is selected}\} \geq P_1^*.$$

PROOF.

$$\begin{aligned}
&P_{\underline{\theta}}\{\pi_{(k)} \text{ is selected}\} \\
&= P_{\underline{\theta}}\{\overline{X}_{(k)} \geq \overline{X}_o + c,\ \overline{X}_{(k)} \geq \overline{X}_{(j)},\ 1 \leq j \leq k-1\} \\
&= P_{\underline{\theta}}\{\frac{\sqrt{N}[(\overline{X}_o - \overline{X}_{(k)}) - (\theta_o - \theta_{[k]})]}{S_o(2(1-\alpha))^{1/2}} \leq \frac{\sqrt{N}(\theta_{[k]} - \theta_o - c)}{S_o(2(1-\alpha))^{1/2}}, \\
&\quad \frac{\sqrt{N}[(\overline{X}_{(j)} - \overline{X}_{(k)}) - (\theta_{[j]} - \theta_{[k]})]}{S_o(2(1-\beta))^{1/2}} \leq \frac{\sqrt{N}(\theta_{[k]} - \theta_{[j]})}{S_o(2(1-\beta))^{1/2}},\ 1 \leq j \leq k-1\} \\
&\geq P\{\frac{U_{ok}}{S_o/\sigma} \leq \frac{\sqrt{N}(\Delta - c)}{S_o(2(1-\alpha))^{1/2}},\ \frac{U_{jk}}{S_o/\sigma} \leq \frac{\sqrt{N}\Delta}{S_o(2(1-\beta))^{1/2}},\ 1 \leq j \leq k-1\} \\
&\qquad \text{if } \underline{\theta} \in \Omega_1
\end{aligned}$$

where

$$\begin{aligned}
U_{jk} &= \sqrt{N}[(\overline{X}_{(j)} - \overline{X}_{(k)} - (\theta_{[j]} - \theta_{[k]})]/\sigma(2(1-\beta))^{1/2}, j = 1, \ldots, k-1 \\
U_{ok} &= \sqrt{N}[(\overline{X}_o - \overline{X}_{(k)}) - (\theta_o - \theta_{[k]})]/\sigma(2(1-\alpha))^{1/2} \\
&(U_{ok}, \ldots, U_{k-1,k}) \sim N_k(0, R^{**}).
\end{aligned}$$

R^{**} has the desired property.

$$\begin{aligned}
&\inf_{\underline{\theta}\in\Omega_1} P_{\underline{\theta}}\{\pi_{(k)} \text{ is selected}\} \\
&= P\{T_o^* \leq \frac{\sqrt{N}(\Delta - c)}{S_o(2(1-\alpha))^{1/2}},\ T_j^* \leq \frac{\sqrt{N}\Delta}{S_o(2(1-\beta))^{1/2}},\ 1 \leq j \leq k-1\} \\
&\geq P\{T_o^* \leq \frac{d(\Delta - c)}{c(2(1-\alpha))^{1/2}},\ T_j^* \leq \frac{d\Delta}{c(2(1-\beta))^{1/2}},\ 1 \leq j \leq k-1\}
\end{aligned}$$

$(T_o^*, \ldots, T_{k-1}^*)$ has the desired distribution. □

REMARK. The values of d in the equation (4.2) or the equation (5.1) can be found by using the tables of Gupta (1963) or Krishnaiah and Armitage (1966).

ACKNOWLEDGMENTS

This research was supported by the National Science Council, Republic of China under the contract NSC 75-0208-M008-08.

BIBLIOGRAPHY

Bechhofer, R. E. and Turnbull, B. W. (1978), Two $(k+1)$-Decision Selection Procedures For Comparing K Normal Means with a Specified Standard, *J.A.S.A.*, Vol. 73, 385–392.

Berger, R. L. (1981), Selecting All Treatments Better Than a Control Using Existing Tables, *Commun. Statist.* A10(20), 2025–2037.

Chen, H. J. (1980), On Selecting a Subset Which Contains All Populations Better Than a Control, *Commun. Statist.* A9(8), 851–864.

Gupta, S. S. (1963), Probability Integrals of Multivariate Normal and Multivariate t, *Ann. Math. Statist.* 34, 792–828.

Gupta, S. S., Nagel, K., and Panchapakesan, S. (1973), On the Order Statistics from Equally Correlated Normal Random Variables, *Biometrika* 60, 403–412.

Krishnaiah, P. R. and Armitage, J. V. (1966), Tables for Multivariate t distribution, *Sankhyā B*, 28, 31–56.

Paulson, E. (1952), On the Comparison of Several Experimental Categories with a Control, *Ann. Math. Statist.* 23, 239–246.

Stein, C. (1945), A Two Sample Test for a Linear Hypothesis Whose Power is Independent of the Variance, *Ann. Math. Statist.* 16, 243–258.

Tong, Y. L. (1977), An Ordering Theorem for Conditionally Independent and Identically Distributed Random Variables, *The Annals of Statist.* Vol. 5, 274–277.

INFERENCE ABOUT THE CHANGE-POINT IN A SEQUENCE OF RANDOM VARIABLES: A SELECTION APPROACH

TACHEN LIANG

Department of Statistics
Purdue University
West Lafayette, Indiana 47907, U.S.A.

S. PANCHAPAKESAN

Department of Mathematics
Southern Illinois University
Carbondale, Illinois 62901, U.S.A.

1. INTRODUCTION

Let $X_1, \ldots, X_k$ be a sequence of independent random variables where X_i has the probability density function $p(x|\theta_i), i = 1, \ldots, k$. The θ_i are unknown. It is assumed that $\theta_1 \leq \ldots \leq \theta_k$ with *exactly one* shift; in other words, there is an unknown integer $m (1 \leq m \leq k-1)$ such that $\theta_1 = \ldots = \theta_m < \theta_{m+1} = \ldots = \theta_k$. Then m is called the *change-point*. Problems of inference about a change-point have been investigated by several authors using different approaches. Page (1955, 1957) and Bhattacharyya and Johnson (1968) have considered testing for a shift using nonparametric methods. Hinkley (1970) used asymptotic arguments based on maximum likelihood estimates and likelihood ratio tests. The problem has also been considered within a Bayesian framework by Chernoff and Zacks (1964), Kander and Zacks (1966), Mustafi (1968), Broemeling (1972, 1974), Sen and Srivastava (1973), Smith (1975), and Raftery and Akman (1986). Broemeling and Magalit (1975) have discussed parametric tests for a shift. Worsley (1986) has investigated confidence regions and tests for a change-point using maximum likelihood methods. Inference about change-points has been studied by econometrists under the general context of structural changes in a model. Some recent papers using a Bayesian approach are Booth and Smith (1982), Diaz (1982), Holbert (1982), and Salazar (1982), which appeared in a special issue of the *Journal of Econometrics* (Volume 19, 1982) edited by Lyle Broemeling. The emphasis and the objectives of the present paper are along the lines of the selection approach of multiple decision problems.

For selecting the true change-point we derive three different selection rules. Section 2 deals with a Bayes rule under a fairly general loss function assuming that the prior distributions of the location of the change-point and the amount of change are independent. A gamma-minimax rule is derived in Section 3.

These two rules select one point as the change-point, while the locally optimal rule derived in Section 4 selects a subset (possibly empty) of the $k-1$ points.

2. BAYES PROCEDURE

Let $\Omega_i = \{\underset{\sim}{\theta} = (\theta_1, \ldots, \theta_k) | \theta_1 = \ldots = \theta_i < \theta_{i+1} = \ldots = \theta_k\}, i = 1, \ldots, k-1$, and $\Omega = \bigcup_{i=1}^{k-1} \Omega_i$. If $\underset{\sim}{\theta} \in \Omega_i$, then i is the change-point. Let $\tau_{ij} = \tau(\theta_i, \theta_j)$ be a measure of separation of θ_j from θ_i, defined so that it is increasing in θ_j for a fixed θ_i and $\tau_{ii} = \tau^*$ for all i. For example, if θ is a location parameter, a natural choice is $\tau_{ij} = \theta_j - \theta_i$ with $\tau^* = 0$. Let T_{ij} be a statistic based on X_i and X_j, suitably defined as a sample measure of the separation τ_{ij}.

Let $Y_i = T_{i,i+1}, i = 1, \ldots, k-1$. We assume that $\underset{\sim}{Y} = (Y_1, \ldots, Y_{k-1})$ has a density $f(\underset{\sim}{y}|\underset{\sim}{\tau})$ depending on the parameter $\underset{\sim}{\tau} = (\tau_{12}, \ldots, \tau_{k-1,k})$. For convenience, let $\alpha_i = \tau_{i,i+1}, i = 1, \ldots, k-1$. For $\underset{\sim}{\theta} \in \Omega_i$, we have correspondingly $\underset{\sim}{\alpha} = (\alpha_1, \ldots, \alpha_{k-1})$ such that $\alpha_i = \tau > \tau^* = \alpha_j$ for all $j \neq i$, where τ is the amount of shift. Thus $\Omega_i = \{\underset{\sim}{\alpha} | \alpha_i > \tau^* = \alpha_j$ for all $j \neq i\}$. In this case, we write $f(\underset{\sim}{y}|\underset{\sim}{\alpha}_i(\tau))$ for $f(\underset{\sim}{y}|\underset{\sim}{\tau})$.

Now, our action space is $\mathcal{A} = \{1, 2, \ldots, k-1\}$. Action i corresponds to the decision that i is the change-point. For given $\underset{\sim}{\alpha} \in \Omega_m$ and action a, the associated loss function is defined by

$$L(\underset{\sim}{\alpha}, a) = \begin{cases} \sum\limits_{i=a}^{m-1} L_{mi}^{(1)}(\underset{\sim}{\alpha}) & \text{if } a < m, \\ \sum\limits_{j=m+1}^{a} L_{mj}^{(2)}(\underset{\sim}{\alpha}) & \text{if } a > m, \\ 0 & \text{if } a = m, \end{cases} \tag{2.1}$$

where

$$L_{mi}^{(1)}(\underset{\sim}{\alpha}) \begin{cases} > 0 & \text{and is nonincreasing in } i \text{ for } i < m, \\ = 0 & \text{for } i \geq m, \end{cases} \tag{2.2}$$

and

$$L_{mj}^{(2)}(\underset{\sim}{\alpha}) \begin{cases} > 0 & \text{and is nondecreasing in } j \text{ for } j > m, \\ = 0 & \text{for } j \leq m. \end{cases} \tag{2.3}$$

Let $p(m) = \Pr\{\underset{\sim}{\alpha} \in \Omega_m\}, m = 1, \ldots, k-1$, and let $g(\tau)$ be the prior density of the amount of shift with support on (τ^*, ∞). It is assumed that the two prior distributions are independent.

A decision rule $\underset{\sim}{\delta} = (\delta_1, \ldots, \delta_{k-1})$ is a measurable mapping from $\mathcal{Y}$, the sample space of $\underset{\sim}{Y}$, to $[0,1]^{k-1}$ such that $0 \leq \delta_i(\underset{\sim}{y}) \leq 1$ and $\sum_{i=1}^{k-1} \delta_i(\underset{\sim}{y}) = 1$ for each $\underset{\sim}{y} \in \mathcal{Y}$. The value of $\delta_i(\underset{\sim}{y})$ is the probability of taking action i given the observation $\underset{\sim}{y}$.

Let $r(\underset{\sim}{\delta}|g,\underset{\sim}{p})$ denote the Bayes risk associated with the rule $\underset{\sim}{\delta}$, where $\underset{\sim}{p} = (p(1),\ldots,p(k-1))$. Then

$$r(\underset{\sim}{\delta}|g,\underset{\sim}{p}) = \sum_{m=1}^{k-1} p(m) \int_{\Omega_m} \int_{\mathcal{Y}} \sum_{a=1}^{k-1} L(\underset{\sim}{\alpha}_m(\tau),a)\ \delta_a(\underset{\sim}{y})\ f(\underset{\sim}{y}|\underset{\sim}{\alpha}_m(\tau))\ g(\tau)\ d\underset{\sim}{y}\ d\tau. \tag{2.4}$$

Now, let

$$\begin{aligned} f_m(\underset{\sim}{y}) &= \int_{\tau^*}^{\infty} f(\underset{\sim}{y}|\underset{\sim}{\alpha}_m(\tau))\ g(\tau)\ d\tau \\ g_m(\tau|\underset{\sim}{y}) &= f(\underset{\sim}{y}|\underset{\sim}{\alpha}_m(\tau))\ g(\tau)/f_m(\underset{\sim}{y}) \\ f(\underset{\sim}{y}) &= \sum_{m=1}^{k-1} p(m)\ f_m(\underset{\sim}{y}) \\ h(m|\underset{\sim}{y}) &= p(m)\ f_m(\underset{\sim}{y})/f(\underset{\sim}{y}) \\ R(a|\underset{\sim}{y}) &= \sum_{m=1}^{k-1} h(m|\underset{\sim}{y}) \int_{\tau^*}^{\infty} L(\underset{\sim}{\alpha}_m(\tau),a)\ g_m(\tau|\underset{\sim}{y})\ d\tau. \end{aligned} \tag{2.5}$$

By the usual interchanging of summations and integrals, it can be easily seen that

$$r(\underset{\sim}{\delta}|g,\underset{\sim}{p}) = \sum_{a=1}^{k-1} \int_{\mathcal{Y}} \delta_a(\underset{\sim}{y})\ R(a|\underset{\sim}{y})\ f(\underset{\sim}{y})\ d\underset{\sim}{y}. \tag{2.6}$$

Then $\underset{\sim}{\delta} = (\delta_1,\ldots,\delta_{k-1})$ is a Bayes rule if $\sum_{a\in A(\underset{\sim}{y})} \delta_a(\underset{\sim}{y}) = 1$, where

$$A(\underset{\sim}{y}) = \{a|R(a|\underset{\sim}{y}) = \min_{1\le a'\le k-1} R(a'|\underset{\sim}{y})\}. \tag{2.7}$$

In order to obtain more insight for implementation of this Bayes rule, let

$$\begin{aligned} S_{m,i}^{(1)}(\underset{\sim}{y}) &= \int L_{m,i}^{(1)}(\underset{\sim}{\alpha}_m(\tau))\ g_m(\tau|\underset{\sim}{y})\ d\tau, \quad i < m \\ S_{m,j}^{(2)}(\underset{\sim}{y}) &= \int L_{m,j}^{(2)}(\underset{\sim}{\alpha}_m(\tau))\ g_m(\tau|\underset{\sim}{y})\ d\tau, \quad j > m \\ D(a|\underset{\sim}{y}) &= R(a+1|\underset{\sim}{y}) - R(a|\underset{\sim}{y}). \end{aligned} \tag{2.8}$$

First, note that using (2.1), we can express $R(a|\underset{\sim}{y})$ as

$$R(a|\underset{\sim}{y}) = \sum_{m=a+1}^{k-1} \sum_{i=a}^{m-1} h(m|\underset{\sim}{y})\ S_{m,i}^{(1)}(\underset{\sim}{y}) + \sum_{m=1}^{a-1} \sum_{j=m+1}^{a} h(m|\underset{\sim}{y})\ S_{m,j}^{(2)}(\underset{\sim}{y}). \tag{2.9}$$

Lemma 2.1. *For fixed m and $\underset{\sim}{y}$,*
(a) $S_{m,i}^{(1)}(\underset{\sim}{y})$ *is nonincreasing in i for $i < m$, and*
(b) $S_{m,j}^{(2)}(\underset{\sim}{y})$ *is nondecreasing in j for $j > m$.*

PROOF. Omitted as it is obvious. □

Lemma 2.2. *For given $\underset{\sim}{y}$, $D(a|\underset{\sim}{y})$ is nondecreasing in a.*

PROOF. From (2.9), it is easy to see that

$$D(a|\underset{\sim}{y}) = \sum_{m=1}^{a} h(m|\underset{\sim}{y})\ S^{(2)}_{m,a+1}(\underset{\sim}{y}) - \sum_{m=a+1}^{k-1} h(m|\underset{\sim}{y})\ S^{(1)}_{m,a}(\underset{\sim}{y}).$$

Therefore,

$$\begin{aligned} &D(a+1|\underset{\sim}{y}) - D(a|\underset{\sim}{y}) \\ &= \sum_{m=1}^{a} h(m|\underset{\sim}{y}) \left[S^{(2)}_{m,a+2}(\underset{\sim}{y}) - S^{(2)}_{m,a+1}(\underset{\sim}{y})\right] + h(a+1,\underset{\sim}{y}) S^{(2)}_{a+1,a+2}(\underset{\sim}{y}) \\ &\quad + \sum_{m=a+2}^{k-1} h(m|\underset{\sim}{y}) \left[S^{(1)}_{m,a}(\underset{\sim}{y}) - S^{(2)}_{m,a+1}(\underset{\sim}{y})\right] + h(a+1|\underset{\sim}{y}) S^{(1)}_{a+1,a}(\underset{\sim}{y}). \end{aligned}$$

Using Lemma 2.1 and the nonnegativity of $h(m|\underset{\sim}{y})$ and $S^{(r)}_{m,i}, r = 1, 2$, it can now be seen that $D(a+1|\underset{\sim}{y}) - D(a|\underset{\sim}{y}) \geq 0$. □

Based on Lemma 2.2, the Bayes rule $\underset{\sim}{\delta}$ can be expressed as follows:

Randomize your decision over the set $B(\underset{\sim}{y}) = \{a|D(a|\underset{\sim}{y}) = 0\}$. If the set $B(\underset{\sim}{y})$ is vacuous, then choose action b where b is the largest a for which $D(a|\underset{\sim}{y}) < 0$. If such a b does not exist, choose action 1.

REMARK 2.1. Because of the monotonicity of $D(a|\underset{\sim}{y})$ in a, the set $B(\underset{\sim}{y})$ is either vacuous or it consists of consecutive members of the set $\{1, 2, \ldots, k-1\}$. We can define a nonrandomizd rule, by taking the action corresponding to the smallest member of $B(\underset{\sim}{y})$.

3. Γ-MINIMAX RULE

In this section, we assume a uniform prior for the change-point, i.e. $p(m) = \frac{1}{k-1}, m = 1, \ldots, k-1$. Further, the loss function is taken to be

$$L(\underset{\sim}{\alpha}_m(\tau), a) = |m - a| L\ I_{[\tau_0,\infty)}(\tau) \tag{3.1}$$

where $L > 0$ is a known constant, $\tau_0 > \tau^*$, and I_B denotes the indicator function of the set B.

Γ-minimax selection rules for selecting the best population has been considered in the literature; see Gupta and Huang (1977), for example. For $0 < \pi_0 < 1$, let

$$\Gamma = \{G| \int_{\tau_0}^{\infty} dG(\tau) \leq \pi_0\} \tag{3.2}$$

A rule $\underset{\sim}{\delta}^\circ$ is a Γ-minimax rule if, for any other rule $\underset{\sim}{\delta}$, $\sup_{G\in\Gamma} r(\underset{\sim}{\delta}, G) \geq \sup_{G\in\Gamma} r(\underset{\sim}{\delta}^\circ, G)$.

We now state and prove a theorem which gives a Γ-minimax rule under certain condition.

Theorem 3.1. *Suppose that there exists a $\tau' \geq \tau_0$ such that, for each $m, 1 \leq m \leq k-1$,*

$$\sup_{\tau\geq\tau_0} \int_{\mathcal{Y}} \sum_{j=1}^{k-1} L(\underset{\sim}{\alpha}_m(\tau), j)\ \delta_j^\circ(\underset{\sim}{y})\ f(\underset{\sim}{y}|\underset{\sim}{\alpha}_m(\tau))\ d\underset{\sim}{y}$$
$$= \int_{\mathcal{Y}} \sum_{j=1}^{k-1} L(\underset{\sim}{\alpha}_m(\tau'), j)\ \delta_j^\circ(\underset{\sim}{y})\ f(\underset{\sim}{y}|\underset{\sim}{\alpha}_m(\tau'))\ d\underset{\sim}{y} \tag{3.3}$$

where

$$\underset{\sim}{\delta}_j^\circ(\underset{\sim}{y}) = \begin{cases} \frac{1}{|M|} & \text{if } L_j(\underset{\sim}{y}) \leq \min_{1\leq i\leq k-1} L_i(\underset{\sim}{y}) \\ 0 & \text{otherwise,} \end{cases}$$

$$L_i(\underset{\sim}{y}) = \sum_{j=1}^{k-1} |i-j| L f(\underset{\sim}{y}|\underset{\sim}{\alpha}_j(\tau')), \tag{3.4}$$

and $|M|$ is the cardinality of the set

$$M = \{j | L_j(\underset{\sim}{y}) = \min_{1\leq i\leq k-1} L_i(\underset{\sim}{y})\}.$$

Then $\underset{\sim}{\delta}^\circ = (\delta_1^\circ, \ldots, \delta_{k-1}^\circ)$ is a Γ-minimax rule.

PROOF. Let G_0 be any distribution such that $G_0(\tau') = 1, G_0(\tau_0-) = G_0(\tau'-) = 1 - \pi_0$. Then $G_0 \in \Gamma$. Therefore, for any rule $\underset{\sim}{\delta}$,

$$\begin{aligned}
&\sup_{G\in\Gamma} r(\underset{\sim}{\delta}|G) \geq r(\underset{\sim}{\delta}|G_0) \\
&= \sum_{m=1}^{k-1} \frac{1}{k-1} \int_{\mathcal{Y}} \sum_{a=1}^{k-1} |m-a| L\delta_a(\underset{\sim}{y})\ \pi_0 f(\underset{\sim}{y}|\underset{\sim}{\alpha}_m(\tau'))\ d\underset{\sim}{y} \\
&= \sum_{a=1}^{k-1} \frac{1}{k-1} \int_{\mathcal{Y}} \pi_0\ \delta_a(\underset{\sim}{y}) L_a(\underset{\sim}{y})\ d\underset{\sim}{y} \\
&\geq \sum_{a=1}^{k-1} \frac{1}{k-1} \int_{\mathcal{Y}} \pi_0 \delta_a^\circ(\underset{\sim}{y}) L_a(\underset{\sim}{y})\ d\underset{\sim}{y} \\
&= \sum_{m=1}^{k-1} \frac{1}{k-1} \int_{\mathcal{Y}} \sum_{a=1}^{k-1} |m-a|\ L\ \delta_a^\circ(\underset{\sim}{y})\ \pi_0\ f(\underset{\sim}{y}|\underset{\sim}{\alpha}_m(\tau'))\ d\underset{\sim}{y} \\
&\geq \sum_{m=1}^{k-1} \frac{1}{k-1} \int_{\tau} \int_{\mathcal{Y}} \sum_{a=1}^{k-1} |m-a| L\ \delta_a^\circ(\underset{\sim}{y})\ f(\underset{\sim}{y}|\underset{\sim}{\alpha}_m(\tau'))\ d\underset{\sim}{y}\ dG(\tau),
\end{aligned}$$

$$
\begin{aligned}
&\text{for any } G \in \Gamma \\
&\geq \sum_{m=1}^{k-1} \frac{1}{k-1} \int_\tau \int_{\underset{\sim}{y}} \sum_{a=1}^{k-1} |m-a|\, L\, \delta_a^\circ(\underset{\sim}{y})\, f(\underset{\sim}{y}|\underset{\sim}{\alpha}_m(\tau))\, d\underset{\sim}{y}\, dG(\tau), \\
&\text{because of (3.3)} \\
&= r(\underset{\sim}{\delta}^\circ, G), \text{ for any } G \in \Gamma.
\end{aligned}
$$

Therefore, $\sup_{G\in\Gamma} r(\underset{\sim}{\delta}|G) \geq \sup_{G\in\Gamma} r(\underset{\sim}{\delta}^\circ|G)$. □

Now, let

$$D^*(a|\underset{\sim}{y}) = L_{a+1}(\underset{\sim}{y}) - L_a(\underset{\sim}{y}), \qquad a = 1, \ldots, k-1. \tag{3.5}$$

Since $D^*(a|\underset{\sim}{y}) = L \sum_{j=1}^{a} f(\underset{\sim}{y}|\underset{\sim}{\alpha}_j(\tau')) - L \sum_{j=a+1}^{k-1} f(\underset{\sim}{y}|\underset{\sim}{\alpha}_j(\tau')), D^*(a|\underset{\sim}{y})$ is nondecreasing in a. Using this property, we can state the Γ-minimax rule in the form:

Randomize your decision over the set $B^*(\underset{\sim}{y}) = \{a|D^*(a|\underset{\sim}{y}) = 0\}$. If the set is vacuous, then choose action b, where b is the largest a for which $D^*(a|\underset{\sim}{y}) < 0$. If such a b does not exist, choose action 1.

A remark similar to Remark 2.1 holds here.

4. LOCALLY OPTIMAL RULE

In this section, a selection rule $\underset{\sim}{\delta}$ is defined by the individual selection probabilities $\delta_i(\underset{\sim}{y}), i = 1, \ldots, k-1$. This results in selection of a subset (possibly empty) of the k-1 points. Further, the selected points need not be consecutive. We restrict ourselves to the class of rules defined by

$$\mathcal{D} = \{\underset{\sim}{\delta}| \lim_{\tau\downarrow\tau^*} E_{\underset{\sim}{\alpha}_m(\tau)}(\delta_m(\underset{\sim}{Y})) = P^* \text{ for } m = 1, \ldots, k-1\}. \tag{4.1}$$

In other words, considering the configurations of $\underset{\sim}{\alpha}$ for any given change-point location, the limiting value of the probability of choosing the true location is P^* as the amount of shift goes to τ^* (corresponding to equi-parameter configuration of $\underset{\sim}{\theta}$). This is similar to the class considered by Gupta, Huang and Nagel (1979) but for a different problem.

We seek a rule $\underset{\sim}{\delta} \in \mathcal{D}$ which is locally optimal in the sense that it maximizes

$$\mathcal{Q} = \sum_{m=1}^{k-1} \lim_{\tau\downarrow\tau^*} \frac{\partial}{\partial\tau} E_{\underset{\sim}{\alpha}_m(\tau)}\, (\delta_m(\underset{\sim}{Y})). \tag{4.2}$$

The quantity $\mathcal{Q}$ reflects the sensitivity of the rule in making the correct decision in a neighborhood of an equi-parameter configuration of $\underset{\sim}{\theta}$. Such a measure in a different context has been used by Huang, Panchapakesan and Tseng (1984).

In deriving a locally optimal rule, we assume usual regularity conditions so that limit operations and differentiations under the integral signs can be carried out. Then, we can write

$$\mathcal{Q} = \sum_{m=1}^{k-1} \int_{\mathcal{Y}} \delta_m(\underset{\sim}{y})\ f_{(m)}\ (\underset{\sim}{y}|\underset{\sim}{\tau}^*)\ d\underset{\sim}{y}$$

where $f_{(m)}\ (\underset{\sim}{y}|\underset{\sim}{\alpha}_m(\tau)) = \frac{\partial}{\partial\tau}\ f(\underset{\sim}{y}|\underset{\sim}{\alpha}_m(\tau))$ and $\underset{\sim}{\tau}^* = (\tau^*, \ldots, \tau^*)$.

Theorem 4.1. *Under regularity conditions, a rule belonging to $\mathcal{D}$ which maximizes $\mathcal{Q}$ is given by*

$$\delta^*_m(\underset{\sim}{y}) = \begin{cases} 1 & \\ \lambda_m & \text{if } \frac{f_{(m)}(\underset{\sim}{y}|\underset{\sim}{\tau}^*)}{f(\underset{\sim}{y}|\underset{\sim}{\tau}^*)} \begin{matrix} > \\ = \\ < \end{matrix} c_m \\ 0 & \end{cases} \tag{4.3}$$

where λ_m and c_m are determined such that

$$\int_{\mathcal{Y}} \delta^*_m(\underset{\sim}{y})\ f(\underset{\sim}{y}|\underset{\sim}{\tau}^*)\ d\underset{\sim}{y} = P^*,\ m = 1, \ldots, k-1. \tag{4.4}$$

PROOF. The proof is straightforward, by noting that for any rule $\underset{\sim}{\delta} \in \mathcal{D}$,

$$\int_{\mathcal{Y}} \delta_m(\underset{\sim}{y}) f(\underset{\sim}{y}|\underset{\sim}{\tau}^*) d\underset{\sim}{y} = P^* \text{ for } m = 1, \ldots, k-1,$$

and that

$$\sum_{m=1}^{k-1} \int_{\mathcal{Y}} (\delta^*_m(\underset{\sim}{y}) - \delta_m(\underset{\sim}{y}))(f_{(m)}(\underset{\sim}{y}|\underset{\sim}{\tau}^*) - c_m f(\underset{\sim}{y}|\underset{\sim}{\tau}^*)) d\underset{\sim}{y} \geq 0.$$

□

EXAMPLE. Suppose we have a sequence of sample means (based on n independent observations) from normal distributions $N(\theta_i, \sigma^2), i = 1, \ldots, k$, where σ^2 is known. We take $Y_i = \bar{X}_{i+1} - \bar{X}_i$, $i = 1, \ldots, k-1$. Then, for $\underset{\sim}{\theta} \in \Omega_m$, $\underset{\sim}{Y}$ has a $(k-1)$-variate normal distribution having mean vector with $\tau = \theta_{m+1} - \theta_m$ as the m-th component and zero everywhere else, and covariance matrix $V = (\sigma_{ij})$, where $\sigma_{ii} = \frac{2\sigma^2}{n}$ and $\sigma_{ij} = -\frac{\sigma^2}{n}$ or 0 according as $|i-j|$ is $=$ or $>$ 1. It is easy to see that

$$\frac{f_{(m)}(\underset{\sim}{y}|\underset{\sim}{\tau}^*)}{f(\underset{\sim}{y}|\underset{\sim}{\tau}^*)} = \sum_{i=1}^{k-1} y_i \sigma^{im}$$

where $V^{-1} = (\sigma^{im})$. By noting that $\sigma^{i1} = \frac{(k-i)n}{k\sigma^2}$ for $1 \leq i \leq k-1$, and $\sigma^{ij} = j\sigma^{i1}$ for $1 \leq j \leq i$, we get

$$\sum_{i=1}^{k-1} y_i\ \sigma^{im} = \frac{(k-m)mn}{k\sigma^2} \left[\frac{\bar{X}_{m+1} + \ldots + \bar{X}_k}{k-m} - \frac{\bar{X}_1 + \ldots + \bar{X}_m}{m} \right].$$

This gives the intuitively appealing rule:

$$\delta_m(\underset{\sim}{y}) = \begin{cases} 1 & \text{if } \frac{\bar{X}_{m+1}+\ldots+\bar{X}_k}{k-m} - \frac{\bar{X}_1+\ldots+\bar{X}_m}{m} > c_m; \\ 0 & \text{otherwise;} \end{cases}$$

where $c_m = \sigma\sqrt{\frac{k}{nm(k-m)}}\ \Phi^{-1}(1-P^*)$.

ACKNOWLEDGMENTS

This research is supported partly by the Office of Naval Research Contract N00014-84-C-0167 at Purdue University.

BIBLIOGRAPHY

Bhattacharyya, G. K. and Johnson, R. A. (1968). Non-parametric tests for shift in location. *Annals of Mathematical Statistics*, 39, 1731–1743.

Booth, N. B. and Smith, A. F. M. (1982). A Bayesian approach to retrospective identification of change-points. *Journal of Econometrics*, 19, 7–22.

Broemeling, L. D. (1972). Bayesian procedures for testing a change in a sequence of random variables. *Metron.*, 30, 1–14.

Broemeling, L. D. (1974). Bayesian inferences about a changing sequence of random variables. *Communications in Statistics*, 3, 243–255.

Broemeling, L. D. and Magalit, H. F. (1975). Some tests for a shift in the mean of a normal distribution occurring at unknown points. *Metron*, 33, 235–243.

Chernoff, H. and Zacks, S. (1964). Estimating the current mean of a normal distribution which is subjected to changes in time. *Annals of Mathematical Statistics*, 35, 999–1018.

Diaz, J. (1982). Bayesian detection of a change of scale parameter in sequences of independent gamma random variables. *Journal of Econometrics*, 19, 23–29.

Gupta, S. S. and Huang, D.-Y. (1977). On some gamma-minimax selection and multiple comparison procedures. *Statistical Decision Theory and Related Topics II*, eds. S. S. Gupta and D. S. Moore, Academic Press, New York, 139–155.

Gupta, S. S., Huang, D.-Y. and Nagel, K. (1979). Locally optimal subset selection procedures based on ranks. *Optimizing Methods in Statistics*, ed. J. S. Rustagi, Academic Press, New York, 251–260.

Hinkley, D. V. (1970). Inference about the change-point in a sequence of random variables. *Biometrika*, 57, 1–17.

Holbert, D. (1982). A Bayesian analysis of a switching linear model. *Journal of Econometrics*, 19, 77–87.

Huang, D.-Y., Panchapakesan, S. and Tseng, S.-T. (1984). Some locally optimal subset selection rules for comparison with a control. *Journal of Statistical Planning and Inference*, 9, 63–72.

Kander, Z. and Zacks, S. (1966). Test procedures for possible changes in parameters of statistical distributions occurring at unknown time points. *Annals of Mathematical Statistics*, 37, 1196–1210.

Mustafi, C. K. (1968). Inference problems about parameters which are subjected to changes over time. *Annals of Mathematical Statistics*, 39, 840–854.

Page, E. S. (1955). A test for a change in a parameter occurring at an unknown point. *Biometrika*, 42, 523–527.

Page, E. S. (1957). On problems in which a change in parameter occurs at an unknown point. *Biometrika*, 42, 248–252.

Raftery, A. E. and Akman, V. E. (1986). Bayesian analysis of a Poisson process with a change-point. *Biometrika*, 73, 85–89.

Salazar, D. (1982). Structural changes in time series models. *Journal of Econometrics*, 19, 147–163.

Sen, A. K. and Srivastava, M. S. (1973). On multivariate tests for detecting change in mean. *Sankhya A*, 35, 173–186.

Smith, A. F. M. (1975). A Bayesian approach to inference about a change-point in a sequence of random variables. *Biometrika*, 62, 407–416.

Worsley, K. J. (1986). Confidence regions and tests for a change-point in a sequence of exponential family random variables. *Biometrika*, 73, 91–104.

ON CONFIDENCE SETS IN MULTIPLE COMPARISONS

GUNNAR STEFANSSON

Marine Research Institute
Reykjavik, Iceland

WOO-CHUL KIM

Seoul National University
Seoul, South Korea

JASON C. HSU

Department of Statistics
The Ohio State University
Columbus, Ohio 43210-1247, U.S.A.

1. A NONSTANDARD PIVOTAL METHOD

The connection between tests and confidence sets is well known (Lehmann, 1986, p. 90).

Lemma 1.1. *If the observable* $\underline{Y}$ *has distribution* $F_{\underline{\theta}}(\underline{y})$ *indexed by* $\underline{\theta} \in \Theta$, *and if* $\{\phi_{\underline{\theta}}(\underline{Y})\colon \underline{\theta} \in \Theta\}$ *is a family of tests such that* $P_{\underline{\theta}}\{\phi_{\underline{\theta}}(\underline{Y}) = 0\} = (\geq) 1 - \alpha$ *for each* $\underline{\theta} \in \Theta$, *then* $P_{\underline{\theta}}\{\underline{\theta} \in C(\underline{Y})\} = (\geq) 1 - \alpha$ *for all* $\underline{\theta}$ *if* $C(\underline{Y}) = \{\underline{\theta}\colon \underline{\theta} \in \Theta, \phi_{\underline{\theta}}(\underline{Y}) = 0\}$.

Somewhat surprisingly, in the usual applications of the pivotal method of deriving confidence sets, this connection has not been fully exploited. Suppose $\underline{Y}$ has a distribution $F(\underline{Y} - \underline{\theta})$, $\underline{\theta} \in \Theta$, then one usually considers a family of equivariant tests $\phi_{\underline{\theta}}(\underline{Y}) = \phi_0(\underline{Y} - \underline{\theta})$, and pivots the full null acceptance region $A = \{\underline{y}\colon \phi_0(\underline{y}) = 0\}$ to obtain $C(\underline{Y}) = -A + \underline{Y}$. However, we will show three examples where, by choosing appropriate non-equivariant tests, new and useful results in multiple comparisons are obtained.

In the first example, a confidence set having coverage probability exactly $1 - \alpha$ which generates the existing constrained Multiple Comparisons with the Best (MCB) and Ranking and Selection results (all of which were conservative) is given. In the second example, a confidence set which generates a stepwise multiple comparisons procedure is given, dispelling a longstanding myth (Lehmann, 1986, p. 388) that stepwise procedures have no confidence sets. In the third example, it is shown that for comparing $k(\geq 2)$ treatments, regardless of k, the sample best treatment can be inferred to be the true best when the two-sample two-sided size-α t-test between the sample best and the sample second best rejects.

The derivations in all three examples assume that a larger treatment effect implies a better treatment. The analogous results when a smaller treatment effect is better are obtained by first taking negatives of all the observations, computing the inference, and then again taking negatives.

2. MULTIPLE COMPARISONS WITH THE BEST (MCB)

In comparing k treatment effects $\theta_1, \ldots, \theta_k$, it has been proposed (Hsu, 1984) that in addition to all-pairwise comparisons (MCA), where the parameters of interest are

$$\theta_i - \theta_j (\text{all } i \neq j),$$

and multiple comparisons with a control (MCC), where the parameters of interest are

$$\theta_i - \theta_k (i = 1, \ldots, k-1),$$

a useful third formulation is multiple comparisons with the best (MCB), where the parameters of interest are

$$\theta_i - \max_{j \neq i} \theta_j (i = 1, \ldots, k).$$

(Here it is assumed that a larger treatment effect implies a better treatment.) The MCB parameters differ from the MCA and MCC parameters in that they are nonlinear.

Suppose $\underline{Y} = (Y_1, \ldots, Y_k)$ has an absolutely continuous distribution $F(\underline{Y} - \underline{\theta})$ where $\underline{\theta} = (\theta_1, \ldots, \theta_k)$ and F is exchangeable. Let $(1), \ldots, (k)$ be (unknown) indices such that

$$\theta_{(1)} \leq \theta_{(2)} \leq \ldots \leq \theta_{(k)}. \tag{2.1}$$

In case of ties, any arrangement is permissible as long as (2.1) is satisfied.

In Hsu (1984), it was shown that if

$$P\{Y_{(k)} - \theta_{(k)} \geq Y_i - \theta_i - d \text{ for all } i \neq (k)\} = 1 - \alpha$$

and $D_i^* = \min(0, Y_i - \max_{j \neq i} Y_j - d)$, $D_i^{**} = \max(0, Y_i - \max_{j \neq i} Y_j + d)$, then

$$P_{\underline{\theta}}\{D_i^* \leq \theta_i - \max_{j \neq i} \theta_j \leq D_i^{**} \text{ for } i = 1, \ldots, k\} \geq 1 - \alpha \tag{2.2}$$

for all $\underline{\theta}$. The confidence statement (2.2) was derived by replacing the event $E = \{Y_{(k)} - \theta_{(k)} \geq Y_i - \theta_i - d \text{ for all } i \neq (k)\}$, which is pivotal but has a not directly observable variable $Y_{(k)}$, by another event which strictly contains E for all $\underline{\theta}$ except when $\theta_{(k)} - \theta_{(k-1)} > d$, in which case the two events are identical. Consequently, the coverage probability (2.2) is strictly greater than $1 - \alpha$ except when $\theta_{(k)} - \theta_{(k-1)} > d$. In other words, the confidence set $[D_1^*, D_1^{**}] \times \ldots \times [D_k^*, D_k^{**}]$ for $(\theta_1 - \max_{j \neq 1} \theta_j, \ldots, \theta_k - \max_{j \neq k} \theta_j)$ is typically conservative rather than exact in terms of coverage probability. So an interesting question has been, "Does an exact MCB confidence set exist?"

Exact MCB confidence sets are obtained by dividing Θ into k regions $\Theta_1 \bigcup \ldots \bigcup \Theta_i$, where $\Theta_i = \{\underline{\theta}\colon \theta_i \geq \max_{j \neq i} \theta_j\}$, and pivoting a different test ϕ_i in each Θ_i.

Let $H = \{1, \ldots, k\}$ and let $g\colon \Theta \to H$ be defined by $\theta_{g(\theta)} = \max_{1 \leq i \leq k} \theta_i$, i.e., $g(\theta)$ is the index of the largest component of $\underline{\theta} = \max_{1 \leq i \leq k} \theta_i$, where we agree to resolve ties for $\max_{1 \leq i \leq k} \theta_i$ is some arbitrary but fixed manner.

For each $\underline{\theta}$, suppose $g(\underline{\theta}) = i$ say, then the test for that $\underline{\theta}$ is defined by

$$\begin{aligned} \phi_{\underline{\theta}}(\underline{Y}) = \phi_i(\underline{Y} - \underline{\theta}) &= 0 \text{ if } Y_i - \theta_i \geq Y_j - \theta_j - d \text{ for all } j \neq i \\ &= 1 \text{ otherwise} \end{aligned} \tag{2.3}$$

where d, as in the definition of D_i^* and D_i^{**}, is chosen such that ϕ_i is a size-α test. We assume $d > 0$. By Lemma 1.1, the set $C(\underline{Y}) = \{\underline{\theta}\colon \phi_{\underline{\theta}}(\underline{Y}) = 0\}$ is an exact $100(1-\alpha)\%$ confidence set for $\underline{\theta}$. One way to think of $C(\underline{Y})$ is that $\underline{\theta} \in C^*(\underline{Y})$, the closure of $C(\underline{Y})$, if and only if there exists i such that $\underline{\theta} \in \Theta_i$ and $\phi_i(\underline{Y} - \underline{\theta}) = 0$. Observe the following.

(a) If $Y_i - \max_{j \neq i} Y_j + d \leq 0$, then $\text{int}(\Theta_i) \bigcap C(\underline{Y}) = \varnothing$.

(b) Since $d > 0$, $\underline{\theta} = (\theta, \ldots, \theta)$ is always accepted by the test $\phi_M(\underline{Y} - \underline{\theta})$ which accepts if $Y_M \geq \max_{j \neq M} Y_j - d$, where M is the random index such that $Y_M = \max_{1 \leq i \leq k} Y_i$. So $C^*(\underline{Y})$ always contains the equi-angular line. In fact, the coverage probability of $C^*(\underline{Y})$, as a function of $\underline{\theta}$, is discontinuous. It is exactly $1 - \alpha$ for each interior point $\underline{\theta}$, higher for any $\underline{\theta}$ on the boundaries of Θ_i, equaling one for any $\underline{\theta} = (\theta, \ldots, \theta)$.

Geometrically, $C^*(\underline{Y})$, being translation invariant, is a cylinder set parallel to and always containing the equi-angular line.

To obtain useful inference on the MCB parameters $\theta_i - \max_{j \neq i} \theta_j$ from $C(\underline{Y})$, we derive upper and lower bounds on $\theta_i - \max_{j \neq i} \theta_j$ for $\underline{\theta} \in C(\underline{Y})$ in the following lemma.

Lemma 2.1. *For $i = 1, \ldots, k$, let*

$$D_i^L = \inf_{\underline{\theta} \in C^*(\underline{Y})} \theta_i - \max_{j \neq i} \theta_j,$$

$$D_i^U = \sup_{\underline{\theta} \in C^*(\underline{Y})} \theta_i - \max_{j \neq i} \theta_j,$$

Then with the notation $-x^- = \min(0,x)$ *and* $x^+ = \max(0,x)$, *we have*

$$D_i^L = -(Y_i - \max_{j\neq i} Y_j - d)^- = D_i^*,$$

$$D_i^U = +(Y_i - \max_{j\neq i} Y_j - d)^+ = D_i^{**}.$$

PROOF. First note that $\theta_i - \max_{j\neq i}\theta_j > 0$ if and only if $\underline{\theta} \in \text{int}(\Theta_i)$. Consider D_i^U. If $Y_i - \max_{j\neq i} Y_j + d > 0$, then (2.3) shows $D_i^U = Y_i - \max_{j\neq i} Y_j + d$. If $Y_i - \max_{j\neq i} Y_j + d \leq 0$, then $D_i^U = 0$ because $\text{int}(\Theta_i) \bigcap C(\underline{Y}) = \emptyset$ but $\underline{0} \in C^*(\underline{Y})$. Thus $D_i^U = (Y_i - \max_{j\neq i} Y_j + d)^+$. Next consider D_i^L. First suppose $Y_i - \max_{j\neq i} Y_j - d < 0$. If $Y_i - \max_{j\neq i} Y_j - d = Y_i - Y_M - d < 0$ for some $M \neq i$, then any $\underline{\theta}$ with $\theta_M - \theta_i = Y_M - Y_i + d$, and $\theta_M = \theta_j$, for all $j \neq M$ or i, is accepted by $\phi_M(\underline{Y} - \underline{\theta})$ which accepts when $Y_M - \theta_M \geq Y_j - \theta_j - d$ for all $j \neq M$. Thus $\inf_{\underline{\theta} \in C^*(\underline{Y})} \theta_i - \max_{j\neq i}\theta_j \leq Y_i - \max_{j\neq i} Y_j - d < 0$. However, if $\underline{\theta} \in \Theta_j$ and $\theta_j - \theta_i > Y_M - Y_i + d > 0$, then $\phi_j(\underline{Y} - \underline{\theta})$ rejects because $Y_j - \theta_j < Y_i - \theta_i - d$ and therefore $\underline{\theta} \notin C(\underline{Y})$. So $\inf_{\underline{\theta} \in C^*(\underline{Y})} \theta_i - \max_{j\neq i}\theta_j = Y_i - \max_{j\neq i} Y_j - d$ if the latter is negative. On the other hand, if $Y_i - \max_{j\neq i} Y_j - d \geq 0$, then $\text{int}(\Theta_j) \bigcap C(\underline{Y}) = \emptyset$ for all $j \neq i$, while $(\theta, \ldots, \theta) \in C(\underline{Y})$ always. Hence, $\inf_{\underline{\theta} \in C(\underline{Y})} \theta_i - \max_{j\neq i}\theta_j = 0$ in this case. Thus $D_i^L = -(Y_i - \max_{j\neq i} Y_j - d)^-$. □

Since the coverage probability of $C(\underline{Y})$ is exactly $1 - \alpha$, we have

Theorem 2.1. *(Hsu 1984) For all* $\underline{\theta}$

$$P_{\underline{\theta}}\{D_i^* \leq \theta_i - \max_{j\neq i}\theta_j \leq D_i^{**} \textit{ for } i = 1, \ldots, k\} \geq 1 - \alpha.$$

Perhaps the most important aspect of the derivation of exact constrained MCB confidence sets is the insight that the "constraints" result from pivoting one-sided tests inwards toward the equi-angular line. In the special case of $k = 2$, the confidence interval was derived by Pratt (1961) as the confidence for $\theta_1 - \theta_2$ having the smallest expected length when $\theta_1 = \theta_2$.

3. STEPWISE MULTIPLE COMPARISONS WITH A CONTROL (MCC)

Consider the problem of comparing k new treatments with a control or standard. Denote the control treatment effect by θ_0 and the new treatment effects by $\theta_1, \ldots, \theta_k$, and the corresponding observables by $Y_0, Y_1, \ldots, Y_k$. We assume that $\underline{Y} = (Y_0, Y_1, \ldots, Y_k)$ is distributed according to $F(\underline{Y} - \underline{\theta})$ with $\underline{\theta} = (\theta_0, \theta_1, \ldots, \theta_k)$.

Suppose the primary purpose of the analysis is to discover as many new treatments that are better than the control as possible, while guaranteeing that the probability of incorrectly asserting any treatment worse than the control as better is no more than α.

The traditional Dunnett (1955) and Gupta and Sobel (1958) method pivots the test

$$\begin{aligned} \phi(\underline{Y} - \underline{\theta}) &= 0 \text{ if } Y_0 - \theta_0 > Y_i - \theta_i - d_k \text{ for } i = 1, \ldots, k \\ &= 1 \text{ otherwise} \end{aligned}$$

where d_k is chosen so that ϕ is of size α, to obtain the confidence statement

$$P\{\theta_i - \theta_0 > \underline{D}_i \text{ for } i = 1, \ldots, k\} = 1 - \alpha \text{ for all } \underline{\theta} \tag{3.1}$$

where $\underline{D}_i = Y_i - Y_0 - d_k$. Thus, if only treatments with positive lower confidence bounds are asserted to be better than the control, then the probability of at least one incorrect assertion is $\leq \alpha$.

However, if the purpose of the analysis is to find new treatments better than the standard, then for any $\underline{\theta}$, it is not necessary to test for θ_i's for which $\theta_i > \theta_0$. This leads us to define the following collection of tests.

Let H be the power set of $\{1, \ldots, k\}$, and let g: $\mathsf{R}^k \to H$ be defined by $g(\underline{\theta}) = \{i\colon 1 \leq i \leq k, \theta_i \leq \theta_0\}$. For any $\underline{\theta}$ such that $g(\underline{\theta}) = B$, define the test for that $\underline{\theta}$ by

$$\begin{aligned} \phi_{\underline{\theta}}(\underline{Y}) = \phi_B(\underline{Y} - \underline{\theta}) &= 0 \text{ if } Y_0 - \theta_0 > Y_i - \theta_i - d_{|B|} \text{ for all } i \in B \\ &= 1 \text{ otherwise} \end{aligned} \tag{3.2}$$

where $d_{|B|}$ is chosen such that ϕ_B is a size α test. Note that $d_{|B|}$ increases with $|B|$, the number of elements in B. (3.2) does not define any test for $\underline{\theta}$ for which $g(\underline{\theta}) = \emptyset$. For any $\underline{\theta}$ such that $\min_i \theta_i - \theta_0 = \delta > 0$, define the test $\phi_{\underline{\theta}}$ by

$$\begin{aligned} \phi_{\underline{\theta}}(\underline{Y}) &= 1 \text{ if } Y_0 + \delta < Y_i - d_1 \text{ for all } i = 1, \ldots, k \\ &= 0 \text{ otherwise} \end{aligned} \tag{3.3}$$

which Lehmann (1952) has shown to be the UMP level α test for H_0: $\min_i \theta_i - \theta_0 = \delta$ against H_a: $\min_i \theta_i - \theta_0 > \delta$ among all level α tests with monotone rejection regions.

By Lemma 1.1, the set $C(\underline{Y}) = \{\underline{\theta}\colon \phi_{\underline{\theta}}(\underline{Y}) = 0\}$ is a $100(1-\alpha)\%$ confidence set for $\underline{\theta}$. Its coverage probability is exactly $1 - \alpha$ for $\underline{\theta}$ with $g(\underline{\theta}) \neq \emptyset$, greater than $1 - \alpha$ for $\underline{\theta}$ with $g(\underline{\theta}) = \emptyset$.

To obtain useful inference from $C(\underline{Y})$, it is convenient to proceed in a stepwise fashion, as follows. Let

$$[1], [2], \ldots, [k]$$

denote the random indices such that

$$Y_{[1]} < Y_{[2]} < \ldots < Y_{[k]}.$$

In the stepwise scheme below, t is the step number, and the random integer T should be thought of as the number of treatments that can be asserted to be better than the control.

Step 1. If $Y_{[k-1+1]} < Y_0 + d_{k-1+1}$, set $T = 0$ and stop;
$\geq$, go to Step 2.

Step t. If $Y_{[k-t+1]} < Y_0 + d_{k-t+1}$, set $T = t - 1$ and stop;
$\geq$, go to Step $t + 1$.

Step k. If $Y_{[k-k+1]} < Y_0 + d_{k-k+1}$, set $T = k - 1$ and stop;
$\geq$, set $T = k$ and stop.

Naik (1975) showed that if one infers $\theta_{[k]}, \ldots, \theta_{[k-T+1]} > \theta_0$ if $T \geq 1$, and infers nothing if $T = 0$, then $\sup_{\underline{\theta}} P_{\underline{\theta}}$ {at least one incorrect inference} $\leq \alpha$. Since $d_{k-t+1} < d_k$ for $t > 1$, the set of treatments inferred to be better than the control by the non-stepwise method is always a subset of the those inferred to be better than the control by the stepwise method, sometimes a proper subset. This type of phenomenon has led some to consider stepwise procedures to be uniform improvements over their non-stepwise counterparts. It has always been believed, however, that the price for this increased sensitivity is the inability for stepwise procedures to provide confidence sets. This belief is false, as a confidence set which implies Naik's 0-1 decisions result is derived below.

Lemma 3.1. *For any $\underline{\theta}$ with $|g(\underline{\theta})| \geq k - T + 1$, we have $\underline{\theta} \notin C(\underline{Y})$.*

PROOF. For any $\underline{\theta}$ with $|g(\underline{\theta})| = k - t + 1 \geq k - T + 1$, we have

$$\max_{j \in g(\underline{\theta})} Y_j - \theta_j \geq \max_{j \in g(\underline{\theta})} (Y_{[k-t+1]} - \theta_j) \geq Y_0 - \theta_0 + d_{k-t+1}$$

by the definition of T and $g(\underline{\theta})$, namely, $j \in g(\underline{\theta})$ if and only if $\theta_j \leq \theta_0$. Thus $\phi_{g(\theta)}(\underline{Y} - \underline{\theta}) = 1$ and $\underline{\theta} \notin C(\underline{Y})$. □

Lemma 3.2. *Suppose $i \in \{[k - T + 1], \ldots, [k]\}$. Then $i \in g(\underline{\theta})$ implies $\underline{\theta} \notin C(\underline{Y})$.*

PROOF. By Lemma 3.1, only $\underline{\theta}$ with $|g(\underline{\theta})| \leq k - T$ need to be considered. Now suppose $i \in \{[k - T + 1], \ldots, [k]\}$, $|g(\underline{\theta})| = k - t + 1 \leq k - T$, and $i \in g(\theta)$. Then

$$Y_i - \theta_i - (Y_0 - \theta_0) - d_{k-t+1} \geq Y_i - Y_0 - d_{k-t+1} \geq Y_i - Y_0 - d_{k-T+1} \geq 0$$

and thus $\phi_{g(\underline{\theta})}(\underline{Y} - \underline{\theta}) = 1$ and $\underline{\theta} \notin C(\underline{Y})$. □

Corollary 3.1. *If* $i \in \{[k-T+1], \ldots, [k]\}$, *then* $\inf_{\underline{\theta} \in C(\underline{Y})} (\theta_i - \theta_0) \geq 0$.

Lemma 3.3. *If* $T \leq k-1$, *then*

$$\inf_{\underline{\theta} \in C(\underline{Y})} \theta_i - \theta_0 = Y_i - Y_0 - d_{k-T} \qquad \text{for } i = [1], \ldots, [k-T],$$
$$\inf_{\underline{\theta} \in C(\underline{Y})} \theta_i - \theta_0 = 0 \qquad \text{for } i = [k-T+1], \ldots, [k].$$

PROOF. By Lemma 3.1, only $\underline{\theta}$ with $|g(\underline{\theta}|)| \leq k-T$ need to be considered. First consider $i \in \{[1], \ldots, [k-T]\}$. For $\underline{\theta}$ such that $i \notin g(\underline{\theta})$, $\theta_i - \theta_0 > 0$ by definition. For $\underline{\theta}$ such that $i \in g(\underline{\theta})$, if $\theta_i - \theta_0 \leq Y_i - Y_0 - d_{[k-T]}$, then $\phi_{g(\underline{\theta})}(\underline{Y} - \underline{\theta}) = 1$ because $Y_i - \theta_i \geq Y_0 - \theta_0 + d_{|g(\theta)|}$ as $d_{[k-T]} \geq d_{|g(\theta)|}$. Thus $\inf_{\underline{\theta} \in C(\underline{Y})} \theta_i - \theta_0 \geq Y_i - Y_0 - d_{[k-T]} < 0$. However, consider $\underline{\theta}(\underline{Y})$ such that $\theta_j - \theta_0 = (Y_j - Y_0 - d_{[k-T]})+ < 0$ for $j = [1], \ldots, [k-T]$ and $\theta_j - \theta_0 = 0+$ for $j = [k-T+1], \ldots, [k]$, then $\phi_{g(\underline{\theta})}(\underline{Y} - \underline{\theta}) = 0$ because $g(\underline{\theta}) = \{[1], \ldots, [k-T]\}$ and $Y_j - \theta_j < Y_0 - \theta_o + d_{[k-T]}$ for $j = [1], \ldots, [k-T]$. Thus $\inf_{\underline{\theta} \in C(\underline{Y})} \theta_i - \theta_0 = Y_i - Y_0 - d_{[k-T]}$. Next consider $i \in \{[k-T+1], \ldots, [k]\}$. By Corollary 3.1, $\inf_{\underline{\theta} \in C(\underline{Y})} \theta_i - \theta_0 \geq 0$. But for the $\underline{\theta}(\underline{Y})$ above, $\theta_i - \theta_0 = 0+$ and $\phi_{g(\underline{\theta})}(\underline{Y} - \underline{\theta}) = 0$. Thus $\inf_{\underline{\theta} \in C(\underline{Y})} \theta_i - \theta_0 = 0$ for $i = [k-T+1], \ldots, [k]$. □

Lemma 3.4. *If* $T = k$, *then*

$$\inf_{\underline{\theta} \in C(\underline{Y})} \theta_i - \theta_0 = \min_i Y_i - Y_0 - d_1 \ \text{for } i = 1, \ldots, k.$$

PROOF. When $T = k$, $\phi_{g(\underline{\theta})}(\underline{Y} - \underline{\theta}) = 1$ for all $\underline{\theta}$ such that $g(\underline{\theta}) \neq \emptyset$. For all $\underline{\theta}$ such that $g(\underline{\theta}) = \emptyset$, it can easily be seen that the test (3.3) gives the bounds $\min_i \theta_i - \theta_0 > \min_i Y_i - Y_0 - d_1$. □

Now define, for $i = 1, \ldots, k$,

$$D_i = \min(0, Y_i - Y_0 - d_{k-T}) \qquad \text{if } T \leq k-1,$$
$$D_i = \min_i (Y_i - Y_0 - d_1) \qquad \text{if } T = k.$$

Then, combining Lemmas 3.3 and 3.4, we have

Theorem 3.1. *For all* $\underline{\theta}$,

$$P_{\underline{\theta}}\{D_i < \theta_i - \theta_0 \ \text{for } i = 1, \ldots, k\} \geq 1 - \alpha.$$

Note that, while the stepwise procedure is better in detecting directional inequalities, in contrast to the $\underline{D}_i$'s in (3.1), the stepwise lower bounds D_i on $\theta_i - \theta_0$ for treatments inferred to be better than the control are 0 rather than positive, except in those lucky cases when all new treatments can be inferred to be better than the control. Thus, the stepwise procedure cannot be considered a uniform improvement over the non-stepwise procedure.

4. COMPARISON WITH THE SAMPLE BEST (CSB)

Suppose $k(>2)$ treatments are being compared. The problem investigated in this section is as follows. When can we infer the *sample* best treatment to be the *true* best treatment? Somewhat surprisingly, for the usual balanced designs, the answer is whenever the sample best is judged better than the sample second best by the *two-sample* two-sided t-test (using the k-sample pooled error variance estimate and degrees of freedom). In fact, we derive a confidence set which contains a positive lower confidence bound on the amount by which the sample best is better than the true best of the other treatments whenever this inference is made.

Consider sampling from k treatments. Let $\theta_1, \ldots, \theta_k$ denote the k treatment effects, and let $Y_1, \ldots, Y_k$ be their least square estimates. Assume $Y_1, \ldots, Y_k$ are independent with joint distribution $\prod_{i=1}^{k} \Phi((y_i - \theta_i)/(\sigma/n^{1/2}))$, where Φ is the standard normal distribution, and σ is unknown. Suppose s^2 is an estimate of σ^2 independent of $Y_1, \ldots, Y_k$ such that vs^2/σ^2 has a χ^2 distribution with v degrees of freedom. We remark that most balanced designs fit into this framework. For example, in a balanced one-way design, $Y_1, \ldots, Y_k$ are the treatment means, and $s^2 = \text{MSE}$ is the usual pooled estimate of σ^2 with $k(n-1)$ degrees of freedom. In a randomized complete blocks design with no interaction, $Y_1, \ldots, Y_k$ are again the sample means, and $s^2 = \text{MSE}$ is the usual estimate of σ^2 with $(k-1)(n-1)$ degrees of freedom.

Again we assume a larger treatment effect implies a better treatment. Our method of deriving the desired confidence set below is to test an exhaustive family of nonlinear contours of $\underline{\theta}$, which is motivated by the following consideration. Let $[1], [2], \ldots, [k]$ denote the random indices such that

$$Y_{[1]} \leq Y_{[2]} \leq \ldots \leq Y_{[k]}.$$

To infer treatment $[k]$, the *sample* best treatment, to be the true best treatment is identical to inferring

$$0 < \theta_{[k]} - \max_{j \neq [k]} \theta_j. \tag{4.1}$$

Therefore, if we can construct a size-α test for

$$H_0\colon\ \theta_i - \max_{j \neq i} \theta_j = \delta \tag{4.2}$$

for *each* $i = 1, \ldots, k$ and $\delta \geq 0$, then by Lemma 1.1, we can determine whether the data allow for the inference (4.1).

The key result of this section shows that the test

$$\begin{aligned} \phi(\underline{Y}) &= 1 \text{ if } Y_{[k]} - Y_{[k-1]} > t_{\alpha/2,v} s(2/n)^{1/2} \\ &= 0 \text{ otherwise} \end{aligned} \tag{4.3}$$

is a level-α test for $H_0\colon \theta_i = \max_{j \neq i} \theta_j$ for any fixed i. Here $t_{\alpha/2,v}$ is the upper $\alpha/2$ quantile of the *univariate* t-distribution with v degrees of freedom. Test for the hypotheses (4.2) with $\delta > 0$ are then easily obtained by translation. Note that in this section dependence on s or σ is suppressed notationally. We first state a lemma.

Lemma 4.1. *Suppose that the joint density of $Y_1, \ldots, Y_k$ is $\prod_{i=1}^{k} f(y_i - \theta_i)$, where $f(y - \theta)$ has monotone likelihood ratio in y and θ. Then, for any fixed $t > 0$, $P_{\underline{\theta}}\{Y_{[k]} - Y_{[k-1]} \geq t\}$ is monotone nonincreasing in $\min_{1 \leq i \leq k} \theta_i$.*

PROOF. By symmetry, assume $\theta_1 \leq \ldots \leq \theta_k$. Then, for any $t > 0$, we have

$$\begin{aligned} P_{\underline{\theta}}\{Y_{[k]} - Y_{[k-1]} \geq t\} &= \sum_{j=1}^{k} P_{\underline{\theta}}\{Y_j \geq \max_{i \neq j} Y_i + t\} \\ &= \sum_{j=1}^{k} \int_{-\infty}^{\infty} \prod_{i=1, i \neq j}^{k} F(y + \theta_j - \theta_i - t) f(y)\, dy \end{aligned}$$

where F is the cdf of f. Therefore,

$$\begin{aligned} &\partial/\partial\theta_1 P_{\underline{\theta}}\{Y_{[k]} - Y_{[k-1]} \geq t\} \\ &= \sum_{j=2}^{k} \int_{-\infty}^{\infty} \prod_{i=2, i \neq j}^{k} F(y + \theta_1 - \theta_i - t) f(y + \theta_1 - \theta_j - t) f(y)\, dy \\ &\qquad - \sum_{j=2}^{k} \int_{-\infty}^{\infty} \prod_{i=2, i \neq j}^{k} F(y + \theta_j - \theta_i - t) f(y + \theta_j - \theta_1 - t) f(y)\, dy \\ &= \sum_{j=2}^{k} \int_{-\infty}^{\infty} \prod_{i=2, i \neq j}^{k} F(y - \theta_i - t)[f(y - \theta_j - t) f(y - \theta_1) \\ &\qquad - f(y - \theta_1 - t) f(y - \theta_j)]\, dy. \end{aligned}$$

Since the expression in the brackets is nonpositive by the monotone likelihood ratio property of $f(y - \theta)$, the result follows. □

Theorem 4.1. *The test $\phi(\underline{Y})$ defined by (4.3) is a level-α test for H_0: $\theta_i = \max_{j \neq i} \theta_j$ for any fixed i.*

PROOF. Let $(1), \ldots, (k)$ be a set of indices such that $\theta_{(1)} \leq \ldots \leq \theta_{(k)}$ with $i = (k)$. By Lemma 4.1, conditionally given s, $\sup_{\underline{\theta} \in H_O} P_{\underline{\theta}}\{\phi(\underline{Y}) = 1\}$ occurs at $-\infty = \theta_{(1)} = \ldots = \theta_{(k-2)}, \theta_{(k-1)} = \theta_{(k)}$. This can be seen by first sending $\theta_{(1)}$ to $-\infty$ and then $\theta_{(2)}$ to $-\infty$ and so on. Therefore, unconditionally the supremum occurs there also. The theorem then follows. □

To obtain a confidence set, we divide the parameter space into contours $\Theta_i(\delta) = \{\underline{\theta}\colon \theta_i = \max_{j \neq i} \theta_j + \delta\}$ for $\delta \geq 0$, and test for all possible contours. Clearly, $\{\Theta_i(\delta)\colon \delta \geq 0, i = 1, \ldots, k\}$ is exhaustive: For a given $\underline{\theta}$, if $\theta_{(k)} > \theta_{(k-1)}$ then $\underline{\theta}$ belongs to a unique $\Theta_i(\delta)$; if $\theta_{(k)} = \theta_{(k-1)}$ then $\underline{\theta}$ belongs to two or more $\Theta_i(\delta)$.

By Theorem 4.1, for each contour $\Theta_i(\delta)$, a level-α test for the composite null hypothesis H_0: $\underline{\theta} \in \Theta_i(\delta)$ is given by $\phi(\underline{Y} - \delta \underline{1}_i)$, where $\underline{Y} - \delta \underline{1}_i = (Y_1, \ldots, Y_i - \delta, \ldots, Y_k)$. Thus by Lemma 1.1, we have

Theorem 4.2. *The set* $C(\underline{Y}) = \{\underline{\theta}\colon \phi(\underline{Y} - \delta \underline{1}_i) = 0$ *for some* $\Theta_i(\delta) \ni \underline{\theta}\}$ *is a (conservative)* $100(1-\alpha)\%$ *confidence set for* $\underline{\theta}$.

The boundaries of $C(\underline{Y})$ are completely described by the following three lemmas, from which it can be seen that $C(\underline{Y})$ is determined by the spacings between $Y_{[k]}, Y_{[k-1]}$, and $Y_{[k-2]}$. As $C(\underline{Y})$ is translation invariant in $\underline{\theta}$ and permutationally invariant in θ_i for $i \neq [k]$ or $[k-1]$, $C(\underline{Y})$ can be graphed exactly using a *contrast* system in two dimensions for $\theta_{[k]}, \theta_{[k-1]}$, and $\theta_i (i \neq [k]$ or $[k-1])$, as shown in Figures 1 and 2. For a given $\underline{Y}$, the confidence set $C(\underline{Y})$ has one of five distinct shapes, depending on the spacings between

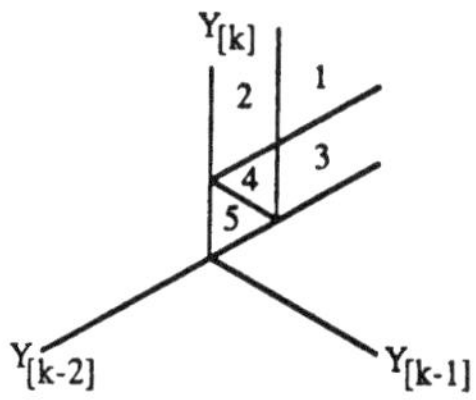

Region	$Y_{[k]} - Y_{[k-1]} - T$	$Y_{[k-1]} - Y_{[k-2]} - T$	$Y_{[k]} - Y_{[k-2]} - T$
1	+	+	+
2	+	-	+
3	-	+	+
4	-	-	+
5	-	-	-

Figure 1. Regions of Sample Space Corresponding to Different Shapes of $C(\underline{Y})$

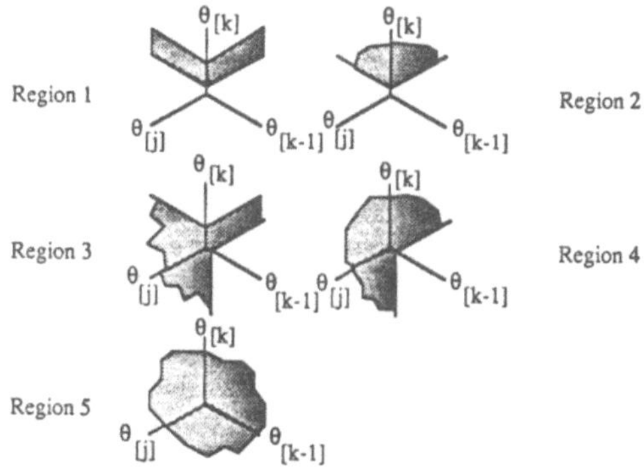

Figure 2. Confidence Sets $C(\underline{Y})$ Corresponding to Regions of the Sample Space in Figure 1

$Y_{[k]}, Y_{[k-1]}$, and $Y_{[k-2]}$. Figure 1 shows the five regions of the sample space ($\underline{Y}$ space) corresponding to the five distinct shapes. Figure 2 shows the confidence set $C(\underline{Y})$ in $\underline{\theta}$ space for a typical $\underline{Y}$ in each of the five regions. In viewing each graph, keep in mind that we are looking at a cross section of $C(\underline{Y})$ perpendicular to the equi-angular line $(\theta, \ldots, \theta)$. The reader should consult Figure 2, and may want to start with Lemma 4.3.

Lemma 4.2. *This lemma describes the boundaries of $C(\underline{Y})$ in $\{\underline{\theta}\colon \theta_{[k]} \geq \theta_j$ for all $j \neq [k]\}$.*

$$\inf_{\delta \geq 0}\{\delta : \phi(\underline{Y} - \delta \underline{1}_{[k]}) = 0\} = (Y_{[k]} - Y_{[k-1]} - T)^+,$$

$$\begin{aligned} \sup\{\delta\colon \phi(\underline{Y} - \delta \underline{1}_{[k]}) = 0\} &= Y_{[k]} - Y_{[k-1]} + T \qquad \text{if } Y_{[k-1]} - Y_{[k-2]} > T, \\ &= +\infty \qquad \text{if } Y_{[k-1]} - Y_{[k-2]} \leq T. \end{aligned}$$

PROOF. For fixed $Y_{[k-1]}, \phi(\underline{Y} - \delta \underline{1}_{[k]}) = 0$ if and only if the largest component of $\underline{Y} - \delta \underline{1}_{[k]}$ is in the interval $Y_{[k-1]} \pm T$. Thus $\inf_{\delta \geq 0}\{\delta\colon \phi(\underline{Y} - \delta \underline{1}_{[k]}) = 0\} = (Y_{[k]} - Y_{[k-1]} - T)^+$. As δ increases past $Y_{[k]} - Y_{[k-1]}, Y_{[k]} - \delta$ becomes the second largest component of $\underline{Y} - \delta \underline{1}_{[k]}$ until $Y_{[k]} - \delta$ reaches $Y_{[k-2]}$. Thus $\sup\{\delta\colon \phi(\underline{Y} - \delta \underline{1}_{[k]}) = 0\} = Y_{[k]} - Y_{[k-1]} + T$ if $Y_{[k-1]} - Y_{[k-2]} > T$; $= +\infty$ if $Y_{[k-1]} - Y_{[k-2]} \leq T$. □

Lemma 4.3. *This lemma describes the boundaries of $C(\underline{Y})$ in $\{\underline{\theta}\colon \theta_{[k-1]} \geq \theta_j$ for all $j \neq [k-1]\}$. The set $\{\delta\colon \delta \geq 0, \phi(\underline{Y} - \delta \underline{1}_{[k-1]}) = 0\} = \emptyset$ if $Y_{[k]} - Y_{[k-1]} > T$. If $Y_{[k]} - Y_{[k-1]} \leq T$, then*

$$\sup_{\delta \geq 0}\{\delta: \phi(\underline{Y} - \delta \underline{1}_{[k-1]}) = 0\} = (Y_{[k-1]} - Y_{[k]} + T) \quad \textit{if } Y_{[k]} - Y_{[k-2]} > T;$$
$$= +\infty \qquad\qquad \textit{if } Y_{[k]} - Y_{[k-2]} \leq T.$$

PROOF. For fixed $Y_{[k]}$, $\phi(\underline{Y} - \delta \underline{1}_{[k-1]}) = 0$ if and only if the second largest component of $\underline{Y} - \delta \underline{1}_{[k-1]}$ is greater than or equal to $Y_{[k]} - T$. The result follows from the fact that $Y_{[k-1]} - \delta$ is the second largest component of $\underline{Y} - \delta \underline{1}_{[k-1]}$ until δ increases to $Y_{[k-1]} - Y_{[k-2]}$. □

Lemma 4.4. *This lemma describes the boundaries of $C(\underline{Y})$ in $\{\underline{\theta}$: $\theta_i \geq \theta_j$ for all $j \neq i\}$ for fixed $i \neq [k]$ or $[k-1]$. For $i \neq [k]$ or $[k-1]$ and any $\delta \geq 0$, $\phi(\underline{Y} - \delta \underline{1}_i) = 0$ if and only if $Y_{[k]} - Y_{[k-1]} \leq T$.*

PROOF. Since $\delta \geq 0$, the largest and second largest component of $\underline{Y} - \delta \underline{1}_i$ remain $Y_{[k]}$ and $Y_{[k-1]}$. □

It is now clear that the confidence intervals for $\theta_i - \max_{j \neq i} \theta_j$ given by $C(\underline{Y})$ can be stated as follows. Let $T = t_{\alpha/2,v} s(2/n)^{1/2}$. For any event E, let $\infty(E) = +\infty$ if E occurs, $-\infty$ otherwise. Let

$$C^*_{[k]} = (Y_{[k]} - Y_{[k-1]} - T) \wedge \infty(Y_{[k]} - Y_{[k-1]} > T),$$
$$C^{**}_{[k]} = (Y_{[k]} - Y_{[k-1]} + T) \vee \infty(Y_{[k-1]} - Y_{[k-2]} \leq T),$$
$$C^*_{[k-1]} = -\infty,$$
$$C^{**}_{[k-1]} = (Y_{[k-1]} - Y_{[k]} + T) \vee \infty(Y_{[k]} - Y_{[k-2]} \leq T),$$
$$C^*_i = -\infty,$$
$$C^{**}_i = (Y_{[k-1]} - Y_{[k]} + T) \vee \infty(Y_{[k]} - Y_{[k-1]} \leq T),$$

for $i \neq [k]$ or $[k-1]$. Here $\wedge$ and $\vee$ are the minimum and maximum functions, respectively. Then, based on the above discussion, we have

Theorem 4.3. *For any $\underline{\theta}$ and σ^2,*

$$P_{\underline{\theta}}\{\theta_i - \max_{j \neq i} \theta_j \in (C^*_i, C^{**}_i) \textit{ for } i = 1, \ldots, k\} \geq 1 - \alpha.$$

REMARK 4.1. Note that $C^*_{[k]} = Y_{[k]} - Y_{[k-1]} - T$ if $Y_{[k]} - Y_{[k-1]} > T$, $-\infty$ otherwise. In practice, the user will likely be most interested in $C^*_{[k]}$ which, when positive, identifies treatment $[k]$ as the best treatment and gives a lower bound on the amount by which it is better than the second best treatment. Thus, if the assertion that treatment $[k]$ is best is made only when $C^*_{[k]} > 0$, i.e., when the "two-sample" t-test between treatments $[k]$ and $[k-1]$ rejects, then the probability of an incorrect assertion is no more than α. Zinger (1961) showed numerical evidence of this last fact, and recently Gutmann and Maymin (1987) proved it. However, their method does not appear capable of producing any confidence bound when σ is unknown.

4.1. Comparisons with Previous Results

Previous results giving similar inference as this section include Bechhofer (1954), Bofinger (1983), Edwards and Hsu (1983), Gupta (1956, 1965), Gutmann and Maymin (1987), Hsu (1981, 1984, 1985). The basic difference is that the method of this section is better in providing a positive lower bound on $\theta_{[k]} - \max_{j \neq [k]} \theta_j$.

Hsu (1984) showed that

$$[-(Y_i - \max_{j \neq i} Y_j - d_{\alpha,k-1,v} s(2/n)^{1/2})^-, (Y_i - \max_{j \neq i} Y_j + d_{\alpha,k-1,v} s(2/n)^{1/2})^+] \quad (4.4)$$

form simultaneous $100(1-\alpha)\%$ confidence intervals for $\theta_i - \max_{j \neq i} \theta_j$, $i = 1, \ldots, k$. Here $-x^- = \min(0, x), x^+ = \max(0, x)$ and $d_{\alpha,k-1,v}$ is the upper α quantile of the maximum of $k-1$ random variables with a multivariate t-distribution with correlation 1/2 and v degrees of freedom. So in particular,

$$D^*_{[k]} = -(Y_{[k]} - Y_{[k-1]} - d_{\alpha,k-1,v} s(2/n)^{1/2})^-$$

is a $100(1-\alpha)\%$ lower confidence bound for $\theta_{[k]} - \max_{j \neq [k]} \theta_j$. Note that $D^*_{[k]}$ equals zero whenever $Y_{[k]} - Y_{[k-1]} > d_{\alpha,k-1,v} s(2/n)^{1/2}$, in which case the inference that treatment $[k]$ is a best treatment is made. (c.f. Remark 4.1) (Both the subset selection inference of Gupta and the indifference zone selection inference of Bechhofer, the later appropriately modified for single-stage experiments with σ unknown, would give this last assertion in this situation. See Hsu (1981).) A comparison of the critical value $t_{\alpha,k-1,v}$ for the lower bound $C^*_{[k]}$ of this section and $d_{\alpha,k-1,v}$ for $D^*_{[k]}$ show for both $\alpha = .05$ and .01,

$$t_{\alpha/2,n} {}^{>}_{<} d_{\alpha,k-1,v} \text{ for } k {}^{\leq}_{>} 3$$

for all degrees of freedom v. So $C^*_{[k]}$ makes the assertion "treatment $[k]$ is best" more (less) often for $k > (\leq) 3$. However, it should be kept clearly in mind that $C^*_{[k]}$ and $D^*_{[k]}$ are opposites in behavior. The previous lower bound $D^*_{[k]}$ is designed to find a treatment close to the true best. ($-\delta < \theta_{[k]} - \max_{j \neq [k]} \theta_j$, where δ is a small positive number.) In those lucky cases where the sample best can be declared to be the true best, $D^*_{[k]}$ is constrained to equal zero, and therefore does not indicate how much the sample best is better than the best of the other treatments. On the other hand, the new lower bound $C^*_{[k]}$ is specifically designed to give a positive lower bound on $\theta_{[k]} - \max_{j \neq [k]} \theta_j$. When the sample best cannot be asserted to be the true best, $C^*_{[k]}$ equals a noninformative $-\infty$. It should also be kept in mind that $D^*_{[k]}$ is but one of the $2k$ confidence bounds (4.4) that can be given simultaneously. In particular, the upper bounds in (4.4) correspond to the subset selection inference of Gupta (1956, 1965).

Bofinger (1983) obtained a lower confidence bound for $\theta_{[k]} - \max_{j \neq [k]} \theta_j$. Edwards and Hsu (1983) and Hsu (1985) gave simultaneous confidence intervals which contain a lower bound on $\theta_{[k]} - \max_{j \neq [k]} \theta_j$. Those lower bounds, unlike $C^*_{[k]}$, are continuous. However, the critical values associated with those bounds are equal to $t_{\alpha/2,\nu}$ for $k = 2$, greater than both $t_{\alpha/2,v}$ and $d_{\alpha,k-1,v}$ for $k > 2$. Gutmann and Maymin (1987) obtained a lower confidence bound for $\theta_{[k]} - \max_{j \neq [k]} \theta_j$ for σ known. However, their conditional approach seems to preclude extension to the σ unknown case.

5. REAL DATA EXAMPLES

To study the effect of smoking on pulmonary health, White and Froeb (1980) measured pulmonary functions in six groups, ranging from nonsmokers to heavy smokers. They were able to get 200 males and females in each of the six groups except for the nonhaling smokers group, which due to the small number of such people available, was limited to 50 males and 50 females. For illustration purposes, we compare the forced vital capacity (FVC) of females using the stepwise multiple comparisons with a control (MCC) method of Section 3, and the forced mid-expiratory flow (FEF) of males using the comparisons with the sample best (CSB) method of Section 4, among the five groups excluding the noninhaling smokers group.

The NS (nonsmoker) group has the highest FVC. Taking it as the control group, a reasonable question is: can we infer that the NS group has the highest population FVC? Denote the population FVC for the NS, ..., HS groups by $\theta_1, \ldots, \theta_5$. For the FVC data, the pooled sample standard deviation $s = MSE^{1/2} = 0.46$ with $5(199) = 995$ degrees of freedom. Choosing $\alpha = .01$, and applying the stepwise and nonstepwise MCC methods of Section 3, we find

Table 1. Forced Vital Capacity (FVC) and Forced Mid-expiratory Flow (FEF) Data

Group number (i)	Group name	Mean FVC (liters)	Std. Dev. FVC (liters)	Mean FEF (liters/sec)	Std. Dev. FEF (liters/sec)
1	NS = nonsmokers	3.35	0.63	3.78	0.79
2	PS = passive smokers	3.23	0.46	3.30	0.77
3	LS = light smokers	3.15	0.39	3.23	0.78
4	MS = moderate smokers	2.80	0.38	2.73	0.81
5	HS = heavy smokers	2.55	0.38	2.59	0.82

Stepwise Lower Confidence Bound	Nonstepwise Lower Confidence Bound
$\theta_1 - \theta_2 > 0.0133$	$\theta_1 - \theta_2 > -0.0069$
$\theta_1 - \theta_3 > 0.0133$	$\theta_1 - \theta_3 > 0.0731$
$\theta_1 - \theta_4 > 0.0133$	$\theta_1 - \theta_4 > 0.4231$
$\theta_1 - \theta_5 > 0.0133$	$\theta_1 - \theta_5 > 0.6731$

The HS (heavy smokers) group has the lowest sample FEF. A reasonable question is: can we infer that the HS group has the lowest population FEF? In other words, if we let $\theta_1, \ldots, \theta_5$ denote the population FEFs for the NS, ..., HS groups respectively, can we infer θ_5 is less than min $\theta_1, \ldots, \theta_4$, and if so, by how much? Since we want to assert θ_5 is the minimum rather than the maximum, we compute the inference for the negatives of the observations. For the FEF data, the pooled sample standard deviation $s = MSE^{1/2} = 0.63$ with $5(199) = 995$ degrees of freedom. Choosing $\alpha = .05$, we find $t_{\alpha/2,v}s(2/n)^{1/2} = (1.96)(0.063) = 0.12$. Applying Theorem 4.3 to the negatives of the observations, we obtain

$$\begin{aligned}
0.02 &< -\theta_5 - \max_{j\neq 5}(-\theta_j) < 0.26 \\
-\infty &< -\theta_4 - \max_{j\neq 4}(-\theta_j) < -0.02 \\
-\infty &< -\theta_i - \max_{j\neq i}(-\theta_j) < -0.02, \quad i = 1, 2, 3.
\end{aligned}$$

Thus, by again taking negatives, we have

$$\begin{aligned}
-0.26 &< \theta_5 - \min_{j\neq 5} -\theta_j < -0.02 \\
0.02 &< \theta_4 - \min_{j\neq 4} -\theta_j < +\infty \\
0.02 &< \theta_i - \min_{j\neq i} -\theta_j < +\infty, \quad i = 1, 2, 3.
\end{aligned}$$

So in particular, at confidence level $1 - \alpha = .95$, we can infer that the HS (heavy smokers) group has the lowest population FEF.

By comparison, it can be verified that none of the following methods: Scheffé, Tukey, Dunnett (with HS as the control group), constrained and unconstrained multiple comparisons with the best, can make this last assertion at $\alpha = .05$, primarily because they are not as specifically designed for this type of inference as the method of this section.

ACKNOWLEDGMENTS

Jason Hsu's research was made possible in part by Grant 1R01 CA41168-01 from the National Cancer Institute.

BIBLIOGRAPHY

Bechhofer, R. E. (1954). A single-sample multiple decision procedure for ranking means of normal populations with known variances. *Annals of Mathematical Statistics* **25**, 16–39.

Bofinger, E. (1983). Multiple comparisons and selections. *Australian Journal of Statistics* **25**, 198–207.

Dunnett, C. W. (1955). A multiple comparison procedure for comparing several treatments with a control. *Journal of the American Statistical Association* **50**, 1096–1121.

Edwards, D. G. and Hsu, J. C. (1983). Multiple comparisons with the best treatment. *Journal of the American Statistical Association* **78**, 965–971. Corrigenda *Journal of the American Statistical Association* **79**, p. 965.

Gupta, S. S. (1956). On a decision rule for a problem in ranking means. Mimeo Series No. 150, Institute of Statistics, University of North Carolina, Chapel Hill.

Gupta, S. S. (1965). On some multiple decision (selection and ranking) rules. *Technometrics* **7**, 225–245.

Gupta, S. S. and Sobel, M. (1958). On selecting a subset which contains all populations better than a standard. *Annals of Mathematical Statistics* **29**, 235–240.

Gutmann, S. and Maymin, Z. (1987). Testing retrospective hypothesis: is the selected population the best? *Annals of Statistics* **15**, 456–461.

Hsu, J. C. (1981). Simultaneous confidence intervals for all distances from the 'best'. *Annals of Statistics* **9**, 1026–1034.

Hsu, J. C. (1984). Constrained two-sided simultaneous confidence intervals for multiple comparisons with the best. *Annals of Statistics* **12**, 1136–1144.

Hsu, J. C. (1985). A method of unconstrained multiple comparisons with the best. *Communications in Statistics A* **14**, 2009–2028.

Lehmann, E. L. (1952). Testing multiparameter hypotheses. *Annals of Mathematical Statistics* **23**, 541–552.

Lehmann, E. L. (1986). *Testing Statistical Hypothesis*, 2nd edition. New York: John Wiley and Sons, Inc.

Naik, U. D. (1975). Some selection rules for comparing p processes with a standard. *Communications in Statistics* **4**, 519–535.

Pratt, J. W. (1961). Length of confidence intervals. *Journal of the American Statistical Association* **56**, 541–567.

White, J. R. and Foreb, H. F. (1980). Small-airways dysfunction in nonsmokers chronically exposed to tobacco smoke. *New England Medical Journal* **302**, 720–723.

Zinger, A. (1961). Detection of best and outlying normal populations with known variances. *Biometrika* **48**, 457–461.

PART 2 - Asymptotic and Sequential Analysis

THE VPRT: OPTIMAL SEQUENTIAL AND NONSEQUENTIAL TESTING

NOEL CRESSIE

Department of Statistics
Iowa State University
Ames, Iowa 50010, U. S. A.

PETER B. MORGAN

Department of Economics
University of Western Ontario
London, Ontario, Canada

1. INTRODUCTION

Wald (1947) proposed a hypothesis testing procedure which is sequential in nature, and thus allows early stopping of an experiment when the evidence for one hypothesis or the other becomes sufficiently strong. He called it the sequential probability ratio test (SPRT), and set it up as follows. Define

$$\Omega = \{\omega_0, \omega_1\}, \tag{1.1}$$

to be the set of "states of nature". A choice is to be made between the two hypotheses,

$$H_0\colon\ \omega = \omega_0, \qquad \text{versus} \qquad H_1\colon\ \omega = \omega_1, \tag{1.2}$$

based on observations on a random variable x.

Suppose the choice between H_0 and H_1 can be made over time, at decision times $t = 1, 2, \ldots$. At t, either the experiment is stopped with one of ω_0 or ω_1 being declared the true state of nature, or more evidence is needed and another observation is collected to be used (along with the previous observations) at decision time $t + 1$. Let

$$y_t = (x_1, x_2, \ldots, x_{t-1}) \tag{1.3}$$

be the collection of independent and identically distributed observations on x, which comprise the data to be used at decision time t. Then rewrite (1.2) as a choice between whether x has distribution function

$$H_0\colon\ F(\cdot\,; \omega_0), \qquad \text{versus} \qquad H_1\colon\ F(\cdot\,; \omega_1). \tag{1.4}$$

Let $f(\cdot\,; \omega)$ be the density or mass function of $F(\cdot\,; \omega)$. Define the likelihood ratio,

$$\lambda_1 \equiv 1, \text{ and } \lambda_t \equiv \frac{\prod_{i=1}^{t-1} f(x_i;\omega_0)}{\prod_{i=1}^{t-1} f(x_i;\omega_1)}; \qquad t = 2,3,\dots . \tag{1.5}$$

Then the SPRT is defined as,

$$\begin{array}{ll} \text{stop sampling and choose } H_1, & \text{if } \lambda_t \,\varepsilon\, [0,\mathrm{L}] \\ \text{stop sampling and choose } H_0, & \text{if } \lambda_t \,\varepsilon\, [\mathrm{U},\infty] \\ \text{continue sampling by collecting} & \\ \quad \text{one more observation } x_t, & \text{if } \lambda_t \,\varepsilon\, (\mathrm{L},\mathrm{U}), \end{array} \tag{1.6}$$

for $t = 2,3,\dots$. Implicit in Wald's definition is a decision to take at least one observation. The quantities L and U of the continue-sampling region (L,U) are shown by Wald to be related to

$$\alpha = \Pr(H_1 \text{ chosen}|H_0 \text{ true}), \qquad \beta = \Pr(H_0 \text{ chosen}|H_1 \text{ true}), \tag{1.7}$$

respectively the probabilities of type I and type II error. Hence for prespecified α and β, L and U can in principle be determined.

Although Wald (1947) showed that with probability one, the SPRT stops in finite time, there was no result to show that this sequential test was an improvement over a fixed-sample-size test with the same α and β. In a landmark paper, Wald and Wolfowitz (1948) demonstrated that of *all* size-α tests having the same power $1-\beta$, the SPRT is a test which requires on average the fewest observations (under either state of nature). Arrow, Blackwell and Girshick's (1949) paper contains the same result, but their emphasis is more on minimizing expected total risk (i.e. on finding a *Bayes* procedure), when losses from making a wrong decision and sampling costs are specified. Their assumptions are more general than Wald and Wolfowitz's, they establish optimality results for the SPRT in the multi-valued decision problem, and their approach foreshadows the use of dynamic programming in constructing optimal decision procedures.

In order to expose the rather specific nature of the Wald-Wolfowitz optimality results, we need to state explicitly their underlying assumptions. First the truncation time T is assumed to be infinite. What if $T < \infty$? Second the cost of each observation is assumed to be a constant c. What if observations have a variable cost structure? Third the loss incurred by continuing to sample is not discounted; i.e. discount factor τ is assumed to be 1. What if $0 \leqq \tau \leqq 1$? Finally, it is assumed that only one observation can be collected at-a-time. What if n_t, the number of observations collected at time t should sampling continue, is not restricted to be 1?

In the rest of this paper we investigate the properties of a sequential decision procedure we call the VPRT (variable-sample-size-sequential probability ratio test) for testing H_0 versus H_1 which is optimal in both senses of being Bayes and minimizing expected total sampling cost for fixed α and β (Cressie and Morgan, 1986a). Section 2 defines the class of sequential decision procedures and in particular the VPRT. Section 3 gives the VPRT for a specific example, as well as the cumulative probabilities of type I and type II error for $t = 1, 2, \ldots T$.

2. SEQUENTIAL TESTING OF TWO HYPOTHESES

Consider a sequential test which allows choosing between H_0: $\omega = \omega_0$ and H_1: $\omega = \omega_1$, at any of $t = 1, 2, \ldots, T$ discrete decision times, where $1 \leqq T \leqq \infty$.

For $n_t \geqq 0$, let $c(n_t)$ be the cost incurred at any decision time t by collecting n_t observations. If $n_t \geqq 1$ observations

$$x_t \equiv (x_{t,1}, \ldots, x_{t,n_t}) \tag{2.1}$$

are asked for at time t, they are received at time $t+1$ and are used at that time (along with previous observations collected) to decide which course of action then should be taken. Assume that $c(n)$ is strictly increasing with n and unbounded above. Let $c_0 \equiv c(0)$ denote the "fixed cost" which is incurred by waiting a period until the next decision point without collecting any observations.

A sequential decision procedure has three components: a terminal decision rule, a sample-size rule, and a stopping rule. Each of these is described briefly here; for a more detailed development see Cressie and Morgan (1986a).

A terminal decision rule δ is a sequence $\{\delta_t\colon\ t = 1, \ldots, T\}$. The quantity δ_t is a function of all available data thus far, viz.

$$y_t = (x_1, \ldots, x_{t-1}), \tag{2.2}$$

and

$$\delta_t(y_t) = \begin{cases} 0, & \text{if } \omega_0 \text{ is chosen} \\ 1, & \text{if } \omega_1 \text{ is chosen;} \quad t = 2, \ldots, T. \end{cases} \tag{2.3}$$

At any decision time t, a choice is made between continuing to take more observations, and stopping to make a terminal decision. The "payoffs" or "rewards" for these terminal decisions are defined by the payoff function

$$U(\omega, \delta_t) \equiv \begin{cases} u_{00}, & \text{if } \delta_t = 0 \text{ and } \omega = \omega_0 \\ u_{01}, & \text{if } \delta_t = 0 \text{ and } \omega = \omega_1 \\ u_{10}, & \text{if } \delta_t = 1 \text{ and } \omega = \omega_0 \\ u_{11}, & \text{if } \delta_t = 1 \text{ and } \omega = \omega_1. \end{cases} \tag{2.4}$$

Typically $u_{00} > u_{10}$, and $u_{11} > u_{01}$ since it is usual to reward correct decisions more than incorrect ones. The units are the same as those for the cost function. Although Section 1 talked about "losses" and "total risk," we have decided to take the more optimistic approach of working with "payoffs" and "net gain". Those familiar with the former should have no trouble thinking of a payoff function as the negative of a loss function. Obviously then, an expected-net-gain-maximizing procedure is a Bayes procedure.

A sample size rule ν is a sequence $\{\nu_t\colon\ t = 1, \ldots, T\}$. This component of a sequential decision procedure is not often seen in sequential analysis articles since one-at-a-time sampling is usually imposed; an exception is Spahn and Ehrenfeld (1974). The quantity $\nu_t \geqq 0$ is a function of all the available data thus far (i.e. y_t) and, if sampling is to continue, it tells the decision maker how many observations are to be asked for at time t, to be received at time $t+1$. It is this generality which allows us to find an optimal ν, and hence to develop decision procedures which dominate the more classical ones (where ν is constrained in some way or other).

A stopping rule S is a sequence $\{S_t\colon\ t = 1, \ldots, T\}$. Each S_t is a function of all the data available at t (i.e. y_t), with

$$S_t(y_t) \equiv \begin{cases} 0, & \text{if sampling continues at } t \\ 1, & \text{if sampling stops at } t, \end{cases} \tag{2.5}$$

for $t = 2, \ldots, T-1$; and $S_T(y_T) \equiv 1$.

The special cases δ_1, ν_1, S_1, do not depend upon data, but are functions of the *prior* $p_1 = \Pr(\omega = \omega_0)$, the *a priori* probability that the true state of nature is ω_0. After data are observed the prior can be updated to give the *posterior* probability that $\omega = \omega_0$, namely p_t. If $\nu_{t-1} > 0$, p_t is

$$p_t = \frac{f(x_{t-1};\omega_0)p_{t-1}}{f(x_{t-1};\omega_0)p_{t-1} + f(x_{t-1};\omega_1)(1-p_{t-1})}; \qquad t = 2, \ldots, T. \tag{2.6}$$

If $\nu_{t-1} = 0$, $p_t = p_{t-1}$; $t = 2, \ldots, T$.

A decision procedure is defined to be the ordered triple $d = (S, \nu, \delta)$. Cressie and Morgan (1986a) find a d^* which maximizes the expected net gain (i.e. is Bayes) over the class of all sequential decision procedures defined above. It should be noted that this class *includes* nonsequential decision procedures, since these are the special case of $S_1(\cdot) \equiv 0$, and $S_t(\cdot) \equiv 1$; $t = 2, \ldots, T$. Cressie and Morgan (1986a) call this optimal procedure d^* the *variable-sample-size-sequential probability ratio test (VPRT)* because of the particular form it takes; see below.

We now set up the backward induction equations which define the VPRT. At time t, the maximum *expected net gain* (taking into account both payoff and cost) of making a terminal decision is

$$\max\{u_{00}\, p_t + u_{01}\,(1-p_t),\quad u_{10}\, p_t + u_{11}\,(1-p_t)\}; \qquad t = 1, 2, \ldots, T, \tag{2.7}$$

which is achieved by using the optimal terminal decision procedure

$$\delta_t^*(p_t) = \begin{cases} 0, & \text{if } u_{00}\, p_t + u_{01}\,(1-p_t) \geqq u_{10}\, p_t + u_{11}\,(1-p_t) \\ 1, & \text{if } u_{00}\, p_t + u_{01}\,(1-p_t) < u_{10}\, p_t + u_{11}\,(1-p_t). \end{cases} \tag{2.8}$$

In particular, at $t = T$ sampling *must* stop and a terminal decision should be taken according to (2.8). This has optimal ENG (expected net gain)

$$V_T^T(p_T) = \max\{u_{00}\, p_T + u_{01}\,(1-p_T), \quad u_{10}\, p_T + u_{11}\,(1-p_T)\}. \tag{2.9}$$

By the optimality principle of dynamic programming (Bellman, 1957), a sequence of maximum ENG functions is given by

$$\begin{aligned} V_t^T(p_t) = \max[&\max\{u_{00}\, p_t + u_{01}\,(1-p_t), \quad u_{10}\, p_t + u_{11}\,(1-p_t)\}, \\ &\max_{n_t \geqq 0}\{-c(n_t) + \tau E_{F_t}(V_{t+1}^T\,(p_{t+1})|p_t,\, n_t)\}], \end{aligned}$$

where $F_t(\cdot) \equiv p_t\; F(\cdot\,;\omega_0) + (1-p_t)\; F(\cdot\,;\omega_1)$, and E_{F_t} denotes expectation with respect to the distribution function F_t; $t = 1, 2, \ldots, T-1$. The optimal procedure that yields these expected net gains, which we call the VPRT, is

$$\delta_t^*(p_t) = \begin{cases} 0, & \text{if } u_{00}\, p_t + u_{01}\,(1-p_t) \geqq u_{10}\, p_t + u_{11}(1-p_t) \\ 1, & \text{if } u_{00}\, p_t + u_{01}\,(1-p_t) < u_{10}\, p_t + u_{11}(1-p_t); \quad t = 1, \ldots, T. \end{cases} \tag{2.10}$$

$$\begin{aligned} \nu_t^*(p_t) &= \arg\max_{n_t \geqq 0}\{-c(n_t) + \tau E_{F_t}(V_{t+1}^T\,(p_{t+1})|p_t,\, n_t)\}; \quad t = 1, \ldots, T-1, \\ \nu_t^*(p_T) &\equiv 0. \end{aligned} \tag{2.11}$$

$$\begin{aligned} S_t^*(p_t) &= \begin{cases} 1, & \text{if } \max\{u_{00}\, p_t + u_{01}(1-p_t),\; u_{10}p_t + u_{11}\,(1-p_t)\} \\ & \qquad \geqq -c(\nu_t^*(p_t)) + \tau E_{F_t}(V_{t+1}^T(p_{t+1})|p_t,\, \nu_t^*(p_t)) \\ 0, & \text{otherwise}; \qquad t = 1, \ldots, T-1, \end{cases} \\ S_T^*(p_T) &\equiv 1. \end{aligned} \tag{2.12}$$

Cressie and Morgan (1986a) show the existence of a sequence

$$\{(p_{tL}, p_{tU}):\; t = 1, 2, \ldots, T-1\}, \tag{2.13}$$

independent of the data and the prior, which are the *continue-sampling intervals* of the VPRT. That is,

$$S_t^*(p_t) = \begin{cases} 1, & \text{if } p_t \leqq p_{tL}, \text{ or } p_t \geqq p_{tU} \\ 0, & \text{if } p_t \in (p_{tL}, p_{tU}). \end{cases} \tag{2.14}$$

These intervals depend upon the cost function, the payoff function, the discount rate τ, the truncation time T, and the hypothesized distributions $F(\cdot\,;\omega_0)$ and $F(\cdot\,;\omega_1)$, and hence can be calculated *before* the sequential test is undertaken.

Moreover the VPRT's terminal decision rule is simply

$$\delta_t^*(p_t) = \begin{cases} 0, & \text{if } p_t \geqq p_{tU} \\ 1, & \text{if } p_t \leqq p_{tL}. \end{cases} \tag{2.15}$$

These two rules depend upon the data and the prior only through the posterior p_t, and hence by (2.11), so does the sample-size rule $\nu_t^*(\cdot)$.

It is now clear why we have used the letters "PRT" standing for "probability ratio test" in our acronym VPRT, representing the expected-net-gain-maximizing decision procedure. Since $p_t = p_1/\{p_1+(1-p_1)/\lambda_t\}$ is an increasing function of λ_t, (2.14) and (2.15) can be rewritten in a manner directly comparable to the SPRT (1.6) with L and U now depending upon t; specifically $L_t = \{p_{tL}(1-p_1)\}/\{(1-p_{tL})p_1\}$ and $U_t = \{p_{tU}(1-p_1)\}/\{(1-p_{tU})p_1\}$. Cressie and Morgan (1986a) show that for discount factor $\tau = 1$, truncation time $T = \infty$, and cost function satisfying $c(0) = 0$, $c(n) > 0$, $c(n) - c(n-1)$ nondecreasing, the VPRT becomes an SPRT (i.e. $\nu_t^* \equiv 1$) and the interval (p_{tL}, p_{tU}) (equivalently (L_t, U_t)) becomes independent of t.

Often the continue-sampling interval (L, U) for the SPRT is approximated because its evaluation has proved to be computationally intensive. One might therefore anticipate that computation of the VPRT's continue-sampling intervals $\{(p_{tL}, p_{tU})\colon\ t = 1, 2, \ldots, T-1\}$, is out of the question. Not so; we are able to compute economically the (p_{tL}, p_{tU}) pairs as follows. The sequence $\{V_t^T(p_t)\colon\ p_t \in [0,1]\}_{t=1}^T$ of maximum expected-net-gain functions can be determined over a suitably fine grid of p_t-values. Cressie and Morgan (1986a) show that if, for $t = 1, \ldots, T-1$, there exists $p_t \in [0,1]$ such that the best action is to sample then the function $h(p_t) = V_t^T(p_t) - \max\{u_{00}\, p_t + u_{01}\, (1-p_t),\ u_{10}\, p_t + u_{11}\, (1-p_t)\}$ has two roots, $p_t = p_{tL}$ and $p_t = p_{tU}$. Locating these zero roots is a relatively simple numerical matter.

Another special case of particular interest, that of the nonsequential tests, arises when $T = 2$ and $p_1 \in (p_{1L}, p_{1U})$. In other words, a sample is asked for at $t = 1$, received at $t = T = 2$, and a decision is then made about H_0 or H_1 based on the sample. The terminal decision rule of the VPRT at $t = T = 2$ is given by

$$\delta_2^* = \begin{cases} 0, & \text{if } \frac{f(x_1;\omega_0)}{f(x_1;\omega_1)} \geqq \frac{(1-p_1)(u_{11}-u_{01})}{p_1(u_{00}-u_{10})} \\ 1, & \text{if } \frac{f(x_1;\omega_0)}{f(x_1;\omega_1)} < \frac{(1-p_1)(u_{11}-u_{01})}{p_1(u_{00}-u_{10})}, \end{cases} \tag{2.16}$$

which is exactly the form of the Neyman-Pearson test (Neyman and Pearson, 1933). As formulated by Neyman and Pearson, the test procedure does not address how to select the sample size but, when viewed as a special case of the VPRT, a sample size of $\nu_1^*(p_1)$ yields an ENG-maximizing (equivalently, Bayes) procedure; see Cressie and Morgan (1986b) for more details.

3. THE VPRT AND ITS PROPERTIES FOR A SPECIFIC EXAMPLE

3.1 THE EXAMPLE

We have chosen a simple example to illustrate the VPRT, and to compare its performance to more classical procedures. The results given are obtained from the computer program referred to above. Suppose $T = 4$; i.e. there are only three decision points available for sampling, since $S_4(\cdot) \equiv 1$. We wish to test H_0 versus H_1, where

$$H_0\colon\ f(x;\omega_0) = \begin{cases} .20; & x = 1 \\ .35; & x = 2 \\ .45; & x = 3 \\ 0\ ; & \text{otherwise,} \end{cases} \tag{3.1}$$

$$H_1\colon\ f(x;\omega_1) = \begin{cases} .60; & x = 1 \\ .30; & x = 2 \\ .10; & x = 3 \\ 0\ ; & \text{otherwise,} \end{cases} \tag{3.2}$$

We shall reward both types of correct decisions equally, and give no reward for an incorrect decision: $u_{00} = u_{11} = 10$, $u_{10} = u_{01} = 0$. We shall discount the expected net gain of deferring a decision between H_0 and H_1 until the next decision period, by $\tau = .9$. Finally, the cost of sampling will include a fixed cost but will have a constant marginal cost: $c(n) = 0.2 + 0.2n$. With these parameters specified, we can now completely determine the continue-sampling intervals and the sample-size rule of the VPRT.

3.2 THE VPRT AND ITS COMPETITORS

We shall now illustrate the dominance of the VPRT over the following sequential decision procedures, each optimal within a more restrictive class.

(i) optimal truncated SPRT (TPRT); i.e. $\nu_1 = \nu_2 = \nu_3 \equiv 1$, and $S_4 \equiv 1$.

(ii) optimal truncated group-sequential procedure (GPRT); i.e. $\nu_1 = \nu_2 = \nu_3 = \nu^*$, and $S_4 \equiv 1$.

(iii) optimal two-stage procedure (2PRT); i.e. $S_3 \equiv 1$.

(iv) optimal one-stage procedure (1PRT); i.e. $S_2 \equiv 1$.

Table 1 gives the continue-sampling intervals for each procedure. Within these three continue-sampling intervals the sample sizes specified by the procedures are inherently different. They are $(\nu_1^*(p_1), \nu_2^*(p_2), \nu_3^*(p_3))$ for the VPRT, (1,1,1) for the TPRT, $(\nu^*(p_1), \nu^*(p_1), \nu^*(p_1))$ for the GPRT, $(\nu_2^*(p_1), \nu_3^*(p_2), 0)$ for the 2PRT, and $(\nu_3^*(p_1), 0, 0)$ for the 1PRT.

The optimal sample size rule for the VPRT is presented in Table 2.

Now compare the expected net gains of the various procedures at $t = 1$, and their probabilities of type I and type II error. These are functions of the prior p_1, as is shown in Table 3.

As to be expected, the ENG for the VPRT dominates those of all the other optimal (in their class) procedures for all values of p_1. The table also shows that the relative type I and type II error probabilities of the procedures can differ considerably even though their relative ENG's are not very different. For example, for $p_1 = 0.45$ the ENG's of the VPRT and the 1PRT differ by only 0.8%, while the 1PRT's type I and type II error probabilities exceed those of the

Table 1. Continue-sampling intervals of the sequential procedures given in (i), (ii), (iii), and (iv). At decision time t, sampling continues if posterior p_t belongs to the interval.

	$t=1$	$t=2$	$t=3$
VPRT	(.310,.686)	(.311,.685)	(.326,.673)
TPRT	(.375,.609)	(.379,.609)	(.387,.609)
GPRT	(.310,.686)	(.311,.685)[1]	(.326,.673)[1]
2PRT	(.311,.685)	(.326,.673)	–
1PRT	(.326,.673)	–	–

[1] These intervals will vary according to the decision maker's prior p_1; shown is the most conservative (widest) interval.

Table 2. The optimal sample size rule is presented as a partition of the VPRT's continue-sampling intervals. At decision time t, if the posterior p_t belongs to any of the subintervals listed for time t then $\nu_t^*(p_t)$, the number associated with the subinterval, is the number of extra observations collected.

t=1

$\nu_1^*(p_1)$		4		5		4	
subintervals	.310		.428		.615		.686

t=2

$\nu_2^*(p_2)$		4		5		4	
subintervals	.311		.436		.608		.685

t=3

$\nu_3^*(p_3)$		5		4		6		5	
subintervals	.326		.558		.582		.645		.673

VPRT by 73% and 102% respectively. The TPRT's performance is well below that of the other four procedures, a direct consequence of the TPRT allowing at most only three observations to be collected. The GPRT's performance in the example is almost identical to that of the VPRT because the GPRT's best sample size (for given p_1) is very similar to the VPRT's at $t = 1$ and $t = 2$. They do differ at $t = 3$ but the probability that either procedure continues to this decision point is close to zero.

Table 3. Expected net gain (ENG) at $t = 1$, probability of type I error (α) and probability of type II error (β), are shown as functions of the prior p_1, for various decision procedures optimal in their own class (see start of Section 3.2 for definitions).

	p_1						
	.35	.40	.45	.50	.55	.60	.65
VPRT							
ENG	6.852	6.784	6.736	6.740	6.761	6.787	6.831
α	.181	.181	.090	.077	.068	.066	.065
β	.052	.052	.049	.105	.111	.115	.158
TPRT							
ENG	–	6.219	6.157	6.095	6.043	6.080	–
α	–	.295	.295	.295	.270	.200	–
β	–	.166	.166	.166	.220	.400	–
GPRT							
ENG	6.847	6.782	6.731	6.738	6.760	6.787	6.825
α	.181	.181	.082	.073	.067	.066	.065
β	.057	.057	.073	.106	.111	.115	.166
2PRT							
ENG	6.839	6.766	6.718	6.734	6.756	6.778	6.820
α	.197	.194	.079	.074	.074	.074	.074
β	.051	.055	.086	.113	.113	.113	.155
1PRT							
ENG	6.730	6.704	6.679	6.653	6.627	6.637	6.694
α	.156	.156	.156	.156	.156	.059	.059
β	.099	.099	.099	.099	.099	.179	.242

3.3 SPENDING α AND β

The cumulative size α_t of a sequential decision procedure is defined by

$$\alpha_t = \sum_{s=2}^{t} \Pr(\text{choose } \omega_1 \text{ at decision time } s|\omega_0); \qquad t = 2,\ldots,T. \qquad (3.3)$$

Obviously $\alpha_T = \alpha$, the probability of the procedure's type I error.

Sequential clinical trials have a need for truncated designs and sampling strategies that allow taking more than one observation at-a-time. While no attempt has been made to find an optimal sample-size rule, group-sequential procedures (amongst others) have been proposed (Pocock, 1977), and their "goodness" assessed by the width of the continue-sampling intervals, as well as by the rate at which the probability of type I error is "spent" (Slud and Wei, 1982). In our terminology, this type I error expenditure is

$$a_t \equiv \alpha_{t+1} - \alpha_t; \qquad t = 1,\ldots,T-1, \qquad (3.4)$$

where $\alpha_1 = 0$.

It is of interest to see how the α-level is spent for the VPRT. Now for $p_1 \in (p_{1L}, p_{1U})$, and $t = 2,\ldots,T-1$,

$$\alpha_t(p_1) = \sum_{s=2}^{t} \Pr(\bigcap_{i=1}^{s-1} \left\{ \frac{p_{iL}(1-p_1)}{(1-p_{iL})p_1} < \lambda_i < \frac{p_{iU}(1-p_1)}{(1-p_{iU})p_1} \right\} \bigcap \left\{ \lambda_s \leqq \frac{p_{sL}(1-p_1)}{(1-p_{sL})p_1} \right\} |\omega_0); \qquad (3.5)$$

and for $t = T$

$$\alpha_T(p_1) = \alpha_{T-1}(p_1) + \Pr(\bigcap_{i=1}^{T-1} \left\{ \frac{p_{iL}(1-p_1)}{(1-p_{iL})p_1} < \lambda_i < \frac{p_{iU}(1-p_1)}{(1-p_{iU})p_1} \right\} \bigcap \left\{ \lambda_T < \frac{\hat{p}(1-p_1)}{(1-\hat{p})p_1} \right\} |\omega_0), \qquad (3.6)$$

where $\hat{p}$ satisfies: $u_{00}\,\hat{p} + u_{01}(1-\hat{p}) = u_{10}\,\hat{p} + u_{11}(1-\hat{p})$. By symmetry, analogous cumulative quantities $\{\beta_t\colon\ t = 2,\ldots,T\}$ for the probabilities of type II errors can be defined. Table 4 presents a_t and $b_t \equiv \beta_{t+1} - \beta_t$ as a function of p_1 and t, for the VPRT in the example analyzed above. The values for the optimal truncated-group-sequential procedure, the VPRT's nearest rival (see Table 1), are also given.

Table 4. Error-spending functions for the VPRT (GPRT is in parentheses).

	p_1			
	.35	.45	.55	.65
a_1	.167(.167)	.059(.059)	.059(.059)	.057(.057)
a_2	.011(.011)	.026(.020)	.007(.006)	.006(.006)
a_3	.003(.003)	.005(.003)	.002(.002)	.002(.002)
b_1	.042(.042)	.032(.064)	.099(.099)	.136(.136)
b_2	.009(.012)	.015(.008)	.009(.009)	.019(.025)
b_3	.001(.003)	.002(.001)	.003(.003)	.003(.005)

Notice that most of the VPRT's α-level is spent early, as is the optimal (optimization is over the constant group size) truncated-group-sequential GPRT procedure's α-level. Given the closeness of the VPRT's and GPRT's performance from Table 1, it should not be surprising that α-levels and β-levels are spent in a similar way. These observations offer further evidence as to why *optimal* group sequential procedures are worth considering in their own right.

The computer program also outputs the stopping-time distribution, its expectation, and the expected total-sampling-cost, under either state of nature, for the various procedures compared in Table 1. A more detailed comparison will appear elsewhere.

ACKNOWLEDGMENTS

In this research Noel Cressie was partially supported by the National Science Foundation under grant SES 84-01460, "On-site Research to Improve Government-Generated Social Science Data Base." The research was partially conducted at the U.S. Bureau of the Census while the author was a participant in the American Statistical Association/National Science Foundation/Census Bureau Research Program.

Peter Morgan was partially supported by the Social Sciences and Humanities Research Council of Canada under grant 410-84-1211, "Learning and the Search for Information" and by financial assistance from the Department of Economics at the University of Western Ontario.

BIBLIOGRAPHY

Arrow, K. J., Blackwell, D., and Girshick, M. A. (1949). Bayes and minimax solutions of sequential decision problems. *Econometrica*, **17**, 213–244.

Bellman, R. (1957). *Dynamic Programming.* Princeton University Press, Princeton, New Jersey.

Cressie, N., and Morgan, P. B. (1986a). The VPRT: a sequential testing procedure dominating the SPRT. Preprint No. 86-17, Statistical Laboratory, Iowa State University.

Cressie, N., and Morgan, P. B. (1986b). Improving upon the Neyman-Pearson approach to testing hypotheses. Preprint No. 86-19, Statistical Laboratory, Iowa State University.

Neyman, J., and Pearson, E. S. (1983). On the problem of the most efficient tests of statistical hypotheses. *Philosophical Transactions of the Royal Society A*, **231**, 140–185.

Pocock, S. J. (1977). Group sequential methods in the design and analysis of clinical trials. *Biometrika*, **64**, 191–199.

Slud, E. V., and Wei, L. J. (1982). Two-sample repeated significance tests based on the modified Wilcoxon statistic. *Journal of the American Statistical Association*, **77**, 862–868.

Spahn, M., and Ehrenfeld, S. (1974). Optimal and suboptimal procedures in group sequential sampling. *Naval Research Logistics Quarterly*, **21**, 53–68.

Wald, A. (1947). *Sequential Analysis.* John Wiley and Sons, New York.

Wald, A., and Wolfowitz, J. (1948). Optimum character of the sequential probability ratio test. *The Annals of Mathematical Statistics*, **19**, 326–339.

AN EDGEWORTH EXPANSION FOR THE DISTRIBUTION OF THE F-RATIO UNDER A RANDOMIZATION MODEL FOR THE RANDOMIZED BLOCK DESIGN

A. W. DAVIS

CSIRO Division of Mathematics and Statistics
Glen Osmond, South Australia, 5064

T. P. SPEED

CSIRO Division of Mathematics and Statistics
Canberra, ACT 2601, Australia

1. INTRODUCTION

The validity of the F-test in the practical situations to which it has been applied has been the subject of much discussion over the years. An early contribution was the paper by Eden and Yates (1933) where, in addition to a survey of work done on the topic up to that point, we find the results of a sampling experiment involving four hypothetical treatments being superimposed upon a set of data arising from the shoot-heights of a particular variety of wheat, as a randomized block design with eight replicates. The data were markedly non-normal, but the empirical distribution over 1,000 samples of Fisher's z-statistic (the forerunner of the usual F-ratio) testing the null hypothesis of no treatment effects, showed excellent agreement with that given by normal theory.

Not long after this Welch (1937) derived the mean and variance of a statistic W closely related to the F-ratio, and independently, Pitman (1938) obtained the first four moments of the same quantity, under the assumption that the observed data were a suitable random permutation of a set $\xi = (\xi_{ij})$ of numbers. Here $i = 1, \ldots, m$ indexes the blocks and $j = 1, \ldots, n$ indexes plots within blocks. Both observed that the closeness of the first two moments of W under the randomization distribution to the values given by normal theory was governed by the coefficient of variation of the m within block sums of squares $\Delta_i = \Sigma_j(\xi_{ij} - \xi_{i\cdot})^2$, $i = 1, \ldots, m$, where $\xi_{i\cdot} = n^{-1}\Sigma_j \xi_{ij}$, and that this coefficient of variation was a simple function of the quantity

$$A = \frac{\Sigma_i \left[\Sigma_j(\xi_{ij} - \xi_{i\cdot})^2\right]^2}{\left[\Sigma_i\Sigma_j(\xi_{ij} - \xi_{i\cdot})^2\right]^2} .$$

Although there has been a good deal more work on this topic since the late 1930's, notably Wald and Wolfowitz (1944), Hoeffding (1952), Wilk (1955), Ogawa (1974), our emphasis is quite different from that adopted by these writers and is much more like the early works. We do not establish precise limit theorems, nor are we concerned with permutation tests; our aim was to see what quantities other than A above govern the extent to which the F-distribution approximates the randomization distribution of the F-ratio. To this end we derive a formal Edgeworth expansion of this distribution following the method introduced in Davis (1976) up to terms in $0(m^{-1})$ where m is the number of blocks, and we find that the relevant correction terms involve the data (ξ_{ij}) only through the quantity A and

$$B = \frac{mn\left[\Sigma_i \Sigma_j (\xi_{ij} - \xi_{i\cdot})^3\right]^2}{\left[\Sigma_i \Sigma_j (\xi_{ij} - \xi_{i\cdot})^2\right]^3} .$$

More fully, we show that if F denotes the F-ratio of treatment to error mean squares, if $P(F)$ and $\Phi(F)$ denote its randomization and normal-theory distribution, respectively, then

$$P(F) - \Phi(F) = a(A)\psi_1(F) + B\psi_2(F) + o(m^{-1})$$

as $m \to \infty$, where $a(A)$ is a function of m, n and A, and ψ_1 and ψ_2 are readily computed functions of m, n and F.

In the final section of this paper we calculate these correction terms for the Eden and Yates data set and compare our approximation to the values obtained by simulation.

2. OUTLINE OF THE METHOD

Let $X = (X_{ij})$ be the array of observations obtained from a randomized block experiment where, as before, $i = 1, \ldots, m$ indexes blocks, and $j = 1, \ldots, n$ indexes plots within blocks. Under the null hypothesis of no treatment effects we will be regarding the array X as simply a random relabelling of an array ξ within blocks and, although it is not necessary and not done in practice, it is convenient to suppose that the blocks are labelled at random as well. Thus (X_{ij}) is a random permutation of (ξ_{ij}) obtained by relabelling the m blocks at random and then, independently within each block, randomly relabelling the n plots as well. There are clearly $m!(n!)^m$ such permutations, assuming (as we will) that all the ξ_{ij} are distinct, and the group of all such permutations is called the *permutation wreath product* of the symmetric group S_n (on plots within blocks) with the symmetric group S_m (on blocks), denoted $S_n wr S_m$: see Bailey et al (1983) for further details.

Let us denote by $k(i,j)$ the treatment allocated to plot j in block i. Our interest will be focussing on the ratio $F = MST/MSE$ of the mean square for treatments to the mean square for error where $MST = SST/(n-1)$ and

$MSE = SSE/(m-1)(n-1)$. Here

$$SST = \Sigma_i \Sigma_{i'} \Sigma_j \Sigma_{j'} t_{ij,i'j'} X_{ij} X_{i'j'} = [t|X \otimes X]$$

where

$$t_{ij,i'j'} = \begin{cases} (n-1)/mn & \text{if } k(i,j) = k(i',j'), \\ -1/mn & \text{otherwise;} \end{cases}$$

while $SSE = \Sigma_i \Sigma_{i'} \Sigma_j \Sigma_{j'} e_{ij,i'j'} X_{ij} X_{i'j'} = [e|X \otimes X]$ with a similar kind of expression for $e_{ij,i'j'}$. We will find this form useful for later computations.

We will be following the formal method outlined in Davis (1976, §2) for deriving Edgeworth expansions of statistics of interest. It is based upon the observation that the Gram-Charlier Type A expansion for the joint density $p(X)$ of $X = (X_{ij})$ may be viewed as the expectation.

$$p(X) = E_Z\{\phi(X|EX + Z,\ DX)\}$$

of a multivariate normal density ϕ with mean $EX + Z$ and dispersion DX, where $Z = (Z_{ij})$ is an array of pseudovariates with $EZ = DZ = 0$ and the same higher-order cumulants as X. Here E and D denote the expectation and dispersion operators, respectively.

This observation can be used to reduce the construction of the Gram-Charlier or Edgeworth expansion of the distribution of a statistic $S(X)$ based upon X to the following two steps:

Step 1. Obtain the density $q(S|Z)$ of S given Z, i.e. in a normal array with mean $EX + Z$, dispersion DX. This is an analytic problem: constructing a normal theory non-central distribution.

Step 2. Compute the 'unconditional' density $q(S)$

$$q(S) = E_Z\{q(S|Z)\}$$

where Z is an array of pseudovariates with $EZ = DZ = 0$, and the same higher order cumulants as X. This is usually a combinatorial problem.

3. GENERALIZED K-STATISTICS AND RANDOMIZATION CUMULANTS

Let $X = (X_{ij}\colon\ i = 1, \ldots, m; j = 1, \ldots, n)$ be an array of random variables whose joint distribution is invariant under the group $G = S_n wr S_m$ of permutations acting on the indices ij as described in §2 above. It follows from results in Speed (1986a, 1986b), hereafter referred to as II and III, whose notation and results we use freely, that the moment generating function (m.g.f.) of X can be written

$$M_X(\theta) = 1 + \sum_{r=1}^{\infty} \frac{1}{r!} \sum_{\rho \in \text{ Hom } (P, \mathcal{P}(\mathbf{r}))} f(\rho)[R_\rho | \theta^{\otimes r}] \tag{3.1}$$

where $\theta = (\theta_{ij})$ is an array of indeterminates with r-fold tensor power $\theta^{\otimes r}$, R_ρ is the relationship tensor labelled by the order preserving map ρ: $P \to \mathcal{P}(\mathbf{r})$ from the partially ordered set $P = \begin{smallmatrix} \bullet\, 1 \\ | \\ \bullet\, 2 \end{smallmatrix}$ into the lattice $\mathcal{P}(\mathbf{r})$ of all (set) partitions of $\mathbf{r} = \{1, \ldots, r\}$, and $f(\rho)$ is the generalized cumulant of the random array X labelled by $\rho = (\rho_1, \rho_2)$. Taking logarithms of (3.1) we find a similar expansion of the cumulant generating function (c.g.f.) of X:

$$K_X(\theta) = \sum_{r=1}^{\infty} \frac{1}{r!} \sum_{\rho \in \text{ Hom } (P, \mathcal{P}(\mathbf{r}))} g(\rho)[R_\rho | \theta^{\otimes r}] \tag{3.2}$$

where the quantities $g(\rho)$ are given by

$$g(\rho) = \sum_{\pi \geq \rho_1} \mu(\pi, \mathbf{r}) \prod_{a=1}^{b(\pi)} f(\rho \cap \pi^a).$$

Here μ is the Möbius function of $\mathcal{P}(\mathbf{r})$ and $\pi = \pi^1 | \ldots | \pi^b$ is an element of $\mathcal{P}(\mathbf{r})$ with $b = b(\pi)$ blocks.

As explained in II and III, when the random array $X = (X_{ij})$ has the randomization distribution obtained by permuting the elements of an array $\xi = (\xi_{ij})$, its generalized cumulants $f(\rho)$ coincide with the generalized k-statistics $k(\rho)$ defined in terms of ξ. We will denote the expressions corresponding to the $g(\rho)$ in this context by $\ell(\rho)$, i.e.

$$\ell(\rho) = \sum_{\pi \geq \rho_1} \mu(\pi, \mathbf{r}) \prod_{a=1}^{b(\pi)} k(\rho \cap \pi^a), \tag{3.3}$$

where $\rho = (\rho_1, \rho_2) \in$ Hom $(P, \mathcal{P}(\mathbf{r}))$ as before.

A number of expressions can be given for the $k(\rho)$, see II and III, Proposition 2.1, and Speed and Silcock (1987b), but the following form is convenient for the asymptotic calculations necessary in §4 below. We use the abbreviation $[\sigma_1, \sigma_2] = [R(\sigma_1, \sigma_2) | \xi]$.

Lemma 3.1. *The following expresion gives* $k(\rho)$, *where* $\rho = (\rho_1, \rho_2)$ *has* $\rho_1, \rho_2 \in \mathcal{P}(\mathbf{r})$ *with* $\rho_1 \geq \rho_2$*:*

$$\sum_{\sigma_1, \sigma_2} [\sigma_1, \sigma_2] \sum_{\tau_1 \geq \rho_2 \vee \sigma_2} \frac{\mu(\tau_1, \rho_1)\mu(\tau_1, \sigma_1)}{(m)_{\tau_1}} \sum_{\tau_2} \frac{\mu(\tau_2, \rho_2)\mu(\tau_2, \sigma_2)}{\prod_{a=1}^{b(\tau_1)} (n)_{\tau_2 \cap \tau_1^a}} \tag{3.4}$$

PROOF. The result follows by putting part (iii) of Proposition 2.1 of II into the same proposition of III and using the following formulae for the Möbius function of the Hom lattice and the generalized incomplete factorial.

$$\mu_H((\tau_1,\tau_2),(\sigma_1,\sigma_2)) = \mu(\tau_1,\sigma_1)\mu(\tau_2,\sigma_2)\varsigma(\sigma_2,\tau_1)$$

where $H =$ Hom $(P, \mathcal{P}(\mathbf{r}))$ and the μ and ς on the RHS refer to $\mathcal{P}(\mathbf{r})$, see III, and

$$(m,n)_{(\tau_2,\tau_1)} = (m)_{\tau_1} \prod_{a=1}^{b(\tau_1)} (n)_{\tau_2 \cap \tau_1^a}.$$

□

Corollary. *The following expression gives $\ell(\rho)$ where $\rho = (\rho_1, \rho_2)$ is as before.*

$$\sum_{\sigma_1,\sigma_2} [\sigma_1,\sigma_2] \sum_{\tau_1 \geq \rho_2 \vee \sigma_2} \mu(\tau_1,\rho_1)\mu(\tau_1,\sigma_1) \left\{ \sum_{\pi \geq \rho_1 \vee \sigma_1} \frac{\mu(\pi,\mathbf{r})}{\prod_{a=1}^{b(\pi)} (m)_{\tau_1 \cap \pi^a}} \right\}$$

$$\times \left\{ \sum_{\tau_2} \frac{\mu(\tau_2,\rho_2)\mu(\tau_2,\sigma_2)}{\prod_{a=1}^{b(\tau_1)} (n)_{\tau_2 \cap \tau_1^a}} \right\} \tag{3.5}$$

PROOF. This follows from (3.3) and (3.4) and the multiplicative property of μ on $\mathcal{P}(\mathbf{r})$. We omit the details. □

4. CALCULATIONS

Throughout this section $X = (X_{ij})$ is a random array obtained by randomly permuting $\xi = (\xi_{ij})$ as explained in §3. In particular the quantities $f_X(\rho)$ and $g_X(\rho)$ in the m.g.f. and c.g.f. of X are given by the expression in Lemma 3.1 and its corollary, coinciding with $k_\xi(\rho)$ and $\ell_\xi(\rho)$ respectively.

4.1 ORDER OF MAGNITUDE OF THE CUMULANTS

We now consider the behaviour of the $g(\rho)$ as the number m of blocks increases. In other words, we regard our underlying array $\xi = (\xi_{ij}\colon\ i = 1,\ldots,m; j = 1,\ldots,n)$ as the initial block of an array $\tilde{\xi} = (\xi_{ij}\colon\ i = 1,2,\ldots, j = 1,\ldots,n)$. Our basic assumption here is

$$\frac{1}{m^{\sigma_1}\ n^{\sigma_2}}[R(\sigma_1,\sigma_2)|\xi] \text{ is bounded as } m \to \infty. \tag{$*$}$$

It is not hard to weaken this assumption by referring to sums of mean-adjusted terms, and then recasting Lemma 3.1 in terms of these sums, but the basic ideas remain the same.

Lemma 4.1. *Let* $\kappa \in \mathcal{P}(\mathbf{s}), p = b(\kappa)$. *Then*

$$m^s \sum_{\lambda \geq \kappa} \frac{\mu(\lambda, \mathbf{s})}{\prod_{a=1}^{b(\lambda)} (m)_{|\lambda^a|}} = 0(m^{-p+1}) \qquad \textit{as } m \to \infty. \tag{4.1}$$

PROOF. We need only consider $p \geq 3$. Write $\lambda^u = \lambda|\dots|\lambda$ (u terms). Then the left hand side of (4.1) is

$$\sum_{\lambda \geq \kappa} \frac{\mu(\lambda, \mathbf{s})}{\{1 + \sum_{0<\phi\leq\lambda} \mu(0,\phi) m^{b(\phi)-s}\}}$$

$$= \sum_{u=1}^{\infty} (-1)^u \sum_{\alpha_1=1}^{s-1} \cdots \sum_{\alpha_u=1}^{s-1} m^{-\sum_{i=1}^{u} \alpha_i} \sum_{\phi \leq \mathbf{s}^u} \mu(0,\phi) \sum_{\lambda \geq \kappa}^{*} \mu(\lambda, \mathbf{s})$$

where $\phi = \phi_1|\dots|\phi_u, b(\phi_i) = s - \alpha_i (i = 1, \dots, u)$, and Σ^* extends over $\lambda \geq \kappa$ such that $\phi \leq \lambda^u$. Clearly

$$\sum_{\lambda \geq \kappa}^{*} \mu(\lambda, \mathbf{s}) = \sum_{\lambda \geq \theta} \mu(\lambda, \mathbf{s}) = \delta(\theta, \mathbf{s}),$$

where θ is the greatest lower bound of all λ such that $\phi \leq \lambda^u$. It is thus sufficient to show that, for ϕ with the property $\sum_{i=1}^{u} \alpha_i \leq p - 2 (u = 1, \dots, p-2)$, there is a λ such that $\phi \leq \lambda^u$ and $b(\lambda) \geq 2$, or more specifically, $b(\lambda) \geq p - \sum_{i=1}^{u} \alpha_i$.

We use induction. If $\sum_{i=1}^{u} \alpha_i = u$, then $\alpha_1 = \dots = \alpha_u = 1$, and each ϕ_i consists of singletons except for a block of size 2. Hence there exists λ such that $\phi_i \leq \lambda (i = 1, \dots, u)$ and $b(\lambda) \geq p - u = p - \sum_{i=1}^{u} \alpha_i$. Assuming the result for $\sum_{i=1}^{u} \alpha_i = g - 1 \geq u$, any ϕ such that $\sum_{i=1}^{u} \alpha_i = g$ may be formed by concatenating two blocks in a $\phi^* = \phi_1^*|\dots|\phi_u^*$ such that $\sum_{i=1}^{u} \alpha_i^* = g - 1 (b(\phi_i^*) = s - \alpha_i^*, i = 1, \dots, u)$. But by assumption there is a λ^* such that $\phi^* \leq (\lambda^*)^u$ and $b(\lambda^*) \geq p - (g-1)$. If necessary, two blocks of λ^* may be concatenated to form a λ with the required property. □

Lemma 4.2. *Under assumption* $(*)$, *if* $b(\rho_1) = b$, *then*

$$\ell_\xi(\rho) = 0(m^{-b+1}) \qquad \textit{as } m \to \infty. \tag{4.2}$$

PROOF. We apply Lemma 4.1 to the π summation in (3.5) with

$$t = b(\rho_1 \vee \sigma_1), \quad \kappa = \mathbf{k}_1|\mathbf{k}_2|\dots|\mathbf{k}_t, \quad s = \sum_{i=1}^{t} k_i = b(\tau_1)$$

where

$$k_i = b(\tau_1 \cap (\rho_1 \vee \sigma_1)^i), \quad i = 1, \dots, t,$$

noting that $\pi \geq \rho_1 \vee \sigma_1$ may be mapped onto $\lambda \geq \kappa$ with $\mu(\pi, \mathbf{r}) = \mu(\lambda, \mathbf{s})$. Then $\ell_\xi(\rho)$ is a sum of terms whose order in m^{-1} is at least

$$\begin{aligned} b(\rho_1 \vee \sigma_1) + b(\tau_1) - b(\sigma_1) - 1 &\geq b(\rho_1 \vee \sigma_1) + b(\rho_1 \wedge \sigma_1) - b(\sigma_1) - 1 \\ &\geq b(\rho_1) - 1. \end{aligned}$$

□

We obtain the c.g.f. $K_Z(\theta)$ of our array of pseudovariates $Z = (Z_{ij})$ by deleting the terms of order $r = 1$ and $r = 2$ in $K_X(\theta)$; of course $M_Z(\theta) = \exp K_Z(\theta)$ as usual. For calculations in §4.3 below we are going to require the order of magnitudes of the terms $f_Z(\rho)$. If ρ_1 has b blocks, then $f_Z(\rho)$ will be a sum of products $\prod_{j=1}^{c} g_X(\rho_j)$, where $(\rho_j)_1$ has r_j blocks, $\sum_1^c r_j = b$ and each ρ_j is at least of third order. By Lemma 4.2 such a term is $0(m^{-(b-c)})$ and this is least when c is greatest. If ρ_1 has p blocks of size 2, then these can at best be grouped into pairs, and so we obtain

Lemma 4.3. *Under assumption* $(*)$, *if* ρ_1 *has* p *blocks of length 2, then*

$$f_Z(\rho) = 0(m^{-[\frac{1}{2}(p+1)]}) \qquad \text{as } m \to \infty. \tag{4.3}$$

4.2 THE F-TEST FOR TREATMENTS

Under the normal theory model with the same mean and second order cumulants as (X_{ij}), and with the array (Z_{ij}) of pseudovariates added, the F-ratio MST/MSE has a doubly non-central F-distribution with $\nu_1 = n - 1$ and $\nu_2 = (m-1)(n-1)$ degrees of freedom, and non-centrality parameters

$$\omega_1 = SST_Z/\sigma^2, \qquad \omega_2 = SSE_Z/\sigma^2$$

where $\sigma^2 = f(12, 12) = [n(n-1)]^{-1}\Sigma_i\Delta_i$. The cumulative distribution function (c.d.f.) is (see Davis (1980)):

$$P(F) = \Phi_{\nu_1,\nu_2}(F) + \phi_{\nu_1,\nu_2}(F) \sum_{k=0}^{\infty} \sum_{\ell=0}^{\infty} \frac{(-\frac{1}{2}\omega_1)^k(-\frac{1}{2}\omega_2)^\ell}{k!\ell!(\frac{1}{2}\nu_1)_k(\frac{1}{2}\nu_2)_\ell} Q_{k\ell}(F) \tag{4.4}$$

where Φ and ϕ are the c.d.f. and density of the normal theory F, and

$$Q_{k\ell}(F) = \begin{cases} (\tfrac{1}{2}\nu_1)_k (\tfrac{1}{2}\nu_2)_\ell F \sum_{i=0}^{k-1} \sum_{j=0}^{\ell-1} (-1)^{i+j} \binom{k-1}{i} \binom{\ell}{j} \dfrac{(\frac{1}{2}\nu_1 + \frac{1}{2}\nu_2)_{i+j} \left(\frac{\nu_1}{\nu_2} F\right)^i}{(\frac{1}{2}\nu_1)_{i+1} (\frac{1}{2}\nu_2)_j \left(1 + \frac{\nu_1}{\nu_2} F\right)^{i+j}} & \text{if } k \geq 1 \\ (\tfrac{1}{2}\nu_2)_\ell F \sum_{j=0}^{\ell-1} (-1)^{j+1} \binom{\ell-1}{j} \dfrac{(\frac{1}{2}\nu_1 + \frac{1}{2}\nu_2)_j}{(\frac{1}{2}\nu_2)_{j+1} \left(1 + \frac{\nu_1}{\nu_2} F\right)^j} & \text{if } k = 0. \end{cases}$$

4.3 Expectations over Z

Since the $Q_{k\ell}$ in (4.4) are 0(1) as $\nu_2 \to \infty$, Davis (1980), Step 2 of the procedure described in §2 requires the behaviour of

$$E_Z \left\{ \frac{\omega_1^k \omega_2^\ell}{(\frac{1}{2}\nu_2)_\ell} \right\} \quad \text{as } m \to \infty.$$

Recalling from §2 that we have written $SST = [t|X \otimes X]$, and similarly for SSE, we can write $SST_Z = [t|Z \otimes Z]$ and $SSE_Z = [e|Z \otimes Z]$ and we are led to consider the order of magnitude of terms of the form

$$f_Z(\rho)[t^{\otimes k} \otimes e^{\otimes \ell}|R(\rho)]/(\tfrac{1}{2}\nu_2)_\ell. \tag{4.5}$$

Suppose that $\rho \in$ Hom $(P, \mathcal{P}(2q))$ and that ρ_1 has b blocks with p of size 2. Then (4.5) is of order at least

$$\left[\tfrac{p+1}{2}\right] + q - b \qquad (q = k + \ell)$$

in m^{-1}. For a given q and $p \leq q, p \neq q - 1$ this is least when b is as large as possible, i.e. $b = p + \left[\frac{2}{3}(q - p)\right]$, in which case the order is

$$\left[\tfrac{1}{2}(p+1)\right] + (q - p) - \left[\tfrac{2}{3}\,(q - p)\right] = \begin{cases} \left[\frac{1}{3}(q + r + 2)\right] & \text{if } p = 2r \\ \left[\frac{1}{3}(q + r + 1)\right] + 1 & \text{if } p = 2r + 1. \end{cases}$$

As can be seen from Table 1, this least order may be achieved for more than one value of p, namely the combinations $p = 0$ or 2 for $q = 2$ and $p = 0$ for $q = 3$, in which cases it is 1. In all other cases the order exceeds 1.

Table 1. Order of Magnitude of (4.5) in m^{-1}

q \ p	0	1	2	3	4	5	6
2	1		1				
3	1	2		2			
4	2	2	2		3		
5	2	3	2	3		3	
6	2	3	3	3	3		3

It is now clear which expectations over Z we need to evaluate in order to obtain all the $0(m^{-1})$ terms in the expansion. They are all stated in the following lemma where we use the notation $CV = s_\Delta/\bar{\Delta}$ where Δ_i is defined in §1, $\bar{\Delta} = m^{-1}\Sigma_i\Delta_i$ and $s_\Delta^2 = (m-1)^{-1}\Sigma_i(\Delta_i - \bar{\Delta})^2$. Recall that B is also defined in §1.

Lemma 4.4. (i) *Case $q = 2$. Here we must consider $p = 0$ or 2, i.e. $\rho_1 = 1234$ or $\rho_1 = 1234$ and 2 similar expressions.*

$$E_Z\{\omega_1^2\} = \frac{-2(n-1)}{m}\left[1 + \frac{m-1}{m}(CV)^2\right] = 0(m^{-1})$$

$$E_Z\{\omega_1\omega_2\} = \frac{-2(m-1)(n-1)}{m}\left[1 - \frac{1}{m}(CV)^2\right] = 0(1)$$

$$E_Z\{\omega_2^2\} = \frac{-2(m-1)^2(n-1)}{m}\left[1 + \frac{1}{m(m-1)}(CV)^2\right] = 0(m)$$

All of these terms are $0(m^{-1})$ when divided by $(\frac{1}{2}\nu_2)_\ell$.

(ii) *Case $q = 3$. Here it is only necessary to consider $p = 0$, i.e. $\rho_1 = 123|456$ and similar expressions.*

$$E_Z\{\omega_1^3\} = \frac{4n\nu_1^3}{m(n)_3}B$$

$$E_Z\{\omega_1^2\omega_2\} = 0$$

$$E_Z\{\omega_1\omega_2^2\} = \frac{4(m)_2 n\nu_1^3}{m^2(n)_3}B$$

$$E_Z\{\omega_2^3\} = \frac{4(m)_3 n\nu_1^3}{m^2(n)_3}B$$

PROOF. The results are derived using the method adopted in Speed and Silcock (1987b §6) which exploits the representation for SST (and SSE) explained in §2 above. We omit the straightforward but tedious calculations. □

5. THE EXPANSION

The results in Lemma 4.4 now enable us to write down the $0(m^{-1})$ expansion of the randomization distribution of the F-ratio MST/MSE.

$$P(F) - \Phi_{\nu_1,\nu_2}(F) = a\psi_1(F) + B\psi_2(F) + o(m^{-1}) \tag{5.1}$$

where

$$a = \left\{\tfrac{1}{2}(n-1) + m^{-1}\right\}(CV)^2 - 1$$

$$\psi_1(F) = \phi_{\nu_1,\nu_2}(F)\frac{\nu_1}{4m}\left\{\frac{Q_{20}}{\left(\frac{1}{2}\nu_1\right)_2} + 2(m-1)\frac{Q_{11}}{\left(\frac{1}{2}\nu_1\right)\left(\frac{1}{2}\nu_2\right)} + (m-1)^2\frac{Q_{02}}{\left(\frac{1}{2}\nu_2\right)_2}\right\}$$

$$= \phi_{\nu_1,\nu_2}(F)\frac{F[\{n-1+2/(m-1)\}F-(n+1)]}{(n+1)\{(m-1)(n-1)+2\}\{1+F/(m-1)\}}$$

$$\psi_2(F) = -\phi_{\nu_1,\nu_2}(F)\frac{n\nu_1^3}{12m(n)_3}\left\{\frac{Q_{30}}{\left(\frac{1}{2}\nu_1\right)_3} + 3(m-1)\frac{Q_{12}}{\left(\frac{1}{2}\nu\right)_1\left(\frac{1}{2}\nu_2\right)_2}\right.$$

$$\left. + (m-1)(m-2)\frac{Q_{03}}{\left(\frac{1}{2}\nu_2\right)_3}\right\}.$$

We note that $(CV)^2$ and the expression A of §1 are related by

$$1 - \frac{(CV)^2}{m} = \frac{m}{m-1}(1-A).$$

6. AN EXAMPLE

We consider the Eden and Yates (1933) data described in §1, used also by Welch (1937) and Pitman (1938). For this data $m = 8$, $n = 4$, $\nu_1 = 3$, $\nu_2 = 21$, and $CV = 1.0354$, $B = 0.010648$.

The expansion (5.1) may be used to approximate the actual significance level of F corresponding to a given nominal (normal theory) level. Table 2 compares the approximation with estimates obtained from a simulation study in which F was compared with the normal theory $100(1-\alpha)$ point $F_{3,21;\alpha}(\alpha = 0.10, 0.05)$ over 10,000 re-randomizations.

Table 2. Actual significance levels of F under the randomization distribution. Comparison of approximation (5.1) with simulation levels (α_{sim}) (standard errors in parentheses).

α	$\psi_1(F_{3,21;\alpha})$	$\psi_2(F_{3,21;\alpha})$	$1-P(F_{3,21;\alpha})$	α_{sim}
0.10	0.004275	0.002123	0.0968	0.0980 (0.0030)
0.05	0.004517	−0.0001512	0.0466	0.0455 (0.0021)

7. CONCLUDING REMARKS

The general approach of this paper can be adapted to other testing problems in the analysis of variance and to different non-normal distributions for the underlying arrays. For example, we can analyse a randomized block design when the underlying array is the usual additive model, consisting of independent random block and plot effects, together with a multiplicative term correlated with each of these effects. Similarly we can analyse random effects analogues of the linear models with a multiplicative term which are closely related to Tukey's one degree of freedom for non-additivity. These results will be reported elsewhere.

BIBLIOGRAPHY

Bailey, R. A., Praeger, C. E., Rowley, C. A., Speed, T. P. (1983). Generalised wreath products of permutation groups. *Proc. London Math. Soc.* (3) **47**, 69–82.

Davis, A. W. (1976). Statistical distributions in univariate and multivariate Edgeworth populations. *Biometrika*, **63**, 661–670.

Davis, A. W. (1980). On the effects of moderate multivariate nonnormality on Wilks's likelihood ratio criterion. *Biometrika*, **67**, 419–427.

Eden, T. and Yates, F. (1933). On the validity of Fisher's z test when applied to an actual sample of non-normal data. *J. Agric. Sci.*, **23**, 6–17.

Hoeffding, W. (1952). The large-sample power of tests based on permutations of observations. *Ann. Math. Statist.*, **23**, 169–192.

Ogawa, J. (1974). *Statistical Theory of the Analysis of Experimental Designs.* Marcel Dekker, Inc.: New York.

Pitman, E. J. G. (1938). Significance tests which may be applied to samples from any populations III. The analysis of variance tests. *Biometrika*, **29**, 322–335.

Speed, T. P. (1986a). Cumulants and partition lattices II. Generalised k-statistics. *J. Aust. Math. Soc. Ser. A*, **40**, 34–53.

Speed, T. P. (1986b). Cumulants and partition lattices III. Multiply-indexed arrays. *J. Aust. Math. Soc. Ser. A*, **40**, 161–182.

Speed, T. P. and Silcock, H. L. (1987a). Cumulants and partition lattices V. Calculating generalized k-statistics. *J. Aust. Math. Soc. Ser. A.* To appear.

Speed, T. P. and Silcock, H. L. (1987b). Cumulants and partition lattices VI. Variances and covariances of mean squares. *J. Aust. Math. Soc. Ser. A.* To appear.

Wald, A. and Wolfowitz, J. (1944). Statistical tests based on permutations of the observations. *Ann. Math. Statist.*, **15**, 358–372.

Welch, B. L. (1937). On the z-test in randomized blocks and Latin squares.

Biometrika, **29**, 21–51.

Wilk, M. B. (1955). The randomization analysis of a generalized randomized block design. *Biometrika*, **42**, 70–79.

ON BAYES SEQUENTIAL TESTS

TZE LEUNG LAI

Department of Statistics
Columbia University
New York, New York 10027, U.S.A.

1. BACKGROUND AND INTRODUCTION

The theory of Bayes sequential tests was initiated by the pioneering work of Wald and Wolfowitz (1948) on the optimality of the sequential probability ratio test (SPRT). Let $X_1, X_2, \ldots$ be i.i.d. with common density f with respect to some measure ν. To test the simple null hypothesis $H_0 : f = f_0$ versus the simple alternative $H_1 : f = f_1$, the SPRT introduced by Wald (1945) stops sampling at stage

$$N = \inf\{n \geq 1 : R_n \geq A \quad \text{or} \quad R_n \leq B\} \tag{1}$$

and rejects H_0 or H_1 according as $R_N \geq A$ or $R_N \leq B$, where

$$R_n = \Pi_{i=1}^n \{f_1(X_i)/f_0(X_i)\}$$

and $A > 1 > B > 0$ are so chosen that $P_0\{R_N \geq A\} = \alpha$ and $P_1\{R_N \leq B\} = \beta$, for prescribed constants $0 < \alpha,\ \beta < 1$. The Wald-Wolfowitz theorem states that this SPRT minimizes both $E_0(T)$ and $E_1(T)$ among all tests whose sample size T has a finite expectation under both H_0 and H_1, and whose error probabilities satisfy

$$P_0[\text{Reject } H_0] \leq \alpha \quad \text{and} \quad P_1[\text{Reject } H_1] \leq \beta. \tag{2}$$

To prove the theorem, a Lagrange-multiplier-type approach to handle the constraints (2) leads to the so-called "auxiliary Bayes problem" of minimizing

$$p\{w_0 P_0[\text{Reject } H_0] + cE_0(T)\} + (1-p)\{w_1 P_1[\text{Reject } H_1] + cE_1(T)\},$$

where $0 < p < 1$ can be regarded as the prior probability in favor of H_0, $c > 0$ as the cost per observation, and w_i as the loss of incorrectly rejecting H_i $(i = 0, 1)$.

The Bayes problem can be decomposed into two components. The first component is the optimal terminal decision rule δ^*, which turns out to be the rule that accepts the H_i with the smaller posterior loss ℓ_n^* when stopping occurs at stage n. The second component of the Bayes test is its stopping rule T^*, which can be studied by the theory of optimal stopping for the stochastic sequence $cn + \ell_n^*$. In the present setting of a two-point prior distribution (simple null versus simple alternative), the optimal stopping rule turns out to be of the form (1). Thus, the Bayes solution is an SPRT and the Wald-Wolfowitz theorem then follows by varying the parameters p, c, w_0, w_1 (cf. Ferguson, 1967).

Although the same ideas can be applied to composite hypotheses, the optimal stopping problems associated with the Bayes sequential tests do not have simple solutions beyond the two-point prior setting considered by Wald and Wolfowitz. There are, therefore, two main directions in the literature on Bayes sequential tests of composite hypotheses. The first is concerned with general qualitative properties of the Bayes tests, while the second provides simple approximations to these tests and their operating characteristics. We now give a brief review of some of the work in each of these direction.

Sobel (1953) studied Bayes sequential tests of $H_0 : \theta \leq \theta_0$ versus $H_1 : \theta \geq \theta_1 (> \theta_0)$ for the parameter θ of the exponential family of densities

$$f_\theta(x) = \exp\{\theta x - \psi(\theta)\} \tag{3}$$

with respect to some nondegenerate measure ν. As is well known, the natural parameter space $\Theta = \{\theta : \int e^{\theta x} d\nu(x) < \infty\}$ is an interval. The loss $\ell(\theta)$, at θ, of accepting the incorrect hypothesis is assumed to be positive and decreasing for $\theta \in \Theta_0 = (-\infty, \theta_0] \cap \Theta$ but increasing for $\theta \in \Theta_1 = [\theta_1, \infty) \cap \Theta$. There is no loss when $\theta_0 < \theta < \theta_1$ (the so-called "indifference zone"). Given a sampling cost of c for each observation and a prior distribution G on Θ, the Bayes risk of a sequential test (T, δ) with stopping rule T and terminal decision rule δ is given by

$$\begin{aligned} r(T,\delta) = c \int_\Theta E_\theta T \, dG + \int_{\Theta_0} \ell(\theta) P_\theta\{\delta \text{ rejects } H_0\} dG \\ + \int_{\Theta_1} \ell(\theta) P_\theta\{\delta \text{ rejects } H_1\} dG. \end{aligned} \tag{4}$$

Sobel (1953) showed that the Bayes test that minimizes (4) is a "generalized sequential probability ratio test" (GSPRT), whose stopping rule is of the form

$$N' = \inf\{n \geq 1 : R_n \geq A_n \text{ or } R_n \leq B_n\}, \tag{5}$$

which is a generalization of the stopping rule (1) of the SPRT. Since $R_n = \Pi_1^n\{f_{\theta_1}(X_i)/f_{\theta_0}(X_i)\}$ is an increasing function of $S_n = \Sigma_1^n X_i$, (5) can also be expressed as

$$N' = \inf\{n \geq 1 : S_n \geq a_n \text{ or } S_n \leq b_n\}. \tag{5$'$}$$

Sobel also showed that such GSPRTs form an essentially complete class.

Brown, Cohen and Strawderman (1979, 1980) extended Sobel's results beyond the exponential family. They gave sufficient conditions for Bayes tests of one-sided hypotheses to have stopping rules of the form (5′) and also developed analogous results for two-sided hypotheses. Moreover, they proved that under very mild conditions Bayes tests form a complete class.

Ray (1965) and later Berk, Brown and Cohen (1981b) gave sufficient conditions under which Bayes tests have bounded stopping times. Berk, Brown and Cohen (1981a) also provided weaker conditions under which the stopping rules of Bayes tests have bounded moment generating functions, and used the result to establish the continuity of the risk functions of Bayes tests.

Moriguti and Robbins (1962) studied the dynamic programming equations for the Bayes test of $H_0 : p < \frac{1}{2}$ versus $H_1 : p > \frac{1}{2}$, with respect to a Beta prior on a Bernoulli parameter p. The posterior distribution of p is also a Beta distribution that can be expressed in terms of the number of successes and failures so far observed. Moriguti and Robbins assumed a loss of $|p - \frac{1}{2}|$ for the wrong decision, and discussed the numerical solution and analytic approximation of the dynamic programming equations. This analytic (diffusion-type) approximation involves a partial differential equation with certain boundary conditions (including a so-called "free boundary" condition), and is closely related to Chernoff's asymptotic theory described below.

There are two different asymptotic theories in the literature on asymptotic approximations to Bayes tests. The first is Schwarz's (1962) theory of asymptotic shapes for testing $H_0 : \theta \leq \theta_0$ versus $H_1 : \theta \geq \theta_1$ for the parameter θ of the exponential family (3). Let $\mathcal{B}(c)$ denote the continuation region of the Bayes rule. Assuming that $\ell(\theta) > 0$ for $\theta \notin (\theta_0, \theta_1)$ and that $G(I) > 0$ for every open interval $I \subset \Theta$, Schwarz's asymptotic theory leads to the following simple limiting continuation region of the Bayes rule: As $c \to 0$,

$$\mathcal{B}(c)/|\log c| \to \{(t, w) : 1 + \min_{i=0,1} (\theta_i w - t\psi(\theta_i)) > \sup_{\theta}(\theta w - t\psi(\theta))\}. \tag{6}$$

Thus, writing $n = t|\log c|$ and $S_n = w|\log c|$, an asymptotic approximation to the Bayes rule is to continue sampling at stage $n + 1$ iff

$$\log c^{-1} + \min_{i=0,1} \{\theta_i S_n - n\psi(\theta_i)\} > \sup_{\theta}\{\theta S_n - n\psi(\theta)\}, \tag{6$'$}$$

or equivalently, to stop sampling at stage

$$\begin{aligned} N_c = \inf\{n \geq 1 : \max[&\Pi_{i=1}^n f_{\hat{\theta}_n}(X_i)/\Pi_{i=1}^n f_{\theta_0}(X_i), \\ &\Pi_{i=1}^n f_{\hat{\theta}_n}(X_i)/\Pi_{i=1}^n f_{\theta_1}(X_i)] \geq c^{-1}\}, \end{aligned} \tag{7}$$

where $\hat{\theta}_n$ is the maximum likelihood estimate of θ. The terminal decision rule δ^* is to accept H_1 (or H_0) if $\Pi_{i=1}^{N_c} f_{\theta_1}(X_i) >$ (or $\leq$) $\Pi_{i=1}^{N_c} f_{\theta_0}(X_i)$.

There are two main steps in Schwarz's derivation of the above asymptotic approximation to the Bayes rule. The first step involves upper and lower bounds for the Bayes continuation region $\mathcal{B}(c)$. Let

$$L(n, x) = \frac{\min_{i=0,1} \int_{\Theta_i} \ell(\theta) \exp\{\theta x - n\psi(\theta)\} dG(\theta)}{\int_{\Theta} \exp\{\theta x - n\psi(\theta)\} dG(\theta)} \tag{8}$$

be the so-called "stopping risk," which is the posterior loss due to the wrong decision of the Bayes test if stopping occurs at stage n and $S_n = x$. Let

$\mathcal{R}(c) = \{(n,x) : L(n,x) \geq c\}$. Schwarz showed that for sufficiently small $c > 0$

$$\mathcal{R}(c) \supset \mathcal{B}(c) \supset \mathcal{R}(3\Delta^{-1}c|\log c|), \tag{9}$$

where $\Delta = \psi(\theta_0) + \psi(\theta_1) - 2\psi((\theta_0+\theta_1)/2)$. The next step is to apply Laplace's asymptotic method to evaluate the integrals in (8), leading to the asymptotic approximation

$$\log L(n, S_n) \sim \min_{i=0,1}\{\theta_i S_n - n\psi(\theta_i)\} - \{\hat{\theta}_n S_n - n\psi(\hat{\theta}_n)\}. \tag{10}$$

Combining (10) with (9) gives the "asymptotic shape" (6′) the Bayes continuation region $\mathcal{B}(c)$.

Let $J(\theta) = \max\{I(\theta,\theta_0), I(\theta,\theta_1)\}$, where $I(\theta,\lambda) = (\theta-\lambda)\psi'(\theta) - (\psi(\theta) - \psi(\lambda))$. Wong (1968) showed that for fixed θ_0 and θ_1, as $c \to 0$,

$$E_\theta N_c \sim |\log c|/J(\theta) \quad \text{for every} \quad \theta,$$

$$\sup_{\theta\leq\theta_0} P_\theta\{\delta^* \text{ accepts } \ H_1\} = o(c|\log c|) = \sup_{\theta\geq\theta_1} P_\theta\{\delta^* \ \text{ accepts } \ H_0\},$$

and the Bayes risk of the test (N_c, δ^*) satisfies

$$r(N_c,\delta^*) \sim c|\log c| \int d\pi(\theta)/J(\theta) \sim \inf_{(T,\delta)} r(T,\delta). \tag{11}$$

Replacing (10) by a second-order asymptotic expansion for $\log L(n, S_n)$ led to subsequent modifications of (7) by Fushimi (1967) and Schwarz (1969). However, it is not clear that these modifications provide actual improvement of the Bayes risks. To develop second-order asymptotic approximations to $\mathcal{B}(c)$, one needs also to know which (if any) of the bounds $\mathcal{R}(c)$ and $\mathcal{R}(3\Delta^{-1}c|\log c|)$ is of the right order of magnitude as $c \to 0$. A partial answer to this question was provided by Lorden (1967) who showed that the rule which stops when the stopping risk is less than a constant times c is Bayes within $O(c)$. Extensions of Schwarz's bounds (9) and asymptotic shape (6) beyond the exponential family to other distributions and to higher dimensions have been considered by Kiefer and Sacks (1963), Schwarz (1968) and Fortus (1979). Woodroofe (1980), however, showed that direct use of Schwarz's asymptotic shape may not lead to Bayes risk efficiency in higher dimensions and gave conditions under which one can obtain an analogue of Wong's result (11) on Bayes risk efficiency for multiparameter exponential families.

A different asymptotic theory in Bayes sequential tests was developed by Chernoff (1961, 1965 a,b) in the context of testing $H_0 : \theta < 0$ versus $H_1 : \theta > 0$ for the mean θ of a normal distribution with unit variance. Instead of assuming an indifference zone (θ_0, θ_1) and a general loss function $\ell(\theta) > 0$ for $\theta \notin (\theta_0, \theta_1)$ as in Schwarz's theory, Chernoff's theory considers the special loss function $\ell(\theta) = |\theta|$ for $\theta \neq 0$ and assumes a normal prior distribution G with mean 0 and variance σ^2. The Bayes terminal decision rule accepts H_0 (or H_1) according as $S_n \leq 0$(or $S_n > 0$) when stopping occurs at stage n. To study the optimal stopping rule, Chernoff (1961) introduced the normalization

$$t = c^{2/3}(n + \sigma^{-2}), \quad w = c^{1/3}S_n, \tag{12}$$

which is different from Schwarz's normalization $t = n/|\log c|$ and $w = S_n/|\log c|$. With the normalization (12) for the problem, Chernoff obtained a limiting continuation region of the form $\{(t, w) : |w| < f(t)\}$ as $c \to 0$. The stopping boundary $f(t)$ arises as the solution of the corresponding continuous-time stopping problem involving the Wiener process, and an asymptotic analysis of the free boundary problem associated with the optimal stopping problem leads to

$$f(t) = \{3t[\log t^{-1} - (\log 8\pi)/3 + o(1)]\}^{\frac{1}{2}} \quad \text{as} \quad t \to 0, \tag{13}$$

$$f(t) \sim (4t)^{-1} \quad \text{as} \quad t \to \infty, \tag{14}$$

(cf. Chernoff, 1965a, Breakwell and Chernoff, 1964). It is interesting to compare this with the boundary $|w| = [(2t)^{\frac{1}{2}} - \Delta t]^+$ in Schwarz's limiting region (6) (with the different normalization $t = n/|\log c|$, $w = S_n/|\log c|$) in the normal case with $\theta_1 = \Delta = -\theta_0$.

There is, therefore, a big difference in the asymptotic solutions as $c \to 0$ between Schwarz's theory, which assumes a fixed indifference zone (θ_0, θ_1), and Chernoff's theory, which assumes (instead of an indifference zone) the special loss function $|\theta|$ for the wrong decision. Not only are the normalizations and limiting regions different in the two theories, but the methods of deriving these results are also very different. Schwarz's theory appears to be considerably simpler than Chernoff's theory, and is derived in the general context of the exponential family instead of Chernoff's special setting of the normal distribution.

A basic problem with Schwarz's theory is whether the simple attractive rule (7) actually provides an adequate approximation to the Bayes rule. The theory rests on the assumption that the indifference zone (θ_0, θ_1) remains fixed as $c \to 0$. To see how this assumption may lead to difficulties in applying the theory, consider the simple case where $\theta_1 = \Delta = -\theta_0$ and the X_i are normal with mean θ and variance 1, so that $J(\theta) = (|\theta| + \Delta)^2/2$. Although c may be very small (say $c = 10^{-20}$) and Δ appears much larger to justify its being "fixed" (say $\Delta = .1$), $|\log c|$ may turn out to be smaller than $1/J(\theta)$ for $|\theta| \leq \Delta$. Since the asymptotic formula for the Bayes risk (11) involves $|\log c|/J(\theta)$, if one uses the "$c \to 0$" approximation in this case, it seems more reasonable not to consider Δ as "fixed" but to also let $\Delta \to 0$, as pointed out by Lai (1986).

It appears somewhat artificial that there are two different kinds of asymptotic approximations to the Bayes tests, depending on whether there is an indifference zone or not. A more natural asymptotic theory should have the property that the approximation in the absence of an indifference zone can be obtained as the limit of the approximations with shrinking indifference zones. One such unified asymptotic theory was recently developed in the context of the 0-1 loss by Lai (1986), whose results will be briefly reviewed in Section 2. An analogue of Chernoff's Wiener process approximation (12) for the 0-1 loss instead of Chernoff's absolute value loss had been introduced earlier by Bickel and Yahav (1972) and Bickel (1973). In Section 3 the results for testing $H_0 : \theta < \theta_0$ versus $H_1 : \theta > \theta_0$ in Lai (1986) are extended from the 0-1 loss to general loss functions. The results developed in Section 3, therefore, also generalize Chernoff's classical theory for the special case of the absolute value loss $|\theta - \theta_0|$ for the wrong decision about a normal mean θ.

2. A UNIFIED ASYMPTOTIC THEORY UNDER THE 0-1 LOSS

To begin with, consider a Wiener process with $E(w(t)) = \mu t$ and Var $(w(t)) = t$ (μ is the drift coefficient). Suppose we want to test sequentially $H : \mu < 0$ versus $K : \mu > 0$. Assume the 0-1 loss, a cost of t for observing the process for a period of length t, and a flat prior (i.e., Lebesgue measure) on $\mu \in (-\infty, \infty)$. Then the Bayes terminal decision rule is to accept H or K according as $w(t) < 0$ or $w(t) > 0$ when stopping occurs at time t. With this terminal decision rule, the posterior loss $L_0(t, w)$ at time t if $w(t) = w$ is observed and we decide to stop can be easily shown to be

$$L_0(t, w) = t + \Phi(-|w|t^{-\frac{1}{2}}), \tag{15}$$

where Φ is the standard normal distribution function. The Bayes stopping rule is of the form $\tau_0 = \inf\{t > 0 : |w(t)| \geq h_0(t)\}$, and an asymptotic analysis of the free boundary problem associated with the optimal stopping problem for the posterior loss (15) shows that

$$h_0(t) = \{2t[\log t^{-1} + \frac{1}{2}\log\log t^{-1} - \frac{1}{2}\log 4\pi + o(1)]\}^{\frac{1}{2}} \quad \text{as} \quad t \to 0, \tag{16}$$

and

$$h_0(t) = \frac{1}{4}(\frac{2}{\pi})^{\frac{1}{2}}\{t^{-\frac{1}{2}} - 5t^{-\frac{5}{2}}/48\pi + \ldots\} \quad \text{as} \quad t \to \infty. \tag{17}$$

The same method can be applied to the problem of testing $H_0 : \mu \leq -\gamma$ versus $H_1 : \mu \geq \gamma$ with the 0-1 loss and indifference zone $(-\gamma, \gamma)$. The posterior loss is given by

$$L_\gamma(t, w) = t + \Phi(-|w|t^{-\frac{1}{2}} - \gamma t^{\frac{1}{2}}), \tag{18}$$

so that (15) corresponds to the special case $\gamma = 0$. The Bayes stopping rule is of the form $\tau_\gamma = \inf\{t > 0 : |w(t)| \geq h_\gamma(t)\}$, and an asymptotic analysis of the associated free boundary problem shows that

(i) for fixed $0 < \gamma < \infty$, as $t \to 0$,

$$h_\gamma(t) = \{2t[\log t^{-1} + \frac{1}{2}\log\log t^{-1} - \frac{1}{2}\log 4\pi + o(1)]\}^{\frac{1}{2}}; \tag{19}$$

(ii) as $\gamma \to \infty$ and $t \to 0$ such that $t = o((\gamma^2 \log \gamma^2)^{-1})$, the asymptotic expansion (19) still holds;

(iii) for every fixed $0 < \rho < 2$, as $\gamma \to \infty$ and $t \to 0$ such that $t \leq \rho\gamma^{-2}\log\gamma^2$,

$$h_\gamma(t) = -\gamma t + \{2t[\log t^{-1} + \frac{1}{2}\log\log t^{-1} + O(1)]\}^{\frac{1}{2}}; \tag{20}$$

(iv) for every $\lambda > 2$, $\sup_{t \geq \lambda\gamma^{-2}\log\gamma^2} h_\gamma(t) \to 0$ as $\gamma \to \infty$.

Comparison of (19) with (16) shows that the optimal boundary for the problem of testing $H_0 : \mu \leq -\gamma$ versus $H_1 : \mu \geq \gamma$ has the same asymptotic behavior for small t as the optimal boundary for testing $H : \mu < 0$ versus $K : \mu > 0$. Moreover, even when $\gamma \to \infty$ (large indifference zones), (19) still holds for $t = o((\gamma^2 \log\gamma^2)^{-1})$ and a modified form (20) also holds for $t \leq (2-\epsilon)\gamma^2\log\gamma^2$. Thus, in the case of 0-1 loss, there is a unified asymptotic theory as $t \to 0$ for continuous-time Bayes sequential tests of H versus K and also of H_0 versus H_1. Further discussions and numerical results on the optimal stopping boundaries h_0 and h_γ are given in Lai (1986).

Now consider the problems of testing (i) $H : \theta < 0$ versus $K : \theta > 0$, and (ii) $H_0 : \theta \leq -\Delta$ versus $H_1 : \theta \geq \Delta$ for the mean θ of i.i.d. normal random variables $X_1, X_2, \ldots$. Define

$$t = cn,\ w(t) = c^{\frac{1}{2}}S_n,\ \mu = c^{-\frac{1}{2}}\theta,\ \gamma = c^{-\frac{1}{2}}\Delta. \tag{21}$$

Since $c^{\frac{1}{2}}\theta n = \mu t$, $w(t)$ is a Wiener process with drift coefficient μ and with t restricted to the set $I_c = \{c, 2c, \ldots\}$. As $c \to 0$, I_c becomes dense in $[0,\infty)$. Moreover, for any prior distribution G on θ such that G has a positive continuous density G', the density function π_c of $\mu = c^{-\frac{1}{2}}\theta$ is given by

$$\pi_c(x) = c^{\frac{1}{2}}G'(c^{\frac{1}{2}}x) \sim c^{\frac{1}{2}}G'(0) \quad \text{as} \quad c \to 0,$$

and therefore the family of probability measures with densities π_c converges to Lebesgue measure (flat prior). This suggests using the flat-prior continuous-time Bayes stopping boundaries h_0 and h_γ as approximations to the Bayes tests of H versus K and of H_0 versus H_1, respectively.

Lai (1986) extended these approximations from the normal distribution to the general exponential family (3) as follows, under the assumption that θ is known to lie in an open interval A with end-points $-\infty \leq a_1 < a_2 \leq \infty$ such that

$$\inf_{a_1-\eta<\theta<a_2+\eta} \psi''(\theta) > 0 \quad \sup_{a_1-\eta<\theta<a_2+\eta} \psi''(\theta) < \infty, \text{ and } \psi'' \text{ is uniformly continuous on } (a_1-\eta, a_2+\eta) \text{ for some } \eta > 0. \tag{22}$$

Since θ is known to lie in A, the maximum likelihood estimate $\hat{\theta}_n$ is obtained from the sample mean $\bar{X}_n = n^{-1}S_n$ by

$$\begin{aligned} \hat{\theta}_n &= (\psi')^{-1}(\bar{X}_n) && \text{if } \psi'(a_1) < \psi'(a_2), \\ &= a_1 && \text{if } \bar{X}_n \le \psi'(a_1), \\ &= a_2 && \text{if } \bar{X}_n \ge \psi'(a_2). \end{aligned} \tag{23}$$

Of primary importance in the extension is the Kullback-Leibler information number

$$\begin{aligned} I(\theta,\lambda) &= E_\theta\{\log[f_\theta(X_1)/f_\lambda(X_1)]\} \\ &= (\theta-\lambda)\psi'(\theta) - (\psi(\theta)-\psi(\lambda)), \quad \text{by (3)}. \end{aligned}$$

Note that in the case of a normal distribution with mean θ and variance 1, $I(\theta,\lambda) = (\theta-\lambda)^2/2$, and $\hat{\theta}_n = \bar{X}_n$ (when $A = (-\infty,\infty)$). Moreover, letting $g_\gamma(t) = (h_\gamma(t)+\gamma t)^2/2t$ for $\gamma \ge 0$, it follows from (21) that

$$\begin{aligned} |w(t)| \ge h_\gamma(t) &\iff (|w(t)|+\gamma t)^2/2t \ge g_\gamma(t) \\ &\iff (|S_n|+\triangle n)^2/2n \ge g_\gamma(cn) \\ &\iff \max\{I(\hat{\theta}_n,\theta_0), I(\hat{\theta}_n,\theta_1)\} \ge n^{-1}g_\gamma(cn), \end{aligned}$$

where $\theta_1 = \triangle = -\theta_0$ in the case $H_0 : \theta \le -\triangle$ versus $H_1 : \theta \ge \triangle$, and $\theta_1 = 0 = -\theta_0$ in the case $H : \theta < 0$ versus $K : \theta > 0$. Hence, the aforementioned approximation to the Bayes stopping rule for H_0 versus H_1 can be expressed in the form

$$N(g,c) = \inf\{n \ge 1 : \max[I(\hat{\theta}_n,\theta_0), I(\hat{\theta}_n,\theta_1)] \ge n^{-1}g(cn)\} \tag{24}$$

with $g = g_\gamma$, while the stopping rule for H versus K can be expressed as

$$T_c = \inf\{n \ge 1 : I(\hat{\theta}_n,\theta_0) \ge n^{-1}g_0(cn)\}. \tag{25}$$

Lai (1986) showed that the stopping rules (24) and (25) are asymptotically optimal, as $c \to 0$, for testing $H_0 : \theta \le \theta_0$ versus $H_1 : \theta \ge \theta_1$ and for testing $H : \theta < \theta_0$ versus $K : \theta > \theta_0$, respectively, for the parameter θ of the exponential family (3), with respect to the 0-1 loss, cost c per observation, and a general class of prior distributions on A. These asymptotic optimality results are described in detail in the next section, where they are further extended to a general class of loss functions.

Note that Schwarz's stopping rule (7) for testing $H_0 : \theta \le \theta_0$ versus $H_1 : \theta \ge \theta_1$ can be expressed in terms the Kullback-Leibler information number $I(\theta,\lambda)$ as

$$\inf\{n \ge 1 : \max[I(\hat{\theta}_n,\theta_0), I(\hat{\theta}_0,\theta_0)] \ge n^{-1}|\log c|\}. \tag{26}$$

Thus, the stopping rule (24) simply modifies Schwarz's rule by using $g(cn)$ in place of the factor $|\log c|$ in (26). Moreover, letting $\theta_1 = \theta_0$ and $g = g_0$ in (24) leads to the stopping rule (25) for testing $H : \theta < \theta_0$ versus $K : \theta > \theta_0$. The stopping rules (24) and (25) therefore provide a unified treatment of testing H_0 versus H_1 (with an indifference zone) and of testing H versus K (without an indifference zone).

The simulation studies of the risk functions of these tests reported in Lai (1986) show that the stopping rules (24) and (25) not only provide approximate Bayes solutions with respect to a large class of priors but also have nearly optimal frequentist properties. In addition, Lai (1986) has also given simple closed-form approximations to the stopping boundaries g_γ $(\gamma \geq 0)$ for easy implementation of these tests.

3. EXTENSIONS TO GENERAL LOSS FUNCTIONS

In this section we extend the results of Lai (1986) on the 0-1 loss to more general loss functions $\ell(\theta)$. First compare the asymptotic behavior (16) of the optimal continuous-time boundary $h_0(t)$ as $t \to 0$ for the 0-1 loss with the asymptotic behavior (13) for Chernoff's boundary $f(t)$ in the case of absolute value loss for testing the sign of the drift μ of a Wiener process. Note that $h_0(t) \sim (2t \log t^{-1})^{\frac{1}{2}}$ while $f(t) \sim (3t \log t^{-1})^{\frac{1}{2}}$ as $t \to 0$. This order of $(t \log t^{-1})^{\frac{1}{2}}$ in fact holds generally for losses of the form $|\mu|^p$ $(p > -1)$ for the wrong decision, as is shown by

Lemma 1. *Let $w(t)$, $t \geq 0$, be a Wiener process with mean function μt and variance function t, where μ is an unknown parameter. Let $p > -1$, $d > 0$. Then the optimal stopping rule that minimizes the Bayes risk*

$$\rho(\tau; p, d) = \int_{-\infty}^{\infty} E(\tau|\mu)d\mu + d\int_{-\infty}^{0} |\mu|^p \; P[w(\tau) > 0|\mu]d\mu + d\int_{0}^{\infty} \mu^p \; P[w(\tau) < 0|\mu]d\mu \qquad (27)$$

is of the form $\tau^* = \inf\{t : |w(t)| \geq b_{p,d}(t)\}$, *where*

$$b_{p,d}(t) = \{t[(p+2)\log t^{-1} + (1-p)\log\log t^{-1} + O(1)]\}^{\frac{1}{2}} \quad as \quad t \to 0, \qquad (28a)$$

$$\lim_{t\to\infty} b_{p,d}(t) = 0. \qquad (28b)$$

The special case $p = 0$, $d = 1$ in Lemma 1 corresponds to the 0-1 loss with the optimal boundary h_0, while the case $p = d = 1$ corresponds to Chernoff's absolute value loss with the optimal boundary f. The proof of Lemma 1 will be given elsewhere. Making use of the optimal stopping boundary $b_{p,d}$, the following theorem provides simple asymptotic approximations to Bayes sequential tests of $H : \theta < \theta_0$ versus $K : \theta > \theta_0$ for the parameter of the exponential family

(3), with respect to a wide range of prior distributions and for very general loss functions.

Theorem 1. *Consider the problem of testing* $H : \theta < \theta_0$ *versus* $K : \theta > \theta_0$ *for the parameter* θ *of the exponential family* (3), *where* $\theta_0 \in A$ *and* $A = (a_1, a_2)$ *satisfies* (22). *Assume a cost* c *for each observation and a loss* $\ell(\theta) \geq 0$ *for the wrong decision such that*

$$\ell(\theta) \sim \xi|\theta - \theta_0|^p \quad as \quad \theta \to \theta_0, \tag{29}$$

for some $\xi > 0$ *and* $p > -1$. *Let* $d = \xi(\psi''(\theta_0)))^{-p/2}$, $B(t) = b^2_{p,d}(t)/2t$, *where* $b_{p,d}(t)$ *is given in Lemma 1. Define the stopping rule*

$$T_c = \inf\{n \geq 1 : \; I(\hat{\theta}_n, \theta_0) \geq n^{-1} B(c^{2/(p+2)} n)\}, \tag{30}$$

where $\hat{\theta}_n$ *is the maximum likelihood estimate defined in* (23). *Let* $\tilde{\delta}$ *be the terminal decision rule that accepts* H *or* K *according as* $\hat{\theta}_{T(c)} \leq \theta_0$ *or* $\hat{\theta}_{T(c)} > \theta_0$. *Let* G *be a prior distribution on* A *having a positive continuous density* G' *in some neighborhood of* θ_0 *and satisfying*

$$\int \ell(\theta) dG(\theta) < \infty. \tag{31}$$

Then as $c \to 0$, *the test* $(T_c, \tilde{\delta})$ *is asymptotically Bayes with respect to* G; *in fact,*

$$\begin{aligned} r(T_c, \tilde{\delta}) &\sim \inf_{(T,\delta)} r(T, \delta) \\ &\sim (\psi''(\theta_0))^{-\frac{1}{2}} G'(\theta_0) \rho(\tau^*; p, \xi/(\psi''(\theta_0))^{p/2}) c^{(p+1)/p+2)}, \end{aligned}$$

where $\rho(\tau; p, d)$ *and* τ^* *are given in Lemma* 1 *and* $r(T, \delta)$ *denotes the Bayes risk (as defined in* (4) *with* $\Theta_0 = (-\infty, \theta_0)$ *and* $\Theta_1 = (\theta_0, \infty)$) *of the test* (T, δ).

The special case $\ell(\theta) = 1$ in Theorem 1, corresponding to the 0-1 loss, has been established by Lai (1986, Theorem 3), whose proof we now modify to establish Theorem 1. A basic idea is the normalization (32) below, which includes the different normalizations (12) and (21) as special cases.

PROOF OF THEOREM 1. Take $M \geq 2$. Let $r = (p+2)^{-1}$. Define

$$w_c(t) = c^r (S_n - n\psi'(\theta_0))/(\psi''(\theta_0))^{\frac{1}{2}} \;\; \text{at} \;\; t = c^{2r} n, \tag{32}$$

for $n \in \{1, 2, ...\}$, $w_c(0) = 0$, and define $w_c(t)$ by linear interpolation at other positive values of t. Then for every $B > 0$, the process $\{w_c(t), 0 \leq t \leq B\}$ under $P_{\theta_0 + \mu c^r (\psi''(\theta_0))^{-\frac{1}{2}}}$ converges weakly to the Wiener process $\{w(t), 0 \leq t \leq B\}$ with drift coefficient μ, the convergence being uniform in $-M \leq \mu \leq M$, as can be shown by an argument similar to the proof of Lemma 4 of Lai (1986). From this weak convergence, it follows by an argument similar to the proof of

Theorem 3 of Lai (1986) that as $c \to 0$,

$$\int_{\theta_0 - Mc^r(\psi''(\theta_0))^{-\frac{1}{2}}}^{\theta_0 + Mc^r(\psi''(\theta_0))^{-\frac{1}{2}}} \{cE_\theta T_c + \ell(\theta) P_\theta(\text{Error})\} dG(\theta)$$

$$\sim c^r(\psi''(\theta_0))^{-\frac{1}{2}} G'(\theta_0) \int_{-M}^{M} \{c^{1-2r} E_{\theta_0 + \mu c^r (\psi''(\theta_0))^{-\frac{1}{2}}}(c^{2r} T_c)$$

$$+ \xi |c^r(\psi''(\theta_0))^{-\frac{1}{2}} \mu|^p P_{\theta_0 + \mu c^r(\psi''(\theta_0))^{-\frac{1}{2}}}(\text{Error})\} d\mu$$

$$\sim c^{1-r}(\psi''(\theta_0))^{-\frac{1}{2}} G'(\theta_0) \{ \int_{-M}^{M} E(\tau^*|\mu) d\mu + \xi(\psi''(\theta_0))^{-p/2}$$

$$\times [\int_{-M}^{0} |\mu|^p P(w(\tau^*) > 0|\mu) d\mu + \int_0^M \mu^p P(w(\tau^*) < 0|\mu) d\mu]\},$$

noting that $r + rp = (p+1)/(p+2) = 1 - r$. The notation $P_\theta(\text{Error})$ above refers to $P_\theta[(T_c, \tilde{\delta})$ accepts the wrong hypothesis]. □

In view of the asymptotic behavior (28) of the stopping boundary $b_{p,d}$, it follows from Lemma 1 of Lai (1985) that as $|\mu| \to \infty$,

$$E(\tau^*|\mu) \sim (p+2)\mu^{-2}(\log \mu^2),$$
$$P[w(\tau^*)\text{sgn}(\mu) < 0|\mu] = O(|\mu|^{-(p+2)}(\log \mu^2)^{-1}).$$

These and the fact that $p > -1$ imply that

$$\int_{-\infty}^{\infty} E(\tau^*|\mu) d\mu < \infty, \quad \int_{-\infty}^{\infty} |\mu|^p P[w(\tau^*)\text{sgn}(\mu) < 0|\mu] d\mu < \infty.$$

Moreover, by Theorem 1 of Lai (1985), $E_\theta T_c = O((\theta - \theta_0)^{-2} \log\{c^{-2r}(\theta - \theta_0)^2\})$ uniformly in $\theta \in A$ with $(\theta - \theta_0)^2 \geq 2c^{2r}, P_\theta\{\hat{\theta}_{T_c} > \theta_0\} = O(\{c^{-2r}(\theta - \theta_0)^2\}^{-(p+2)/2})$ uniformly in $\theta \in A$ with $\theta - \theta_0 \leq -c^r, P_\theta\{\hat{\theta}_{T_c} < \theta_0\} = O(\{c^{-2r}(\theta - \theta_0)^2\}^{-(p+2)/2})$ uniformly in $\theta \in A$ with $\theta - \theta_0 \geq c^r$, (cf. also Lemma 5 of Lai (1986)). Making use of these estimates and (31), we can proceed in the same way as the proof of Theorem 3 of Lai (1986) to complete the proof.

We next consider the problem of testing $H_0 : \theta \leq \theta_0$ versus $H_1 : \theta \geq \theta_1 (> \theta_0)$. The argument in Section 6 of Lai (1986) dealing with the 0-1 loss can be easily extended to a general loss function $\ell(\theta)$ that is bounded away from 0 outside the indifference zone (θ_0, θ_1). This is the content of

Theorem 2. *Let G be a prior distribution on A and let $r(T, \delta)$ be the Bayes risk (4) of a test (T, δ) of $H_0 : \theta \in \Theta_0 = (-\infty, \theta_0]$ versus $H_1 : \theta \in \Theta_1 = [\theta_1, \infty)$ with cost c per observation and loss $\ell(\theta)$ for the wrong decision such that* (31) *holds and there exists $\Delta > 0$ for which*

$$\ell(\theta) \geq \Delta \quad \textit{for all} \quad \theta \notin (\theta_0, \theta_1). \tag{33}$$

Let g be a nonnegative function on $(0,\infty)$ such that

$$g(t) \sim \log t^{-1} \text{ and } g(t) \geq \log t^{-1} + \beta \log\log t^{-1} \text{ for some } \beta > -\frac{1}{2}, \text{ as } t \to 0. \tag{34}$$

Define the stopping rule $N(g,c)$ by (24), and let δ^* be the terminal decision that accepts H or K according as $\hat{\theta}_{N(g,c)} \leq \theta^*$ or $\theta_{N(g,c)} > \theta^*$, where $\theta \in (\theta_0, \theta_1)$ is such that $I(\theta^*, \theta_0) = I(\theta^*, \theta_1)$.

(i) *Assume that $G([\theta_0 - t, \theta_0]) > 0$ and $G([\theta_1, \theta_1 + t]) > 0$ for all $t > 0$ and that for $\rho > 0$ and $\epsilon > 0$,*

$$G([x,y]) \leq \rho(y-x) \text{ for all } x,y \in [\theta_0 - \epsilon, \theta_0] \cup [\theta_1, \theta_1 + \epsilon] \text{ with } x < y.$$

Then for fixed θ_0 and θ_1, as $c \to 0$,

$$r(N(g,c),\delta^*) \sim c|\log c| \int_A dG(\theta)/J(\theta) \sim \inf_{(T,\delta)} r(T,\delta),$$

where $J(\theta) = \max\{I(\theta,\theta_0),\ I(\theta,\theta_1)\}$.

(ii) *Assume that G has a positive continuous density G' in some neighborhood of θ_0. Then as $c \to 0$ and $\theta_1 \to \theta_0$ such that $(\theta_1 - \theta_0)^2/c \to \infty$,*

$$\begin{aligned} r(N(g,c),\delta^*) &\sim (8G'(\theta_0)/\psi''(\theta_0))c(\theta_1-\theta_0)^{-1}\log[(\theta_1-\theta_0)^2/c] \\ &\sim \inf_{(T,\delta)} r(T,\delta). \end{aligned}$$

BIBLIOGRAPHY

Berk, R. H., Brown, L. D., and Cohen, A. (1981a). Properties of Bayes sequential tests. *Ann. Statist.* **9** 678–682.

Berk, R. H., Brown, L. D., and Cohen, A. (1981b). Bounded stopping times for a class of sequential Bayes tests. *Ann. Statist.* **9** 834–845.

Bickel, P. J. (1973). On the asymptotic shape of Bayes sequential tests of $\theta \leq 0$ versus $\theta > 0$ for exponential families. *Ann. Statist.* **1** 231–240.

Bickel, P. J. and Yahav, Y. A. (1972). On the Wiener process approximation to Bayesian sequential testing problems. *Proc. Sixth Berkeley Symp. Math. Statist. Probab.* **1** 57–84. Univ. California Press.

Breakwell, J. and Chernoff, H. (1964). Sequential tests for the mean of a normal distribution II (large t). *Ann. Math. Statist.* **35** 162–173.

Brown, L. D., Cohen, A., and Strawderman, W. E. (1979). Monotonicity of Bayes sequential tests. *Ann. Statist.* **7** 1222–1230.

Brown, L. D., Cohen, A. and Strawderman, W. E. (1980). Complete classes for sequential tests of hypotheses. *Ann. Statist.* **8** 377–398.

Chernoff, H. (1961). Sequential tests for the mean of a normal distribution. *Proc. Fourth Berkeley Symp. Math. Statist. Probab.* **1** 79–91. Univ. California Press.

Chernoff, H. (1965a). Sequential tests for the mean of a normal distribution III (small t). *Ann. Math. Statist.* **36** 28–54.

Chernoff, H. (1965b). Sequential tests for the mean of a normal distribution IV (discrete case). *Ann. Math. Statist.* **36** 55–68.

Ferguson, T. (1967). *Mathematical Statistics – A Decision Theoretic Approach.* Academic Press, New York.

Fortus, R. (1979). Approximations to Bayesian sequential tests of composite hypotheses. *Ann. Statist.* **7** 579–591.

Fushimi, M. (1967). On the rate of convergence of asymptotically optimal Bayes tests. *Rep. Stat. Appl. Res. JUSE* **14** 1–17.

Kiefer, J. and Sacks, J. (1963). Asymptotically optimum sequential inference and design. *Ann. Math. Statist.* **34** 705–750.

Lai, T. L. (1985). Boundary crossing problems for sample means. To appear in *Ann. Probability.*

Lai, T. L. (1986). Nearly optimal sequential tests of composite hypotheses. To appear in *Ann. Statist.*

Lorden, G. (1967). Integrated risk of asymptotically Bayes sequential tests. *Ann. Math. Statist.* **38** 1399–1422.

Moriguti, S. and Robbins, H. (1962). A Bayes test of $p \leq \frac{1}{2}$ versus $p > \frac{1}{2}$. *Rep. Stat. Appl. Res., JUSE* **9** 39–60.

Ray, S. N. (1965). Bounds on the maximum sample size of a Bayes sequential procedure. *Ann. Math. Statist.* **36** 859–878.

Schwarz, G. (1962). Asymptotic shapes of Bayes sequential testing regions. *Ann. Math. Statist.* **33** 224–236.

Schwarz, G. (1968). Asymptotic shapes for sequential tests of truncation parameters. *Ann. Math. Statist.* **39** 2038–2043.

Schwarz, G. (1969). A second-order approximation to optimal sampling regions. *Ann. Math. Statist.* **40** 313–315.

Sobel, M. (1953). An essentially complete class of decision functions for certain standard sequential problems. *Ann. Math. Statist.* **24** 319–337.

Wald, A. (1945). Sequential tests of statistical hypotheses. *Ann. Math. Statist.* **16** 117–186.

Wald, A. and Wolfowitz, J. (1948). Optimum character of the sequential probability ratio test. *Ann. Math. Statist.* **19** 326–339.

Wong, S. P. (1968). Asymptotically optimum properties of certain sequential tests. *Ann. Math. Statist.* **39** 1244–1263.

Woodroofe, M. (1980). On the Bayes risk incurred by using asymptotic shapes. *Comm. Statist. Ser. A.* **9** 1727–1748.

STOCHASTIC SEARCH IN A SQUARE AND ON A TORUS

S. P. Lalley

Department of Statistics
Purdue University
West Lafayette, Indiana 47907-2005, U.S.A.

H. E. Robbins

Department of Statistics
Columbia University
New York, New York 10027, U.S.A.

1. INTRODUCTION

Princess and Monster (cf. Isaacs (1965)) is a two-person zero-sum game with the following rules. The two players are restricted to a bounded, connected, two-dimensional region Ω. The Monster (M) has maximum speed 1, the Princess (P) has maximum speed $v < 1$. Neither player obtains any information about the position or movements of the other until the first time the distance between the two is $\leq \epsilon$; at this time M captures P and the game ends. The payoff to P is the time elapsed before capture occurs.

This game is a crude model for naval operations involving a submarine (P) and a surface ship (M) attempting to locate the submarine with active sonar. In such operations the parameter 2ϵ (the "sweep width") is typically small relative to the dimensions of Ω.

The princess and monster game is too complex to admit simple minimax strategies even for simple regions Ω. S. Gal (1979), (1980) has shown, however, that it is possible to find strategies for both players that are "asymptotically minimax" as $\epsilon \to 0$, and has proved that the value $V(\epsilon)$ of the game satisfies

$$\lim_{\epsilon \to 0} 2\epsilon V(\epsilon) = |\Omega|, \tag{1.1}$$

where $|\Omega|$ denotes the area of Ω. (Some restrictions on Ω are necessary for this to hold.) Gal's strategy for P (section 2 below) is simple and does not change with ϵ. His strategy for M, however, is rather complicated, and changes drastically with ϵ. For small ϵ his strategy calls for M to search for relatively long periods of time in small regions of Ω, and therefore his strategy would work poorly if the rules of the game were changed to allow P occasional (but infrequent) information concerning the whereabouts of the M.

The purpose of this paper is to present simple strategies for M (section 3 below) that are asymptotically minimax but do not change with ϵ, and do not "localize" M in Ω for long periods of time. We shall only consider the case where Ω is a square (or torus). The search plans and the main results generalize easily to rectangles, parallelograms, and triangles. We believe that good strategies for more general regions may be obtained by breaking up such regions into congruent rectangles and somehow piecing together the rectangular strategies, but we do not yet know how to do this in such a way that the new strategies will not suffer from the same localization problem as does Gal's.

2. GAL'S STRATEGIES

A. The Princess

Let T be large but small compared to $\epsilon^{-1}|\Omega|$. Since M's maximum speed is 1, the maximum area it can search in time T is $2\epsilon T + \pi\epsilon^2 \sim 2\epsilon T$, no matter what strategy it uses. Therefore, if P chooses a point in Ω at random (with the uniform distribution on Ω) and stays there, the probability of capture by time T is no greater than $2\epsilon T/|\Omega|$.

Gal's strategy for P is as follows. She chooses P_1, P_2, ...i.i.d. with the uniform distribution on Ω; she starts at P_1, stays there T time units, then moves to P_2 at full speed, stays there T time units, etc. Assuming that T is large (e.g., $T = \epsilon^{-\frac{1}{2}}$), the probability that the princess will be caught in transit from P_i to P_{i+1} for some i is miniscule compared to the probability that she will be captured at P_i for some i, and may be neglected. Therefore, the probability that P is captured before time nT is no more than $1-(1-2\epsilon T/|\Omega|)^n$ (approximately), and consequently the expected capture time is no less than $|\Omega|/2\epsilon$ (approximately).

One may wonder why P should move at all. If P were simply to choose a point in Ω at random and stay there indefinitely, and if M knew this, then M could search the entire area Ω in time $|\Omega|/2\epsilon$ (approximately), capturing P *for certain* by time $|\Omega|/2\epsilon$: the *expected* capture time would be $|\Omega|/4\epsilon$. Evidently, the advantage of movement for P is that if forces M to retrace its steps from time to time.

B. The Monster

Obtaining *lower* bounds for capture probabilities is much more difficult than obtaining upper bounds. Gal's strategy for M is correspondingly more complicated and less natural. We shall describe it only for the case where Ω is a square (of side one).

Partition the square into disjoint rectangles with sides 1 and α, where α is small but α/ϵ is large (i.e., $\alpha \to 0$ and $\epsilon/\alpha \to 0$ as $\epsilon \to 0$). The number of rectangles in the partition is α^{-1}. M's strategy is as follows. It chooses one of the α^{-1} rectangles at random and moves back and forth along lines parallel to the *long* edges N times, where N is large (specifically, $N \to \infty$ and $\epsilon N \to 0$ as $\epsilon \to 0$). The N passes through the rectangle are made on line segments chosen so that the projections on the short edge of the rectangle are i.i.d. uniform on the short edge. After N passes, the monster chooses another of the α^{-1} rectangles and makes N passes through it, and so on.

The reason for making such a large number of passes through each narrow rectangle before moving on to the next is that this makes the time spent on the boundary of the square (i.e., the time *between* successive passes) negligible compared to the time making the passes. This is essential to Gal's proof of the asymptotic minimaxity of this strategy. But it forces M to restrict its movements to small areas for large intervals of time. If the rules of the game were changed so that P got a glimpse of M's position once every N units of time or so, then she might be able to elude it indefinitely if it used Gal's strategy.

Gal's (1979) main result is that if M uses the strategy described above, then no matter what strategy P uses the expected time until capture is no greater than $|\Omega|/2\epsilon$ (approximately). This, together with the arguments outlined above in connection with P strategy, implies (1.1).

3. SEMI-MARKOV SEARCH PLAN

Henceforth we shall assume that Ω is either a square or a torus (a square with opposite sides identified). There are two reasons for studying the princess and monster game on a torus. First, the torus has a simple geometry: it is a compact surface without boundary and it has a large group of symmetries (since it has a group structure). Second, any square is a quotient space of a torus (see below): this enables us to "lift" continuous paths in a square to continuous paths on a torus.

For our purposes it is convenient to regard the torus $\mathcal{T}$ as a quotient group $\mathcal{R}^2/\mathcal{Z}^2$, where $\mathcal{Z}^2 = \{(m,n) : m,n \epsilon \mathcal{Z}\}$. The natural projection from $\mathcal{R}^2$ to $\mathcal{T}$ will be denoted by $\tilde{\pi}$. The torus $\mathcal{T}$ inherits a metric and an invariant measure from $\mathcal{R}^2$: these will be referred to as the *natural metric* and the *Lebesgue measure* on $\mathcal{T}$. A *geodesic* with respect to the natural metric is just the projection onto $\mathcal{R}^2/\mathcal{Z}^2$ of a straight line in $\mathcal{R}^2$ (Figure 1). The *uniform distribution* on $\mathcal{T}$ is just Lebesgue measure.

It will also be convenient to regard the square $\mathcal{S}$ of side $\frac{1}{2}$ as a quotient space $\mathcal{T}/\sim$ of the torus. The equivalence relation $\sim$ is defined as follows: if $t_1+\mathcal{Z}^2, t_2+\mathcal{Z}^2$ are two distinct elements of $\mathcal{T} = \mathcal{R}^2/\mathcal{Z}^2$, then $t_1+\mathcal{Z}^2 \sim t_2+\mathcal{Z}^2$ iff $t_1 + t_2 + (\alpha, 0) \epsilon \mathcal{Z}^2$ for some $\alpha \epsilon \mathcal{R}$ or $t_1 + t_2 + (0, \alpha) \epsilon \mathcal{Z}^2$ for some $\alpha \epsilon \mathcal{R}$. In geometrical terms, two points of $\mathcal{T}$ are identified if one can be obtained from

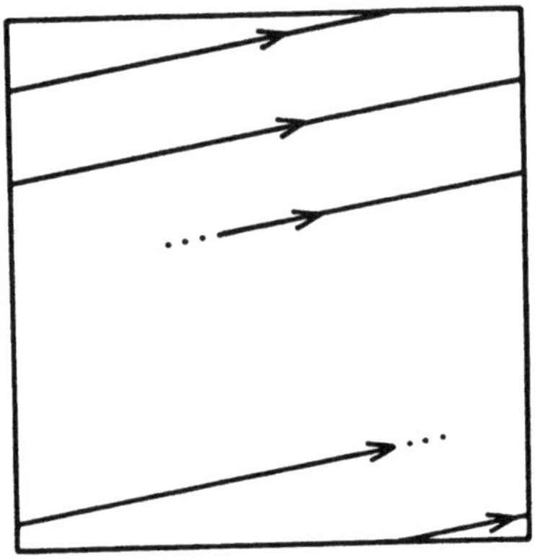

Figure 1.

the other by reflections through the midlines of $\mathcal{T}$ (Figure 2a). The projection mapping from $\mathcal{T}$ to $\mathcal{S} = \mathcal{T}/\sim$ that sends each point of $\mathcal{T}$ to its $\sim$ equivalence class will be denoted by π. Observe that the natural metric and the Lebesgue measure on $\mathcal{T}$ project to the usual Euclidean metric and Lebesgue measure on $\mathcal{S}$. Notice also that each continuous path in $\mathcal{S}$ that does not touch the boundary of $\mathcal{S}$ lifts (by π^{-1}) to four distinct paths in $\mathcal{T}$ (Figure 2b).

Henceforth, points of $\mathcal{T}$ will be denoted by the letter x and points of $\mathcal{S}$ will be denoted by the letter y. Lebesgue measure on $\mathcal{T}$ will be denoted by dx and Lebesgue measure on $\mathcal{S}$ will be denoted by dy.

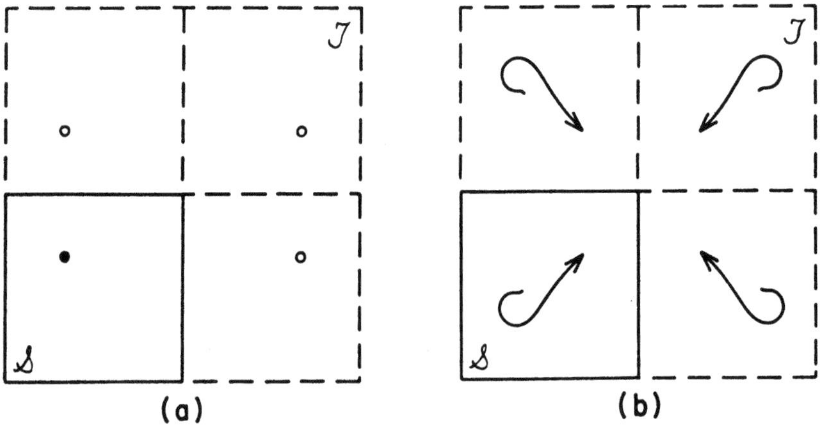

Figure 2.

Our semi-Markov search plan for the monster in $\mathcal{T}$ is as follows. Let $\tilde{V}_1, \tilde{V}_2, \ldots$ be i.i.d. random vectors in $\mathcal{R}^2$ with the uniform distribution on the circumference of the circle of radius ρ centered at the origin, where $0 < \rho \leq \frac{1}{2}$ is fixed. Let $V_i = \tilde{\pi}(\tilde{V}_i)$ be the projection of $\tilde{V}_i$ onto $\mathcal{T}$, and let $X(0)$ be uniformly distributed on $\mathcal{T}$ and independent of $V_1, V_2, \ldots$. For $n \geq 0$ and $0 \leq t \leq \rho$, define

$$X(n\rho + t) = X(0) + \sum_{k=1}^{n} V_i + tV_{n+1}/\rho \tag{3.1}$$

(here $tV_{n+1}/\rho = \tilde{\pi}(t\tilde{V}_{n+1}/\rho)$; the addition is carried out in the group $\mathcal{T}$, so $X(t)$ is a path in $\mathcal{T}$). The point $X(t)$ describes the position of M at time $t \geq 0$. The process $X(t), t \geq 0$, is a semi-Markov process in $\mathcal{T}$. Some of its ergodic properties are described in section 4.

A semi-Markov search plan for M in $\mathcal{S}$ is obtained from $X(t)$ by projection. Thus,

$$Y(t) = \pi(X(t)) \tag{3.2}$$

describes the position of M at time $t \geq 0$. The process $Y(t), t \geq 0$, is a semi-Markov process in $\mathcal{S}$.

Let $\mathcal{F}$ be the set of all piecewise continuously differentiable paths $z(t)$ in $\mathcal{T}$ with $|z'(t)| \leq v < 1$. For $z \epsilon \mathcal{F}$ and $\epsilon > 0$, define

$$T_\epsilon(z) = \inf\{t : \text{ dist } (X(t), z(t)) \leq \epsilon\}. \tag{3.3}$$

(Here dist refers to the natural metric on $\mathcal{T}$.)

Theorem 1. $\limsup_{\epsilon \to 0} \sup_{z \epsilon \mathcal{F}} \epsilon E T_\epsilon(z) \leq \frac{1}{2}$.

In views of the arguments in section 2A, this shows that the semi-Markov search plan $X(t)$ is asymptotically minimax for M. In section 4 we will prove that as $t \to \infty$, the distribution of $X(t)$ conditional on the value of $X(0)$ converges to the uniform distribution on $\mathcal{T}$. It can be shown that if ρ is not too small this convergence is fairly rapid. Consequently, *infrequent* information for P concerning the position and direction of the monster would not enable her to lengthen the expected capture time appreciably. Thus, the semi-Markov plan, in addition to being asymptotically minimax, is robust to certain changes in the rules of the game favoring P.

For $0 < \delta < \frac{1}{4}$, let $\mathcal{F}_\delta$ be the set of all piecewise continuously differentiable paths $z(t)$ in $\mathcal{S}$ such that $|z'(t)| \leq v < 1$ and dist $(z(t), \partial\mathcal{S}) \geq \delta$ for all t, where $\partial\mathcal{S}$ denotes the boundary of $\mathcal{S}$. For $z \epsilon \mathcal{F}_\delta$, define

$$T_\epsilon(z) = \inf\{t \geq 0 : \text{ dist } (Y(t), z(t)) \leq \epsilon\}. \tag{3.4}$$

Theorem 2. *If $\rho < 2\delta$, then*

$$\limsup_{\epsilon \to 0} \sup_{z \in \mathcal{F}_\delta} \epsilon E T_\epsilon(z) \leq 1/8.$$

This is not quite as satisfactory as Theorem 1, because $\mathcal{F}_\delta$ does not include all the pure strategies available to P. Nevertheless, Theorem 2 does provide search strategies that are asymptotically minimax but do not suffer from localization. In particular, if M follows the strategy $Y(t)$ suitable for a slightly expanded square $(1+\delta)\mathcal{S}$, with ρ suitably small, then the maximum of its risk (as $\epsilon \to 0$) over all allowable pure strategies for P in $\mathcal{S}$ is no greater than $(1+\delta)^2$ times the minimax value. (In practice it would be pointless for M to search outside $\mathcal{S}$, so some simple modification should be made at the boundary.) Unfortunately, efficiency and robustness pull in opposite directions: in order to make $(1+\delta)^2 - 1$ small one must choose ρ small, but the smaller ρ, the longer the "relaxation time," i.e., the amount of time it takes the conditional distribution of $Y(t)$ given $Y(0)$ to become approximately uniform.

We suspect that a large class of semi-Markov search strategies are asymptotically minimax for P and M game. The crucial properties of a good search strategy, we believe, are that it should have a uniform stationary distribution and good ergodicity properties. Some of the arguments below use only these properties of our strategies; however, the arguments of section 5 rely heavily on the symmetry properties of $\mathcal{T}, \mathcal{S}, X(t)$, and $Y(t)$.

The rest of the paper is devoted to the proofs of Theorems 1 and 2.

4. ERGODICITY OF THE SEARCH PROCESS

Let $X(t) = X_\rho(t)$ be the search process defined by (3.1). By hypothesis $X(0)$ is uniformly distributed on $\mathcal{T}$ and is independent of the random vectors $V^{(1)}, V^{(2)}, \ldots$. Since the uniform distribution is invariant under translations of $\mathcal{T}$, for any $x \in \mathcal{T}$ the random vector $x + X(0)$ is uniformly distributed on $\mathcal{T}$ and is independent of $V_1, V_2, \ldots$. Consequently, for any $x \in \mathcal{T}$ the stochastic processes $X(t)$ and $x + X(t)$ are identical in law, and it follows that for every $t \geq 0$ the random vector $X(t)$ is uniformly distributed on $\mathcal{T}$. Since $Y(t) = \pi(X(t)), Y(t)$ is uniformly distributed on $\mathcal{S}$ for every $t \geq 0$.

Consider the discrete-time process

$$X(n\rho) = X(0) + \sum_{i=1}^{n} V_i, \quad n = 0, 1, 2, \ldots.$$

Since $X(0)$ is uniformly distributed on $\mathcal{T}$ and is independent of $V_1, V_2, \ldots,\ X(\rho)$ is uniformly distributed and is independent of $V_2, V_3, \ldots$. Hence, $X(n\rho)$, $n =$

$0, 1, 2, \ldots$, is a stationary Markov chain on $\mathcal{T}$. It follows that the same is true of the process $Y(n\rho)$, $n = 0, 1, 2, \ldots$.

Recall that the random vectors V_i, $i = 1, 2, \ldots$, are i.i.d. on $\mathcal{T}$. Their common distribution is the projection onto $\mathcal{T}$ of the uniform distribution on the circumference of the circle of radius ρ centered at the origin. It is easily verified that for sufficiently large k the k-fold convolution of this distribution with itself has a continuously differentiable density with respect to Lebesgue measure on $\mathcal{T}$. (In fact, the Fourier transforms of these convolutions may be written explicitly in terms of Bessel functions: cf. Spitzer (1976), Ch. 2, exercise 15.) Let $f_k(x)$ denote this density.

Proposition 1. $\lim_{k \to \infty} \sup_{x \in \mathcal{T}} |f_k(x) - 1| = 0$.

PROOF. Let k be so large that $f_k(x)$ is continuously differentiable on $\mathcal{T}$. Since $f_k(x)$ is a probability density, its zeroth Fourier coefficient is 1, and all the others are strictly less than one in absolute value. Since $f_k(x)$ is C^1, it has a uniformly convergent Fourier series.

For $n \geq 1$, the Fourier coefficients of $f_{nk}(x)$ are the nth powers of the Fourier coefficients of $f_k(x)$. Hence, for large n the dominant term in the Fourier series of $f_{nk}(x)$ is the zeroth, and it follows that $\lim_{n \to \infty} f_{nk}(x) = 1$ uniformly for $x \in \mathcal{T}$. The proposition now follows routinely. □

For $x_0, x \in \mathcal{T}$ and $y_0, y \in \mathcal{S}$, define

$$g_n(x_0, x) = P(X(n\rho) \in dx | X(0) = x_0)/dx, \tag{4.1}$$

$$h_n(y_0, y) = P(Y(n\rho) \in dy | Y(0) = y_0)/dy, \tag{4.2}$$

provided n is large enough that these densities exist. Observe that $g_n(x_0, x) = f_n(x - x_0)$, and that h_n is the projection onto $\mathcal{S}$ (under $\pi : \mathcal{T} \to \mathcal{S}$) of the density g_n, i.e., if $x_0 \in \pi^{-1}(\{y_0\})$,

$$h_n(y_0, y) = \sum_{x \in \pi^{-1}(\{y\})} g_n(x_0, x).$$

Therefore, Proposition 1 implies that as $n \to \infty$,

$$g_n(x_0, x) \to 1 \tag{4.3}$$

and

$$h_n(y_0, y) \to 4 \tag{4.4}$$

uniformly for $x_0, x \in \mathcal{T}$ and $y_0, y \in \mathcal{S}$.

Recall that $\mathcal{F}$ is the set of all piecewise continuously differentiable paths $z(t)$ in $\mathcal{T}$ with $|z'(t)| \leq v < 1$, and that $\mathcal{F}_\delta$ is the set of all piecewise continuously differentiable paths $z(t)$ in $\mathcal{S}$ with $|z'(t)| \leq v$ and such that dist $(z(t), \partial\mathcal{S}) \geq \delta$ for all $t \geq 0$. Also, for $z \in \mathcal{F}$, $T_\epsilon = T_\epsilon(z)$ is the first time $t \geq 0$

that dist $(z(t), X(t)) \leq \epsilon$; for $z \epsilon \mathcal{F}_\delta, T_\epsilon = T_\epsilon(z)$ is the first time $t \geq 0$ that dist $(z(t), Y(t)) \leq \epsilon$.For $z \epsilon \mathcal{F}$ and $\epsilon > 0$ define

$$G_n(x; \epsilon, z) = P(X(n\rho) \epsilon dx; T_\epsilon(z) > n\rho)/dx, \tag{4.5}$$

and for $z \epsilon \mathcal{F}_\delta$ define

$$H_n(y; \epsilon, z) = P(Y(n\rho) \epsilon dy;\ T_\epsilon(z) > n\rho)/dy. \tag{4.6}$$

Proposition 2. *As* $\epsilon \to 0$

$$\sup_{n \geq 0} \sup_{z \epsilon \mathcal{F}} \int_{\mathcal{T}} |G_n(x; \epsilon, z) - P(T_\epsilon(z) > n\rho)| dx \to 0 \tag{4.7}$$

and

$$\sup_{n \geq 0} \sup_{z \epsilon \mathcal{F}_\delta} \int_S |H_n(y; \epsilon, z) - P(T_\epsilon(z) > n\rho)| dy \to 0. \tag{4.8}$$

For the proof we will need

Lemma 1. *As* $\epsilon \to 0$

$$\sup_{n \geq 1} \sup_{z \epsilon \mathcal{F}} P((n-1)\rho \leq T_\epsilon(z) \leq n\rho) \to 0$$

and

$$\sup_{n \geq 1} \sup_{z \epsilon \mathcal{F}_\delta} P((n-1)\rho \leq T_\epsilon(z) \leq n\rho) \to 0.$$

PROOF. It suffices to show that

$$\sup_{z \epsilon \mathcal{F}} P(T_\epsilon(z) \leq \rho) \to 0 \tag{4.9}$$

and

$$\sup_{z \epsilon \mathcal{F}_\delta} P(T_\epsilon(z) \leq \rho) \to 0 \tag{4.10}$$

as $\epsilon \to 0$, because the process $X(t), t \geq 0$, has the same distribution as $X(t + n\rho), t \geq 0$ (and hence, since $Y(t) = \pi(X(t))$, the process $Y(t), t \geq 0$, has the same distribution as $Y(t + n\rho), t \geq 0$).

To prove (4.9) and (4.10) it suffices to show that there exist constants $c(\epsilon) \to 0$ as $\epsilon \to 0$ such that for each w on the circle of radius ρ centered at the origin,

$$\sup_{z \epsilon \mathcal{F}} P(T_\epsilon(z) \leq \rho | V_1 = \tilde{\pi}(w)) \leq c(\epsilon) \tag{4.11}$$

and

$$\sup_{z \epsilon \mathcal{F}_\delta} P(T_\epsilon(z) \leq \rho | V_1 = \tilde{\pi}(w)) \leq c(\epsilon). \tag{4.12}$$

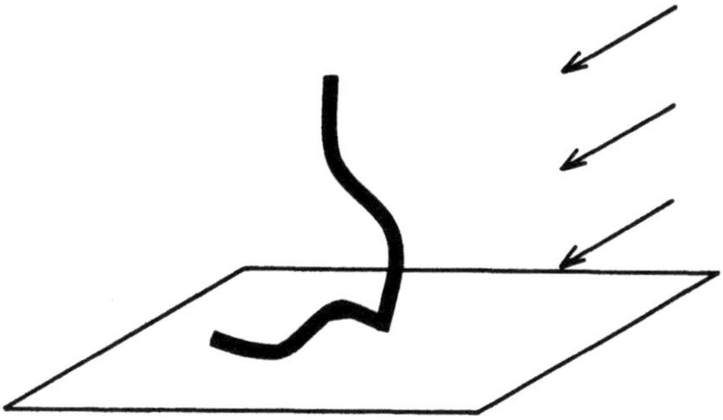

Figure 3.

For $z \epsilon \mathcal{F}$, let $A(z, \epsilon)$ be the set of all $x \epsilon \mathcal{T}$ such that dist $(x + \tilde{\pi}(tw/\rho), z(t)) \leq \epsilon$ for some $0 \leq t \leq \rho$; then

$$P(T_\epsilon(z) \leq \rho | V_1 = \tilde{\pi}(w)) = \text{area } (A(z, \epsilon)).$$

The region $A(z, \epsilon)$ may be thought of as the "projection" of the ϵ-tube around the path $(t, z(t))$ in $[0, \rho] \times \mathcal{T}$ onto the "base" $\{0\} \times \mathcal{T}$ through the vector $(\rho, -w)$. In other words, $A(z, \epsilon)$ is the shadow that the ϵ-tube around $(t, z(t))$, $0 \leq t \leq \rho$, would cast if light were shining down along rays parallel to $(\rho, -w)$ (Figure 3). Since $z(t)$ is piecewise continuously differentiable with $|z'(t)| \leq v$, the area of the shadow $A(z, \epsilon)$ is no greater than $C \times \epsilon$, where the constant C depends only on v (not on z or w). This proves (4.11). A similar argument proves (4.12). (Any path $z(t)$ in $\mathcal{S}$ is the image of four paths in $\mathcal{T}$ under the mapping $\pi : \mathcal{T} \rightarrow \mathcal{S}$. Therefore, the same argument may be repeated almost verbatim, the only difference being that there are now four shadows instead of just one.) □

PROOF OF PROPOSITION 2 Let $\eta > 0$ be small. By (4.3), for k sufficiently large

$$|g_k(x_0, x) - 1| < \eta. \tag{4.13}$$

By Lemma 1, for all $\epsilon > 0$ sufficiently small

$$\sup_{n \geq k} \sup_{z \epsilon \mathcal{F}} P((n-k)\rho \leq T_\epsilon(z) \leq n\rho) < \eta. \tag{4.14}$$

For $z \epsilon \mathcal{F}$ and $n \geq k$

$$\begin{aligned} G_n(x; \epsilon, z) = 1 &- P(X(n\rho) \epsilon dx; T_\epsilon(z) < (n-k)\rho)/dx \\ &- P(X(n\rho) \epsilon dx; (n-k)\rho \leq T_\epsilon(z) \leq n\rho)/dx \end{aligned} \tag{4.15}$$

because $X(n\rho)$ is uniformly distributed on $\mathcal{T}$. By the semi-Markov property of $X(t)$, for $n \geq k$

$$P(X(n\rho)\epsilon dx; T_\epsilon(z) < (n-k)\rho)/dx$$
$$= \int_{x_0 \epsilon \mathcal{T}} g_k(x_0, x) P(X((n-k)\rho)\epsilon dx_0;\ T_\epsilon(z) < (n-k)\rho).$$

Combining this with (4.13) we obtain, for $n \geq k$,

$$\int_{x\epsilon\mathcal{T}} |P(X(n\rho)\epsilon dx; T_\epsilon(z) < (n-k)\rho)/dx - P(T_\epsilon(z) < (n-k)\rho)|dx < \eta. \tag{4.16}$$

It now follows from (4.14)-(4.16), together with the triangle inequality, that

$$\int_{x\epsilon\mathcal{T}} |G_n(x;\epsilon,z) - P(T_\epsilon(z) > n)|dx < 3\eta,$$

and since $\eta > 0$ is arbitrary, this proves (4.7). The proof of (4.8) is virtually the same. □

5. ONE-STEP DETECTION PROBABILITIES

In this section we obtain lower bounds for the probability of detection in one step. These bounds are the key to establishing the asymptotic minimaxity of the semi-Markov search policies $X(t)$ and $Y(t)$.

Proposition 3. *For any $z\epsilon\mathcal{F}$, any $\rho \leq \frac{1}{2}$, and any $\epsilon \leq 1/2$,*

$$P\{dist(X(t), z(t)) \leq \epsilon, \quad some \quad 0 \leq t \leq \rho\} \geq 2\rho\epsilon. \tag{5.1}$$

Proposition 4. *Let $\rho < 2\delta < \frac{1}{2}$. For any $z\epsilon\mathcal{F}_\delta$ and any $\epsilon \leq 1/2$,*

$$P\{dist(Y(t), z(t)) \leq \epsilon, \quad some \quad 0 \leq t \leq \rho\} \geq 8\rho\epsilon. \tag{5.2}$$

The rest of this section is devoted to the proofs of these propositions. The arguments rely rather heavily on symmetry properties of $\mathcal{T}$ and $\mathcal{S}$, especially the group structure of $\mathcal{T}$ and the group invariance (in distribution) of the stochastic process $X(t)$.

We begin with some elementary properties of closed geodesics on $\mathcal{T}$. Recall that a *geodesic* in $\mathcal{T}$ is just a straight line in the square $[0,1] \times [0,1]$ that, upon reaching one of the four sides, "jumps" to the corresponding point of the opposite side and continues in the same direction (Figure 1). A *closed geodesic* is a geodesic which goes through the same point of $\mathcal{T}$ at least twice, and hence is periodic. A geodesic is closed iff its slope is rational (Figure 4).

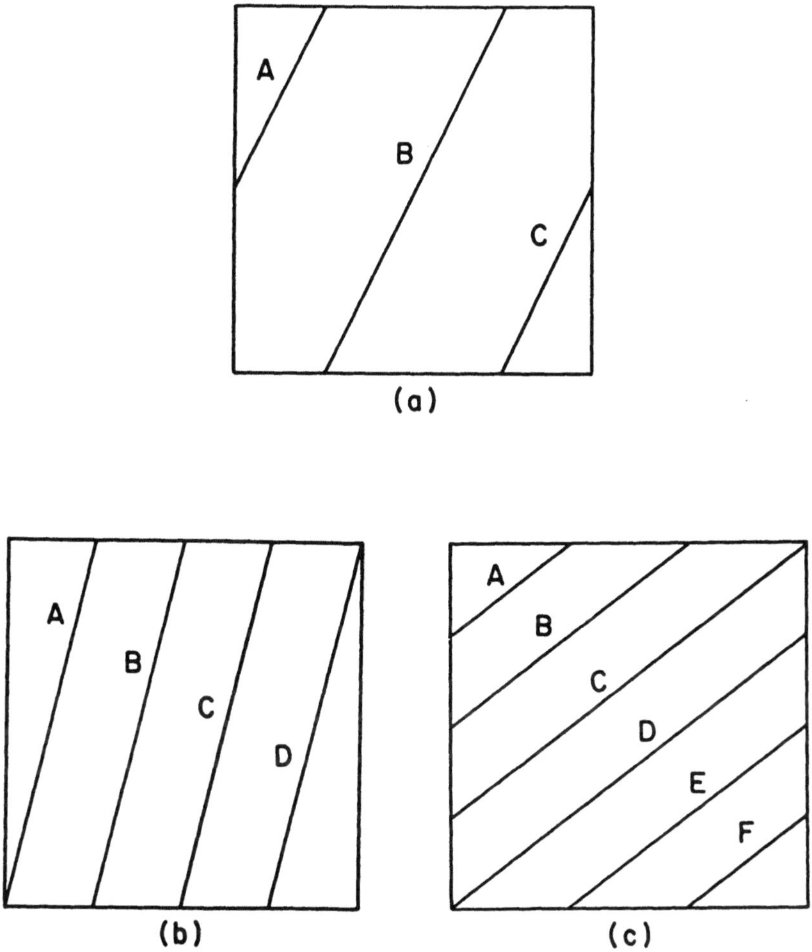

Figure 4.

If the unit square $[0,1] \times [0,1]$ with opposite sides identified is used as a model for $\mathcal{T}$, then a closed geodesic appears as a finite collection of equally spaced parallel line segments in the square (Figure 4). Some of these segments connect opposite sides of the square (B in Figure 4a; A, B, C, D in Figure 4b; C, D in Figure 4c); others connect adjacent sides (A, C in Figure 4a; A, B, C, D in Figure 4c). We define a *cycle* to be either a segment connecting opposite

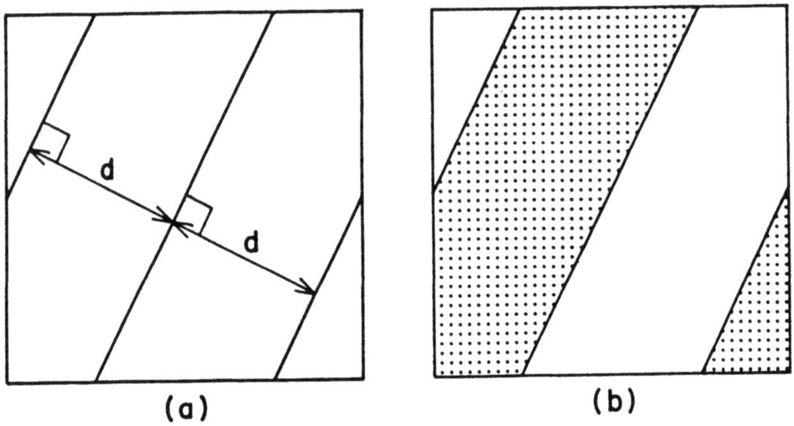

Figure 5.

sides or the union of two successive segments, each of which connects adjacent sides. (In Figure 4a, the cycles are B, $A \cup C$; in Figure 4b, the cycles are A, B, C, D; in Figure 4c, the cycles are $A \cup E$, $B \cup F, C, D$.) Observe that for any closed geodesic, all of the cycles have the same length.

Fix a closed geodesic Γ; let N be the number of cycles, ℓ the length of a cycle, L the total length of Γ, and d the distance between neighboring segments of Γ (cf. Figure 5a). Then

$$L = N\ell, \tag{5.3}$$

$$d = 1/L. \tag{5.4}$$

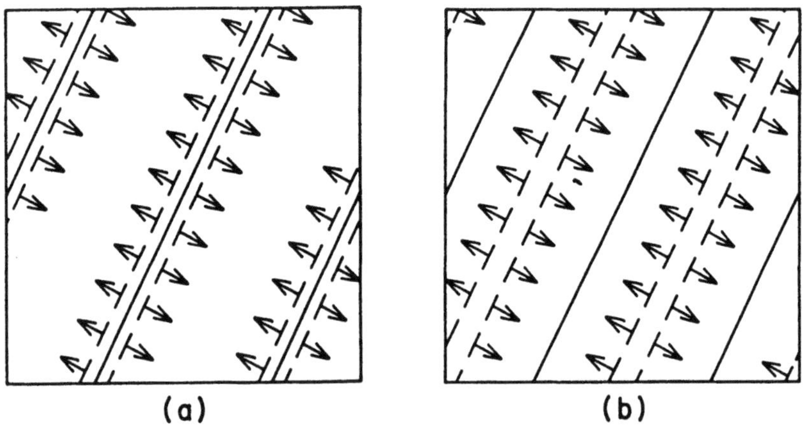

Figure 6.

Equation (5.3) follows from the fact that all cycles of Γ have the same length ℓ. Equation (5.4) follows from a computation of the area of $[0,1] \times [0,1]$: the area (1) is the sum of the areas of the N disjoint "parallelograms" into which the cycles of Γ partition the square, and each such parallelogram has area ℓd (Figure 5b: the shaded region counts as a single "parallelogram").

Imagine now a shock wave emanating from Γ and travelling orthogonally to Γ at speed 1 in each of the two orthogonal directions (Figure 6). Initially the wave front consists of a pair of parallel bands surrounding Γ (Figure 6a). These bands travel in opposite directions, away from Γ. At time $d/2$, the bands of the wave front travelling in opposite directions will meet and cross (Figure 6b). As time goes on, the wave front will pass through itself repeatedly: in fact, this will happen at times $d/2, 2d/2, 3d/2, \ldots$. Thus, by time $r > 0$ the wave front will have passed through itself $[[2r/d]] = [[2rL]]$ times (here $[[\cdot]]$ denotes greatest integer).

Let $z \epsilon \mathcal{F}$. Since $|z'(t)| \leq v < 1$, the path $z(t)$ moves more slowly than the wave front. At time $t = 0, z(t)$ is located either on Γ or between two bands of the wave front. These two bands move toward each other, meeting at time $d/2$; hence, $z(t)$ must pass through the wave front at least once in the time interval $[0, d/2]$. Similarly, $z(t)$ must pass through the wave front at least once in each of the time intervals $[d/2, 2d/2], [2d/2, 3d/2], \ldots$. Note that if $z(t)$ meets the wave front at one of the times $t = 0, d/2, 2d/2, \ldots$, then it simultaneously crosses two bands of the wave front, which may be counted as a double cross. With this convention, it follows that *for any $r > 0$ $z(t)$ must cross the wave front at least $[[2rL]]$ times in the time interval $[0, r]$.*

We may now prove Proposition 3. Recall that the "first step" of the process $X(t)$ is the path

$$X(t) = X(0) + tV_1/\rho, \quad 0 \leq t \leq \rho, \tag{5.5}$$

where $X_\rho(0)$ is uniformly distributed on $\mathcal{T}$ and $V_1 = \tilde{\pi}(\tilde{V}_1)$ where $\tilde{V}_1$ is uniformly distributed on the circle of radius ρ centered at zero. We are interested in the probability that the path (5.5) comes within ϵ of the path $z \epsilon \mathcal{F}$.

The distribution of the path (5.5) (conditional on $\pm V_1$) is invariant under translations of $\mathcal{T}$, since $X(0)$ is uniformly distributed on $\mathcal{T}$ and since the uniform distribution is invariant under translations. Also, the distribution of the path (5.5) (conditional on $\pm V_1$) is the same as that of

$$\hat{X}(t) = X(0) - tV_1/\rho, \quad 0 \leq t \leq \rho,$$

since the distribution of $\tilde{V}_1/\rho$ is rotation-invariant. Therefore, for any x_1, $x_2, \ldots, x_n \epsilon \mathcal{T}$, each of the paths

$$X^{(i)}(t) = x_i + X(0) + tV_1/\rho, \quad 0 \leq t \leq \rho, \tag{5.6}$$

and

$$\hat{X}^{(i)}(t) = x_i + X(0) - tV_1/\rho, \quad 0 \le t \le \rho, \tag{5.7}$$

has the same distribution as the path (5.5). Consequently, to prove Proposition 3 it suffices to show that for each $\eta > 0$ and almost every value of $\pm V_1$, there exist $x_1, x_2, \ldots, x_n$ such that the proportion of the $2n$ paths (5.6) and (5.7) that come within ϵ of $z(t)$ is greater than $2\rho\epsilon - \eta$.

With probability one, the vector V_1 has irrational slope. Consequently, the geodesic through $X(0)$ perpendicular to V_1 is not closed. However, there exists a sequence of closed geodesics Γ_m through $X(0)$ such that as $m \to \infty$ the angle between Γ_m and V_1 approaches $\pi/2$; moreover, if L_m is the length of Γ_m then as $m \to \infty$, $L_m \to \infty$, because for large m, Γ_m closely approximates the geodesic through $X_\rho(0)$ perpendicular to V_1. Fix Γ_m, where m is large. Mark points $x_i + X(0)$, $i = 1, 2, \ldots, n$ on Γ_m in such a way that

$$\begin{aligned} \text{dist}(x_i, x_{i+1}) &= 2\epsilon, \quad i = 1, 2, \ldots, n-1, \\ \text{dist}(x_n, x_1) &\le 2\epsilon. \end{aligned}$$

(The distances are measured according to the metric on $\mathcal{T}$.) Observe that $n = [[L_m/2\epsilon]] + 1$ or $n = [[L_m/2\epsilon]]$.

We argue now that with $x_1, x_2, \ldots, x_n$ defined as in the preceding paragraph, the proportion of the $2n$ paths (5.6) and (5.7) which come within ϵ of $z(t)$ is greater than $2\rho\epsilon - \eta$, provided m is sufficiently large. Think of the paths (5.6) and (5.7) as being the paths of $2n$ men pushing brooms of width 2ϵ. These brooms fit together to form a single wave front emanating from the closed geodesic Γ_m. Observe that the speed of propagation of this wave is not quite 1, because Γ_m is not quite perpendicular to the paths (5.6) and (5.7); nevertheless, by choosing m sufficiently large, the speed of propagation may be made greater than $1 - \lambda$ for any $\lambda > 0$. It follows from the previous discussion that $z(t)$ crosses the wave front at least $[[2\rho(1-\lambda)L_m]]$ times in the time interval $[0, \rho]$. Since $|z'(t)| \le v < 1$, the broom pushers move faster than $z(t)$; hence, $z(t)$ cannot meet the same broom more than once during $[0, \rho]$, since $\rho < \frac{1}{2}$. It follows that the number of broom pushers who pass within ϵ of $z(t)$ during the time interval $[0, \rho]$ is at least $[[2\rho(1 - \lambda)L_m]]$. Since the number of broom pushers is $2n < L_m/\epsilon + 2$, if follows that the proportion of the $2n$ paths (5.6) and (5.7) that pass within ϵ of $z(t)$ is at least

$$[[2\rho(1 - \lambda)L_m]]/(L_m\epsilon^{-1} + 2).$$

As $m \to \infty$, $L_m \to \infty$, hence this proportion can be made arbitrarily close to $2\epsilon\rho - 2\epsilon\rho\lambda$. Since $\lambda > 0$ is arbitrary, it follows that the proportion may be made arbitrarily close to $2\epsilon\rho$. This proves Proposition 3.

The proof of Proposition 4 is a variation on the same theme. Recall that the square $\mathcal{S} = [0, \frac{1}{2}] \times [0, \frac{1}{2}]$ may be thought of as the torus $\mathcal{T}$ with identification of points that correspond by reflections through the lines $z_1 = \frac{1}{2}$ and $z_2 = \frac{1}{2}$, and that with these identifications the process $X(t)$ maps into $Y(t)$. Any path $z \epsilon \mathcal{F}_\delta$ corresponds to four distinct paths in $\mathcal{T}$ by reflections through $z_1 = \frac{1}{2}$ and $z_2 = \frac{1}{2}$; these paths do not come any closer than δ to the boundaries of the four reflections of $\mathcal{S}$ in $\mathcal{T}$ (Figure 2b).

The argument used in proving Proposition 3 may now be repeated. Since the path $z \epsilon \mathcal{F}_\delta$ corresponds to four distinct paths in $\mathcal{T}$, the number of times $z(t)$ crosses the wave front will be four times the number of crossings in the proof of Proposition 3. Moreover, since $\rho < 2\delta$ and since the four images of $z(t)$ in $\mathcal{T}$ never come closer than δ to one of the boundaries, all crossings of the wave front by an image of $z(t)$ for $0 \le t \le \rho$ correspond to different broom pushers. The inequality (5.2) follows easily.

6. ASYMPTOTIC MINIMAXITY OF THE SEARCH PLANS

We shall now prove Theorem 1. Let $z \epsilon \mathcal{F}$ and $T_\epsilon = T_\epsilon(z)$; then

$$ET_\epsilon \le \rho \sum_{n=0}^{\infty} P(T_\epsilon > n\rho). \tag{6.1}$$

Using the results of sections 4 and 5, we will obtain upper bounds for the terms in the series that are independent of z.

Recall that Proposition 1 implies that $g_n(x_0, x) \to 1$ as $n \to \infty$, uniformly for $x_0, x \epsilon \mathcal{T}$, where $g_n(x_0, x)$ is defined by (4.1) (cf. (4.3)). Hence there exists $k \ge 1$ such that

$$g_{k-1}(x_0, x) \ge \frac{1}{2} \qquad \forall\, x_0, x \epsilon \mathcal{T}.$$

Lemma 2. *For any $m \ge 0$, $0 < \rho \le \frac{1}{2}$, and $0 < \epsilon \le 1$*

$$P(T_\epsilon > (m+k)\rho | T_\epsilon > m\rho) \le 1 - \rho\epsilon.$$

PROOF. Let $F = \{\text{dist}(X(t), z(t)) \le \epsilon \text{ for some } t \epsilon [(m+k-1)\rho, (m+k)\rho]\}$. Then

$$\begin{aligned} &P(T_\epsilon \le (m+k)\rho | T_\epsilon > m\rho) \\ &\quad \ge \int\int_{x, x' \epsilon \mathcal{T}} P(F | X((m+k-1)\rho) = x) \\ &\qquad\qquad \times g_{k-1}(x', x) P(X(m\rho) \epsilon dx' | T_\epsilon > m\rho) dx \\ &\quad \ge \frac{1}{2} \int_{x \epsilon \mathcal{T}} P(F | X((m+k-1)\rho) = x) dx \end{aligned}$$

$$= \frac{1}{2} P(F)$$
$$\geq \rho\epsilon,$$

by Proposition 3. □

Lemma 3. *For each $\eta > 0$ there exist constants $C(\epsilon) \uparrow 1$ as $\epsilon \to 0$ such that*

$$P(T_\epsilon \leq (m+1)\rho | T_\epsilon > m\rho) \geq C(\epsilon) 2\rho\epsilon$$

provided that

$$P(T_\epsilon > m\rho) \geq \eta.$$

The constants $C(\epsilon)$ may depend on η but do not depend on z.

PROOF OF LEMMA 3. Let $F = \{\text{dist}(X(t), z(t)) \leq \epsilon \text{ for some } t \epsilon [m\rho, (m+1)\rho]\}$. Then

$$P(T_\epsilon \leq (m+1)\rho; T_\epsilon > m\rho) = \int_{x \epsilon \mathcal{T}} P(F | X(m\rho) = x) G_m(x; \epsilon, z) dx,$$

where $G_m(x; \epsilon, z)$ is defined by (4.5). Proposition 3 implies that

$$P(F) = \int_{x \epsilon \mathcal{T}} P(F | X(m\rho) = x) dx \geq 2\rho\epsilon.$$

The lemma now follows from Proposition 2. □

PROOF OF THEOREM 1. Choose $\eta > 0$ small, and let $C(\epsilon)$ be the constants provided by Lemma 3. Let

$$n_* = n_*(\eta, \epsilon) = [[\log \eta / \log(1 - C(\epsilon) 2\rho\epsilon)]];$$

$$\eta \leq (1 - C(\epsilon) 2\rho\epsilon)^{n_*}.$$

Consequently,

$$P(T_\epsilon > n\rho) \leq (1 - C(\epsilon) 2\rho\epsilon)^n \quad \forall\, n \leq n_*,$$

because Lemma 3 implies that $P(T_\epsilon > (n+1)\rho | T_\epsilon > n\rho) \leq 1 - C(\epsilon) 2\rho\epsilon$ if $P(T_\epsilon > n\rho) \geq \eta$. On the other hand, Lemma 2 implies that

$$P(T_\epsilon > n\rho) \leq (1 - \rho\epsilon)^{[[n/k]]} \quad \forall\, n > n_*.$$

Thus, by (6.1),

$$ET_\epsilon \leq \rho\{\sum_{n=0}^{n_*} (1 - C(\epsilon) 2\rho\epsilon)^n + \sum_{n=n_*+1}^{\infty} (1 - \rho\epsilon)^{[[n/k]]}\}$$
$$\leq \rho\{(C(\epsilon) 2\rho\epsilon)^{-1} + k(1 - \rho\epsilon)^{[[n_*/k]]} (\rho\epsilon)^{-1}\}$$
$$= (C(\epsilon) 2\epsilon)^{-1} + k(1 - \rho\epsilon)^{[[n_*/k]]} \epsilon^{-1}.$$

By choosing $\eta > 0$ very small we can make the second term arbitrarily small relative to the first term. Since $C(\epsilon) \uparrow 1$ as $\epsilon \downarrow 0$, we have therefore obtained an upper bound for ET_ϵ that is asymptotic to $1/2\epsilon$ as $\epsilon \downarrow 0$. □

The proof of Theorem 2 is almost identical to that of Theorem 1. The only changes are that in proving the analogue of Lemma 2, one must use (4.4) in place of (4.3) and Proposition 4 in place of Proposition 3; and in proving the analogue of Lemma 3 one must use (4.8) in place of (4.7) and Proposition 4 in place of Proposition 3 (note that the lower bound in the analogue of Lemma 3 will be $C(\epsilon)8\rho\epsilon$ instead of $C(\epsilon)2\rho\epsilon$).

ACKNOWLEDGMENTS

This research was supported by NSF grant DMS 82-01723 and AFOSR grant 840136.

BIBLIOGRAPHY

Gal, S. (1979). Search games with mobile and immobile hider. *SIAM J. Control Optim.* **17**, 99–122.

Gal, S. (1980). *Search Games.* Academic Press, New York.

Issacs, R. (1965). *Differential Games.* Wiley, New York.

Spitzer, F. (1976). *Principles of Random Walk*, 2nd ed. Springer-Verlag, New York.

DISTINGUISHED STATISTICS, LOSS OF INFORMATION AND A THEOREM OF ROBERT B. DAVIES

LUCIEN LE CAM

Department of Statistics
University of California
Berkeley, California 94720, U.S.A.

GRACE L. YANG

Department of Mathematics
University of Maryland
College Park, Maryland 20742, U.S.A.

1. INTRODUCTION

This is a sequel to a paper by R. B. Davies (1985) and a paper by Le Cam and Yang (1985). Davies (1985) addressed himself to a question that can be paraphrased as follows: Let $\mathcal{E}_n = \{P_{\theta,n}; \theta \in \Theta_n\}$ be experiments given by measures $P_{\theta,n}$ on σ-fields $\mathcal{A}_n$. Let $\mathcal{F}_n$ be the experiment formed by the restrictions $Q_{\theta,n}$ of the $P_{\theta,n}$ to a sub-field $\mathcal{B}_n \subset \mathcal{A}_n$. Suppose that Θ_n is a subset of a given Euclidean space and that the $\mathcal{E}_n$ satisfy the so called LAN assumptions.

What other conditions are needed to insure that the $\mathcal{F}_n$ also satisfy the LAN assumptions, or at least have log likelihoods approximable by a sum of two random terms, the first linear and the second quadratic in θ?

Davies gives sufficient conditions. They involve an asymptotic normality requirement for the conditional distributions given $\mathcal{B}_n$ of the log likelihoods of the $P_{\theta,n}$.

Le Cam and Yang (1985) treated cases where the observations $\{x_1, \ldots, x_n\}$ are independent or Markovian and where the passage to subfields is carried out on the individual x_j's. No conditional distributions were involved. However the technique could not yield the result obtained by Davies in the application to branching processes of his Theorem 4.1.

Since the study of approximations for conditional distributions is often delicate we took another look at the situation to see, among other things, whether one could use instead approximations for joint distributions. More precisely we looked for conditions under which one could take limits of joint distributions and then look at conditional distributions in these limits.

It turns out that this can be done if the statistics used are "distinguished" according to a definition of Le Cam (1986), Chapter 7, Section 3, page 104.

This property is often, but not always, readily verifiable. It held trivially for the likelihood ratio approximations of Le Cam and Yang (1985). It must hold for the pair (μ_n, τ_n) of Theorem 4.1 of Davies (1985) and, as shown in Section 3 below, can be checked without much effort for the branching process case of Davies (1985).

Another aspect of the situation is that both Davies (1985) and Le Cam and Yang (1985) rely heavily on contiguity restrictions. These may be quite appropriate for the restricted experiments $\mathcal{F}_n$ on the subfields, but they cannot be defended easily for the parent experiments $\mathcal{E}_n$. A case in point would be the estimation of variances in the branching processes of Davies (1985), as will be shown in Section 3.

The paper is organized as follows. Section 2 contains our main result on passages to subfields in the form of Theorem 2. It says that, under appropriate restrictions on the statistics involved, one can take limits $\{D_{\theta,\infty}; \theta \in \Theta\}$ of families of distributions $\{D_{\theta,n}; \theta \in \Theta\}$ and deduce the asymptotic behavior of the likelihood ratios of the $D_{\theta,n}$ from that of the likelihood ratios for the $D_{\theta,\infty}$. The passage to subfields can also be carried out on the $D_{\theta,\infty}$. It is also shown (Lemma 2) that a requirement similar to that imposed in Davies' Theorem 4.1 on conditional distributions does imply the requirements of our Theorem 2.

The results do not make any reference to asymptotic normality or contiguity. They can be used more generally.

Section 3 returns to the branching processes. It uses arguments similar to those of Duby and Rouault in (1982). However these authors needed to let the size of the initial population increase to infinity. We show briefly that their results remain valid "locally" with initial population size $M_0 = 1$ and that the statistics of Davies (1985) are "distinguished" there. It is also shown that an apparently minor departure from the assumptions makes them lose this property. As already mentioned, the inappropriateness of contiguity restrictions on the parent experiments $\mathcal{E}_n$ is indicated.

Further applications of Theorem 2 will await another publication.

2. STABLE AND DISTINGUISHED SEQUENCES OF STATISTICS

The objects named in the title are described in some detail in Le Cam (1986), chapter 7, pages 99-108. After recalling their definitions we proceed to give a result relative to the possibility of interchanging two limit procedures as described in the introduction.

Let $\mathcal{E} = \{P_\theta; \theta \in \Theta\}$ be an experiment given by probability measures P_θ on a σ-field $\mathcal{A}$ of subsets of a set Ω. Let $\mathcal{X}$ be a compact space. By "statistics with values in $\mathcal{X}$" will be meant a Markov kernel $\omega \to F_\omega$ that assigns to each $\omega \in \Omega$ a probability measure F_ω on $\mathcal{X}$. These probability measures will be identified with positive linear functionals on the space $C(\mathcal{X})$ of continuous real functions on $\mathcal{X}$. This is sufficient for all purposes if the family $\mathcal{E}$ is dominated. Otherwise we shall extend the definition, admitting as statistics the more general "transitions" of Le Cam (1986). However the main arguments involve only finite sets Θ. Hence no difficulty will occur.

The restriction to *compact* $\mathcal{X}$ is a matter of convenience. In applications, $\mathcal{X}$ will often be Euclidean and not compact. One can always compactify as follows: Take a separating algebra Γ of bounded numerical functions on $\mathcal{X}$, imbed $\mathcal{X}$ in the cartesian produce $\mathbb{R}^\Gamma$ and take its closure there. (The choice of Γ does matter).

In the rest of this section the objects $\mathcal{X}$ and Θ will be kept fixed. One will consider nets or sequences $\{\mathcal{E}_\nu\}$ of experiments $\mathcal{E}_\nu = \{P_{\theta,\nu}; \theta \in \Theta\}$ and statistics S_ν with values in $\mathcal{X}$. This gives rise to families of distributions $\mathcal{D}_\nu = \{D_{\theta,\nu}; \theta \in \Theta\}$ where $D_{\theta,\nu}$ is the distribution of S_ν under $P_{\theta,\nu}$, that is the image $S_\nu P_{\theta,\nu}$ of $P_{\theta,\nu}$ by the kernel S_ν. The expectation of a $\gamma \in C(\mathcal{X})$ for this distribution will be written $\gamma D_{\theta,\nu}$.

Definition 1. Let $\{\mathcal{D}_\nu\}$ be indexed by the directed set $\{\nu\}$. The $\mathcal{D}_\nu$ converge vaguely to $\mathcal{D}_\infty = \{D_{\theta,\infty}; \theta \in \Theta\}$ if for every $\theta \in \Theta$ and $\gamma \in C(\mathcal{X})$ the expectations $\gamma D_{\theta,\nu}$ converge to $\gamma D_{\theta,\infty}$.

Note that, for the vague convergence, the set of all families of distributions indexed by Θ is a compact space. Thus every directed set of such families will have cluster points. That is why we have assumed compactness for $\mathcal{X}$.

A family $\mathcal{D}_\nu = \{D_{\theta,\nu}; \theta \in \Theta\}$ is also an experiment in its own right. In Le Cam (1986), page 100 we introduced a concept of "stability" for nets $\{\mathcal{D}_\nu\}$ in terms of a correspondence between vague convergence as families of distributions and weak convergence as experiments. It is a property that is not very strong in appearance, but is shown in Le Cam to be equivalent to the following approximation property, to be called finite approximability.

Let $\{s,t\}$ be a two point subset of θ and let $\mu_\nu = D_{s,\nu} + D_{t,\nu}$. Let f_ν be the Radon-Nikodym density of $D_{t,\nu}$ with respect to μ_ν.

Definition 2. The net $\{\mathcal{D}_\nu\}$ is stable if for every two point set $\{s,t\} \subset \Theta$ and every $\varepsilon > 0$, there is a finite subset $\{\gamma_j, j \in J\}$ of $C(\mathcal{X})$ and a $\nu(\varepsilon)$ such that $\nu \geq \nu(\varepsilon)$ implies

$$\inf_{j \in J} \int |f_\nu - \gamma_i| d\mu_\nu < \varepsilon.$$

A formulation involving finite subsets of θ instead of two point sets in given in Le Cam (1986), Theorem 1, page 101, together with several other equivalent conditions. For the two points sets $\{s,t\}$ another equivalent statement is as follows

Lemma 1. *The net $\{D_{s,\nu}, D_{t,\nu}\}$ is stable if and only if for every subnet $\{\nu(\xi)\}$ such that $[D_{s,\nu(\xi)}, D_{t,\nu(\xi)}]$ converges vaguely to a limit $[D_{s,\infty}, D_{t,\infty}]$ one has*

$$\liminf \int f_{\nu(\xi)}(1 - f_{\nu(\xi)})d\mu_{\nu(\xi)} \geq \int f_\infty(1 - f_\infty)d\mu_\infty$$

with $\mu_\infty = D_{s,\infty} + D_{t,\infty}$ and $f_\infty = \frac{dD_{t,\infty}}{d\mu_\infty}$.

For a proof, see Le Cam (1986), Chapter 7 Theorem 1, page 101 and below. (An equivalent statement can also be written in terms of Hellinger affinities $\int \sqrt{f_\nu(1 - f_\nu)}\, d\mu_\nu$).

To state our next result, it will be convenient to look at a particular pair $[D_{0,\nu}, D_{1,\nu}]$ and consider that the elements of $\mathcal{X}$ are random variables X_ν with possible distributions $D_{i,\nu}$, $i = 0, 1$. Each $D_{i,\nu}$ yields a joint distribution for the pair (f_ν, X_ν). Convergence of such joint distributions will be the vague convergence on $C([0,1] \times \mathcal{X})$.

Theorem 1. *Let $[D_{0,\nu}, D_{1,\nu}]$ converge vaguely to $[D_{0,\infty}, D_{1,\infty}]$. Let $\mu_\nu = D_{0,\nu} + D_{1,\nu}$ and $f_\nu = dD_{1,\nu}/d\mu_\nu$. The following conditions are all equivalent:*

1) $\liminf_\nu \int f_\nu(1 - f_\nu)d\mu_\nu \geq \int f_\infty(1 - f_\infty)d\mu_\infty$.
2) *For every $\varepsilon > 0$ there is a continuous function ϕ such that $0 \leq \phi \leq 1$ and such that $\int |f_\nu - \phi| d\mu_\nu < \varepsilon$ for all sufficiently large ν.*
3) *The joint distribution of (f_ν, X_ν) under $D_{i,\nu}$, $i = 0, 1$, converges to that of (f_∞, X_∞) for $D_{i,\infty}$.*

If any one of those three conditions hold and if, in addition, the set of discontinuities of f_∞ has μ_∞ measure zero, then $\int |f_\nu - f_\infty| d\mu_\nu$ tends to zero.

PROOF. Consider a decision problem in which it is desired to select a point z in the interval [0,1] with a loss function $W_i(z) = (i - z)^2$. A decision procedure is a Markov kernel $K(dz, x)$. For $[D_{0,\nu}, D_{1,\nu}]$ it has a risk sum equal to

$$\int \int [z^2(1 - f_\nu) + (1 - z)^2 f_\nu] K(dz, x)\mu_\nu(dx)$$
$$= \int f_\nu(1 - f_\nu)d\mu_\nu + \int \int (z - f_\nu)^2 K(dz, x)\mu_\nu(dx).$$

It follows that the minimum risk sum is $r_\nu = \int f_\nu(1 - f_\nu)d\mu_\nu$, attained by the nonrandomized estimate f_ν.

Let $\phi \in C(\mathcal{X})$ be such that $0 \leq \phi \leq 1$. It can be used to estimate z yielding a risk sum

$$R_\nu(\phi) = \int \phi^2 dD_{0,\nu} + \int (1-\phi)^2 dD_{1,\nu}$$
$$= r_\nu + \int |f_\nu - \phi|^2 d\mu_\nu.$$

Do this also for the limit pair $[D_{0,\infty}, D_{1,\infty}]$ obtaining a risk sum

$$R_\infty(\phi) = r_\infty + \int |f_\infty - \phi|^2 d\mu_\infty.$$

Note that, according to the first expression for $R_\nu(\phi)$ one will have $\lim_\nu R_\nu(\phi) = R_\infty(\phi)$ for $\phi \in C(\mathcal{X})$. Now select a $\phi \in C(\mathcal{X})$, $0 \le \phi \le 1$, so that $\int |f_\infty - \phi|^2 d\mu_\infty < \varepsilon^2$. One can write

$$\int |f_\nu - \phi|^2 d\mu_\nu = [R_\nu(\phi) - R_\infty(\phi)] - (r_\nu - r_\infty) + \int |f_\infty - \phi|^2 d\mu_\infty.$$

Since $R_\nu(\phi) - R_\infty(\phi) \to 0$ and since $\liminf(r_\nu - r_\infty) \ge 0$, this implies that for ν large one will have $\int (f_\nu - \phi)^2 d\mu_\nu < \varepsilon^2$. Thus (1) $\Rightarrow$ (2). To show that (2) $\Rightarrow$ (3) it is enough to show that for all $\gamma_1 \in C[0,1]$ and $\gamma_2 \in C(\mathcal{X})$, $\|\gamma_i\| \le 1$, the expectations $E_{i,\nu}\gamma_1(f_\nu)\gamma_2(X_\nu)$ for $D_{i,\nu}$ converge to the corresponding expectations for $D_{i,\infty}$. However, take a ϕ as done above and consider $E_{i,\nu}[\gamma_1(f_\nu) - \gamma_1(\phi)]\gamma_2(X_\nu)$. This does not exceed

$$\int |\gamma_1(\phi) - \gamma_1(f_\nu)| d\mu_\nu \le \mu_\nu[|f_\nu - \phi| > \varepsilon^{1/2}] + Osc(\gamma_1, \varepsilon)$$

where $Osc(\gamma_1, \varepsilon)$ is the maximum oscillation of γ_1 on sets of diameter $\varepsilon^{1/2}$.

By Chebyshev's inequality, the first term does not exceed ε for all ν sufficiently large. By taking ε sufficiently small, one can make $Osc(\gamma_1, \varepsilon)$ as small as one please. Thus $\int |\gamma_1(f_\nu) - \gamma_1(\phi)| d\mu_\nu$ can be made arbitrarily small. The same applies to $\int |\gamma_1(f_\infty) - \gamma_1(\phi)| d\mu_\infty$ and the conclusion follows by the triangle inequality.

That (3) implies (1) is immediate. It even implies $\lim r_\nu = r_\infty$. Hence the equivalence of (1) (2) and (3).

To obtain the final statement, note that if the set of discontinuities of f_∞ has μ_∞ measure zero, then $\int |f_\infty - \phi| d\mu_\nu$ tends to $\int |f_\infty - \phi| d\mu_\infty < \varepsilon$. Thus $\int |f_\nu - f_\infty| d\mu_\nu \le \int |f_\nu - \phi| d\mu_\nu + \int |\phi - f_\infty| d\mu_\nu$ will eventually be inferior to 2ε.

This concludes the proof of the theorem. □

REMARK. Life would be more pleasant if, under the equivalent conditions (1) (2) (3) of the theorem, the integrals $\int |f_\nu - f_\infty| d\mu_\nu$ would always converge to zero. This is not the case as shown by the following example.

Let ψ be a function defined on [0,1], equal to zero at zero and to $\psi(x) = \sin(\frac{c}{x})$ for $x \in (0,1]$. Select c so that $\int_0^1 \sin(\frac{c}{x})dx = 0$. This is possible. There is a value close to 2π with that property.

Let δ_0 be the probability measure concentrated at zero and let λ be the Lebesgue measure on [0,1]. Let $\mu_\infty = \delta_0 + \lambda$ and let $f_\infty = \frac{1}{2} + \frac{1}{4}\psi$. For integer values ν, let $\varepsilon_\nu = \frac{c}{2\pi\nu}$. Take $\phi_\nu(x) = \frac{1}{2}$ for $0 \le x \le \varepsilon_\nu$ and let $\phi_\nu(x) = f_\infty(x)$ for $\varepsilon_\nu < x \le 1$. Let $\mu_\nu = \lambda_\nu + \lambda$ where λ_ν is the Lebesgue measure on $[0, \varepsilon_\nu]$ multiplied by ε_ν^{-1}. Then $\int \phi_\nu d\mu_\nu$ may differ from unity but one can make an f_ν such that $\int f_\nu d\mu_\nu = 1$ by letting $f_\nu = \phi_\nu$ on $[0, \varepsilon_\nu]$ and $f_\nu = k_\nu \phi_\nu$ in $(\varepsilon_\nu, 1]$ for constants k_ν that tend to unity. Then the conditions (1) (2) (3) of the theorem are satisfied. However

$$\int |f_\nu - f_\infty| d\mu_\nu \ge \frac{1}{4} \int_1^\infty |\sin\left(\frac{2\pi\nu y}{c}\right)| \frac{dy}{y^2}$$

does not tend to zero.

To go further we need an additional definition as follows.

Consider experiments $\mathcal{E}_\nu$ as before with statistics S_ν whose distributions form families $\mathcal{D}_\nu$.

Definition 3. The statistics S_ν are pairwise asymptotically sufficient if for every two point set $\{s, t\} \subset \Theta$ the distance between the binary experiments $\{P_{s,\nu}, P_{t,\nu}\}$ and $\{D_{s,\nu}, D_{t,\nu}\}$ tends to zero as ν increases indefinitely. The S_ν are called *distinguished*, for $\mathcal{E}_\nu$, if they are pairwise asymptotically sufficient and if their distributions are stable.

With this we can pass to our main goal which is to obtain relations between pairs of statistics (S_ν, T_ν).

Consider then experiments $\mathcal{E}_\nu = \{P_{\theta,\nu}; \theta \in \Theta\}$, a pair $(\mathcal{X}, \mathcal{Y})$ of compact spaces and pairs (S_ν, T_ν) where S_ν is a statistic with values in $\mathcal{X}$ and T_ν is statistic with values in $\mathcal{Y}$. Both are defined on $\mathcal{E}_\nu$. Under $P_{\theta,\nu}$ the pair (S_ν, T_ν) has a joint distribution $D_{\theta,\nu} = (S_\nu, T_\nu)P_{\theta,\nu}$ considered as a positive linear functional on the space $C(\mathcal{X} \times \mathcal{Y})$. Vague convergence of distributions will be pointwise convergence on $C(\mathcal{X} \times \mathcal{Y})$.

For simplicity we shall consider only the case where Θ is the two point set $\{0, 1\}$. Then one can let $m_\nu = P_{0,\nu} + P_{1,\nu}$ and $\mu_\nu = D_{0,\nu} + D_{1,\nu}$. Each joint measure on $\mathcal{X} \times \mathcal{Y}$ has a marginal on $\mathcal{X}$. This will be noted by a "prime" so that for instance the marginal of μ_ν on $\mathcal{X}$ is μ'_ν. The corresponding marginals on $\mathcal{Y}$ will have double primes, so that the $\mathcal{Y}$=marginal of μ_ν is μ''_ν.

With these notations one can introduce likelihood ratios such as

$$f_\nu = \frac{dD_{1,\nu}}{d\mu_\nu}, \ f'_\nu = \frac{dD'_{1,\nu}}{d\mu'_\nu}, \ f''_\nu = \frac{dD''_{1,\nu}}{d\mu''_\nu}$$

and $f^*_\nu = \frac{dP_{1,\nu}}{dm_\nu}$.

Note that f_ν is a function of the pair (x, y) and that, for instance, f'_ν is a function of x only, conditional expectation of f_ν given x for the measure μ_ν.

If the $D_{i,\nu}$ have vague limits $D_{i,\infty}$, the pair $(D_{0,\infty}, D_{1,\infty})$ defines corresponding densities f_∞, f'_∞, f''_∞. It will be convenient to say that $D_{i,\infty}$ is the distribution of a pair (X, Y) on $\mathcal{X} \times \mathcal{Y}$.

With this notation one can state the following.

Theorem 2. *Assume that $D_{i,\nu}$ converges vaguely to $D_{i,\infty}$. Assume in addition the following:*

(A) *For the experiments $\mathcal{E}_\nu$ the S_ν are distinguished.*

(B) *The net of pairs $(D''_{0,\nu}, D''_{1,\nu})$ is stable.*

Then for $(D_{0,\infty}, D_{1,\infty})$ the X coordinate is sufficient and the joint distributions of $(f_\nu, f'_\nu, f''_\nu, S_\nu, T_\nu)$ for $D_{i,\nu}$ converge to that of

$$(f_\infty, f'_\infty, f''_\infty, X, Y) \text{ for } D_{i,\infty}.$$

PROOF. The proof follows closely the argument used for Theorem 1. Thus we shall just give a short sketch. A first remark is that one may as well assume that S_ν is sufficient for $\mathcal{E}_\nu$. Indeed let $\overline{f}_\nu = f'_\nu S_\nu$ be the image of f'_ν by S_ν. Let $Q_{1,\nu}$ be the measure whose density with respect to m_ν is $\overline{f}_\nu$. Since S_ν is asymptotically sufficient the inequalities of Le Cam (1986), page 72, show that $\int |f^*_\nu - \overline{f}_\nu| dm_\nu \to 0$. Equivalently $\|P_{1,\nu} - \|Q_{1,\nu}\|^{-1} Q_{1,\nu}\| \to 0$. Then replace $P_{0,\nu}$ by $Q_{0,\nu} = m_\nu - \|Q_{1,\nu}\|^{-1} Q_{1,\nu}$. For the pair $(Q_{0,\nu}, Q_{1,\nu})$ the statistics S_ν are sufficient and the properties A and B are preserved.

Now, as in the proof of Theorem 1, find functions $\phi' \in C(\mathcal{X})$ and $\phi'' \in C(\mathcal{Y})$ such that $\int |f'_\infty - \phi'| d\mu'_\infty < \varepsilon^2$ and $\int |f''_\infty - \phi''| d\mu''_\infty < \varepsilon^2$. Then $\int |f'_\nu - \phi'| d\mu'_\nu < \varepsilon^2$ and $\int |f''_\nu - \phi''| d\mu''_\nu < \varepsilon^2$ for all sufficiently large ν and one can repeat the argument that yielded property (3) in Theorem 1. This gives the result for the systems $(f'_\nu, f''_\nu, S_\nu, T_\nu)$. The extra part f_ν need no special attention since $\int |f_\nu - \tilde{f}_\nu| d\mu_\nu \to 0$ for a function $\tilde{f}_\nu$ defined on $\mathcal{X} \times \mathcal{Y}$ by writing $\tilde{f}_\nu(x, y) = f'_\nu(x)$. □

The result as stated follows.

REMARK 1. The theorem can be extended without difficulty, except for more complex notation, to finite sets Θ instead of two point sets.

REMARK 2. Note well that the result is not only that the $(f_\nu, f'_\nu, f''_\nu, S_\nu, T_\nu)$ have limiting distributions but that the distributions are those given by the likelihood ratios in the limits $(D_{0,\infty}, D_{1,\infty})$.

REMARK 3. The result does not remain valid under condition (A) alone, even if T_ν is also exactly sufficient. The stability of the paris $(D_{0,\nu}, D_{1,\nu})$ is not inherited by their marginals $(D''_{0,\nu}, D''_{1,\nu})$.

An example can be constructed as follows. Let $\mathcal{X}$ be the two point set $\{0, 1\}$ and let $\mathcal{Y}$ be the interval [0,1]. Let Ω be [0,1] with its Lebesgue measure

λ. Let λ_i be the Lebesgue measure on the segment $i \times \mathcal{Y}$ in $\mathcal{X} \times \mathcal{Y}$. Let $h_\nu(\omega)$ be the νth binary digit of ω in Ω. Consider an experiment $\mathcal{E}_\nu$ where for $\theta = 0$ the distribution of $\omega \in \Omega$ is λ and where for $\theta = 1$ the distribution of ω has a density $2h_\nu$ with respect to λ. Map the experiment on $\mathcal{X} \times \mathcal{Y}$ by the statistic $\omega \to [h_\nu(\omega), \omega]$. Here our $D_{i,\nu}$ have limits on $\mathcal{X} \times \mathcal{Y}$ in the sense that $\int \gamma dD_{i,\nu} \to \int \gamma dD_{i,\infty}$ for every bounded measurable function γ on $\mathcal{X} \times \mathcal{Y}$. Each coordinate map is sufficient. Yet both $D''_{0,\infty}$ and $D''_{1,\infty}$ are identical to the Lebesgue measure on [0,1]. Condition A is satisfied. Condition B is not. The systems $(f_\nu, f'_\nu;, f''_\nu)$ have a limiting distribution, but it is not that of $(f_\infty, f'_\infty, f''_\infty)$.

A natural example of a similar situation will be encountered in Section 3.

If one can establish that our condition (B) holds, our Theorem 2 can be used to bypass the requirement of convergence of conditional distributions used by Davies (1985) in Theorem 4.1. There the role of our S_ν is played by Davies $\tilde{\Delta}_n$ and the role of T_ν is played by the pair (μ_n, τ_n).

In the set up of Theorem 2 conditional distributions can be used to establish (B) as follows.

Consider the joint distributions $D_{i,\nu}$ as distributions of the pair of coordinates (X_ν, Y_ν) in $\mathcal{X} \times \mathcal{Y}$. Let us say that the conditional distributions of X_ν given Y_ν are *finitely approximable* if for every $\gamma \in C(\mathcal{X})$ and every $\varepsilon > 0$ there is a finite set $\{\psi_j; j \in J\}$ of elements of $C(\mathcal{Y})$ such that, taking conditional expectations for μ_ν one has

$$\inf_{j \in J} E|E[(\gamma(X_\nu)|Y_\nu] - \psi_j(Y_\nu)| < \varepsilon$$

for all sufficiently large ν.

Lemma 2. *Consider the situation described for Theorem 2. Let (A) be satisfied and assume that the conditional distributions of X_ν given Y_ν for μ_ν are finitely approximable. Then condition (B) holds.*

PROOF. As in Theorem 2, one can assume that X_ν is exactly sufficient. Let then $\overline{f}_\nu(x, y) = f'_\nu(x)$. For μ_ν the conditional expectations satisfy

$$E(\overline{f}_\nu|Y_\nu) = E(f'_\nu|Y_\nu) = f''_\nu(Y_\nu).$$

By assumption (A) and Theorem 1, for each $\varepsilon > 0$ there is a $\phi \in C(\mathcal{X})$ such that $\int |f'_\nu - \phi| d\mu'_\nu < \varepsilon$ for all large ν this will give

$$E|Ef''_\nu(Y_\nu) - E(\phi|Y_\nu)| < \varepsilon.$$

By the assumption of the lemma, the $E(\phi|Y_\nu)$ can be approximated by elements of $C(\mathcal{Y})$. Therefore the finite approximability also holds for f''_ν and the f''_ν satisfy the condition given in Definition 2. This gives the desired result. □

According to this lemma, the conditions (A) (B) of Theorem 2 are implied by those used by Davies in his Theorem 4.1. Note, however, that under the other restrictions imposed there, the type of convergence of conditional distributions required by Davies is a necessary requirement for his conclusion. This is so because of the properties of Laplace transforms defined on open sets. It would not be necessary if one would work with a finite Θ, as done here.

In conclusion, Theorem 2 offers an alternate approach to that of Davies in a fairly general framework. The stability property (B) is often easy to verify (as in the cases treated in Le Cam and Yang (1985), where it is entirely obvious). In the branching process example used by Davies (1985), the stability is obvious once one has established the required asymptotic sufficiency. Some additional remarks on this case follow.

3. SOME REMARKS ON BRANCHING PROCESSES

In this section we consider a branching process constructed as follows. There is a probability measure p on the integers $\{1,2,3\ldots\}$. If the nth generation contains m_n individuals, each one of them, independently of the rest, produces a progeny whose size is taken at random from the distribution p, leading to a total population M_{n+1} for the $(n+1)^{st}$ generation.

If ξ is a variable with distribution p, it will be assumed that $E\xi = \mu > 1$ and that $\sigma^2 = E(\xi - \mu)^2$ is finite and non zero. There is a huge literature on inference for such processes. See for instance the references in Davies (1985) and Duby and Rouault (1982). Thus we shall content ourselves with a few simple remarks.

Davies (1985) assumes that p depends on a single parameter $\theta = \mu$ and studies the asymptotic behavior of the experiment $\mathcal{F}_n$ in which one observes $M_0 = 1$, $M_1, \ldots, M_n$. This is done through his Theorem 4.1 by introducing a larger experiment $\mathcal{E}_n$ in which one observes not only the total size of the population but also the entire genealogical tree for each individual in it. Note that here the initial population has size $M_0 = 1$. One can also start with a population of size z as done by Duby and Rouault (1982) and let both n and z go to infinity. Naturally, the asymptotics will be different for such a case.

Davies did include a short remark about estimating parameters other than the mean μ. As will be explained below a technique different from that of his Theorem 4.1 becomes necessary.

Technically it should be noted that the results of Davies depend very strongly on the fact that he works in contiguous neighborhoods of a fixed distribution p. By contrast the results of Duby and Rouault (1982) are uniform over a certain class $\mathcal{P}$ of measures p, one of the important features being that it is proved that, asymptotically, the experiment can be parametrized through the pair (μ, σ^2)

independently of all other features of p.

As said above the result of Duby and Rouault (1982) depends materially on the fact that the initial population size z goes to infinity. Here are some simple remarks on what happens if M_0 is kept equal to unity.

To begin consider numbers μ, σ^2, b_0 and ε_0 with $0 < \sigma^2 \leq b_0 < \infty$ and $\varepsilon_0 > 0$ with $\mu > 1$. Let $\mathcal{P}_0$ be the class of all probability measures on the integers $\{1, 2, \ldots\}$ that satisfy the following requirements

1) $E\xi = \mu$
2) $E(\xi - \mu)^2 = \sigma^2$
3) $E|\xi - \mu|^3 \leq b_0$
4) $\sup_k p(k)p(k+1) \geq \varepsilon_0$.

Lemma 3. *Let μ, σ^2, b_0 and ε_0 be fixed. Consider an experiment $\mathcal{F}_\nu$ in which one observes the entire process $M_0 = 1, M_1, M_2, \ldots, M_n \ldots$. Assume that the possibilities for the distributions of progeny are either p_ν or q_ν, both in $\mathcal{P}_0$. Let P_ν and Q_ν be the corresponding distributions for the process.*

Then if $\|p_\nu - q_\nu\| \to 0$ so does $\|P_\nu - Q_\nu\|$.

PROOF. Consider the behavior of M_{n+1} conditionally given the size M_n of the nth generation. According to local limit theorems, for which see for instance Petrov (1975), Chapter 7, pages 187 to 216, there are constants $b < \infty$ and $c \in (0,1)$ such that for any $p \in \mathcal{P}_0$ one has

$$\Sigma_k |P[M_{n+1} = k|M_n] - \frac{1}{\sigma\sqrt{2\pi M_n}} \exp\left\{-\frac{1}{2}\frac{(k - \mu M_n)}{\sigma^2 M_n}\right\}|$$
$$\leq b\left[\frac{1}{M_n^{1/3}} + M_n^{2/3} c^{M_n}\right].$$

For any $p \in \mathcal{P}_0$ the series $\sum_n M_n^{-1/3}$ and $\sum_n M_n^{2/3} c^{M_n}$ are almost surely convergent, with uniform bounds as p varies in $\mathcal{P}_0$. Thus, for pairs (p,q) in $\mathcal{P}_0$ the tail of the process $\{M_n\}$ will be uninformative. For our $\mathcal{F}_\nu$, if $\|p_\nu - q_\nu\|$ tends to zero, the distance between the two possible distributions of $\{M_1, \ldots, M_n\}$ also tends to zero for each fixed n. The result follows. □

REMARK 1. With $M_0 = 1$ if $\|p_\nu - q_\nu\|$ does not tend to zero the experiment $\mathcal{F}_\nu$ can retain a great deal of information. For instance it could happen that p_ν and q_ν are disjoint so that the first generation allows one to find out which one of the two is operating.

However, Lemma 3 or a slight modification can still be used to show that the asymptotic parametrizability by μ and σ^2 only, described in Duby and Rouault (1982) or Swensen (1980), remains, valid "locally", see below.

If one maintains conditions (1) (2) and (3) but removes condition (4) of the definition of $\mathcal{P}_0$ the situation can be entirely different. There are pairs (p,q)

satisfying (1) (2) and (3) such that $p(k) > 0$ for all $k \geq 1$ but $q(2k+1) \equiv 0$ for all integers k. (For the measure q every birth is a twin birth). What happens then in the process $M_0 = 1, M_1, \ldots, M_n$ may be complex, but what happens in one generation is easy to see.

Consider experiments $\mathcal{G}_n$ where one observes M_{n+1} conditionally given the size $M_n = m_n$ of the nth generation. Assume that the distribution of the progeny is one of two possibilities $f_n = (1-\alpha_n)q + \alpha_n p$ or q, where α_n is some number $a_n \in [0,1]$.

If $m_n \alpha_n \to \infty$, then the (conditional) distribution of M_{n+1} under f_n differs little from what it would be under p, even if $\alpha_n \to 0$. The conditional experiment $\mathcal{G}_n$ does not degenerate, nor does it become perfect. The same is true if $m_n \to \infty$ and $m_n \alpha_n$ tends to a finite non zero limit. However if $m_n \to \infty$ and $m_n \alpha_n \to 0$, the experiment $\mathcal{G}_n$ degenerates.

This suggests, but does not quite prove, that experiments $\mathcal{F}_n$ in which one observes $M_0 = 1, M_1, \ldots, M_n$ and where the distribution of the progeny is either f_n or q will become perfect if $\mu^{r_n}\alpha_n \to \infty$ for some sequence r_n such that $n - r_n \to \infty$. This would be so even if $\alpha_n \to 0$ so that $\|f_n - q\| \to 0$.

A proof that the $\mathcal{F}_n$ do tend to the perfect experiment is easy. Assume $M_n = m_n$ and let ξ_j; $j = 1, 2, \ldots, m_n$ be the sizes of the individual progenies from M_n. Let $\eta_j = \xi_j$ modulo 2. Then $\Sigma \eta_j$ is a binomial variable with m_n trials and probability of success $\alpha_n \pi$ with $\pi = p\{x \text{ is odd}\}$. If $m_n \alpha_n \pi$ is large the probability that $\Sigma_j \eta_j$ is odd will be close to $1/2$. Thus, if $\mu^{r_n}\alpha_n \to \infty$, with $n - r_n \to \infty$, the probability that some M_k, $k \leq n$ will be odd will tend to unity under f_n but will be exactly zero under q.

Under such conditions, note that the conditional distribution of $\frac{M_{n+1} - \mu M_n}{\sigma\sqrt{M_n}}$ is asymptotically normal $\mathsf{N}(0,1)$ for all f_n. Thus the statistics $(M_{n+1} - \mu m_n)/\sigma\sqrt{m_n}$ cannot be stable for the experiments $\mathcal{G}_n$. This is very much the same thing as the behavior of the Y_ν in the example of Remark 3, following Theorem 2.

In the class $\mathcal{P}_0$ the expectation μ and variance σ^2 of the progeny distribution were kept constant. One can use the local limit theorems to obtain results in which μ and σ^2 vary. To do this, consider fixed numbers $0 \leq \varepsilon_0 < b_0 < \infty$ and a sequence $\{\varepsilon_n\}$, $\varepsilon_n > 0$, $\varepsilon_n \to 0$. Let $\mathcal{P}_1$ be the class of distributions $p = \mathcal{L}(\xi)$ such that

1) $1 + \varepsilon_0 \leq E\xi \leq b_0$
2) $\varepsilon_0 \leq E(\xi - \mu)^2 \leq b_0$
3) $E|\xi - \mu|^3 \leq b_0$
4) $\sup_k p(k)p(k+1) \geq \varepsilon_0$.

Let $\mathcal{P}_n$ be any subset of $\mathcal{P}_1$ that has a total variation norm diameter at most ε_n.

The approximation to likelihood ratios described by Duby and Rouault

(1982) is still valid on $\mathcal{P}_n$ if one observes $M_0 = 1, M_1, \ldots, M_n$. That approximation involves only the parameters μ and σ^2 and the expressions

$$\sum_{k=0}^{n-1} \frac{(M_{k+1} - \mu M_k)^2}{\sigma^2 M_k}.$$

We shall not enter into the details, for which see Duby and Rouault (1982), but end on a remark concerning what happens in a class such as $\mathcal{P}_n$ defined here if one keeps μ fixed but vary σ.

Let p_n and q_n be two elements of P_n with the same expectation μ but respective variances σ_n^2 and τ_n^2. Let h_{k+1}^2 be the conditional square Hellinger distance

$$h_{k+1}^2 = \frac{1}{2}\Sigma_j |[P(M_{k+1} = j|M_k, p_n)]^{1/2} - [P(M_{k+1} = j|M_k, q_n)]^{1/2}|^2.$$

Up to terms of order $M_k^{-1/3}$ this is approximately $1 - [1 - \frac{(\sigma_n - \tau_n)^2}{\sigma_n^2 + \tau_n^2}]^{1/2}$.

Let P_n be the distribution of $M_0 = 1, M_1, \ldots, M_n$ under p_n. Let Q_n be the corresponding distribution under q_n. It follows as in Lemma 3 that if $n(\sigma_n - \tau_n)^2$ remains bounded then the sequences $\{P_n\}$ $\{Q_n\}$ will be contiguous.

The experiment (P_n, Q_n) behaves asymptotically as if one observed a variable $T_n = \theta\chi_n^2$ where χ_n^2 is a chi-square with n degrees of freedom and θ is allowed to take value σ_n^2 or τ_n^2.

To apply Davies' Theorem 4.1 one would look at the bigger experiment in which the genealogies of the individuals in the nth generation are all observed. This would give other measures, say P_n^* and Q_n^*. To insure contiguity for $\{P_n^*\}$, $\{Q_n^*\}$ one needs to have that $\mu^n(\sigma_n - \tau_n)^2$ remain bounded. Davies' Theorem 4.1 could then be applied, but the reduced experiment (P_n, Q_n) would tend to the trivial uninformative experiment. Thus the note about estimating other parameters at the end of Section 5 of Davies (1985) should be given more weight that its short length seems to suggest.

We would also like to attract the reader's attention to the fact that the conditions (4) imposed on our classes $\mathcal{P}_0$ and $\mathcal{P}_1$ are not visible in the formulations of Davies (1985) or Swensen (1980), but an analogous condition is implied.

The apparent lack of need for our condition (4) in the papers cited is due to the fact that the authors work with alternatives q_n such that $\|q_n - p\| \to 0$ for a fixed p, independent of n. A similar remark applies to the third moment requirement. As we have seen above a condition such as (4) is needed to obtain uniform results on classes such as $\mathcal{P}_n$. It is also needed for the uniform results of Duby and Rouault (1982).

ACKNOWLEDGMENTS

Research partially supported by National Science Foundation Grant DMS-8403239.

BIBLIOGRAPHY

Davies, R. B. (1985). "Asymptotic inference when the amount of information is random" *Proc. Berkeley Conference in Honor of J. Neyman and J. Kiefer* Vol. II, Wadsworth Inc. pp. 841–864.

Duby, C. and Rouault, A. (1982). "Estimation non paramétrique de l'espérance et de la variance de la loi de reproduction d'un processus de ramification" *Ann. Inst. Henri Poincaré*, Vol. XVIII #2 pp. 149–163.

Le Cam, L. (1986). *Asymptotic methods in statistical decision theory.* Springer Verlag, N.Y. xxvi + 742 pages.

Le Cam, L. and Yang, G. L. "On the preservation of local asymptotic normality under information loss" Technical report No. 53, Nov 1985 submitted to *Ann. Statist.*

Petrov, V. V. (1975). *Sums of independent random variables.* Springer Verlag, N.Y.

Swensen, A. R. (1980). "Asymptotic inference for a class of stochastic processes" Ph.D. Thesis U.C. Berkeley.

PROPHET INEQUALITIES FOR THRESHOLD RULES FOR INDEPENDENT BOUNDED RANDOM VARIABLES

ESTER SAMUEL-CAHN

Department of Statistics
Hebrew University
Jerusalem, Israel

1. INTRODUCTION

Let X_i, $i = 1, 2, \ldots$ be independent and uniformly bounded random variables. In order to simplify notation we can (by proper shift of location, and change of scale) and shall, assume that $0 \le X_i \le 1$. Let T_n be the set of all stopping rules t for $X_1, \ldots, X_n$, i.e. satisfying $P(t \le n) = 1$, and let T be the set of all stopping rules, i.e. $P(t < \infty) = 1$. Denote $U_n = E[X_1 \vee \ldots \vee X_n] = E[\max_{i=1,\ldots,n} X_i]$ and $U = E[\vee_{i=1}^{\infty} X_i]$ and set $V_n = V_n(X_1, \ldots, X_n) = \sup\{EX_t; t \in T_n\}$, $V = V(X_1, X_2, \ldots) = \sup\{EX_t; t \in T\}$. Hill and Kertz (1981) and Hill (1983) show that for any such X_is

$$U_n \le 2V_n - V_n^2 \tag{1.1}$$

and the same inequality holds without the subscript n. Hill (1983) shows that for any (x, y) satisfying $0 \le x \le 1$, $x \le y \le 2x - x^2$ there exists a sequence $X_1, \ldots, X_n$ as above, such that the U_n and V_n values of $X_1, \ldots, X_n$ satisfy $V_n = x$, $U_n = y$.

In the present paper we consider threshold stopping rules, $t(c)$, defined for each constant threshold c for the sequence $X_1, \ldots, X_n$, as follows:

$$t(c) = \inf\{i \le n;\ X_i \ge c\}, \qquad = n \qquad \text{otherwise.} \tag{1.2}$$

We generalize the previous results for threshold rules. Threshold rules were introduced for the ratio prophet problem for independent nonnegative random variables in Samuel-Cahn (1984), and are also considered for negatively dependent random variables in Rinott and Samuel-Cahn (1987). They are simpler to handle both in theory and applications than optimal rules leading to $V(X_1, \ldots, X_n)$. Let $T_n^* =$ set of all threshold rules for $X_1, \ldots, X_n$. Set $E^+X_{t(c)} = E[X_{t(c)} I(X_{t(c)} \ge c)] \le EX_{t(c)}$, and let

$$Q_n(X_1, \ldots, X_n) = \sup\{EX_t\colon t \in T_n^*\},\ W_n(X_1, \ldots, X_n) = \sup\{E^+X_t\colon t \in T_n^*\} \tag{1.3}$$

and let $c_n = \sup\{c; E^+X_{t(c)} \ge c\}$. By the left continuity in c of $E^+X_{t(c)}$ it follows that c_n also satisfies $E^+X_{t(c_n)} \ge c_n$. To abbreviate, set $A_n = E^+X_{t(c_n)}$. For every sequence $X_1, \ldots, X_n$ we clearly have (omitting the dependence on

$X_1, \ldots, X_n)$,

$$c_n \le A_n \le W_n \le Q_n \le V_n \le U_n. \tag{1.4}$$

Our main purpose is to prove the following strengthening of (1.1).

Theorem 1.1. *Let* $0 \le X_i \le 1$, $i = 1, \ldots, n$ *be independent, and*

$$c_n = \sup\{c; E^+ X_{t(c)} \ge c\}.$$

Then for every $c_n \le c \le 1$,

$$E[X_1 \vee \ldots \vee X_n - c]^+ \le (1-c)c_n. \tag{1.5}$$

Clearly (1.5) is true if c_n is replaced by any greater value, and in particular by any of the other values in (1.4). Also the inequality remains true if the $+$ in the left hand side of (1.5) is removed. Replacing c and c_n by V_n in (1.5) and removing the $+$, yields (1.1).

2. PROOFS AND FURTHER RESULTS

Given X_i define $Y_i(c)$ for any $0 \le c < 1$ by

$$Y_i(c) = \begin{cases} 1 & \text{with probability } p_i(c) \\ c & \qquad\qquad\quad q_i(c) \\ 0 & \qquad\qquad\quad w_i(c) \end{cases} \tag{2.1}$$

where

$$p_i(c) = (1-c)^{-1} \int_{c \le x \le 1} (x-c)\,dF_{X_i}(x) \text{ and}$$

$$q_i(c) = (1-c)^{-1} \int_{c \le x \le 1} (1-x)\,dF_{X_i}(x),$$

where F_{X_i} is the distribution function of X_i, and $p_i(c) + q_i(c) + w_i(c) = 1$. For $c = 1$ define $Y_i(1) = 1$ with probability $p_i(1) = P(X_i = 1)$ and $Y_i(1) = 0$ with probability $w_i(1) = 1 - p_i(1)$. Set $q_i(1) = 0$. This definition ensures $EX_i I(X_i \ge c) = EY_i(c) I(Y_i(c) \ge c) = EY_i(c)$ and $P(X_i \ge c) = P(Y_i(c) \ge c)$. We have

Lemma 2.1. *Let* $0 \le X_i \le 1$, $i = 1, \ldots, n$ *be independent, and let* $Y_i(c)$, $i = 1, \ldots, n$ *be independent, defined by* (2.1). *Then*

$$E^+ X_{t(c)} = E^+ Y(c)_{t(c)} \tag{2.2}$$

and for any $0 \le c \le 1$

$$E[X_1 \vee \ldots \vee X_n - c]^+ \le E[Y_1(c) \vee \ldots \vee Y_n(c) - c]^+ \tag{2.3}$$

REMARK. It should be noticed that our notation is deficient, since the rule $t(c)$ depends on the sequence of random variables to which it is applied. Thus a more careful notation would be $t(c; X_1, \ldots, X_n)$, and correspondingly $X_{t(c;X_1,\ldots,X_n)}$. We chose the simpler notation, with the convention that $t(c)$ is the threshold stopping rule for the sequence of which it is used as a subscript.

PROOF. (2.2) is immediate by the definition, and (2.3) follows from the independence of the random variables and the convexity of the function $h(z) = E[X \vee z - c]^+$. □

(For a similar definition and more detailed proofs, see Hill and Kertz (1981)).

Lemma 2.2. *Let $c > c_n$. Then*

$$E^+Y(c)_{t(1)} < c. \tag{2.4}$$

PROOF. Since here we consider a fixed c we shall simplify notation and write Y_i, p_i, q_i, w_i instead of $Y_i(c)$, $p_i(c)$, $q_i(c)$, $w_i(c)$. Note that

$$E^+X_{t(c)} = E^+Y_{t(c)} = c \sum_{i=1}^{n} q_i \prod_{j=1}^{i-1} w_j + \sum_{i=1}^{n} p_i \prod_{j=1}^{i-1} w_j$$

and

$$E^+Y_{t(1)} = 1 - \prod_{i=1}^{n} (w_i + q_i).$$

Since $c > c_n$ we therefore have

$$c > \frac{\sum_{i=1}^{n} p_i \prod_{j=1}^{i-1} w_j}{1 - \sum_{i=1}^{n} q_i \prod_{j=1}^{i-1} w_j}. \tag{2.5}$$

The lemma will therefore follow if we show

$$(1 - \prod_{i=1}^{n}(w_i + q_i))(1 - \sum_{i=1}^{n} q_i \prod_{j=1}^{i-1} w_j) \leq \sum_{i=1}^{n} p_i \prod_{j=1}^{i-1} w_j. \tag{2.6}$$

Notice that $1 - \sum_{i=1}^{n} q_i \prod_{j=1}^{i-1} w_j = \prod_{j=1}^{n} w_j + \sum_{i=1}^{n} p_i \prod_{j=1}^{i-1} w_j$. Substituting this in (2.6), and simplifying, we get that (2.6) is equivalent to

$$(\prod_{j=1}^{n} w_j)[1 - \prod_{i=1}^{n}(w_i + q_i)] \leq [\prod_{k=1}^{n}(w_k + q_k)] \sum_{i=1}^{n} p_i \prod_{j=1}^{i-1} w_j. \tag{2.7}$$

Substituting $1 - \prod_{i=1}^{n} (w_i + q_i) = \sum_{i=1}^{n} p_i \prod_{k=1}^{i-1} (w_k + q_k)$ in the left hand side of (2.7), and rearranging terms, (2.7) is equivalent to

$$\sum_{i=1}^{n} p_i \{ (\prod_{j=1}^{n} w_j) [\prod_{k=1}^{i-1} (w_k + q_k)] - (\prod_{j=1}^{i-1} w_j) [\prod_{k=1}^{n} (w_k + q_k)] \} \leq 0. \tag{2.8}$$

Hence it suffices to show that for each i the value of the expression in the curly bracket is nonpositive. If $(\prod_{j=1}^{i-1} w_j) (\prod_{k=1}^{i-1} (w_k + q_k))$ equals zero, this is obvious. Otherwise divide by this expression, to get that one only must show that

$$\prod_{j=1}^{n} w_j - \prod_{j=1}^{n} (w_j + q_j) \leq 0.$$

The latter is obvious, since $0 \leq w_j \leq w_j + q_j$. □

PROOF OF THEOREM 1.1. Notice that by (2.3),

$$E[X_1 \vee \ldots \vee X_n - c]^+ \leq E[Y_1(c) \vee \ldots \vee Y_n(c) - c]^+ = (1-c) E^+ Y(c)_{t(1)}, \tag{2.9}$$

where $E^+ Y(c)_{t(1)} = 1 - \prod_{i=1}^{n} (1 - p_i(c))$. But, as is easily verified by its definition, $p_i(a)$ is nonincreasing in a and hence for a sequence $\{a_k\}$ $c \geq a_k > c_n$, $a_k \searrow c_n$ using (2.4) we have

$$E^+ Y(c)_{t(1)} \leq E^+ Y(a_k)_{t(1)} \leq \lim_{k \to \infty} E^+ Y(a_k)_{t(1)} \leq c_n. \tag{2.10}$$

By (2.9) and (2.10) the theorem follows. □

Let $0 \leq X_i \leq 1$, $i = 1, 2, \ldots$ and consider the infinite stopping problem. Let $c_\infty = \sup\{c\colon E^+ X_{t(c)} \geq c; t \,\varepsilon\, T^*\}$, where T^* is the set of all threshold rules for the infinite sequence. Corresponding to Theorem 1.1 we have

Theorem 2.3. *For any* $c_\infty \leq c \leq 1$,

$$E[\vee_{i=1}^{\infty} X_i - c]^+ \leq (1 - c) c_\infty. \tag{2.11}$$

PROOF. We use the notation $E_n^+ X_{t(c)}$ to stand for the value in the n-dimensional problem, i.e. $E_n^+ X_{t(c)} = E \sum_{i=1}^{n} X_i I(x_i \geq c)\ P(X_j < c,\ j = 1, \ldots, i-1)$. Then $E_n^+ X_{t(c)}$ is nondecreasing in n for every c. Hence $\{c_n\}$ is a nondecreasing sequence, with $\lim_{n \to \infty} c_n \leq c_\infty$. Since $c_\infty \leq c \leq 1$, it follows that $c_n \leq c \leq 1$ for all n, and since $\lim_{n \to \infty} E[\vee_{i=1}^{n} X_i - c]^+ = E[\vee_{i=1}^{\infty} X_i - c]^+$, (2.11) is an immediate consequence of (1.5). □

Corresponding to Theorem 2.3 of Hill (1983) we have:

Theorem 2.4. *Let* $S = \{(x,y): 0 \le y \le 1,\ 0 \le x \le (1-y)y\}$. *For* $n = 1,2,\dots,\infty$ *the set* S *is exactly the set of all values* $y = c_n$ $x = E[\vee_{i=1}^{n} X_i - c_n]^+$ *obtainable for some independent* $X_1,\dots,X_n$, $0 \le X_i \le 1$.

PROOF. In view of Theorems 1.1 and 2.3 we need only show that each value in S is attainable. It suffices to show this for $n = 1$, since one can always take $X_2 = X_3 = \dots \equiv 0$. Let $(x,y) \in S$. If $y = 0$ take $X_1 = 0$. Otherwise let

$$X_1 = \begin{cases} y^2/(y-x) & \text{with prob. } (y-x)/y \\ 0 & \text{otherwise.} \end{cases}$$

Then it is immediate that $0 < y^2/(y-x) \le 1$ and $0 < (y-x)/y \le 1$, and $c_1 = y$, $E[X_1 - c_1]^+ = x$. □

REMARKS

A: If for all i $a \le X_i \le a + d$ then the statement corresponding to Theorem 1.1 should read

$$E[X_1 \vee \dots \vee X_n - c]^+ \le [1 - d^{-1}(c-a)](c_n - a) \text{ for all } c_n \le c \le a+d, \quad (2.12)$$

where $c_n = \sup\{c\colon E^+(X-a)_{t(c-a)} \ge c - a\}$. (For $a \ge 0$ this is the same as $\sup\{c\colon E^+X_{t(c)} \ge c\}$, but for $a \le 0$ one would trivially get c_n nonnegative by the latter definition.) The other theorems should be changed correspondingly.

B: Assume the X_i to be independent, nonnegative, but not necessarily bounded, and to avoid trivialities assume $EX_i < \infty$ for all i, and $P(X_1 \vee \dots \vee X_n \ge 0) > 0$. Krengel and Sucheston (1978) were the first to consider the ratio prophet inequality, for such random variables. With the notation of the introduction, they show that $U_n \le 2V_n$ and $U \le 2V$. In Samuel-Cahn (1984) the similar ratio threshold prophet-inequality was shown to hold for such X_i, and it was considered further in Rinott-Samuel (1987). More specifically, let b be the unique constant satisfying $E \sum_{i=1}^{n} [X_i - b]^+ = b$. Then clearly

$$E(X_1 \vee \dots \vee X_n) \le 2b \quad (2.13)$$

and Rinott-Samuel (1987) show that $E^+X_{t(b)} > b$, and hence the ratio prophet inequality

$$E(X_1 \vee \dots \vee X_n) < 2W_n(X_1,\dots,X_n) \le 2Q_n(X_1,\dots,X_n) \quad (2.14)$$

holds, and a similar weak inequality holds for the infinite case. (2.14) with weak inequality can also be easily deduced from (2.12), by taking $a = 0$ and truncating the X_i at d. Denote the truncated X_i by $X_i(d)$, and denote the corresponding c_n-value by $c_n(d)$, and the value for the untruncated random variables by c_n. Eliminating the + in the left hand side of (2.12) and taking $c = c_n(d)$ we get $E[X_1(d) \vee \dots \vee X_n(d)] \le 2c_n(d)$ and letting $d \to \infty$ we get

$$E[X_1 \vee \dots \vee X_n] \le 2c_n \le 2W_n \le 2Q_n \quad (2.15)$$

and a similar statement can be deduced for an infinite sequence. Notice, however, that (2.13) is stronger than (2.15), since $E^+X_{t(b)} > b$ implies that $b \leq c_n$, and indeed in many examples, such as the one below, $b < c_n$. Let $X_1 = \mu$, $X_2 = 1$ or 0 with probabilities μ and $1 - \mu$ respectively. Here $c_2 = \mu$ and $b = 2\mu/(2 + \mu)$.

BIBLIOGRAPHY

Hill, T. P. (1983). Prophet inequalities and order selection in optimal stopping problems. *Proc. Amer. Math. Soc.* **88**, 131–137.

Hill, T. P. and Kertz, R. P. (1981). Additive comparisons of stop rule and supremum expectations of uniformly bounded independent random variables. *Proc. Amer. Math. Soc.* **83**, 582–585.

Krengel, U. and Sucheston, L. (1978). On semiamarts, amarts, and processes with finite value. *Adv. in Probab.* **4**, 197–266. Dekker, New York.

Rinott, Y. and Samuel-Cahn, E. (1987). Comparisons of optimal stopping values and prophet inequalities for negatively dependent random variables To appear in *Ann. Statist.*, **15**.

Samuel-Cahn, E. (1984). Comparison of threshold stop rules and maximum for independent nonnegative random variables. *Ann. Probab.* **12**, 1213–1216.

WEAK CONVERGENCE OF THE AALEN ESTIMATOR FOR A CENSORED RENEWAL PROCESS

THOMAS SELLKE

Department of Statistics
Purdue University
West Lafayette, Indiana 47907-2005, U.S.A.

1. INTRODUCTION

Suppose that $X_1, X_2, \ldots$ are the lifetimes of a sequence of machine components. We assume that the X_i's are i.i.d. with unknown continuous c.d.f. F and cumulative hazard function $\Lambda(v) = -\log\{1 - F(v)\}$. The components will be put on test one at a time and will be replaced with a new component either upon failure or at some preplanned replacement (or "censoring") time, whichever comes first. The censoring time for the nth component is allowed to depend upon observations of the previous $n-1$ components. Bather (1977) considers this situation under the assumption that replacing a failed component costs c_1 dollars, while the planned replacement of a working component costs c_2 dollars, with $c_1 > c_2 > 0$. It is intuitively plausible that the expected cost of replacement per unit time over a long period should be minimized by a policy of replacing components whenever they reach a certain fixed age a_0. That this is so is shown by Berg (1976). However, the optimal age of replacement a_0 clearly depends upon the unknown F as well as on c_1 and c_2. Bather's (1977) goal is to minimize long term average cost in the absence of prior knowledge of F. Since it is necessary to have information about the tail of F in order to estimate a_0 consistently, there is at each stage a conflict between the choice of trying to minimize short term costs by replacing the current component at an estimate of a_0 and the choice of not replacing the current component until it fails in order to gain information about the tail of F, thereby making possible better estimation of a_0 in the future. Bather's procedure specifies a sequence of numbers $1 = p_1 \geq p_2 \geq p_3 \ldots$ such that $p_n \downarrow 0$ and $\Sigma p_n = \infty$. The nth component is allowed to continue on until failure with probability p_n, and is otherwise censored at a current estimate of a_0. Since infinitely many components will not be subject to replacement until they fail, a crude policy is to consistently estimate F based only on observations from these components and to use this estimate of F to estimate a_0. This approach discards all of the information available from the majority of components, which are either censored or potentially censored. The sequential replacement policy actually proposed by Bather (1977) uses data from all previous components to estimate a_0, but for technical reasons he is forced to modify the past data by imposing a "fake" censoring scheme which does throw out some of the available information. He specifies a fixed sequence $b_0 > b_1 > \ldots, b_n \downarrow 0$, and if the actual censoring age for the nth component was a_n, then he pretends that the nth component was censored instead at age b_j, where $b_j \leq a_n < b_{j+1}$.

It would be more natural to use the Kaplan-Meier estimator (or the asymptotically equivalent Aalen estimator) based on all past data to estimate F. However, the peculiar nature of the censoring here prevents the usual proofs of consistency and weak convergence from applying. When censoring times are independent of component lifetimes, there is an obvious martingale structure which makes possible straightforward proofs of weak convergence to a Gaussian process. (See, for example, Gill (1980).) The fact that censoring here is allowed to depend on previous lifetimes destroys this martingale structure.

This paper will prove a weak convergence result which applies to censored renewal processes of the above form. The proof is based on the fact that an approximation of the Aalen estimator of $\Lambda(v)$ has a certain orthogonal martingale structure when this approximation is viewed as a process in a two-dimensional time scale. The theorem is not the most general possible, but generalizations in various directions (noncontinuous F's, triangular arrays, etc.) seem straightforward.

2. NOTATION AND THE THEOREM

In what follows, calendar time will be denoted by t and u, and age time will be denoted by s and v. The assumption that components are necessarily put on test only one at a time will be dropped, and it will be permitted for components to be put *back* on test after having been removed from use. Repair of failed components will also be allowed, with the assumption that the distribution of remaining time on test until the next failure is the same as if no previous failure had occurred, i.e., past failures of repaired components do not affect future performance.

The situation of interest here is most easily modelled using the machinery of counting processes. The reference is Gill (1980). Let $Y_i(t)$ be the indicator of the event that the ith component is on test at time t. Define

$$Z_i(t) = \int_0^t Y_i(u)\, du, \tag{1}$$

so that $Z_i(t)$ is the length of time that component i has been on test by calendar time t. Let $N_i(t)$ be the number of failures that component i has suffered at or before time t. Let $\mathcal{F}_t$, $t \geq 0$, be a complete, right-continuous filtration, where $\mathcal{F}_t$ represents knowledge at time t. For each i, $Y_i(t)$ is assumed to be left-continuous and $\mathcal{F}_t$-predictable. Thus, for each ω, $Y_i(t)$ is the indicator function of a union of disjoint intervals which are open on the left and closed on the right. The counting processes $N_i(t)$ are assumed to be $\mathcal{F}_t$-adapted with right-continuous paths. They are zero at time zero, integer valued, nondecreasing, and have jumps of size +1. The process $N_i(t)$ are connected with the continuous cumulative hazard function Λ by the assumption that

$$M_i(t) = N_i(t) - \Lambda\{Z_i(t)\} \tag{2}$$

is an $\mathcal{F}_t$-martingale. Independence of the different components is reflected in the assumption that, with probability one, no two counting processes jump simultaneously. Let $A_i(t) = \Lambda\{Z_i(t)\}$ = amount of accumulated hazard to which component i has been subjected by time t. It follows from Theorem 2.3.1, of Gill (1980) that

$$\langle M_i, M_j\rangle(t) = A_i(t)1\{i = j\}. \tag{3}$$

Here, $\langle M_i, M_j\rangle$ is the "predictable covariation process" of M_i and M_j, so that the product M_iM_j is a martingale for $i \neq j$, i.e., M_i and M_j are orthogonal martingales. The "predictable variance process" of M_i is

$$\langle M_i\rangle = \langle M_i, M_i\rangle = A_i, \tag{4}$$

so that $M_i^2 - A_i$ is a martingale.

The Aalen estimator of Λ based on the data available at time t is a function of the amount of time that each component has been in use and of the ages at which failures have occurred. Define

$$N_i(t, s) = \int_0^s 1\{Z_i(u) \leq s\}\ N_i(du) \tag{5}$$

to be the number of failures suffered by component i before calendar time t and age s. Set $N(t, s) = \sum_i N_i(t, s)$. Define

$$R(t, s) = \sum_i 1\{Z_i(t) \geq s\} \tag{6}$$

to be the number of components which have been on test for at least s units of time by calendar time t. Thus, $R(t, s)$ is the size of the "risk set" at age s and time t. The Aalen estimator of $\Lambda(s)$ based on data available at time t is defined by

$$\hat{\Lambda}(t, s) = \int_0^s \frac{N(t, dv)}{R(t, v)}. \tag{7}$$

(Interpret $\frac{0}{0}$ as 0.)

Note that $N_i(t, s)$ is, as a function of t for fixed s, simply a stopped version of $N_i(t)$ and is therefore itself a counting process. Define

$$A_i(t,s) = \Lambda\{Z_i(t) \ \wedge \ s\} \quad \text{and} \quad A(t,s) = \sum_i A_i(t,s),$$

so that $A_i(t,s)$ is the amount of accumulated hazard to which component i has been exposed before age s and time t. Then

$$M_i(t,s) = N_i(t,s) - A_i(t,s) \tag{8}$$

is, as a function of t for fixed s, a stopped version of $M_i(t)$ and therefore also an $\mathcal{F}_t$-martingale. It follows from (3) and (8) that for $0 \le s_1 \le s_2$,

$$\langle M_i(\cdot,s_1), M_j(\cdot,s_2)\rangle(t) = A_i(t,s_1)1\{i=j\}. \tag{9}$$

If we define

$$M(t,s) = \sum_i M_i(t,s) = N(t,s) - A(t,s),$$

then it follows from (9) that

$$\langle M(\cdot,s_1), M(\cdot,s_2)\rangle(t) = A(t, s_1 \wedge s_2). \tag{10}$$

and that, for $0 \le s_1 \le s_2 \le s_3$

$$\langle M(\cdot,s_1), M(\cdot,s_3) - M(\cdot,s_2)\rangle(t) = 0. \tag{11}$$

Thus, $M(t,s_1)$ and $M(t,s_3) - M(t,s_2)$ are orthogonal $\mathcal{F}_t$-martingales. This orthogonality structure of $M(t,s)$ will play only a minor direct role in the next section. However, a related two-dimensional process $L(t,s)$ will turn out to have the same sort of orthogonal martingale structure as $M(t,s)$, and this will be central to the proof of the following theorem.

Theorem. *Let r: $[0,a] \to (0,\infty)$ be a fixed, left-continuous, nonincreasing function. Suppose that*

$$\frac{R(t,s)}{t} \xrightarrow{P} r(s) \tag{12}$$

as $t \to \infty$, for continuity points s of r, and for $s = 0$ and $s = a$. Assume also that the cumulative hazard function Λ is continuous on $[0,a]$ and that $\Lambda(a) < \infty$. Then as $t \to \infty$,

$$t^{\frac{1}{2}}\{\hat{\Lambda}(t,\cdot) - \Lambda(\cdot)\} \xrightarrow{w} B\left(\int_0^{(\cdot)} \frac{\Lambda(dv)}{r(v)}\right), \tag{13}$$

where B is standard Brownian motion and the weak convergence is in the space of functions $D[0,a]$ with the Skorokhod topology.

3. PROOF OF THE THEOREM

I will assume for convenience that for each t under consideration, $R(t,0)$ (and therefore $R(t,s)$, $0 \le s \le a$) is bounded above by $2r(0)t$, a.s. Condition (12) does not imply this, but the general case is easily reduced to this one by a trivial localization argument.

Note that

$$A(t,dv) = R(t,v)\Lambda(dv). \tag{14}$$

If $R(t,s) > 0$, it follows from (14) that

$$\begin{aligned} t^{\frac{1}{2}}\{\hat{\Lambda}(t,s) - \Lambda(s)\} &= t^{\frac{1}{2}} \int_0^s \frac{N(t,dv) - A(t,dv)}{R(t,v)} \\ &= t^{-\frac{1}{2}} \int_0^s \{t/R(t,v)\}\ M(t,dv). \end{aligned} \tag{15}$$

Define

$$L(t,s) = \int_0^s \{r(v)\}^{-1}\ M(t,dv). \tag{16}$$

Assumption (12) in the Theorem makes it plausible that (15) should, with high probability, be approximately equal to $t^{-\frac{1}{2}}L(t,s)$ for all $s \in [0,a]$ when t is large. The proof that this is so is postponed until the end of this section. Modulo this asymptotic equivalence, it suffices to show that

$$t^{-\frac{1}{2}}\ L(t,\cdot) \xrightarrow{w} B\{\int_0^{(\cdot)} r(v)^{-1}\ \Lambda(dv)\} \tag{17}$$

as $t \to \infty$. The proof of convergence of finite dimensional distributions in (17) depends upon the orthogonal martingale structure alluded to in the previous section.

Lemma 1. *For each fixed $s \in [0,a]$, $L(t,s)$ is a square integrable $\mathcal{F}_t$-martingale, and for $0 \le s_1 \le s_2 \le a$*

$$\langle L(\cdot,s_1), L(\cdot,s_2)\rangle(t) = \int_0^{s_1} \frac{R(t,v)}{r^2(v)}\ \Lambda(dv).$$

PROOF.

$$\begin{aligned} L(t,s) &= \sum_i \int_0^s r(v)^{-1} M_i(t,dv) \\ &= \sum_i \int_0^t r\{Z_i(u)\}^{-1}\ M_i(du,s) \end{aligned}$$

(This last equality is "obvious", but its proof is a bit tedious and therefore omitted.) Since the integrands $r\{Z_i(u)\}^{-1}$ are bounded and predictable, it follows from Gill (1980), page 10, that $L(t,s)$ is a square-integrable martingale in t, and that for $0 \le s_1 \le s_2 \le a$

$$\begin{aligned}\langle L(\cdot,s_1), L(\cdot,s_2)\rangle(t) &= \sum_i \int_0^t r\{Z_i(u)\}^{-2}\, A_i(du,s_1)\\ &= \sum_i \int_0^{s_1} r(v)^{-2}\, A_i(t,dv) \quad (\text{"obvious"})\\ &= \int_0^{s_1} r(v)^{-2}\, A(t,dv).\end{aligned}$$

Now use (14). □

The convergence of finite dimensional distributions in (17) is an immediate consequence of Lemma 2.

Lemma 2. *Fix $0 = s_0 < s_1 < \ldots < s_k = a$, and set $c_i = \int_{s_{i-1}}^{s_i} r(v)^{-1}\Lambda(dv)$. Let $B_1, \ldots, B_k$ be independent standard Brownian motions, and let $W^\infty(u) = [c_1B_1(u), \ldots, c_kB_k(u)]$ be a k-dimensional diffusion. For each t, let $W^t(u)$ be the k-dimensional stochastic process with time index u and with i^{th} coordinate function*

$$W_i^t(u) = t^{-\frac{1}{2}}\{L(tu,s_i) - L(tu,s_{i-1})\}. \tag{18}$$

Then as $t \to \infty$, $W^t \xrightarrow{w} W^\infty$ in the Skorokhod topology on $(D[0,\infty))^k$.

PROOF. By Lemma 1,

$$\langle W_i^t, W_j^t\rangle(u) = 1\{i=j\} \int_{s_{i-1}}^{s_i} \frac{R(tu,v)}{tr^2(v)}\, \Lambda(dv), \tag{19}$$

so that by (12),

$$\langle W_i^t, W_j^t\rangle(u) \xrightarrow{P} 1\{i=j\}c_iu \tag{20}$$

as $t \to \infty$ for each i,j,u. Also, the jumps of W_i^t are bounded in size by $t^{-\frac{1}{2}}r(a)^{-1}$. The claim follows from the martingale central limit theorem 2.4.1. of Gill (1980). □

Tightness in (17) will be proven using the tightness criterion (12.51) of Billingsley (1968). This tightness criterion only applies to continuous random elements in $D[0,a]$, however, so it will be necessary to define a continuous (in s) modification $\tilde{L}(u,s)$ of $L(u,s)$. The criterion (12.51) will be verified for $t^{-\frac{1}{2}}\tilde{L}(t,s)$, and then we'll see that

$$\sup_{0\le s\le a} t^{-\frac{1}{2}}|\tilde{L}(t,s) - L(t,s)| \xrightarrow{P} 0.$$

In order to conform with Billingsley's notation, define

$$\tilde{F}(s) = \int_0^s r(v)^{-1}\, \Lambda(dv).$$

Now fix t, and partition $[0, a]$ into $k = k_t = [t^{\frac{2}{3}}]$ pieces with partition points $0 = v_0 < v_1 < \ldots < v_k = a$ so that

$$2^{-\frac{1}{2}}\tilde{F}(a)t^{-\frac{2}{3}} \leq \tilde{F}(v_{i+1}) - \tilde{F}(v_i) \leq 2\tilde{F}(a)t^{-\frac{2}{3}}. \tag{21}$$

For $v_i \leq s < v_{i+1}$, define

$$\tilde{L}(u, s) = L(u, v_i) + \frac{\tilde{F}(s) - \tilde{F}(v_i)}{\tilde{F}(v_{i+1}) - \tilde{F}(v_i)}\{L(u, v_{i+1}) - L(u, v_i)\}. \tag{22}$$

Lemma 3.

$$E[\{t^{-\frac{1}{2}}\, \tilde{L}(t, s_2) - t^{-\frac{1}{2}}\, \tilde{L}(t, s_1)\}^4] \leq C\{\tilde{F}(s_2) - \tilde{F}(s_1)\}^2 \tag{23}$$

for $0 \leq s_1 \leq s_2 \leq a$, C a suitable constant not depending on t, and t sufficiently large.

(23) is a version of the tightness criterion (12.51) in Billingsley (1968). The tool used to prove Lemma 3 will be the Burkholder-Davis-Gundy inequality for continuous-time martingales quoted in the Appendix.

PROOF. It follows from Lemma 1 that if $v_i < s_1 \leq v_{i+1} \leq v_j \leq s_2 < v_{j+1}$, then

$$\begin{aligned}\langle \tilde{L}(\cdot, s_2) - \tilde{L}(\cdot, s_1)\rangle(t) = &\int_{v_{i+1}}^{v_j} \frac{R(t,v)}{r^2(v)}\, \Lambda(dv) \\ &+ \left\{\frac{\tilde{F}(s_2) - \tilde{F}_{(v_j)}}{\tilde{F}(v_{j+1}) - \tilde{F}(v_j)}\right\}^2 \int_{v_j}^{v_{j+1}} \frac{R(t,v)}{r^2(v)}\, \Lambda(dv) \\ &+ \left\{\frac{\tilde{F}(v_{i+1}) - \tilde{F}(s_1)}{\tilde{F}(v_{i+1}) - \tilde{F}(v_i)}\right\}^2 \int_{v_i}^{v_{i+1}} \frac{R(t,v)}{r^2(v)}\, \Lambda(dv).\end{aligned} \tag{24}$$

The assumption that $R(t, 0) \leq 2r(0)t$ implies

$$\int_x^y \frac{R(t,v)}{r^2(v)}\, \Lambda(dv) \leq 2t\frac{r(0)}{r(a)}\, \{\tilde{F}(y) - \tilde{F}(x)\}.$$

Applying this to (24) yields

$$\langle \tilde{L}(\cdot, s_2) - \tilde{L}(\cdot, s_1)\rangle(t) \leq 2t\frac{r(0)}{r(a)}\{\tilde{F}(s_2) - \tilde{F}(s_1)\}. \tag{25}$$

A very similar calculation shows that (25) also holds if $v_i \leq s_1 \leq s_2 \leq v_{i+1}$.

In order to apply the Burkholder-Davis-Gundy inequality, we need to get a bound on the jumps of the martingale (in u)

$$t^{-\frac{1}{2}}\{\tilde{L}(u, s_2) - \tilde{L}(u, s_1)\}. \tag{26}$$

Claim. *The jumps of (26) are bounded in size by* $\{\tilde{F}(s_2) - \tilde{F}(s_1)\}^{\frac{1}{2}}$ *for* t *sufficiently large.*

PROOF OF THE CLAIM: The jumps of (26) are always bounded by $t^{-\frac{1}{2}}\, r(a)^{-1}$. This is more than enough if $\tilde{F}(s_2) - \tilde{F}(s_1) > t^{-\frac{3}{4}}$. If $\tilde{F}(s_2) - \tilde{F}(s_1) \le t^{-\frac{3}{4}}$ and $v_i \le s_1 < s_2 < v_{i+1}$, then the interpolation used in the definition of $\tilde{L}(u,s)$ reduces the bound to

$$t^{-\frac{1}{2}} r(a)^{-1} \frac{\tilde{F}(s_2) - \tilde{F}(s_1)}{\tilde{F}(v_2) - \tilde{F}(v_1)} < 2t^{\frac{1}{6}} r(a)^{-1}\, \{\tilde{F}(s_2) - \tilde{F}(s_1)\}. \tag{27}$$

(The inequality follows from (21).) The right-hand side of (27) is less than $\{\tilde{F}(s_2) - \tilde{F}(s_1)\}^{\frac{1}{2}}$ if $\tilde{F}(s_2) - \tilde{F}(s_1) \le t^{-\frac{3}{4}}$ and t is sufficiently large. If $\tilde{F}(s_2) - \tilde{F}(s_1) \le t^{-\frac{3}{4}}$ and $s_1 \le v_i \le s_2$, then a similar argument applies. □

By the Burkholder-Davis-Gundy inequality, (25) and the Claim imply that Lemma 3 holds with $C = c_0 \left[\left\{ 2\frac{r(0)}{r(a)} \right\}^{\frac{1}{2}} + 1 \right]^4$ for t large enough. □

Lemma 4. *As* $t \to \infty$,

$$\sup_{0 \le s \le a} t^{-\frac{1}{2}} |\tilde{L}(t,s) - L(t,s)| \xrightarrow{P} 0.$$

PROOF. For any i,

$$\begin{aligned} \langle L(\cdot, v_{i+1}) - L(\cdot, v_i) \rangle (t) &= \int_{v_i}^{v_{i+1}} \frac{R(t,v)}{r^2(v)}\, \Lambda(dv) \\ &\le 2 + \frac{r(0)}{r(a)}\, \{\tilde{F}(v_{i+1}) - \tilde{F}(v_i)\} \\ &\le 4\frac{r(0)}{r(a)} t^{\frac{1}{3}}. \end{aligned}$$

By Freedman's (1975) tail inequality for bounded-jump martingales (quoted in the Appendix),

$$P\{|L(t,v_{i+1}) - L(t,v_i)| \ge t^{\frac{2}{5}}\} \le 2\exp\left[\frac{-t^{\frac{4}{5}}}{2\{r(a)^{-1} t^{\frac{2}{5}} + 4r(0) t^{\frac{1}{3}} / r(a)\}} \right].$$

(Take $a = t^{\frac{2}{5}}$, $b = 4r(0)t^{\frac{1}{3}}/r(a)$, and $K = r(a)^{-1}$ in Freedman's theorem.) Since $t^{\frac{2}{3}}$ times the bound on the right-hand side goes to 0 as $t \to \infty$, we have

$$P\{ \sup_{1 \le i \le [t^{\frac{2}{3}}]} |L(t,v_{i+1}) - L(t,v_i)| \le t^{\frac{2}{5}} \} \longrightarrow 1 \tag{28}$$

as $t \to \infty$. Since for $v_i \leq s \leq v_{i+1}$,

$$L(t,s) - L(t,v_i) = \int_{v_i}^{s} r(v)^{-1} M(t,dv)$$
$$= \int_{v_i}^{s} r(v)^{-1} N(t,dv) - \int_{v_i}^{s} r(v)^{-1} R(t,v)\Lambda(dv),$$

$L(t,s) - L(t,v_i)$ is bounded below by the negative of

$$\int_{v_i}^{v_{i+1}} r(v)^{-1} R(t,v)\Lambda(dv) < 2\, t\, r(0)\{\tilde{F}(v_{i+1}) - F(v_i)\}$$
$$\leq 4\, r(0)\, t^{\frac{1}{3}}, \quad \text{by (21)}.$$

If the event in (28) holds, then

$$\int_{v_i}^{v_{i+1}} r(v)^{-1} N(t,dv) = L(t,v_{i+1}) - L(t,v_i) + \int_{v_i}^{v_{i+1}} r(v)^{-1} R(t,v)\Lambda(dv)$$
$$< t^{\frac{2}{5}} + 4r(0)t^{\frac{1}{3}}. \tag{29}$$

Since the left-hand side of (29) is an upper bound for $L(t,s) - L(t,v_i)$, $v_i \leq s \leq v_{i+1}$, this implies that

$$\sup_{0 \leq s \leq a} t^{-\frac{1}{2}} |\tilde{L}(s,t) - L(s,t)| \leq 2t^{-\frac{1}{2}}\{t^{\frac{2}{5}} + 4\, r(0)t^{\frac{1}{3}}\}$$

whenever the event in (28) holds. This proves Lemma 4. □

Lemma 4 and Lemma 2 imply the convergence of finite dimensional distributions in

$$t^{-\frac{1}{2}}\tilde{L}(t,\cdot) \xrightarrow{w} B\{\int_0^{(\cdot)} r(v)^{-1}\Lambda(dv)\}, \tag{30}$$

and Lemma 3 implies tightness. Together, (30) and Lemma 4 imply (17).

It remains to prove

Lemma 5. *As $t \to \infty$,*

$$\sup_{0 \leq s \leq a} |t^{-\frac{1}{2}} L(t,s) - t^{\frac{1}{2}}\{\hat{\Lambda}(t,s) - \Lambda(s)\}| \xrightarrow{P} 0.$$

PROOF. It follows from (16) and (17) that

$$t^{-\frac{1}{2}} M(t,\cdot) \xrightarrow{w} B\{\int_0^{(\cdot)} r(v)\Lambda(dv)\} \tag{31}$$

on $D[0,a]$. Alternatively, this can be proven by repeating the previous arguments of this section, with $M(t,s)$ in place of $L(t,s)$ and with formula (10) taking the place of Lemma 1.

Define, for $0 \le s \le a$,

$$G(s) = r(s)^{-1}, \quad G_t(s) = t/R(t,s), \quad \text{and} \quad \tilde{B}_t(s) = t^{-\frac{1}{2}} M(t,s).$$

If $R(t,a) > 0$, it follows from (15) and (16) that

$$\begin{aligned} t^{-\frac{1}{2}} L(t,s) - t^{\frac{1}{2}}\{\hat{\Lambda}(t,s) - \Lambda(s)\} &= \int_0^s \{r(v)^{-1} - t/R(t,v)\} t^{-\frac{1}{2}} M(t,dv) \\ &= \int_0^s \{G(v) - G_t(v)\} \tilde{B}_t(dv). \end{aligned} \tag{33}$$

The salient properties of G, G_t, and $\tilde{B}_t$ needed to prove Lemma 5 are:

(a) G is an increasing, left-continuous function with $0 < G(0) \le G(a) = r(a)^{-1}$.

(b) G_t is a non-negative, left-continuous random function for which

$$P\{G_t(a) < 2r(a)^{-1}\} \longrightarrow 1 \text{ as } t \to \infty, \tag{34}$$

and

$$G_t(s) \xrightarrow{P} G(s) \text{ as } t \to \infty \tag{35}$$

for continuity points s of G, and for $s = 0$, $S = a$.

(c) For each t, $\tilde{B}_t(\cdot)$ is a right-continuous process of almost surely bounded variation. Let

$$w_t(\delta) = \sup_{|s_2 - s_1| < \delta} |\tilde{B}_t(s_2) - \tilde{B}_t(s_1)|$$

be the modulus of continuity of $\tilde{B}_t$. Then by (31), for each $\varepsilon > 0$, there exists t_0 and $\delta > 0$ such that $t \ge t_0$ implies

$$P\{w_t(\delta) > \varepsilon\} < \varepsilon. \tag{36}$$

Let $\varepsilon > 0$. Choose t_0 and δ so that (36) holds. Then choose continuity points $s_1, \ldots, s_{k-1}$ of G so that

$$0 = s_0 < s_1 < \ldots < s_k = a \quad \text{and} \quad s_{i+1} - s_i < \delta.$$

By (35) and (34), $t \ge t_0$ will imply

$$P\{\sup_{0 \le i \le k} |G_t(s_i) - G(s_i)| \ge 1/k\} < \varepsilon \tag{37}$$

and

$$P\{G_t(a) < 2r(a)^{-1}\} < \varepsilon \tag{38}$$

if t_0 is made sufficiently larger. Note that the event in (38) implies $R(t,a) > 0$.

Assume that the events in (36), (37), and (38) hold. (This is obviously true with probability $> 1 - 3\varepsilon$ for $t \ge t_0$). A bound for the absolute value of (33) will now be found. Fix $s \in [0, a]$, and let $s_j \leq s < s_{j+1}$. Rewrite (33) as

$$\sum_{i=0}^{j-1} \int_{s_i}^{s_{i+1}} \{G(v)-G_t(v)\} d\{\tilde{B}_t(v)-\tilde{B}_t(s_i)\} + \int_{s_j}^{s} \{G(v)-G_t(v)\} d\{\tilde{B}_t(v)-\tilde{B}_t(s_j)\}. \tag{39}$$

Integration by parts changes (39) into

$$\begin{aligned} &\sum_{i=0}^{j-1} \{G(s_{i+1}) - G_t(s_{i+1})\}\{\tilde{B}_t(s_{i+1}) - \tilde{B}_t(s_i)\} \\ &\quad - \sum_{i=0}^{j-1} \int_{s_i}^{s_{i+1}} \{\tilde{B}_t(v) - \tilde{B}_t(s_i)\}\{G(dv) - G_t(dv)\} \\ &\quad + \{G(s) - G_t(s)\}\{\tilde{B}_t(s) - \tilde{B}_t(s_j)\} \\ &\quad - \int_{s_j}^{s} \{\tilde{B}_t(v) - \tilde{B}_t(s_j)\}\{G(dv) - G_t(dv)\}. \end{aligned} \tag{40}$$

The events in (36) and (37) imply that (40) is bounded in absolute value by

$$\sum_{i=0}^{j-1} \varepsilon/k + 3\{G(a) + G_t(a)\}\varepsilon$$

which by (38) is bounded by $\varepsilon + 9r(a)^{-1}\varepsilon$. Since this bound is valid for all s with probability $> 1 - 3\varepsilon$ whenever $t \geq t_0$, Lemma 5 follows. □

APPENDIX

The following Burkholder-Davis-Gundy inequality for continuous-time martingales can be found in Lenglart, Lepingle, and Pratelli (1980).

BURKHOLDER-DAVIS-GUNDY INEQUALITY

Let $t \geq 0$, be a continuous-time, square integrable martingale satisfying $M(0) = 0$ and with predictable variance process $\langle M \rangle(t)$. Suppose the jumps of M are bounded in size by a real number $K \geq 0$, and let $M^*(t) = \sup_{u \leq t} |M(u)|$. Then there exists a universal constant c_0 such that

$$E[\{M^*(\infty)\}^4] \leq c_0 \; E[\{\langle M \rangle(\infty)^{\frac{1}{2}} + K\}^4].$$

The following result is a continuous-time version of Freedman's (1975) tail inequality for martingales (Proposition 2.1). Although Freedman only considers discrete-time martingales, his proof is easily adapted to the continuous-time case.

TAIL INEQUALITY FOR BOUNDED-JUMP MARTINGALES

Suppose $M(t)$, $t \geq 0$, is a locally square integrable martingale with jumps *bounded in size by K. Then for all positive real numbers a and b,*

$$P\{|M(t)| \geq a \quad \text{and} \quad \langle M\rangle(t) \leq b \quad \text{for some} \quad t \geq 0\} \leq 2\exp\left[-\frac{a^2}{2(Ka+b)}\right].$$

ACKNOWLEDGMENTS

The techniques of this paper were originally developed to study the evolution of the partial likelihood function for the proportional hazards model in staggered-entry clinical trials. I am very grateful to Richard Gill for pointing out to me that these methods apply here.

BIBLIOGRAPHY

Bather, J. A. (1977). On the sequential construction of an optimal age replacement policy. *Proc. 41st Session ISI*, New Delhi, 253–266.

Berg, M. (1976). A proof of optimality for age replacement policies. *J. Appl. Prob.* **13**, 751–759.

Billingsley, P. (1968). *Weak Convergence of Probability Measures.* Wiley, New York.

Freedman, D. A. (1975). On tail probabilities for martingales. *Ann. Probab.* **3**, 100–118.

Gill, R. D. (1975). *Censoring and Stochastic Integrals.* Mathematical Centre Tracts 124, Mathematisch Centrum, Amsterdam.

Lenglart, E., Lepingle, D., Pratelli, M. (1980). Présentation unifiće de certaines inéqualités de la théorie des martingales. *Seminaire de Probabilités XIV 1978/79*, Springer Verlag Lecture Notes in Mathematics, No. 784, 26–48.

SEQUENTIAL STEIN-RULE MAXIMUM LIKELIHOOD ESTIMATION: GENERAL ASYMPTOTICS

PRANAB KUMAR SEN

Biostatistics Department
School of Public Health
University of North Carolina
Chapel Hill, North Carolina 27514, U.S.A.

1. INTRODUCTION

In the classical parametric theory, the *maximum likelihood estimators* (MLE) play a central role. Asymptotic properties and optimality of the MLE have been studied extensively in the literature; we may refer to the classical text by Ibragimov and Hasminskii (1981) for an excellent account of this asymptotic theory. In a general multi-parameter model, however, the classical MLE may not be fully *risk-efficient* (with respect to some specific *risk functions*), and suitable *shrinkage versions* dominating the classical MLE (known as the *Stein-rule estimators*) have been worked out for various models; we may refer to Berger (1985) for a detailed account of this theory (mostly, for the multi-normal family of distributions). In this context, both the *preliminary test estimation* (PTE) and Stein-rule estimation theory provide alternative estimators possessing some desirable properties. The Stein-rule estimators generally dominate the classical versions, but neither the PTE nor the Stein-rule estimator dominates the other. For the MLE, the asymptotic theory of PTE has been considered in a general framework in Sen (1979). The asymptotic distribution theory of the Stein-rule MLE has also been recently studied [Sen (1986)] in a systematic manner, and the relative performance of these versions of the MLE has been considered. All these relate to a traditional non-sequential (i.e., fixed sample size) setup. In the *minimum risk estimation* problem, generally, the risk function involves other unknown (nuisance) parameters, and hence, *sequential estimation* procedures are generally advocated to provide (at least, asymptotically) efficient solutions. For the multivariate normal mean vector, sequential shrinkage estimation procedures have recently been considered by Ghosh, Nickerson and Sen (1987). Under the protection of the multi-normality, these procedures exhibit dominance of the sequential shrinkage estimators over the classical sequential estimator, considered by Ghosh, Sinha and Mukhopadhyay (1976) and others. Improvements over an alternative shrinkage (sequential) procedure by Takada (1984) and two-stage sequential Stein-rule estimators by Ghosh and Sen (1983) have also been discussed there. The current study is centered around the Stein-rule sequential MLE in a general setup, and, extends the results of Sen (1986) to the sequential Stein-rule MLE, and in this context, a comparative picture of the sequential PT MLE has also been drawn. Our study is greatly facilitated by the incorporation of the notion of *asymptotic distributional risk* (ADR) of plausible estimators, and in this vein, the results considered here are the direct extensions of the asymptotic efficiency results on MLE (in the fixed sample or sequential case) to the shrinkage estimation problem.

Along with the preliminary notions, the proposed PT MLE and Stein-rule MLE are considered in Section 2; the notion of ADR is also introduced in the same section. Section 3 deals with the ADR of the different versions of the MLE in the sequential case. Allied asymptotic risk-efficiency results are presented in Section 4. some general remarks are made in the concluding section.

2. THE PRELIMINARY NOTIONS AND THE PROPOSED SEQUENTIAL ESTIMATORS

Let X_i, $i \geq 1$ be a sequence of independent and identically distributed (i.i.d.) random vectors (r.v.) with a distribution function (d.f.) $F(x, \underset{\sim}{\theta})$, $x \in E^m$ and $\underset{\sim}{\theta} = (\theta_1, \ldots, \theta_r)' \in \Omega C E^r$, for some $m \geq 1$ and $r \geq 1$. We assume that $F(x, \underset{\sim}{\theta})$ admits a density function $f(x, \underset{\sim}{\theta})$ (with respect to some sigma-finite measure μ). Then the *log-likelihood function* is defined by

$$\log L_n(\underset{\sim}{\theta}) = \sum_{i=1}^{n} \log f(X_i, \underset{\sim}{\theta}), \ \underset{\sim}{\theta} \in \Omega, \ n \geq 1. \tag{2.1}$$

We define the restricted parameter space ω as

$$\omega = \{\theta : \ \underset{\sim}{h}(\underset{\sim}{\theta}) = \underset{\sim}{h}_1(\underset{\sim}{\theta}), \ldots, h_p(\underset{\sim}{\theta})) = 0\}, \text{ for some } p \leq r. \tag{2.2}$$

Then, an *unrestricted MLE* (UMLE) and a *restricted MLE* (RMLE) of the true parameter $\underset{\sim}{\theta}$, denoted by $\tilde{\underset{\sim}{\theta}}_n$ and $\hat{\underset{\sim}{\theta}}_n$, respectively, are defined as the solutions of

$$\log L_n(\tilde{\underset{\sim}{\theta}}_n) = \sup_{\underset{\sim}{\theta} \in \Omega} \log L_n(\underset{\sim}{\theta}); \ \log L_n(\hat{\underset{\sim}{\theta}}_n) = \sup_{\underset{\sim}{\theta} \in \omega} \log L_n(\underset{\sim}{\theta}). \tag{2.3}$$

We assume that Ω is a convex, compact subset of E^r, and $\log F(x, \underset{\sim}{\theta})$ is almost everywhere (a.e.) differentiable with respect to $\underset{\sim}{\theta}$ (at least twice), these derivatives are dominated by some integrable functions, and the second order derivative matrix has the continuity property in the first mean. Further, the density f has a finite Kullback-Leibler information. Let then

$$\underset{\sim}{B}_{\underset{\sim}{\theta}} = ((\int \int_{E^m} (\partial/\partial\theta_j) \log f(x, \underset{\sim}{\theta})(\partial/\partial\underset{\sim}{\theta}_\ell) \log f(x, \underset{\sim}{\theta}) dF(x, \underset{\sim}{\theta})))_{j,\ell=1,\ldots,r}. \tag{2.4}$$

Then, we assume that $\underset{\sim}{B}_{\underset{\sim}{\theta}}$ is continuous in $\underset{\sim}{\theta}$ in some neighborhood of the true parameter point $\underset{\sim}{\theta}_o$ and $\underset{\sim}{B}_{\underset{\sim}{\theta}_o}$ is positive definite (p.d.). We also assume that

$\underset{\sim}{h}(\underset{\sim}{\theta})$ possesses continuous first and second order derivatives with respect to $\underset{\sim}{\theta}$, for every $\underset{\sim}{\theta} \in \Omega$. Let then

$$\underset{\sim}{H}_{\underset{\sim}{\theta}} = (((\partial/\partial\underset{\sim}{\theta})\underset{\sim}{h}(\underset{\sim}{\theta}))) \text{ (of order } r \times p), \tag{2.5}$$

and assume that

$$\underset{\sim}{H}_{\underset{\sim}{\theta}_o} \text{ is of rank } p(\leq r). \tag{2.6}$$

Further, we define the *likelihood score function* as

$$\underset{\sim}{\Lambda}_n(\underset{\sim}{\theta}) = n^{-1/2}(\partial/\partial\underset{\sim}{\theta}) \log L_n(\underset{\sim}{\theta}); \; \underset{\sim}{\Lambda}_n^o = \underset{\sim}{\Lambda}_n(\underset{\sim}{\theta}_o). \tag{2.7}$$

Regularity conditions pertaining to the asymptotic normality and optimality of the classical MLE have been studied in increasing generality in Cramér (1946), Huber (1967), Hájek (1970), Le Cam (1970), Inagaki (1970) and others. We shall mainly follow here the treatment of Sen (1979, 1986) and omit details of these discussions. Then, we have under θ_o, for large n,

$$n^{1/2}(\tilde{\underset{\sim}{\theta}}_n - \underset{\sim}{\theta}_o) = \underset{\sim}{B}_{\underset{\sim}{\theta}_o}^{-1}\underset{\sim}{\Lambda}_n^o + o_p(1) \overset{\mathcal{D}}{\sim} \mathcal{N}(\underset{\sim}{0}, \underset{\sim}{B}_{\underset{\sim}{\theta}_o}^{-1}). \tag{2.8}$$

Suppose now that having recorded $X_1, \ldots, X_n$, the loss incurred in estimating $\underset{\sim}{\theta}$ by an estimator $\underset{\sim}{T}_n$ is

$$u(\underset{\sim}{T}_n, \underset{\sim}{\theta}; \underset{\sim}{A}, c, n) = (\underset{\sim}{T}_n - \underset{\sim}{\theta})'\underset{\sim}{A}(\underset{\sim}{T}_n - \underset{\sim}{\theta}) + cn, \tag{2.9}$$

where $\underset{\sim}{A}$ is a known $r \times r$ p.d. matrix and $c(> 0)$ is the known cost per unit sample. Then, the *risk* (at $\underset{\sim}{\theta}_o$) is given by

$$R_n(c, \underset{\sim}{A}) = Eu(\underset{\sim}{T}_n, \underset{\sim}{\theta}_o, \underset{\sim}{A}, c, n) = Tr(\underset{\sim}{A}[E\{\underset{\sim}{T}_n - \underset{\sim}{\theta}_o)(\underset{\sim}{T}_n - \underset{\sim}{\theta}_o)'\}]) + cn. \tag{2.10}$$

For the UMLE, we gather from (2.8) and (2.10) that under appropriate regularity conditions, for large n, the risk (under θ_o) is given by

$$R_n(\tilde{\underset{\sim}{\theta}}_n; c, \underset{\sim}{A}) = n^{-1}Tr(\underset{\sim}{A}\underset{\sim}{B}_{\underset{\sim}{\theta}_o}^{-1}) + cn + o(n^{-1}). \tag{2.11}$$

Thus, if $\underset{\sim}{B}_{\underset{\sim}{\theta}_o}$ were known and c is small, the risk in (2.11) would have a minimum value $2cn_c^*$), where the *optimal sample size* n_c^* is given by

$$n_c^* = [Tr(\underset{\sim}{A}\underset{\sim}{B}_{\underset{\sim}{\theta}_o}^{-1})/c]^{1/2}. \tag{2.12}$$

In practice, neither $\underset{\sim}{\theta}_o$ nor the information matrix $\underset{\sim}{B}_{\underset{\sim}{\theta}_o}$ is generally known. Hence, this optimal sample size is not known. We may, however, define

$$\hat{\underset{\sim}{B}}_n = -n^{-1}(\partial^2/\partial\underset{\sim}{\theta}\partial\underset{\sim}{\theta}') \log L_n(\underset{\sim}{\theta})|_{\underset{\sim}{\theta}=\tilde{\underset{\sim}{\theta}}_n}, \tag{2.13}$$

and formulate a *stopping variable* N_c, by letting

$$N_c = \inf\{n \geq n_o : \; n^2 \geq \{Tr(\underset{\sim}{A}\hat{\underset{\sim}{B}}_n^{-1}) + n^{-a}\}/c\}, \; c > 0, \tag{2.14}$$

where $a(> 0)$ is a suitable positive number and n_o is a positive integer $(> r)$. For the multinormal mean estimation problem, $\hat{\underset{\sim}{B}}_n$ in (2.13) reduces to some simple functions of the sample covariance matrix, and alternative formulations of the stopping number in (2.14) has already been considered by Ghosh, Sinha and Mukhopadhyay (1976) and others. Their formulation relates to a particular choice of the matrix $\underset{\sim}{A}$ (having null elements outside the upper left-corner $p \times p$ block). In this as well as the univariate sequential estimation problem, the factor n^{-a} in (2.14) plays a specific role: A very early stopping may take place (without this factor) whenever $Tr(\underset{\sim}{A}\hat{\underset{\sim}{B}}_n^{-1})$ is very small (which may happen with a positive probability), and this factor avoids that possibility and makes mathematical manipulations easier.

With the stopping variable N_c, defined by (2.14), a sequential version of the UMLE is given by $\tilde{\underset{\sim}{\theta}}_{N_c}$. We shall see later on that under appropriate regularity conditions, the asymptotic risk of this sequential UMLE may also be approximated by $(2cn_c^*)$, so that it is asymptotically risk-efficient.

The RMLE, defined in (2.3), generally, behaves better (at least asymptotically) than the UMLE when $\underset{\sim}{\theta}_o$ is contained in ω. However, this picture does not hold when $\underset{\sim}{\theta}_o \notin \omega$. Towards this study, we assume that the following matrix (of order $(p+r) \times (p+r)$)

$$\begin{pmatrix} \underset{\sim}{B}_{\underset{\sim}{\theta}_o} & -\underset{\sim}{H}_{\underset{\sim}{\theta}_o} \\ -\underset{\sim}{H}'_{\underset{\sim}{\theta}_o} & \underset{\sim}{0} \end{pmatrix} \text{ is p.d.,} \tag{2.15}$$

and let

$$\begin{pmatrix} \underset{\sim}{P}_{\underset{\sim}{\theta}_o} & \underset{\sim}{Q}_{\underset{\sim}{\theta}_o} \\ \underset{\sim}{Q}'_{\underset{\sim}{\theta}_o} & \underset{\sim}{R}_{\underset{\sim}{\theta}_o} \end{pmatrix} = \begin{pmatrix} \underset{\sim}{B}_{\underset{\sim}{\theta}_o} & -\underset{\sim}{H}_{\underset{\sim}{\theta}_o} \\ -\underset{\sim}{H}'_{\underset{\sim}{\theta}_o} & \underset{\sim}{0} \end{pmatrix}^{-1}. \tag{2.16}$$

Then [viz., Sen (1986)], for $\underset{\sim}{\theta}_o \in \omega$, as $n \to \infty$,

$$n^{1/2}(\hat{\underset{\sim}{\theta}}_n - \underset{\sim}{\theta}_o) = \underset{\sim}{P}_{\underset{\sim}{\theta}_o}\underset{\sim}{\Lambda}_n^o + \underset{\sim}{o}_p(1) \ \overset{\mathcal{D}}{\sim} \ \mathcal{N}(\underset{\sim}{0}, \underset{\sim}{P}_{\underset{\sim}{\theta}_o}), \tag{2.17}$$

where

$$\underset{\sim}{I}_r - \underset{\sim}{B}_{\underset{\sim}{\theta}_o}\underset{\sim}{P}_{\underset{\sim}{\theta}_o} \text{ is idempotent (of rank } p(\leq r)). \tag{2.18}$$

Thus, for $\underset{\sim}{\theta}_o \in \omega$, the asymptotic risk of the RMLE is given by

$$\begin{aligned} R(\hat{\underset{\sim}{\theta}}_n; c, \underset{\sim}{A}) &= n^{-1}Tr(\underset{\sim}{A}\underset{\sim}{P}_{\underset{\sim}{\theta}_o}) + cn + o(n^{-1}) \\ &= n^{-1}Tr(\underset{\sim}{A}\underset{\sim}{B}_{\underset{\sim}{\theta}_o}^{-1}) - n^{-1}Tr(\underset{\sim}{A}\underset{\sim}{B}_{\underset{\sim}{\theta}_o}^{-1}(\underset{\sim}{I} - \underset{\sim}{B}_{\underset{\sim}{\theta}_o}\underset{\sim}{P}_{\underset{\sim}{\theta}_o})) + cn + o(n^{-1}) \\ &= R(\tilde{\underset{\sim}{\theta}}_n; c, \underset{\sim}{A}) - n^{-1}Tr(\underset{\sim}{A}\underset{\sim}{B}_{\underset{\sim}{\theta}_o}^{-1}(\underset{\sim}{I} - \underset{\sim}{B}_{\underset{\sim}{\theta}_o}\underset{\sim}{P}_{\underset{\sim}{\theta}_o})) + o(n^{-1}) \\ &\leq R(\tilde{\underset{\sim}{\theta}}_n; c, \underset{\sim}{A}) + o(n^{-1}). \end{aligned} \tag{2.19}$$

This picture may not hold when $\underset{\sim}{\theta}_o \notin \omega$. We shall make more comments on that in a later section. With the stopping time N_c defined in (2.14), we denote the sequential version of the RMLE by $\hat{\underset{\sim}{\theta}}_{N_c}$.

For testing the null hypothesis H_0: $\underset{\sim}{\theta}_o \in \omega$, the classical (log-) likelihood ratio test statistic is given by

$$\mathcal{L}_n = -2\log\{L_n(\hat{\theta}_n)/L_n(\tilde{\theta}_n)\}. \tag{2.20}$$

The null hypothesis is then accepted or rejected according as $\mathcal{L}_n$ is $\geq$ or $>$ $\ell_{n,\alpha}$, the level $\alpha(0 < \alpha < 1)$ critical value of $\mathcal{L}_n$; $\ell_{n,\alpha}$ is generally taken as $\chi^2_{p,\alpha}$, the upper $100\alpha\%$ point of the central chi square d.f. with p degrees of freedom (DF). The PT MLE $\hat{\theta}_n^{PT}$ is defined then as

$$\hat{\theta}_n^{PT} = \begin{cases} \hat{\theta}_n, & \text{if } \mathcal{L}_n \leq \ell_{n,\alpha} \\ \tilde{\theta}_n, & \text{if } \mathcal{L}_n > \ell_{n,\alpha} \end{cases} \tag{2.21}$$

with the stopping number N_c defined as in (2.14), we denote the sequential version of the PT MLE by $\hat{\theta}_{N_c}^{PT}$.

To introduce the Stein-rule MLE, we consider the usual James-Stein (1961) rule and define [as in Sen (1986)]

$$\hat{\theta}_n^S = \tilde{\theta}_n + (\hat{\theta}_n - \tilde{\theta}_N)(p-2)/\mathcal{L}_n. \tag{2.22}$$

It is possible to consider a more general form of shrinkage MLE (and we shall comment on that in the concluding section). However, for the risk-efficiency picture, this simple form provides a basis for other complicated forms, and hence, for the sake of simplicity of presentation, we shall mainly consider this form. The sequential version of (2.22) with the stopping number defined in (2.14) is denoted by $\hat{\theta}_{N_c}^S$.

While all these estimators are comparable under the null hypothesis H_0: $\theta_o \in \omega$, when θ_o lies outside ω, asymptotically, the PT MLE and the Stein-rule MLE both become equivalent (in probability) to the UMLE. Thus, to have meaningful comparisons, we confine ourselves to local alternatives, for which the different versions have possibly non-equivalent asymptotic risks. Note that n_c^* defined in (2.12) is $0(c^{-1/2})$ and it goes to ∞ as $c \downarrow 0$. To retain the asymptotic structure of the proposed sequential versions of the MLE, we therefore consider the case where c goes to 0, and, for this case, we consider the following sequence $\{K_c\}$ of local alternatives:

$$K_c : \ h(\theta_o) = c^{1/4}\gamma, \ \gamma \text{ a real vector in } E^P, \ c \downarrow 0. \tag{2.23}$$

It is well known that in the usual (parametric or nonparametric) case, a shrinkage estimator leads to an effective reduction of risk only when the actual parameter point lies near the pivot, and the choice of (2.22) is geared to this consideration too. For more detailed discussions of such local alternatives in shrinkage estimation (in the non-sequential case), we may refer to Sen (1984) and Sen and Saleh (1985). (2.23) is an analogue of the usual Pitman alternative in the sequential case. In passing, we may note that we can conceive of a sequence $\{\underset{\sim}{\theta}_o^{(c)}\}$ of parametric points, such that $\underset{\sim}{h}(\underset{\sim}{\theta}_o^{(c)}) = \underset{\sim}{0}$ and

$$\lim_{c\downarrow 0} c^{-1/4}(\underset{\sim}{\theta}_o^{(c)} - \underset{\sim}{\theta}_o) = \underset{\sim}{\gamma}^* \text{ exists, where } \underset{\sim}{\gamma} = \underset{\sim}{H}_{\underset{\sim}{\theta}_o}\underset{\sim}{\gamma}^*. \tag{2.24}$$

In general, the MLE are non-linear functions of the sample observations, and hence, the computation of the moment in (2.10) may need additional regularity conditions. On the top of that in the PT MLE and the Stein-rule MLE, the likelihood ratio statistic introduces further complications for such moment computations. To overcome this problem, we take recourse to the *asymptotic distributional risk (ADR)* of the sequential versions of the MLE. For a suitable sequential estimator $\underset{\sim}{\theta}_{N_c}^*$ of $\underset{\sim}{\theta}_o$, we denote the asymptotic d.f. (under $\{K_c\}$ in (2.23)) by

$$G^*(\underset{\sim}{x}) = \lim_{c\downarrow 0} P\{c^{-1/4}(\underset{\sim}{\theta}_{N_c}^* - \underset{\sim}{\theta}_o) \le \underset{\sim}{x}|K_c\}, \tag{2.25}$$

and assume that G^* exists and is non-degenerate. Then, corresponding to the loss function in (2.9), the ADR of $\underset{\sim}{\theta}_{N_c}^*$, at the point $\underset{\sim}{\theta}_o$, is defined by

$$\begin{aligned} R^*(\underset{\sim}{\theta}_{N_c}^*; c, \underset{\sim}{A}) &= c^{1/2}[Tr(\underset{\sim}{A}[\int\int \underset{\sim}{x}\underset{\sim}{x}' dG^*(\underset{\sim}{x})]) + \lim_{c\downarrow 0} c^{1/2}\tilde{E}(N_c|K_c)] \\ &= c^{1/2}[Tr(\underset{\sim}{A}\underset{\sim}{V}^*) + \xi^*], \text{ say,} \end{aligned} \tag{2.26}$$

where $\underset{\sim}{V}^*$ stands for the mean product matrix for the d.f. G^* and $\tilde{E}(\cdot)$ refers to the expected value computed from the asymptotic distribution of $c^{1/4}(N_c - n_c^*)$ under $\{K_c\}$, so that

$$\begin{aligned} \xi^* &= \lim_{c\downarrow 0}\{c^{1/2}\tilde{E}(N_c|K_c)\} = \lim_{c\downarrow 0}\{c^{1/2}n_c^* + c^{1/4}\tilde{E}\{c^{1/4}(N_c - n_c^*)|K_c\}\} \\ &= [Tr(\underset{\sim}{A}\underset{\sim}{B}_{\underset{\sim}{\theta}_o}^{-1})]^{1/2} + \lim_{c\downarrow 0}[0(c^{1/4})] = [Tr(\underset{\sim}{A}\underset{\sim}{B}_{\underset{\sim}{\theta}_o}^{-1})]^{1/2}. \end{aligned} \tag{2.27}$$

The risk thus computed [in (2.26)-(2.27)] from the asymptotic distributions of the (sequential) estimator as well as the stopping number provides a meaningful and adaptable way for comparing the performance characteristics of the proposed sequential versions of the MLE under the usual regularity conditions. For parallel results in the non-sequential case, we have a comparatively simpler picture (as the stopping number becomes a degenerate r.v.), and, we may refer to Sen (1986), for some details.

3. ADR OF THE PROPOSED SEQUENTIAL VERSIONS OF THE MLE

Keeping in mind the *Mahalanobis distance* for the UMLE and looking at (2.8), it seems quite plausible to choose $\underset{\sim}{A} = \underset{\sim}{B}_{\underset{\sim}{\theta}_o}$. Of course, in (2.14), $\underset{\sim}{A}$ is assumed to be a given p.d. matrix, while $\underset{\sim}{B}_{\underset{\sim}{\theta}_o}$ is generally unknown. This apparent anomaly may easily be eliminated by considering an arbitrary $\underset{\sim}{A}$. However, since our basic goal is to study the related *asymptotic risk-efficiency* (ARE) results (in Section 4), this particular choice of $\underset{\sim}{A}$ seems to be very appropriate, and it also leads to simplifications in the mathematical manipulations. The results to follow with minor modifications hold for an arbitrary $\underset{\sim}{A}$ too.

First, we show that under the assumed regularity conditions, under $\underset{\sim}{\theta}_o$,

$$(c/r)^{1/2}N_c \to 1 \text{ almost surely (a.s.), as } c \downarrow 0. \tag{3.1}$$

In this context, we may note that in (2.13) and (2.14) we have made use of the UMLE (and not the RMLE), so that for (3.1), it suffices to prove the result under $\underset{\sim}{\theta}_o$. We denote the characteristic roots of $\underset{\sim}{B}^{-1}_{\underset{\sim}{\theta}_o}\hat{\underset{\sim}{B}}_n$ by b_{nj}, $j = 1,\ldots,r$, respectively, so that $Tr(\underset{\sim}{B}_{\underset{\sim}{\theta}_o}\hat{\underset{\sim}{B}}{}_n^{-1}) = \sum_{j=1}^{r} b_{nj}^{-1}$. Then, note that by (2.14), for every $c > 0$ and $n \geq n_0$,

$$\begin{aligned}P\{N_c > n\} &= P\{cm^2 - m^{-a} \leq \sum_{j=1}^{r} b_{mj}^{-1}, \text{ for every } m \in [n_o, n]\}\\ &\leq P\{cn^2 - n^{-a} \leq \sum_{j=1}^{r} b_{nj}^{-1}\},\end{aligned} \tag{3.2}$$

where we note that $(n_c^*)^2 \sim (r/c)$, as $c \downarrow 0$. Then, for every $\varepsilon > 0$, whenever $n \geq n_c^*(1+\varepsilon)$, $cn^2 - n^{-a} > r(1+2\varepsilon) + o(1)$. Thus, if we are able to show that

$$\max\{|b_{nj} - 1| : 1 \leq j \leq r\} \to 0 \text{ a.s., as } n \to \infty, \tag{3.3}$$

then, $P\{N_c > n_c^*(1+\varepsilon)\} \to 0$ as $c \downarrow 0$ (for every $\varepsilon > 0$). A similar treatment holds for the lower tail $P\{N_c < n_c^*(1-\varepsilon)\}$. Thus, the problem reduces to showing that (3.3) holds. Towards this note that on defining $\hat{\underset{\sim}{B}}{}_n^o$ as in (2.13) with $\tilde{\underset{\sim}{\theta}}_n$ being replaced by $\underset{\sim}{\theta}_o$, we have $\hat{\underset{\sim}{B}}_n = \hat{\underset{\sim}{B}}{}_n^o + (\hat{\underset{\sim}{B}}_n - \hat{\underset{\sim}{B}}{}_n^o)$, so that on $[\|\tilde{\underset{\sim}{\theta}}_n - \underset{\sim}{\theta}_o\| < \delta]$, $\delta > 0$,

$$\begin{aligned}\|\hat{\underset{\sim}{B}}_n - \hat{\underset{\sim}{B}}{}_n^o\| &\leq \sup_{\|\underset{\sim}{t}\| < \delta} \left\| \begin{matrix}\text{difference of the second derivative}\\ \text{matrices evaluated at } \underset{\sim}{\theta}_o \text{ and } \underset{\sim}{\theta}_o + \underset{\sim}{t}\end{matrix} \right\|\\ &\leq n^{-1}\sum_{i=1}^{n}\{\sup_{\|\underset{\sim}{t}\| < \delta} \|(\partial^2/\partial\underset{\sim}{\theta}\partial\underset{\sim}{\theta}')\log f(X_i,\underset{\sim}{\theta})|_{\underset{\sim}{\theta}_o}\\ &\qquad - (\partial^2/\partial\underset{\sim}{\theta}\partial\underset{\sim}{\theta}')\log f(X_i,\underset{\sim}{\theta})|_{\underset{\sim}{\theta}_o+\underset{\sim}{t}}\|\}\\ &= n^{-1}\sum_{i=1}^{n} U_i(\delta), \text{ say,}\end{aligned} \tag{3.4}$$

where the $U_i(\delta)$ are independent r.v., and, by the assumed continuity property (in the first mean) of the second order derivative matrix, we have

$$E[U_i(\delta)|\underset{\sim}{\theta}_o] \downarrow 0 \text{ as } \delta \downarrow 0. \tag{3.5}$$

Now, by the Khintchine strong law of large numbers and (3.5), we conclude that $n^{-1}\sum_{i=1}^{n} U_i(\delta)$ converges a.s. to 0 as $n \to \infty$ (and δ is made to converge to 0). On the other hand, the assumed regularity conditions insure the strong consistency of the UMLE, so that $\tilde{\theta}_n \to \theta_o$ a.s., as $n \to \infty$. Thus, $\hat{B}_n - \hat{B}_n^o \to 0$ a.s., as $n \to \infty$. On the other hand, by the Khintchine strong law of large numbers, as $n \to \infty$, $\hat{B}_n^o = -\frac{1}{n}\sum_{i=1}^{n}(\partial^2/\partial\theta\partial\theta')\log f(X_i,\theta)|_{\theta_o}$ converges a.s. to its expectation i.d., B_{θ_o}. Therefore, $\hat{B}_n \to B_{\theta_o}$ a.s., as $n \to \infty$, so that $B_{\theta_o}^{-1}\hat{B}_n \to I_r$ a.s., as $n \to \infty$. This ensures that all the characteristic roots of $B_{\theta_o}^{-1}\hat{B}_n$ converge a.s. to 1 as $n \to \infty$, and this proves (3.3).

Now, in the context of repeated significance testing, *local asymptotic Wiener* property of the MLE's (in the unrestricted as well as restricted cases) and the associated likelihood ratio test statistics have been studied in detail by So and Sen (1981) and others. A direct consequence of this local asymptotic Wiener property is that for each of these sequence of estimators and statistics, the classical Anscombe (1952) condition (on uniform continuity in probability with respect to $\{n^{-1/2}\}$) holds. As such, by an appeal to (3.1) and the results in So and Sen (1981), we conclude that under the assumed regularity conditions, as $c \downarrow 0$,

$$c^{-1/4}\|\tilde{\theta}_{N_c} - \tilde{\theta}_{n_c^*}\| \to 0, \text{ in probability,} \tag{3.6}$$

$$c^{-1/4}\|\hat{\theta}_{N_c} - \hat{\theta}_{n_c^*}\| \to 0, \text{ in probability,} \tag{3.7}$$

$$\mathcal{L}_{N_c} - \mathcal{L}_{n_c^*} \to 0, \text{ in probability.} \tag{3.8}$$

Note that (3.6)-(3.8) hold under $\{K_c\}$ (and H_0 as well). As such, using (2.21), (2.22), (3.6)-(3.8) and the Slutzky theorem, it follows that as $c \downarrow 0$,

$$c^{-1/4}\|\hat{\theta}_{N_c}^{PT} - \hat{\theta}_{n_c^*}^{PT}\| \to 0, \text{ in probability,} \tag{3.9}$$

$$c^{-1/4}\|\hat{\theta}_{N_c}^{S} - \hat{\theta}_{n_c^*}^{S}\| \to 0, \text{ in probability,} \tag{3.10}$$

This shows that for the proposed sequential versions of the MLE, the asymptotic distribution theory agrees with the non-sequential counterparts when for the sample size (in the later case), we take n_c^*. This asymptotic equivalence makes it easier for us to use the results in Sen (1986) to derive the parallel ones in the sequential case. These ADR results in the non-sequential case are treated in Section 3 [viz., (3.18)-(3.23)] of Sen (1986). Whereas, there, the normalizing factor in the asymptotic distribution was $n^{1/2}$, here in (2.25), as have $c^{-1/4}$. Secondly, here, we have $n_c^* \sim (r/c)^{1/2}$, so that replacing $n_c^{*1/2}$ by $c^{-1/4}$ results in a scale factor adjustment by $r^{1/4}$. Further, by (2.27) and the specific choice of A, we have $\xi^* = r^{1/2}$. Finally, by (2.23) and the non-sequential counterpart of it in Sen (1986) [see his (3.8)], we have the same scale factor $r^{1/4}$, connecting the two cases. Thus, noting that the dispersion matrix B^* in (2.26) (in the sequential case) is $r^{1/2}$ times the corresponding matrix in the non-sequential case [viz., (3.17)-(3.18) of Sen (1986)], we may use (2.26), (2.27) and the ADR results in Sen (1986) along with the modifications stated above. With these adjustments, for the specific choice of $\underset{\sim}{A} = \underset{\sim}{B}_{\underset{\sim}{\theta}_o}$, the expressions for the ADR in Sen (1986) simplify, and we arrive at the following:

$$R^*(\tilde{\underset{\sim}{\theta}}_{N_c}; c, \underset{\sim}{B}_{\underset{\sim}{\theta}_o}) = 2(cr)^{1/2}, \ \forall \underset{\sim}{\gamma} \in E^p, \tag{3.11}$$

$$\begin{aligned} R^*(\hat{\underset{\sim}{\theta}}_{N_c}; c, \underset{\sim}{B}_{\underset{\sim}{\theta}_o}) &= (c/r)^{1/2}[Tr(\underset{\sim}{B}_{\underset{\sim}{\theta}_o}\underset{\sim}{P}_{\underset{\sim}{\theta}_o}) + (\underset{\sim}{\gamma}^{*\prime}\underset{\sim}{B}_{\underset{\sim}{\theta}_o}\underset{\sim}{\gamma}^*)] + (cr)^{1/2} \\ &= (cr)^{1/2}[2 - r^{-1}\{Tr(\underset{\sim}{I} - \underset{\sim}{B}_{\underset{\sim}{\theta}_o}\underset{\sim}{P}_{\underset{\sim}{\theta}_o}) - (\underset{\sim}{\gamma}^{*\prime}\underset{\sim}{B}_{\underset{\sim}{\theta}_o}\underset{\sim}{\gamma}^*)\}] \\ &= (cr)^{1/2}[2 - r^{-1}\{Tr(\underset{\sim}{I} - \underset{\sim}{B}_{\underset{\sim}{\theta}_o}\underset{\sim}{P}_{\underset{\sim}{\theta}_o}) - \Delta\}], \end{aligned} \tag{3.12}$$

$$\begin{aligned} R^*(\hat{\underset{\sim}{\theta}}{}^{PT}_{N_c}; c, \underset{\sim}{B}_{\underset{\sim}{\theta}_o}) &= (cr)^{1/2} + (c/r)^{1/2}\{r - Tr(\underset{\sim}{I} - \underset{\sim}{B}_{\underset{\sim}{\theta}_o}\underset{\sim}{P}_{\underset{\sim}{\theta}_o})\Pi_{p+2}(\chi^2_{p,\alpha}; \Delta) \\ &\qquad \Delta[2\Pi_{p+2}(\chi^2_{p,\alpha}; \Delta) - \Pi_{p+4}(\chi^2_{p,\alpha}; \Delta)]\} \\ &= (cr)^{1/2}\{2 - r^{-1}(Tr(\underset{\sim}{I} - \underset{\sim}{B}_{\underset{\sim}{\theta}_o}\underset{\sim}{P}_{\underset{\sim}{\theta}_o})\Pi_{p+2}(\chi^2_{p,\alpha}; \Delta) + \\ &\qquad \Delta[2\Pi_{p+2}(\chi^2_{p,\alpha}; \Delta) - \Pi_{p+4}(\chi^2_{p,\alpha}; \Delta)])\}, \end{aligned} \tag{3.13}$$

where $\Delta = (\underset{\sim}{\gamma}^{*\prime}\underset{\sim}{B}_{\underset{\sim}{\theta}_o}\underset{\sim}{\gamma}^*)$ and $\Pi_q(x; \delta)$ stands for the non-central chi square distribution function with q degrees of freedom and noncentrality parameter δ. Finally,

$$\begin{aligned} R^*(\hat{\underset{\sim}{\theta}}{}^{S}_{N_c}; c, \underset{\sim}{B}_{\underset{\sim}{\theta}_o}) &= (cr)^{1/2} + (c/r)^{1/2}\{r + (p-2)^2[rE(\chi^{-4}_{p+2}(\Delta)) \\ &\quad + \Delta E(\chi^{-4}_{p+4}(\Delta))] - 2(p-2)rE(\chi^{-2}_{p+2}(\Delta)) + 4(p-2)\Delta E(\chi^{-4}_{p+4}(\Delta))\} \\ &= (cr)^{1/2}\{2 - 2(p-2)E(\chi^{-2}_{p+2}(\Delta)) + 4r^{-1}(p-2)\Delta E(\chi^{-4}_{p+4}(\Delta)) \\ &\quad + (p-2)^2[E(\chi^{-4}_{p+2}(\Delta)) + r^{-1}\Delta E(\chi^{-4}_{p+4}(\Delta)]\}, \end{aligned} \tag{3.14}$$

where, for every $r(\geq 0)$ and $q(> 2r)$,

$$E(\chi_q^{-2r}(\delta)) = \int_0^\infty x^{-r} d\Pi_q(x, \delta), \qquad \delta \geq 0, \tag{3.15}$$

stands for the negative rth moment of a r.v. having the distribution $\Pi_q(x, \delta)$. We shall find these expressions convenient for further analysis in the next section.

4. ASYMPTOTIC DOMINANCE AND RISK-EFFICIENCY RESULTS

In the light of the ADR results in Section 3, we shall study the asymptotic dominance of some versions of the sequential MLE over the others (along with some related asymptotic risk-efficiency results). In this context, the key role is played by the sequence of local alternatives in (2.23), and this is in line with the usual asymptotic theory of sequential point estimation [as has been treated in detail in Chapter 10 of Sen (1981)]. Comparing (3.11) and (3.14), we obtain that

$$R^*(\hat{\theta}^S_{N_c}; c, B_{\theta_o})/R^*(\tilde{\theta}_{N_c}; c, B_{\theta_o}) = 1 - (p-2)E(\chi^{-2}_{p+2}(\Delta)) + \frac{1}{2}(p-2)^2 E(\chi^{-4}_{p+2}(\Delta)) + \frac{1}{2}(p-2)(p+2)r^{-1}\Delta E(\chi^{-4}_{p+4}(\Delta)). \quad (4.1)$$

Under the null hypothesis H_0, we have $\Delta = 0$ and (4.1) reduces to $1 - (p-2)/p + \frac{1}{2}(p-2)^2/p(p-2) = \frac{1}{2} + p^{-1})(< 1)$ for every $p > 2$; the larger is the value of p, the greater is the reduction in the ADR of the Stein-rule MLE (under H_0). It also follows from the results of Sen (1986) [after noting that $r > p > 2$] that the right hand side of (4.1) is less than 1 for all γ^* such that $\Delta < \infty$, and as $\Delta \to \infty$, (4.1) converges to 1. This shows that in the light of the ADR, the Stein-rule sequential MLE dominates the usual sequential MLE. Comparing (3.12) and (3.11), we obtain that

$$R^*(\hat{\theta}_{N_c}; c, B_{\theta_o})/R^*(\tilde{\theta}_{N_c}; c, B_{\theta_o}) \text{ is } \gtreqless 1 \text{ according as } \Delta \gtreqless Tr(I - B_{\theta_o} P_{\theta_o}). \quad (4.2)$$

Thus, under the null hypothesis (and for smaller values of Δ), the restricted MLE (in the sequential case) dominates the other (in the light of the ADR). The RMLE has the discouraging picture that as $\Delta \to \infty$, the ADR in (3.12) tends to $+\infty$, and this unbounded risk makes it a poor competitor.

Comparing (3.12) and (3.14), we obtain that under H_0,

$$\begin{aligned} R^*(\hat{\theta}_{N_c}; c, B_{\theta_o})/R^*(\hat{\theta}_{N_c}; c, B_{\theta_o}) &= \frac{2 - 2(p-2)/p + (p-2)^2/p(p-2)}{2 - [Tr(I - B_{\theta_o} P_{\theta_o})]/r} \\ &= [r(p+2)]/[r(p+2) - \{p^2 - r(p-2)\}]. \end{aligned} \quad (4.3)$$

Thus, (4.3) exceeds one whenever $p^2 > r(p-2)$, and the opposite inequality holds when $p^2 < r(p-2)$. The situation is very much comparable to the non-sequential case treated in Sen (1986), and hence, avoiding details, we conclude that usually, under H_0, the sequential Stein-rule MLE may not dominate the RMLE. When the null hypothesis is not true, the ADR of the RMLE increases with Δ, while the Stein-rule MLE has a bounded ADR, and hence, at least outside a closed ellipsoid (in γ^*) with 0 as the origin, the Stein-rule MLE would have a smaller ADR than the RMLE. Thus, though the Stein-rule sequential MLE may not dominate the RMLE, it possesses a more robust ARE picture.

Similarly, if we compare (3.13) and (3.14), we obtain that under H_0,

$$R^*(\hat{\theta}^S_{N_c}; c, B_{\theta_o})/R^*(\hat{\theta}^{PT}_{N_c}; c, \beta_{\theta_o}) = \frac{2 - 2(p-2)/p + (p-2)^2/p(p-2)}{2 - (p/r)\Pi_{p+2}(\chi^2_{p,\alpha}; 0)}$$

$$= [r(p+2)]/[p(2r - p\Pi_{p+2}(\chi^2_{p,\alpha};0))]. \qquad (4.4)$$

Now, (4.4) is $>$ or $<$ 1 according as $\Pi_{p+2}(\chi^2_{p,\alpha};0)$ is $>$ or $<$ $r(p-2)/p^2$. Note that $\Pi_{p+2}(\chi^2_{p,\alpha};0) < \Pi_p(\chi^2_{p,\alpha};0) = 1-\alpha$, and larger is the value of p, the smaller is the difference between the two. Thus, generally, when α is small and p is not very small, (4.4) exceeds 1, so that the Stein-rule sequential MLE may not dominate the PT MLE (under H_0). However, as we move away from the null space ω, the ADR of each of these two sequential estimators becomes larger. In the case of the PT MLE, the ADR first increases, attains a maximum [somewhat larger than $2(cr)^{1/2}$] and then converges to $2(cr)^{1/2}$ as the noncentrality parameter Δ tends to $+\infty$. On the other hand, the ADR of the sequential Stein-rule MLE is monotone in the noncentrality parameter Δ, it never exceeds the value $2(cr)^{1/2}$ and it converges to its upper asymptote $2(cr)^{1/2}$ as $\Delta \to \infty$. Consequently, outside a closed ellipsoid (in γ^*) with $\underset{\sim}{0}$ as the origin, the ADR of the sequential Stein-rule MLE is smaller than that of the PT MLE, although their ratio is never too large and it converges to 1 as $\Delta \to \infty$. The situation is quite similar to the non-sequential case, treated in detail in Sen (1986) with the only change that the ARE is much closer to 1 [because of the second term on the right hand side of (2.26) which remains the same for both the versions]. As such, we may refer to Sen (1986) for some details of this ARE study.

We conclude this section with some general remarks on the ADR based ARE results in the sequential case. Under the null hypothesis H_0: $\underset{\sim}{\theta}_o \in \omega$, the ARE results are quite clearly interpretable and reveal a clear picture on the relative performances of the different versions. For a fixed alternative (i.e., $\underset{\sim}{\theta}_o \notin \omega$), the asymptotic distributions of the different versions of sequential MLE's reduce to that of the UMLE (excepting the case of the RMLE which becomes degenerate). As such, there remains very little interest in the study of their ARE. The sequence of alternatives in (2.23) is the right choice for the study of the ADR and ARE results, and under such alternatives, we have discriminating results casting light on the relative performances of the different versions of the sequential MLE. In this context, the interpretation of the 'small' values of c remains the same as in the classical sequential point estimation problem (where, of course, such alternatives do not enter into the formulation of the asymptotic risk, although the expected sample size is $0(c^{-1/2})$). The other difference between the sequential and non-sequential cases is the actual values of the ARE for the different versions. This is due to the second factor (i.e., ξ^*) in (2.26) which is generally the same for all the versions of the MLE. As such, the ADR in the sequential case is the sum of the ADR in the non-sequential case and a second term (constant) of the same order of magnitude. This makes the ARE in the sequential case closer to 1 than in the non-sequential case.

Thus, in the non-sequential case, under H_0, the ARE of the Stein-rule MLE with respect to the UMLE is $p/2$, and this can be very large for large values of p, while, in the sequential case, the parallel ARE values is $2p/(p+2)$ which can not exceed 2. Similarly, for the non-null case, the ARE of the sequential Stein-rule MLE with respect to the sequential case. A similar picture holds for the other versions too. This difference between the sequential and non-sequential cases can be attributed to the difference of their corresponding risk functions. However, the relative picture (of the ARE) remains very much comparable in the two cases.

5. SOME GENERAL REMARKS

The sequential Stein-rule MLE $\hat{\theta}^S_{N_c}$, defined by (2.22) and (2.14), is motivated by the particular choice of $A = B_{\theta_o}$. In line with general shrinkage estimators, we may consider a more general shrinkage MLE as

$$\theta^*_n = \hat{\theta}_n + (I - q_n d_n \mathcal{L}_n^{-1} A^{-1} S_n^{-1})(\tilde{\theta}_n - \hat{\theta}_n), \tag{5.1}$$

where $\{q_n\}$ is a sequence of positive numbers converging to a positive constant q with $q \in (0, 2(p-2))$, d_n = smallest characteristic root of AS_n, and S_n is a BAN estimator of $B^{-1}_{\theta_o} - P_{\theta_o}$. Such an estimator has also been studied in Sen (1986), and using our stopping variable N_c in (2.14), we may as well consider the sequential shrinkage MLE $\theta^*_{N_c}$. Like in the case of $\hat{B}_n$, the strong consistency of S_n may be established along the same line, so that parallel to (3.10), we have

$$c^{-1/4} \| \theta^*_{N_c} - \theta^*_{n^*_c} \| \to 0, \text{ in probability, as } c \downarrow 0. \tag{5.2}$$

As such, the ADR and ARE results studied in Sen (1986) can directly be incorporated to show that in the light of the ADR, $\theta^*_{N_c}$ dominates the UMLE $\tilde{\theta}_{N_c}$, where we do not have to take $A = B_{\theta_o}$. The ARE picture of this shrinkage (sequential) MLE with respect to the sequential RMLE and PT MLE remains very much comparable to the one presented in Section 4 [see (4.3), (4.4) and the discussions following those]. However, the expressions for these relative risks, for an arbitrary A, are generally more complicated than the ones considered in Section 4.

The sequential versions of the different MLE's are all based on the common stopping number N_c, defined in (2.14), and this has been motivated by the classical sequential estimation problem (without appealing to the Stein-rule philosophy). The asymptotic theory presented here remains intact for any other well defined stopping variable (say, N^*_c), such that N^*_c increases indefinitely as $c \downarrow 0$. In such a case, we need to find out the corresponding sequence $\{n^*_c\}$

for which (under $\underset{\sim}{\theta}_o$) as $c \downarrow 0$, $N_c^*/n_c^* \xrightarrow{P} 1$, and in the ADR results, we have to consider the obvious modifications with the adjustment for this sample size. There remains the issue of finding an (asymptotically) optimal stopping rule in this context, and this will be considered in a future publication.

The choice of the matrix $\underset{\sim}{A}$ in (2.9) is of some importance too. In the usual linear models, $\underset{\sim}{B}_{\underset{\sim}{\theta}_o}$ in (2.4) usually involves a known matrix $\underset{\sim}{B}_o$ and an unknown scalar factor σ^{-2}. In such a case, $\underset{\sim}{A}$ can naturally be taken as $\underset{\sim}{B}_o$, and the stopping variable in (2.14) is then based on the sequence of estimators of the scale factor σ^2. For the normal theory case, optimal estimators of σ^2 are known, and hence, some further strengthening of the asymptotic theory (for moderately small values of c) can be made with some high power mathematics. However, in the general case, such a simplification may not be possible, and the asymptotic theory presented here may be used to provide reasonable approximations to the actual solutions. Our moderate experience with some Monte Carlo studies reveals that under (2.23), for moderate values of c, the asymptotic theory provides good approximations (much better, if c is still smaller).

ACKNOWLEDGMENTS

Thanks are due to the referee for his very useful comments on the paper.

This work was partially supported by the Office of Naval Research, Contract No. N00014-83-K-0387.

BIBLIOGRAPHY

Anscombe, F. J. (1952). Large sample theory of sequential estimation. *Proc. Cambridge Phil. Soc.* **48**, 600–607.

Berger, J. O. (1985). *Statistical Decision Theory*, 2nd Ed. Springer-Verlag, New York.

Cramér, H. (1946). *Mathematical Methods of Statistics.* Princeton Univ. Press, New Jersey.

Ghosh, M., Nickerson, D. M. and Sen, P. K. (1987). Sequential shrinkage estimation. *Ann. Statist.*

Ghosh, M. and Sen, P. K. (1983). On two-stage James-Stein estimators. *Sequential Anal.* **2**, 359–367.

Ghosh, M., Sinha, B. and Mukhopadhyay, N. (1976). Multivariate sequential point estimation. *Jour. Multivar. Anal.* **6**, 281–294.

Hájek, J. (1970). A characterization of limiting distributions of regular estimates. *Zeit. Wahrsch. verw. Geb.* **14**, 323–330.

Huber, P. J. (1967). The behavior of maximum likelihood estimators under non-standard conditions. *Proc. 5th Berkeley Symp. Math. Statist. Prob.* **1**, 221–233.

Ibragimov, I. A. and Hasminskii, R. Z. (1981). *Statistical Estimation: Asymptotic Theory.* Springer-Verlag, New York.

Inagaki, N. (1970). On the limiting distribution of a sequence of estimators under uniformity property. *Ann. Inst. Statist. Math.* **22**, 1–13.

James, W. and Stein, C. (1981). Estimation with quadratic loss. *Proc. 4th Berkeley Symp. Math. Statist. Prob.* **1**, 361–380.

Le Cam, L. M. (1970). On the assumptions used to prove asymptotic normality of maximum likelihood estimators. *Ann. Math. Statist.* **41**, 802–828.

Sen, P. K. (1979). Asymptotic properties of maximum likelihood estimators based on conditional specifications. *Ann. Statist.* **7**, 1019–1033.

Sen, P. K. (1981). *Sequential Nonparametrics: Invariance Principles and Statistical Inference.* John Wiley, New York.

Sen, P. K. (1984). A James-Stein detour of U-statistics. *Commun. Statist. Theor. Meth.* **A13**, 2725–2747.

Sen, P. K. (1986). On the asymptotic distributional risks of shrinkage and preliminary test versions of maximum likelihood estimators. *Sankhya*, Ser. **A48** 354–371.

Sen, P. K. and Saleh, A. K. Md. E. (1985). On some shrinkage estimators of multivariate location. *Ann. Statist.* **13**, 272–281.

So, Y. C. and Sen, P. K. (1981). Repeated significance tests based on likelihood ratio statistics. *Commun. Statist. Theor. Meth.* **A10**, 2149–2176.

Takada, Y. (1984). Inadmissibility of a sequential estimation rule of the mean of a multivariate normal distribution. *Sequential Anal.* **3**, 267–271.

FIXED PROPORTIONAL ACCURACY IN THREE STAGES

MICHAEL WOODROOFE

Department of Statistics
The University of Michigan
Ann Arbor, Michigan 48109, U.S.A.

1. INTRODUCTION

Let $x_1, x_2, \ldots$ denote independent random variables which are normally distributed with an unknown mean $\theta \in R = (-\infty, \infty)$ and unit variance; let Q denote a positive, twice continuously differentiable function on R. A sequential confidence interval for θ with width proportional to $\sqrt{Q(\theta)}$ is sought; that is, given $0 < \gamma < 1$ and $h > 0$, a stopping time $t = t(x_1, x_2, \ldots)$ and an estimator $\theta_t = \theta_t(x_1, \ldots, x_t)$ are sought for which

$$P_\theta\{| \theta_t - \theta |< h\sqrt{Q(\theta)}\} \approx \gamma \tag{1}$$

for all θ. Small values of h are of special interest here, and $\approx$ is interpreted to mean equality up to $o(h^2)$. Woodroofe (1986a) considered this problem and established (1) in a very weak sense for a fully sequential procedure, one which monitors the data continuously. The main result of this paper is to show that (1) holds (in a conventional sense) for a three stage procedure, one which requires monitoring the data only three times. For simplicity, θ is estimated by the sample mean,

$$\theta_n = \bar{X}_n = (X_1 + \ldots + X_n)/n, \; n = 1, 2, \ldots, \tag{2}$$

although more general estimators are amenable to similar analyses, as in Woodroofe (1986a). Thus, the problem is to find a stopping time t for which (1) holds with θ_t as in (2).

To get a rough idea of the sample size required, let Φ denote the standard normal distribution function. If a sample of fixed (non-random) size n were taken, then the left side of (1) would be $2\Phi(h\sqrt{[nQ(\theta)]}\} - 1$, which exceeds $\gamma \iff$

$$n > N := c^2/h^2Q(\theta), \tag{3}$$

where $c = \Phi^{-1}[(1+\gamma)/2]$, and $:=$ indicates a definition. Of course, N depends on θ, which is unknown.

To describe the sequential procedure, let q denote a twice continuously differentiable function on $\mathcal{R}$; let

$$Q_n = \max\{1/n, \; \min\{n, Q(\bar{X}_n)\}\}$$

and

$$q_n = \max\{-n, \ \min\{n, q(\bar{X}_n)\} \text{ for } n = 1, 2, \ldots,$$

and let $m = m(h)$ denote a positive integer valued function for which

$$hm + (1/m)\log(1/h) \to 0 \qquad \text{as } h \to 0. \tag{4}$$

The procedure first takes an initial sample of size m; it then takes a second sample (if necessary) to bring the sample size to

$$s = \max\{m, \ \lfloor ac^2/h^2 Q_m \rfloor), \tag{5a}$$

where $0 < a < 1$ and $\lfloor x \rfloor$ denotes the greatest integer which does not exceed x; the procedure then takes a third sample (if necessary) to bring the total sample size to

$$t = \max\{s, \lfloor c^2/h^2 Q_s + q_s \rfloor\}. \tag{5b}$$

The best results are obtained when m and $t - s$ are small compared to s. Then the three samples may reasonable be called "the pilot study," "the main study," and "fine tuning." The main results determine asymptotic expansions for the coverage probability in (1) and the expected sample size $E_\theta(t)$ as $h \to 0$; and a corollary shows how q may be chosen so that (1) holds, up to $o(h^2)$.

Theorem 1. *As $h \to 0$;*

i) *$t/N \to 1$ in probability;*

ii) *$t* := (t - N)/\sqrt{N}$ is asymptotically normal with mean 0 and variance $Q'(\theta)^2/aQ(\theta)^2$;*

iii) *t^{*2} is uniformly integrable for all $\theta \in R$. Moreover, if $Q'(\theta) \neq 0$, then*

iv) $\lim E_\theta(t - N) = q(\theta) - 1/2 - (1/2a)Q''(\theta)/Q(\theta)$.

Theorem 2. *If $Q'(\theta) \neq 0$, then as $h \to 0$,*

$$P_\theta\{| \bar{X} - \theta | < h\sqrt{Q(\theta)}\} = \gamma + h^2 R(\theta)\phi(c)/c + o(h^2),$$

where

$$R = \{Q[q - 1/2] - (Q'^2/4Q)[c^4 + (1/a - 2)c^2 + (1/a - 1)] - (Q''/2)[c^2 + 1/a + 1]\}.$$

Corollary 1. *If $q = 1/2 + [c^4 + (1/a - 2)c^2 + (1/a - 1)]Q'^2/4Q^2 + [c^2 + 1/a + 1]Q''/2Q$, then*

$$P_\theta\{| \bar{X} - \theta | < h\sqrt{Q(\theta)}\} = \gamma + o(h^2).$$

as $h \to 0$ for all θ.

Corollary 2. *With q as in Corollary 1,*

$$E(t - N) = [c^4 + (1/a - 2)c^2 + (1/a - 1)]Q'^2/4Q^4 + [c^2 + 1]Q''/2Q + o(1) \qquad \textit{as } h \to 0.$$

The corollaries are obvious consequences of the theorems.

Corollary 2 is of interest, since it computes the regret, the expected number of additional observations needed to attain (1) with the three stage procedure, compared to a lucky guess with the fixed sample size procedure. This difference depends on the design parameter a and attains its infimum as $a \to 1$. The expected sample size is not highly sensitive to a, however.

Theorems 1 and 2 are proved in Sections 3 and 4.

The procedure (5) is similar to one introduced by Hall (1981) for fixed width interval estimation of the mean of a normal distribution with unknown variance. The analysis of the coverage probability here is quite different from Hall's, however, since the stopping rule and estimator are more highly dependent in this case.

2. PRELIMINARY LEMMAS

In the proofs of Theorems 1 and 2, there is no loss of generality in supposing that

$$\theta = 0 \text{ and } Q(0) = 1, \tag{6}$$

since this may be arranged by a simple reparameterization. Accordingly, (6) is assumed throughout the next three sections, and P is written for P_0. Observe that then $N = c^2/h^2$.

To begin, observe that

$$P\{\sup_{n \geq m} \mid \bar{X}_n \mid > \delta\} \leq 2\exp(-m\delta^2/2)$$

for any $\delta > 0$, by Bernstein's inequality for martingales. So, for any $\varepsilon > 0$, there is a $\delta > 0$ for which

$$P\{\sup_{n \geq m} \mid Q_n - 1 \mid > \varepsilon\} \leq 2\exp(-m\delta^2/2), \tag{7}$$

by the continuity of Q.

The first lemma details some properties of s.

Lemma 1. *For any* $\varepsilon > 0$ *and* $p > 0$,

$$P\{\mid s/N - a \mid \geq \varepsilon\} = 0(h^p),$$
$$\lim E \mid s/N - a \mid^p = 0,$$

and

$$\lim E \mid N/s - 1/a \mid^p = 0, \qquad \textit{as } h \to 0.$$

PROOF. If $0 < \varepsilon < a/2$ and $N > 2\max\{m/a, 1/\varepsilon\}$, then $| Q_m - 1 | < \varepsilon/4$ implies $| s/N - a | < \varepsilon$. So, the first assertion follows directly from (7). Since, $m \leq s \leq Nm$, in any case, by the definition of Q_n, $n \geq 1$,

$$E | s/N - a |^p \leq \varepsilon^p + m^p P\{| s/N - a | > \varepsilon\}$$

for all $h > 0$. The second assertion now follow, by letting $h \to 0$ and $\varepsilon \to 0$ (in that order); and the third may be established similarly. □

In the next lemma, $\mathcal{D}_n$ denotees the sigma-algebra generated by $X_1, \ldots, X_n$; that is,

$$\mathcal{D}_n = \sigma\{X_1, \ldots, X_n\},$$

for all $n = 1, 2, \ldots$. Observe that s and t are stopping times with respect to $\mathcal{D}_n$, $n \geq 1$, for which s is $\mathcal{D}_m$ measurable and t is $\mathcal{D}_s$-measurable.

Lemma 2. *The conditional distribution of $\bar{X}_s$ given $\mathcal{D}_m$ is normal with mean $m\bar{X}_m/s$ and variance $(s-m)/s^2$.*

The distribution of $\sqrt{s}\bar{X}_s$ converges weakly to the standard normal distribution; and all positive powers of $\sqrt{s}\bar{X}_s$ are uniformly integrable.

PROOF. The first assertion follows by writing

$$\bar{X}_s = m\bar{X}_m/s + (1/s) \sum_{i=m+1}^{s} X_i$$

and using the independence of $\mathcal{D}_m$ and X_{m+i}, $i = 1, 2, \ldots$.

The asymptotic normality of $\sqrt{s}\bar{X}_s$ follows directly from Anscombe's (1952) Theorem and Lemma 1. For the uniform integrability, let v_p denote the pth absolute moment of the standard normal distribution for $p > 0$. If $p > 0$ then

$$\begin{aligned} E\{| \sqrt{s}\bar{X}_s |^p\} &= E\{E[| \sqrt{s}\bar{X}_s |^p | \mathcal{D}_m]\} \\ &\leq 2^{p-1} E\{| m\bar{X}_m/\sqrt{s} |^p + v_p\} \\ &\leq 2^p v_p \end{aligned}$$

for all $h > 0$. So, all moments of $| \sqrt{s}\bar{X}_s |$ remain bounded as $h \to 0$ and, therefore, all powers of $| \sqrt{s}\bar{X}_s |$ are uniformly integrable. See, for example, Loeve (1963, p. 164). □

Lemma 3.

i) $\sqrt{m}[(N/s - 1/a) - Q'(0)\bar{X}_m/a] \to 0$ *in probability;*
ii) *all positive powers* $| \sqrt{m}(N/s - 1/a) |$ *are uniformly integrable; and*
iii) $\lim NE(\bar{X}_s) = Q'(0)/a$.

PROOF. The first assertion follows directly from (5a), Lemma 2, and the Mean Value Theorem.

For the second, let $\delta > 0$ be so small that $| Q(\theta) - 1 | < 1 - a/4$ for all

$|\theta| < \delta$; and let $A = A(h)$ be the event

$$A = \{|\bar{X}_m| < \delta\}. \tag{8}$$

Then A is $\mathcal{D}_m$-measurable and $P(A') = 0(h^p)$ as $h \to 0$ for all $p > 0$, where $'$ denotes complement. Using the mean value theorem again, it is easily seen that $|\sqrt{m}(N/s - 1/a)| \leq \sqrt{m}/N + K\sqrt{m}|\bar{X}_m|$ on A for all small h for some constant K; and

$$\int_{A'} |\sqrt{m}(N/s - 1/a)|^p \, dP \leq (2mN)^p P(A'),$$

which tends to zero as $h \to 0$.

For the final assertion, condition on $\mathcal{D}_m$ and use $E(\bar{X}_m) = 0$ to write

$$NE(\bar{X}_s) = NE(m\bar{X}_m/s) = E[(m\bar{X}_m)(N/s - 1/a)].$$

The latter integrand converges in distribution to $Q'(0)Z^2/a$ as $h \to 0$, where Z has the standard normal distribution, and is uniformly integrable by ii). So, the latter expectation converges to $Q'(0)/a$, as asserted. □

3. THE PROOF OF THEOREM 1

The first assertion of Theorem 1 follows directly from the Law of Large Numbers by an argument similar to that of Lemma 2.

For the others, let $\delta > 0$ be so small that $|Q(\theta) - 1| < 1 - a/4$ whenever $|\theta| < \delta$, as in Lemma 3; and $B = B(h)$ denote the event

$$B = \{|\bar{X}_n| \leq \delta \qquad \text{for all } n = m, \ldots, s\}.$$

Then B is $\mathcal{D}_s$-measurable; $P(B') = 0(h^p)$ for all $p > 0$; and B implies that

$$t = \lfloor N/Q_s + q_s \rfloor = N/Q_s + q_s - < N/Q_s + q_s >$$

for all sufficiently small $h > 0$, where $\lfloor y \rfloor$ and $< y >$ denote the integer and fractional parts of y. In particular, B implies

$$t^* = (t - N)/\sqrt{N} = \sqrt{N}[1/Q_s - 1] + o_p(1),$$

which is asymptotically normal with mean 0 and variance $Q'(0)^2/a$ as $h \to 0$, by Lemmas 1 and 2 and the Mean Value Theorem. This establishes the second assertion of Theorem 1.

For the uniform integrability, observe that $t \leq N/Q_s + s \leq 2Ns$ $w.p.1$ for all small $h \to 0$; so $|t^*| \leq 2\sqrt{Ns}$ and

$$\int_{B'} |t^*|^p \, dP \leq (2\sqrt{N})^p \sqrt{P(B')}\sqrt{E(s^{2p})},$$

which approaches zero as $h \to 0$ for all $p > 0$ by (7) and Lemma 1. If B occurs,

then

$$| t^* | \leq K\sqrt{N} | \bar{X}_s | + K/\sqrt{N}$$

for some constant K. The second term on the right rpesents no difficulty. For the first,

$$E\{| \sqrt{N}\bar{X}_s |^p\} \leq \sqrt{E\{(N/s)^{2p}\}}\sqrt{E\{| \sqrt{s}\bar{X}_s |^{2p}\}},$$

which remains bounded as $h \to 0$ for any $p > 0$. Uniform integrability follows.

Finally, consider $E(t)$. As above, it is easily seen that

$$\int_{B'} t dP \to 0 \qquad \text{as } h \to 0.$$

Next,

$$\begin{aligned}\int_B (t - N) dP &= \int_B N[1/Q_s - 1] dP + \int_B q_s dP \\ &- \int_B < N/Q_s + q_s > dP \\ &= I_1 + I_2 - I_3,\end{aligned}$$

say, for all sufficiently small $h > 0$. These three integrals are considered separately.

First, a Taylor series expansion of $1/Q$ shows that

$$\begin{aligned}I_1 &= -\int_B NQ'(0)\bar{X}_s dP + \int_B (1/2)N(1/Q)''(Y)\bar{X}_s^2 dP \\ &= I_{11} + I_{12'}\end{aligned}$$

say, where Y denotes an intermediate point with $| Y | \leq | \bar{X}_s |$. Let Z denote a standard normal random variable. Then the integrand in I_{12} converges in distribution to $[2Q'^2 - Q''](0) \times Z^2/2a$; and this integrand is uniformly integrable, since both $(N/s)^2$ and $s^2\bar{X}_s^4$ are by Lemmas 1 and 2. So,

$$\lim I_{12} = [2Q'^2 - Q''](0)/2a, \qquad \text{as } h \to 0.$$

To estimate I_{11}, first observe that $\int_{B'} N | \bar{X}_s | dP \to 0$, as $h \to 0$, as above. So, by Lemma 3,

$$\begin{aligned}I_{11} &= -NQ'(0)E(\bar{X}_s) + o(1) \\ &= -Q'(0)^2/a + o(1),\end{aligned}$$

and

$$\lim I_1 = -Q''(0)/2a \qquad \text{as } h \to 0$$

It is easily seen that $I_2 \to q(0)$. So, to complete the proof, it suffices to show that $I_3 \to 1/2$. For this, it suffices to show that the fractional part $< n/Q_s + q_s >$ has an asymptotic uniform distribution on $[0, 1)$. A slightly strong statement is proved in Lemma 4 below.

Lemma 4. *The distribution of $< N/Q_s + q_s >$ converges weakly to the uniform distribution on $[0,1)$; and $< n/Q_s + q_s >$ is asymptotically independent of $\sqrt{s}\bar{X}_s$.*

PROOF. Let $W = W_h = N/Q_s + q_s$ and let $Z_s = \sqrt{s}\bar{X}_s$. Then it suffices to show that the expected value of $g(< W >)k(Z_s)$ converges to $(\int_0^1 g du)(\int_{-\infty}^{\infty} k d\Phi)$ for all bounded continuous functions g and k on R. In fact, it suffices to verify this when k is a positive continuously differentiable function with compact support and g is a periodic function with unit period for which $\int_0^1 g(u)du = 0$. Given such functions g and k, let $I > 0$ be so large that the support of k is contained in $[-I, I]$; let $\delta > 0$ be so small that $| Q'(\theta) |> 0$ for all $| \theta |\leq \delta$; and let h be so small that $I/\sqrt{m} < \delta$ and $| Q | +1/ | Q |< m >| q |$ on $| \theta |\leq \delta$. Let

$$w(z) = N/Q(z/\sqrt{s}) + q(z/\sqrt{s})$$

for all z and fixed s, so that $W = w(Z_s)$. Then $| w'(z) |> 0$ for all $| z |< I$ and all sufficiently small h, so that the restriction of w to $[-I, I]$ has an inverse function v; and differentiation shows that

$$v'(x)^2 + | v''(x) |\leq Km/N$$

for $w(-I) < x < w(I)$ for some constant K, independent of s, h, and x. Let e denote the conditional density of Z_s, given $\mathcal{D}_m$, normal with mean $m\bar{X}_m/\sqrt{s}$ and variance $(1 - m/s)$; and let $f = (ke)ov | v' |$, where o denotes composition. Then the conditional expectation of $g(< W >)k(Z_s)$, given $\mathcal{D}_m$, is

$$\begin{aligned} E\{g(< W >)k(Z_s) \mid \mathcal{D}_m\} &= \int_{-\infty}^{\infty} gow(z)k(z)e(z)dz \\ &= \int_{-\infty}^{\infty} g(x)f(x)dx \\ &= \int_{-\infty}^{\infty} -G(y)f'(y)dy, \end{aligned}$$

where G denotes the indefinite integral of g (from 0). Since the absolute value of the latter integral is at most Km/N $w.p.1$ for some constant K, it follows that $E\{g(< W >)k(Z_s)\} \to 0$ to complete the proof. □

4. PROOF OF THEOREM 2

The details are supplied only in the case that $a = 1/2$ since the algebra is substantially simpler in this case. The following lemma is used repeatedly.

Lemma 5. *For any real b and μ, any positive σ and π, and any non-negative* *integer* j,

$$(1/\sigma\pi)\int_{-\infty}^{\infty}\phi[(b-x)/\pi]\phi[(x-\mu)/\sigma]x^j\,dx$$
$$= (1/\pi')\phi[(b-\mu)/\pi']\int_{-\infty}^{\infty}(\sigma'y+\mu')^j\phi(y)dy,$$

where $\pi'^2 = \sigma^2+\pi^2$, $\mu' = (b\sigma^2+\mu\pi^2)/\pi'^2$, *and* $\sigma' = \sigma\pi/\pi'$.

PROOF. The lemma is elementary and well-known. □

To begin with the proof of Theorem 2, let $A = A(h)$ denote the ($\mathcal{D}_m$-measurable) event (8); and observe A implies that $|\ s - N/2\ | < N/4$ for all small $h > 0$. Then

$$P\{\bar{X}_t < h\} = P\{\bar{X}_t < h, A\} + o(h^2)$$
$$= \int_A P\{\bar{X}_t < h \mid \mathcal{D}_s\}dP + o(h^2)$$

as $h \to 0$. Since the conditional distribution of $\bar{X}_t$ given $\mathcal{D}_s$ is normal with mean $s\bar{X}_s/t$ and variance $(t-s)/t^2$, as in Lemma 2, the conditional probability may be written

$$P\{\bar{X}_t < h \mid \mathcal{D}_s\} = \Phi o w(t),$$

where

$$w(t) = w(t; s, \bar{X}_s, h) = [ht - s\bar{X}_s]/\surd(t-s)$$

and o denotes composition. The dependence of w on s, $\bar{X}_s$, and h is suppressed in the notation. So

$$P\{\bar{X}_t < h\} = \int_A \Phi o w(N)dP$$
$$+ \int_A \{\Phi o w(t) - \Phi o w(N)\}dP + o(h^2). \tag{9}$$

Since the conditional distribution of $s\bar{X}_s$ given $\mathcal{D}_m$ is normal with mean $m\bar{X}_m$ and variance $s - m$, the conditional expectation of $\Phi o w(N)$ given $\mathcal{D}_m$ is the convolution of two normal distributions. Using Lemma 5, this simplifies to $\Phi[(hN - m\bar{X}_m)/\surd(N-m)]$ on A. Then, using Lemma 5 again,

$$\int_A \Phi o w(N)dP = \int_A \Phi[(hN - m\bar{X}_m)/\surd(N-m)]dP$$
$$= \int \Phi[(hN - m\bar{X}_m)/\surd(N-m)]dP + o(h^2)$$
$$= \Phi(c) + o(h^2) \qquad \text{as } h \to 0. \tag{10}$$

So, it remains to consider the difference $\Phi o w(t) - \Phi o w(N)$.

A simple Taylor series expansion shows that

$$\int_A \{\Phi ow(t) - \Phi ow(N)\}dP = \int_A (\Phi ow)'(N)(t-N)dP + (1/2)\int_A (\Phi ow)''(t\#)(t-N)^2 dP = I_1 + I_2$$

say, where $t\#$ denotes an intermediate point with $| t\# - N | < | t - N |$. These two terms are considered separately.

The (partial) derivatives of w (with respect to t) are

$$w'(t) = h/\sqrt{(t-s)} - (ht - s\bar{X}_s)/2\sqrt{(t-s)^3}$$

and

$$w''(t) = -h/\sqrt{(t-s)^3} + 3(ht - s\bar{X}_s)/4\sqrt{(t-s)^5};$$

and the derivatives of Φow are $\phi(w)w'$ and $\phi(w)[w'' - ww'^2]$. Let Z denote a standard normal random variable. Then

$$t/N \to 1, \qquad s/N \to 1/2, \text{in probability}$$
$$\sqrt{s}\bar{X}_s \Rightarrow Z, \text{ and } (t-N)/\sqrt{N} \Rightarrow -\sqrt{2}Q'(0)Z,$$

where $\Rightarrow$ denotes convergence of the joint distributions. It follows easily that

$$w(N) \Rightarrow c' - Z, \qquad Nw'(N) \Rightarrow Z,$$

and

$$N^2 w''(N) \Rightarrow c' - 3Z,$$

where $c' = \sqrt{2}c$; and the same limits are obtained if N is replaced by t or $t\#$. So,

$$N(\Phi ow)''(t\#)(t-N)^2 = \{N^2(\Phi ow)''(t\#)\}\{(t-N)^2/N\} \Rightarrow \phi(c'-Z)\{(c'-3Z) - (c'-Z)Z^2\}2Q'(0)^2 Z^2.$$

As in the proof of Theorem 1, it is easily seen that the left side is uniformly integrable on A. So, using Lemma 5,

$$NI_2 \to Q'(0)^2 E\{\phi(c'-Z)[c'Z^2 - 3Z^3 - c'Z^4 + Z^5]\} = -Q'(0)^2\phi(c)[c^5 + 4c^3 + 5c]/8. \tag{11}$$

To estimate I_1 write, as in the proof of Theorem 1,

$$I_1 = \int_A (\Phi ow)'(N)N[1/Q_s - 1]dP + \int_A (\Phi ow)'(N)(t - N/Q_s)dP = I_{11} + I_{12},$$

say. Since $N(\Phi ow)'(N)$ is uniformly integrable on A, by Lemma 2, it is easily seen that

$$NI_{12} \to [q(0) - 1/2]E\{\phi(c' - Z)Z\} = [q(0) - 1/2]\phi(c)c/2 \qquad (12)$$

by Lemmas 4 and 5. Similarly, writing

$$\begin{aligned} I_{11} &= -Q'(0) \int_A N(\Phi o w)'(N)\bar{X}_s dP \\ &\quad + (1/2) \int_A N(\Phi o w)'(N)(1/Q)''(Y)\bar{X}_s^2 dP \\ &= J + J^{\#}, \end{aligned} \qquad (13)$$

say, where $| Y | < | \bar{X}_s |$, and using Lemma 5 again, it is easily seen that

$$\begin{aligned} NJ^{\#} &\to [2Q'(0)^2 - Q''(0)]E\{\phi(c' - Z)Z^3\} \\ &= [2Q'(0)^2 - Q''(0)]\phi(c)(c^3 + 3c)/4. \end{aligned} \qquad (14)$$

The remaining integral J is more challenging. To estimate it, write

$$\begin{aligned} N(\Phi o w)'(N)\bar{X}_s &= C_1 \phi o w(N) s\bar{X}_s/\surd(N - s) \\ &\quad + C_2 \phi o w(N)(s\bar{X}_s)^2/\surd(N - s), \end{aligned}$$

where

$$C_1 = hN(N - 2s)/[2s(N - s)]$$

and

$$C_2 = N/[2s(N - s)].$$

The conditional expectations of these two terms given $\mathcal{D}_m$ are of the form considered in Lemma 5 with $b = hN$, $\pi = \surd(N - s)$, $\sigma = \surd(s - m)$, and $\mu = m\bar{X}_m$. Thus, with $\pi' = \surd(N - m)$, $\sigma' = \surd[(s - m)(N - s)/(N - m)]$, and $\mu' = [(s - m)hN + (N - s)m\bar{X}_m]/(N - m)$, the following relation holds on A:

$$\begin{aligned} E\{N(\Phi o w)'(N)\bar{X}_s \mid \mathcal{D}_m\} &= C_1 \pi'^{-1}\phi[(b - \mu)/\pi']\mu' \\ &\quad + C_2 \pi'^{-1}\phi[(b - \mu)/\pi'](\sigma'^2 + \mu'^2) \\ &= \{1/\surd(N - m)\}\phi[(hN - m\bar{X}_m)/\surd(N - m)]\{\sum_{j=0}^{2} B_j (m\bar{X}_m)^j\}, \end{aligned} \qquad (15)$$

where

$$\begin{aligned} B_0 &= c^2 N(N - 2s)(s - m)/[2s(N - s)(N - m)] \\ &\quad + N(s - m)/[2s(N - m)] + c^2 N^2 (s - m)^2/[2s(N - s)(N - m)^2] \\ B_1 &= hN(N - 2s)/[2s(N - m)] + (s - m)hN^2/[s(N - m)^2] \end{aligned} \qquad (16)$$

and

$$B_2 = N(N - s)/[2s(N - m)^2].$$

Let

$$J_i = \int_A [1/\sqrt{(N-m)}]\phi[(hN - m\bar{X}_m)/\sqrt{(N-m)}]B_i(m\bar{X}_m)^i dP$$

for $i = 0, 1, 2$, so that $J = -Q'(0)\{J_0 + J_1 + J_2\}$.

It is easily seen that $| B_2 | < 2/N$ on A for all sufficiently small h; and it follows easily that

$$J_2 \leq (2/\sqrt{N^3})[m(N-m)/N + (mh)^2] = o(h^2). \tag{17}$$

For the middle term ($j = 1$) in the sum, one may write

$$B_1 = h[N/2s - r_1],$$

where

$$| r_1 | \leq 4m/N \qquad \text{on } A.$$

So,

$$\begin{aligned} J_1 &= h\{\int_A [1/\sqrt{(N-m)}]\phi[(hN - m\bar{X}_m)/\sqrt{(N-m)}](m\bar{X}_m)dP \\ &+ \int_A [1/\sqrt{(N-m)}]\phi[(hN - m\bar{X}_m)/\sqrt{(N-m)}][N/2s - 1](m\bar{X}_m)dp \\ &+ \int_A [1/\sqrt{(N-m)}]\phi[(hN - m\bar{X}_m)/\sqrt{(N-m)}]r_1 m\bar{X}_m dP\} \\ &= J_{11} + J_{12} + J_{13'} \end{aligned}$$

say. Here the first integral J_{11} may be estimated using Lemma 5 and (7) and is easily seen to be of smaller order of magnitude than h^2 as $h \to 0$. The last integral J_{13} may be estimated as in (15), using Schwarz' Inequality, and is also $o(h^2)$. For J_{12},

$$\phi[(hN - m\bar{X}_m)/\sqrt{(N-m)}][N/2s - 1](m\bar{X}_m) \Rightarrow \phi(c)Q'(0)Z^2,$$

where Z has the standard normal distribution, and is uniformly integrable on A, by Lemma 3. So,

$$NJ_{12} \to cE\{\phi(c)Q'(0)Z^2\} = c\phi(c)Q'(0). \tag{18}$$

Similarly, B_0 may be written in the form

$$B_0 = (1 + c^2)/2 + r_0,$$

where

$$| r_0 | \leq Km/N$$

for some constant K; and, using Lemma 5 and (7), one finds that

$$J_0 = (1/\sqrt{N})(1 + c^2)\phi(c)/2 + o(1/N) \qquad \text{as } h \to 0. \tag{19}$$

Combining equations (9)-(19) now shows that

$$P\{\bar{X}_t < h\} = \Phi(c) - \phi(c)R_1/\sqrt{N} - c\phi(c)R_2/N + o(1/N),$$

where

$$R_1 = Q'(0)(1+c^2)/2$$

and

$$R_2 = Q'(0)^2(c^4+1)/8 + Q''(0)(c^2+3)/4 - [q(0) - 1/2]/2\}.$$

Finally, replacing c by $-c$ and subtracting shows that the coverage probability (at $\theta = 0$) is

$$P\{| \bar{X}_t |< h\} = \gamma - 2R_2c\phi(c)/N + o(1/N)$$

as $h \to 0$, to complete the proof.

5. REMARKS

If $Q'(\theta) = 0$ in Theorems 1 and 2, then the fractional part no longer has a uniform distribution, and its expectation may oscillate. The other parts of the expansions are unaffected, however.

It may be shown that the expansion of Theorem 2 holds in the very weak sense of Woodroofe (1986a), even near points where $Q'(\theta) = 0$.

There is theoretical interest in letting $a = a(h) \to 1$ as $h \to 0$, since this value minimizes the regret calculated in Corollary 2. It may be shown that Theorems 1 and 2 remain valid if $a \to 1$ sufficiently slowly as $h \to 0$. Then the regret of the three stage procedure is the same as that of a fully sequential procedure.

With (primarily notational) complications, the limits in Theorems 1 and 2 may be shown to hold uniformly on compacts (in θ). The details are omitted here, but supplied by Woodroofe (1986b) for a closely related problem. When so extended, the results here are stronger than those of Woodroofe (1986a).

It seems likely that the procedures developed here and by Woodroofe (1986a) may be extended to one parameter exponential families. Meslem (1986) takes an important step in this direction.

BIBLIOGRAPHY

Anscombe, F. J. (1952). Large sample theory of sequential estimation. *Proc. Camb. Philos. Soc.*, Vol. 48, 600–607.

Hall, P. (1981). Asymptotic theory of triple sampling for sequential estimation of a normal mean. *Ann. Statist.*, Vol. 9, 1229–1238.

Loeve, M. (1963). *Probability Theory.* Van Nostrand.

Meslem, A. (1986). *Asymptotic expansions for fixed width confidence intervals.* Technical Report, University of Michigan.

Woodroofe, M. (1986a). *Confidence intervals with fixed proportional accuracy.* Technical Report, University of Michigan. To appear in *J.S.P.I.*.

Woodroofe, M. (1986b). *Asymptotically optimal sequential point estimation in three stages.* Technical Report, University of Michigan. To appear in *New Perspectives in Theoretical and Applied Statistics.*

PART 3 - Estimation and Testing

DOMINATING INADMISSIBLE TESTS IN EXPONENTIAL FAMILY MODELS

HUSSEIN AL-RAWWASH

Department of Statistics
Yarmouk University
Jordan

JOHN I. MARDEN

Department of Statistics
University of Illinois
Urbana, Illinois 61820, U.S.A.

1. INTRODUCTION

Given an inadmissible hypothesis testing procedure, it is natural to wonder what tests dominate it. Unfortunately, most of the necessary conditions for a test to be admissible found in the literature are not constructive. In this paper we present an approach to the construction of dominating tests in exponential family problems when the inadmissible test violates requisite convexity and/or monotonicity conditions.

We assume that we have a complete p-dimensional exponential family model. The sample space is $\mathcal{X}$, a Borel subset of $\mathbb{R}^p$, the dominating measure ν is a Borel probability measure on $\mathcal{X}$, and the family of probability measures we consider is indexed by $\boldsymbol{\theta} \in \Theta$, where

$$\Theta = \{\boldsymbol{\theta} \in \mathbb{R}^p | \int_{\mathcal{X}} \exp\{\boldsymbol{\theta} \circ x\}\nu(dx) < \infty\} \tag{1.1}$$

is the natural parameter space. Here, $\boldsymbol{\theta} \circ x = \sum \theta_i x_i$. The probability measure corresponding to $\boldsymbol{\theta}$ has density

$$\exp\{\boldsymbol{\theta} \circ x - \psi(\boldsymbol{\theta})\} \text{ with respect to } \nu. \tag{1.2}$$

Our problem tests

$$H_0 : \boldsymbol{\theta} \in \Theta_0 \text{ versus } H_A : \boldsymbol{\theta} \in \Theta_A, \tag{1.3}$$

where Θ_0 and Θ_A are disjoint subsets of Θ, and $\Theta_0 \subseteq \text{int}\Theta$.

A *test* is a measurable function ϕ: $\mathcal{X} \to [0,1]$, where $\phi(x)$ represents the probability of rejecting H_0 when x is observed. A test ϕ *dominates* another test ϕ_0 if

$$E_{\boldsymbol{\theta}}\phi \leq E_{\boldsymbol{\theta}}\phi_0 \quad \forall \boldsymbol{\theta} \in \Theta_0$$

and

$$E_\theta\phi \geq E_\theta\phi_0 \quad \forall\theta \in \Theta_A \tag{1.4}$$

and ϕ ***strictly*** dominates ϕ_0 if (1.4) holds, as well as a strict inequality holding for some $\theta' \in \Theta_0 \cup \Theta_A$. A test is *inadmissible* if there exists another test which dominates it, and otherwise it is *admissible*.

Some of the results in the literature on testing $\theta_1 = \ldots = \theta_q = 0$ for fixed q are collected in the theorem below. The parameters $\theta_{q+1}, \ldots, \theta_p$ are called nuisance parameters.

Theorem 1.1 *(Birnbaum, Matthes, Truax, Eaton). Suppose for given q that in (1.3) we have*

$$\Theta_0 = \{\theta \in \Theta | \theta_1 = \ldots = \theta_q = 0\} \text{ and } \Theta_A = \{\theta \in \Theta | (\theta_1, \ldots, \theta_q) \in V\}/\Theta_0, \tag{1.5}$$

where V is a given closed convex cone in $\mathbb{R}^q$. A necessary condition for a test to be admissible is that it equals a.e. $[\nu]$ one of the form

$$\phi(x) = \begin{cases} 1 & \text{if } s_q \notin A(t_q) \\ 0 & \text{if } s_q \in \operatorname{int} A(t_q) \end{cases} \tag{1.6}$$

where

$$s_q = (x_1, \ldots, x_q),\ t_q = (x_{q+1}, \ldots, x_p), \tag{1.7}$$

and for each t_q, $A(t_q)$ is a closed convex V-decreasing subset of $\mathbb{R}^q$. [A set $A \subseteq \mathbb{R}^q$ is said to be V-decreasing if for every u, $w \in \mathbb{R}^q$,

$$\begin{aligned} &u \in A \text{ and } w \leq u[V] \Rightarrow w \in A; \\ &w \leq u[V] \Leftrightarrow w \circ v \leq u \circ v \forall v \in V. \end{aligned} \tag{1.8}$$

Birnbaum (1955) presented, and Matthes and Truax (1967) corrected and extended, the proof for $p = q$ and $V = \mathbb{R}^p$, i.e., when there are no nuisance parameters and the V-decreasing property is vacuous. Matthes and Truax also proved the result for $q < p$ and $V = \mathbb{R}^p$. Birnbaum presented the result for $p = q = 2$ and $V =$ the nonnegative orthant in $\mathbb{R}^2$. Eaton (1970) then proved the theorem in its stated generality.

Our objective is to take a test ϕ_0 which is not essentially of the appropriate form (1.6), and find a test ϕ which strictly dominates it. Our approach is motivated by Section 4 of Matthes and Truax (1967), which deals with (1.5), $q = 1$, $V = \mathbb{R}$. In this case ϕ in (1.6) can be written

$$\phi(s_1, t_1) = \begin{cases} 1 & \text{if } s_1 < a_1(t_1) \text{ or } s_1 > a_2(t_1) \\ c_i(t_1) & \text{if } s_1 = a_i(t_1),\ i = 1, 2 \\ 0 & \text{if } a_1(t_1) < s_1 < a_2(t_1) \end{cases} \tag{1.9}$$

for some functions a_1, a_2, c_1 and c_2. Given any test ϕ_0, there is a test which we call $c\phi_0$ which satisfies (1.9) and dominates ϕ_0, strictly so unless $\phi_0 = c\phi_0$ a.e.

$[\nu]$. The test $c\phi_0$ is the unique test ϕ which satisfies (1.9),

$$E_0(\phi(S_1,T_1)|T_1=t_1)=E_0(\phi_0(S_1,T_1)|T_1=t_1), \tag{1.10}$$

and

$$E_0(S_1\phi(S_1,T_1)|T_1=t_1)=E_0(S_1\phi_0(S_1,T_1)|T_1=t_1), \tag{1.11}$$

for each t_1. [The uniqueness depends on the version of the conditional distribution of S_1 given $T_1=t_1$ chosen, and the convention that for each t_1, $c_i(t_1)=0$ if $\nu(S_1=a_i(t_1)|T_1=t_1)=0$, $i=1,2$.] The test $c\phi_0$ is the "convexification" of ϕ_0. See Ferguson (1967), Section 5.3, for the case $p=q=1$. When we have (1.5) with $q=1$ and $V=[0,\infty)$, the test (1.6) satisfies

$$\phi(s_1,t_1)=\begin{cases}1 & \text{if } s_1>a(t_1)\\ c(t_1) & \text{if } s_1=a(t_1)\\ 0 & \text{if } s_1<a(t_1)\end{cases} \tag{1.12}$$

for some functions a and c. For any test ϕ_0, there is a test which we call $m\phi_0$ which satisfies (1.12) and dominates ϕ_0, strictly so unless $m\phi_0=\phi_0$ a.e. $[\nu]$. The test $m\phi_0$ is the unique (as before) test ϕ satisfying (1.10) and (1.12). The test $m\phi_0$ is the "monotonization" of ϕ_0. Brown, Cohen and Strawderman (1976) present this monotonization in a more general setting, i.e., when the density of S given T has strict monotone likelihood ratio. Our notation was suggested by that in Van Houwelingen and Verbeek (1985).

Now consider (1.5) with $q>1$, $V=\mathrm{R}^q$. A test satisfying (1.6) will also satisfy (1.9), and if a test ϕ_0 violates (1.9), $c\phi_0$ will still strictly dominate ϕ_0. Note that any one of $x_1,\ldots,x_q$ could play the role of s_1 in the above construction, as well as any linear combination of $x_1,\ldots,x_q$. This observation is the basis of our method. A test ϕ_0 which violates (1.6) may also violate (1.9) for a particular choice of s_1. If so, we obtain ϕ_1 which strictly dominates ϕ_0 using the construction. The test ϕ_1 may then violate (1.9) for another choice of s_1, hence we can find ϕ_2 which strictly dominates ϕ_1. We continue until we obtain a test which can no longer be improved in this manner, or at least which appears close enough to such a test. When V is strictly contained in R^q, it behooves us to use the monotonization construction for certain choices of s_1. In Section 2 we present the constructions explicitly, and in Section 3 formalize the procedure of "choosing an s_1". But now we illustrate our approach with an example.

EXAMPLE 1.1. Let X_1 and X_2 be independent, $X_i\sim$ Binomial $(10,p_i)$, $i=1,2$. We test

$$H_0:\ (p_1,p_2)=\left(\frac{1}{2},\frac{1}{2}\right)\ \text{versus}\ H_A:\ (p_1,p_2)\in(0,1)^2/\left\{\left(\frac{1}{2},\frac{1}{2}\right)\right\}, \tag{1.13}$$

which is (1.5) with $p=q=2$, $V=\mathrm{R}^2$, and $\theta_i=ln\ p_i(1-p_i)^{-1}$, $i=1,2$. Now

Table 1.1. The tests ϕ_0, ϕ_1, ϕ_2, ϕ_4, and ϕ_5. The "." represents a value strictly between 0 and 1. The actual values, given in the order as if reading a page of English, are

ϕ_1: .79, .56, .47, .56, .47, .47, .56, .47, .56, .79;

ϕ_2: .79, .56, .47, .76, .61, .56, .61, .47, .76, .76, .47, .61, .56, .61, .76, .47, .56, .79;

ϕ_4: .55, .21, .60, .12, .12, .60, .21, .55;

ϕ_5: .37, .37, .37, .37, .37, .37, .37, .37.

x_2 ↑											
10	1	1	1	1	0	0	0	1	1	1	1
9	1	1	1	1	0	0	0	1	1	1	1
8	1	1	1	1	0	0	0	1	1	1	1
7	1	1	1	1	0	0	0	1	1	1	1
6	0	0	0	0	0	0	0	0	0	0	0
5	0	0	0	0	0	0	0	0	0	0	0
4	0	0	0	0	0	0	0	0	0	0	0
3	1	1	1	1	0	0	0	1	1	1	1
2	1	1	1	1	0	0	0	1	1	1	1
1	1	1	1	1	0	0	0	1	1	1	1
0	1	1	1	1	0	0	0	1	1	1	1
x_1 →	0	1	2	3	4	5	6	7	8	9	10

ϕ_0

1	1	1	1	1	1	1	1	1	1	1
1	1	1	1	1	1	0	1	1	1	1
1	1	.	.	.	0	0	1	1	1	1
1	1	.	0	0	0	0	1	1	1	1
1	1	.	0	0	0	0	0	0	0	1
1	1	0	0	0	0	0	0	0	1	1
1	0	0	0	0	0	0	0	.	1	1
1	1	1	1	0	0	0	0	.	1	1
1	1	1	1	0	0	.	.	.	1	1
1	1	1	1	0	1	1	1	1	1	1
1	1	1	1	1	1	1	1	1	1	1

ϕ_1

1	1	1	1	1	1	1	1	1	1	1
1	1	1	1	1	1	1	1	1	1	1
1	1	.	.	.	0	.	.	1	1	1
1	1	.	0	0	0	0	0	.	1	1
1	1	.	0	0	0	0	0	.	1	1
1	1	0	0	0	0	0	0	0	1	1
1	1	.	0	0	0	0	0	.	1	1
1	1	.	0	0	0	0	0	.	1	1
1	1	1	.	.	0	.	.	.	1	1
1	1	1	1	1	1	1	1	1	1	1
1	1	1	1	1	1	1	1	1	1	1

ϕ_2

1	1	1	1	1	1	1	1	1	1	1
1	1	1	1	1	1	1	1	1	1	1
1	1	1	1	.	0	.	1	1	1	1
1	1	1	0	0	0	0	0	1	1	1
1	1	.	0	0	0	0	0	.	1	1
1	1	0	0	0	0	0	0	0	1	1
1	1	.	0	0	0	0	0	.	1	1
1	1	1	0	0	0	0	0	1	1	1
1	1	1	1	.	0	.	1	1	1	1
1	1	1	1	1	1	1	1	1	1	1
1	1	1	1	1	1	1	1	1	1	1

ϕ_4, ϕ_5

t_2 is void, so that (1.6) shows that a test must have a convex acceptance region to be admissible. We start with the level $\alpha = .118$ test ϕ_0 defined by

$$\phi_0(x) = \begin{cases} 1 & \text{if } x_1 \text{ and } x_2 \in \{0,1,2,3,7,8,9,10\} \\ 0 & \text{if } x_1 \text{ or } x_2 \in \{4,5,6\}. \end{cases} \tag{1.14}$$

This test, based on the union-intersection principle, is created to have good power when both p_1 and p_2 are far from $\frac{1}{2}$. From Table 1.1, it is clear that ϕ_0 does not have a convex acceptance region. It does, however, satisfy (1.9) with $(s_1, t_1) = (x_1, x_2)$ and with $(s_1, t_1) = (x_2, x_1)$, but if we set $(s_1, t_1) = (x_1 + x_2, x_1 - x_2)$, then the test $\phi_0'(s_1, t_1) \equiv \phi_0((s_1 + t_1)/2, (s_1 - t_1)/2)$ defined on the (s_1, t_1)-space violates (1.9). Now $c\phi_0'$ can be constructed to satisfy (1.9), (1.10) and (1.11) with ϕ_0' in place of ϕ_0, and by defining ϕ_1 on the x-space by $\phi_1(x_1, x_2) = c\phi_0'(x_1 + x_2, x_1 - x_2)$ as in Table 1.1, we have that ϕ_1 strictly dominates ϕ_0. Now ϕ_1 violates (1.9) with $(s_1, t_1) = (x_1 - x_2, x_1 + x_2)$. Test ϕ_2 is then constructed to strictly dominate ϕ_1. This test ϕ_2 violates (1.9) with $(s_1, t_1) = (x_1, x_2)$, so we find $\phi_3 = c\phi_2$. Finally, ϕ_4 is constructed by taking $(s_1, t_1) = (x_2, x_1)$. The test ϕ_4 satisfies (1.6), and in fact can be shown to be admissible.

We could stop here, but note that the problem (1.13) and test (1.14) are invariant under the group G generated by the following symmetries:

$$x_1 \leftrightarrow x_2, \quad x_1 \leftrightarrow 10 - x_1, \quad x_2 \leftrightarrow 10 - x_2. \tag{1.15}$$

Thus, esthetically, we would like our final dominating test to be G-invariant as well. The test ϕ_5 constructed by symmetrizing ϕ_4,

$$\phi_5(x) = \frac{1}{8} \sum_{g \in G} \phi_4(gx),$$

is invariant and strictly dominates ϕ_0. The test ϕ_5 is also admissible, and in fact happens to be the likelihood ratio test and Pearson χ^2 test for level $\alpha = .118$. Table 1.2 contains values of $1000E_{(p_1, p_2)}(\phi_5 - \phi_0)$ (recall the symmetries (1.15).) Note that ϕ_5 performs much better than ϕ_0 if only one p_i is far from $\frac{1}{2}$, and somewhat better even when both p_1 and p_2 are far from $\frac{1}{2}$.

The sequence of tests in the above example converged to an admissible test in four steps. One might wonder in general what convergence results can be obtained. In Section 4 we consider such results. We show that any such sequence of tests does converge in the weak* sense. Also, when $\mathcal{X}$ is countable,

Table 1.2. Improvement of ϕ_5 over ϕ_0. Values are $1000 \times E_{(p_1, p_2)}(\phi_5 - \phi_0)$.

	.50	.55	.60	.65	.70	.75	.80	.85	.90	.95
.95										2
.90									17	12
.85								46	40	46
.80							70	72	90	111
.75						77	89	120	164	204
.70					67	85	125	184	254	314
.65				73	65	105	169	253	347	426
.60			23	40	73	129	210	314	427	522
.55		6	17	39	81	147	240	356	481	586
.50	0	3	15	40	84	154	251	371	500	609

a sequence is given which converges to a test which can no longer be improved using the convexification or monotonization algorithm, but may not satisfy condition (1.6). We conjecture that a similar sequence will converge to a test of the form (1.6) when ν is absolutely continuous with respect to Lebesgue measure on $\mathbb{R}^p$. We note that Van Houwelingen and Verbeek (1985) have a related result in which a sequence of tests obtained by alternating monotonization and symmetrization converges to a monotone symmetric test.

2. CONVEXIFICATION AND MONOTONIZATION

Take (s_1, t_1) as in (1.7), and drop the subscripts. Let $\mathcal{G} \subseteq \mathbb{R}$ and $\mathcal{T} \in \mathbb{R}^{p-1}$ be the ranges of s and t, respectively. Choose a set $\{\nu(\cdot|t)|t \in \mathcal{T}\}$ of conditional probability distributions on $\mathcal{G}$ of S given $T = t$ arising from ν, and let $\{F(\cdot;t)|t \in \mathcal{T}\}$ be the corresponding set of distribution functions. Thus from (1.2), we have that the conditional distribution of X given $T = t$ when $\theta \in \Theta$ obtains has density

$$\exp\{\theta_1 s - \psi(\theta;t)\} \text{ with respect to } \nu(\cdot|t), \tag{2.1}$$

and if ν^* is the marginal distribution of $\mathcal{T}$ arising from ν, then T has the density

$$\exp\{(\theta_2, \ldots, \theta_p) \circ t - \psi^*(\theta_2, \ldots, \theta_p)\} \text{ with respect to } \nu^*. \tag{2.2}$$

Given a test ϕ_0, we will construct a test $c\phi_0$ which satisfies (1.9), (1.10), and (1.11). The method is nearly identical to that in the Appendix of Lehmann and Scheffé (1955), where the uniformly most powerful unbiased test for (1.5) with $q = 1$ and $V = \mathbb{R}$ is obtained. See also Problem 14 on page 153 of Lehmann (1959).

Define

$$\alpha(t) = E_0(\phi_0(S,T)|T = t) \text{ and } \beta(t) = E_0(S\phi_0(S,T)|T = t) \tag{2.3}$$

for $t \in \mathcal{T}$. Since α and β are conditional expectations, they are measurable functions on $\mathcal{T}$. (A "measurable function on $\mathcal{B}$" is a Borel measurable function from $\mathcal{B}$ to $\mathbb{R}$, or to $\mathbb{R} \cup \{-\infty, +\infty\}$.) The inverse function

$$F^{-1}(y;t) \equiv \inf\{u|F(u;t) \geq y\} \tag{2.4}$$

is measurable on $[0,1] \times \mathcal{T}$. Thus the following functions are measurable on $\mathcal{A} \equiv \{(y,t) \in [0,1] \times \mathcal{T}|y \leq \alpha(t)\}$:

$$\begin{aligned} \overline{a}(y;t) &= F^{-1}(y;t); \overline{a}_2(y;t) = F^{-1}(1 - \alpha(t) + y;t); \\ \overline{c}_1(y;t) &= (y - F^{-1}(\overline{a}_1(y;t)-;t))/\nu(\{\overline{a}_1(y;t)\}|t); \\ \overline{c}_2(y;t) &= 1 - (1 - \alpha(t) + y - F^{-1}(\overline{a}_2(y;t)-;t))/\nu(\{\overline{a}_2(y;t)\}|t), \end{aligned} \tag{2.5}$$

where we take $\overline{c}_i(y;t) = 0$ if $\nu(\{\overline{a}_i(y;t)\}|t) = 0$, $i = 1, 2$. Thus we have that the following function is measurable on $\mathcal{G} \times \mathcal{A}$:

$$\overline{\phi}(s;y,t) = \begin{cases} 1 & \text{if } s < \overline{a}_1(y;t) \text{ or } s > \overline{a}_2(y;t) \\ \overline{c}_i(y;t) & \text{if } s = \overline{a}_i(y;t),\ i = 1,2 \\ 0 & \text{if } \overline{a}_1(y;t) < s < \overline{a}_2(y;t). \end{cases} \tag{2.6}$$

Note that by (2.5),

$$E_0(\overline{\phi}(S;y,t)|T = t) = \alpha(t) \text{ for } (y,t) \in \mathcal{A}. \tag{2.7}$$

Since $0 \in \text{int}\Theta$, $E_0(S|T = t)$ exists for each $t \in \mathcal{T}$, so that the function

$$H(y;t) \equiv E_0(S\overline{\phi}(S;y,t)|T = t) - \beta(t) \tag{2.8}$$

is measurable on $\mathcal{A}$. As in Lehmann and Scheffé (1955), H is continuous in y, $H(\alpha(t);t) \geq 0$, and $H(0;t) \leq 0$ for each $t \in \mathcal{T}$. Furthermore, it can be shown that H is strictly decreasing in y for each $t \in \mathcal{T}$. Thus the function $y(t)$ on $\mathcal{T}$ given by

$$H(y(t);t) = 0 \tag{2.9}$$

is well-defined and measurable. Now we are ready to present $c\phi_0$.

Lemma 2.1. *The test $c\phi_0$ on $\mathcal{X} \subseteq \mathcal{G} \times \mathcal{T}$ defined by*

$$c\phi_0(s,t) = \overline{\phi}(s;y(t),t), \tag{2.10}$$

where $\overline{\phi}$ is given in (2.6) *and $y(t)$ in* (2.9)*, satisfies* (1.9), (1.10), *and* (1.11).

PROOF. The measurability of $c\phi_0$ follows from the measurability of $\overline{\phi}$ and y. The form (1.9) holds by setting a_1, a_2, c_1 and c_2 equal to their barred counterparts in (2.5) with $y(t)$ instead of y. Equation (1.10) follows from (2.7), and (1.11) follows from (2.8) and (2.9). □

The construction $m\phi_0$ is given without proof in the next lemma. See Brown, Cohen and Strawderman (1976) for the result extended to more general densities.

Lemma 2.2. *The test $m\phi_0$ defined on $\mathcal{X}$ given by $m\phi_0(s,t) = \overline{\phi}(s;0,t)$ satisfies* (1.10).

REMARK 2.3. For each t, there is an essentially $[\nu(\cdot|t)]$ unique test ϕ_t on $\mathcal{G}$ which satisfies (1.9), (1.10) and (1.11). Thus if ϕ_0 satisfies (1.9), $c\phi_0 = \phi_0$ a.e. $[\nu]$. Similarly, if ϕ_0 satisfies (1.12), $m\phi_0 = \phi_0$ a.e. $[\nu]$.

We close this section with a result we will use to prove dominance results.

Lemma 2.4.

(a) *If $\nu(\{x|\phi_0(x) \neq c\phi_0(x)\}) > 0$ then*

$$E_\theta c\phi_0 \left\{ \begin{matrix} = \\ > \end{matrix} \right\} E_\theta \phi_0 \text{ as } \theta_1 \left\{ \begin{matrix} = \\ \neq \end{matrix} \right\} 0. \tag{2.11}$$

(b) *If* $\nu(\{x|\phi_0(x) \neq m\phi_0(x)\}) > 0$, *then*

$$E_\theta m\phi_0 \left\{\begin{smallmatrix}\geq\\=\\<\end{smallmatrix}\right\} E_\theta\phi_0 \text{ as } \theta_1 \left\{\begin{smallmatrix}\geq\\=\\<\end{smallmatrix}\right\} 0. \tag{2.12}$$

PROOF. By Lemma 2.1, we have that for each $t \in \mathcal{T}$, (1.10) and (1.11) hold with $\phi = c\phi_0$. We can then use the Generalized Neyman Pearson Lemma to show that

$$E_\theta(c\phi_0|T = t) > E_\theta(\phi_0|T = t) \text{ if } \theta_1 \neq 0 \text{ and } t \in D, \tag{2.13}$$

where

$$D = \{t \in \mathcal{T}|\nu(\{s|c\phi_0(s,t) \neq \phi_0(s,t)\}) > 0\}, \tag{2.14}$$

and equality holds in (2.13) for $t \notin D$. [See Lehmann (1959), Theorem 3.5, with $f_1(s) = 1$, $f_2(s) = s$, and $f_3(s) = \exp(\theta \cdot s)$.] Thus (2.11) follows by taking expectation over t on both sides of (2.13) with density given in (2.2). The strict inequality holds since $\nu(\{x|\phi_0(x) \neq c\phi_0(x)\}) > 0$ implies that $\nu^*(D) > 0$. Equation (2.12) can be proved similarly. □

3. USING LINEAR COMBINATIONS OF $X_1, \ldots, X_Q$ AS S_1

Let C be the unit sphere in $\mathbb{R}^p$, and for each $\gamma \in C$, let Γ_γ be an arbitrarily chosen orthogonal matrix whose first row is γ. (We are taking elements of $\mathbb{R}^k$ to be row vectors.) For fixed $\gamma \in C$, consider the mapping

$$\begin{aligned}\Gamma_\gamma : \quad & \mathcal{X} \to \mathcal{X}' \\ & x \mapsto x\Gamma_\gamma^t,\end{aligned} \tag{3.1}$$

where $\mathcal{X}'(\subseteq \mathbb{R}^p)$ is the range of the mapping. Given a test ϕ on $\mathcal{X}$, let ϕ' be the test on $\mathcal{X}'$ defined by

$$\phi'(x') = \phi(x'\Gamma_\gamma). \tag{3.2}$$

In the following definitions, $s = x_1'$ and $t = (x_2', \ldots, x_p')$.

Definition 3.1. A test ϕ on $\mathcal{X}$ is γ-convex (γ-monotone) if ϕ' on $\mathcal{X}'$ is of the form (1.9) ((1.12)), and is *essentially* γ-convex (γ-monotone) if it equals a.e. $[\nu]$ a test which is γ-convex (γ-monotone).

It is useful to note that a test ϕ is γ-convex if and only if its restriction to each line parallel to γ in $\mathcal{X}$ yields a test with convex acceptance region. That is, if $a^{(i)}\gamma + m \in \mathcal{X}$, $i = 1,2,3$, for $a^{(1)} < a^{(2)} < a^{(3)}$ and $m \in \mathbb{R}^p$, then

$$\phi(a^{(1)}\gamma + m) < 1 \text{ and } \phi(a^{(3)}\gamma + m) < 1 \Rightarrow \phi(a^{(2)}\gamma + m) = 0. \tag{3.3}$$

Similarly, a test ϕ is γ-monotone if and only if for any $a^{(i)}\gamma + m$, $i = 1,2$, as above,

$$\phi(a^{(2)}\gamma + m) < 1 \Rightarrow \phi(a^{(1)}\gamma + m) = 0. \tag{3.4}$$

Definition 3.2. Given a test ϕ_0 on $\mathcal{X}$, define the γ-convex test $c_\gamma\phi_0$ on $\mathcal{X}$ by $c_\gamma\phi_0(x) = c\phi_0'(x\Gamma_\gamma^t)$, and the γ-monotone test $m_\gamma\phi_0$ on $\mathcal{X}$ by $m_\gamma\phi_0(x) = m\phi_0'(x\Gamma_\gamma^t)$.

The next result follows from Lemma 2.4 since the mapping (3.1) preserves the exponential family structure. The transformed parameter is $\theta' \equiv \theta\Gamma_\gamma^t$, so that $\theta_1' = \gamma \circ \theta$.

Lemma 3.3. *If ϕ_0 is not essentially γ-convex, then*

$$E_\theta c_\gamma\phi_0 \left\{ \begin{matrix} = \\ > \end{matrix} \right\} E_\theta\phi_0 \text{ as } \gamma \circ \theta \left\{ \begin{matrix} = \\ \neq \end{matrix} \right\} 0. \tag{3.5}$$

If ϕ_0 is not essentially γ-monotone, then

$$E_\theta m_\gamma\phi_0 \{\gtreqless\} E_\theta\phi_0 \text{ as } \gamma \circ \theta \{\gtreqless\} 0. \tag{3.6}$$

Now we given conditions under which $c_\gamma\phi_0$, or $m_\gamma\phi_0$, dominates ϕ_0.

Definition 3.4. The problem (1.3) is *γ-two-sided* if $\gamma\circ\Theta_0 = \{0\}$ and $\gamma\circ\Theta_A/\{0\} \neq \emptyset$, and is *$\gamma$-one-sided* if $\gamma\circ\Theta_0 \subseteq (-\infty, 0]$, $\gamma\circ\Theta_A \subseteq [0, \infty)$, and $\gamma\circ(\Theta_A \cup \Theta_A) \neq \emptyset$.

Lemma 3.5. *If problem (1.3) is γ-two-sided (γ-one-sided) and ϕ_0 is not essentially γ-convex (γ-monotone), then $c_\gamma\phi_0(m_\gamma\phi_0)$ strictly dominates ϕ_0.*

PROOF. Follows directly from Lemma 3.3, Definition 3.4, and the definition of strictly dominates (1.4). □

REMARK 3.6. Many problems (1.3) are γ-two-sided, or γ-one-sided, for a multitude of vectors $\gamma \in C$. For example, (1.5) is γ-two-sided for all $\gamma \in C_q \equiv \{\gamma \in C | \gamma_{q+1} = \ldots = \gamma_p = 0\}$, and γ-one-sided for all $\gamma \in C_q(V) \equiv \{\gamma \in C_q | -(\gamma_1, \ldots, \gamma_q) \in V^\perp\}$, where $V^\perp$ is the cone dual to V in $\mathbb{R}^q$. This observation, together with Lemma 3.5, leads to the following rival to Theorem 1.1.

Theorem 3.7. *If* (1.5) *holds, then a necessary condition for a test ϕ_0 to be admissible is that ϕ_0 be essentially γ-convex (γ-monotone) for all $\gamma \in C_q(C_q(V))$.*

The proof(s) of Theorem 1.1 in the indicated references used limits of proper Bayes tests when $p = q$, and a complicated measurable piecing together of sets $A(t_q)$ in (1.6) when $q < p$. The proof of Theorem 3.7 has a completely different flavor, and to us is much easier. In general, though, Theorem 3.7 is weaker than Theorem 1.1. For example, take $p = q = 2$, $V = \mathbb{R}^2$, and $\mathcal{X} = \{(0,0), (1,0), (2,-1), (2,1)\}$. Then test ϕ_0 defined by $\phi_0(1,0) = 1$, $\phi_0(x) = 0$ otherwise, is γ-convex for all $\gamma \equiv C$, but clearly violates (1.6). Now when $\mathcal{X}$ is convex, and ν and Lebesgue measure on $\mathbb{R}^p$ are mutually absolutely continuous, the two theorems can be shown to be equivalent. That is, a test ϕ_0 satisfies the conditions in Theorem 3.7 if and only if it is essentially of the form (1.6). See Al-Rawwash (1986) for details.

4. ITERATIVE SCHEMES

For problem (1.3), let $C^{(2)}$ be the set of vectors $\gamma \in C$ for which it is γ-two-sided, and $C^{(1)}$ be the set of vectors for which it is γ-one-sided. See Definition 3.4. Given a test ϕ_0 and sequence $\{\gamma_i\} \subseteq C^{(1)} \cup C^{(2)}$, we construct a sequence $\{\phi_i\}$ of tests defined by

$$\phi_i = d_{\gamma_i}\phi_{i-1},\ i \geq 1, \text{ where } d_\gamma = \begin{cases} c_\gamma & \text{if } \gamma \in C^{(2)}/C^{(1)} \\ m_\gamma & \text{if } \gamma \in C^{(1)} \end{cases}. \tag{4.1}$$

See Definition 3.2. [If $\gamma \in C^{(1)} \cap C^{(2)}$, then m_γ is a stronger construction than c_γ.] By Lemma 3.5 we know that each ϕ_i dominates ϕ_{i-1}, strictly so unless $\phi_i = \phi_{i-1}$ a.e. $[\nu]$. Our first result proves that the sequence converges.

Lemma 4.1. *Suppose that the family of densities* (1.2) *for* $\theta \in \Theta_0 \cup \Theta_A$ *is complete. Then for any test* ϕ_0 *and sequence* $\{\gamma_i\} \subseteq C^{(1)} \cup C^{(2)}$*, the sequence* $\{\phi_i\}$ *given in* (4.1) *has a weak* limit* ϕ*, that is, for any bounded continuous function* f *on* $\mathbb{R}^p$,

$$\int_{\mathcal{X}} \phi_i(x)f(x)\nu(dx) \to \int_{\mathcal{X}} \phi(x)f(x)\nu(dx). \tag{4.2}$$

Furthermore, the test ϕ *dominates* ϕ_0*, strictly so unless* ϕ_0 *is essentially* γ_i*-convex (*γ_i*-monotone) for all* $\gamma_i \in C^{(2)} (\gamma_i \in C^{(1)})$, $i \geq 1$.

PROOF. The class of tests is compact with respect to the weak* topology, hence to prove that $\{\phi_i\}$ converges it is enough to show that any two convergent subsequences have a common limit. Let $\{\phi_{n(i)}\}$ and $\{\phi_{m(i)}\}$ be such subsequences, and ϕ_N and ϕ_M their respective limits. For each i, we can find i' and i'' such that $n(i) \leq m(i') \leq n(i'')$. Hence by (4.1) and Lemma 3.5,

$$E_\theta(\phi_{n(i)}) \left\{\substack{\geq \\ = \\ \leq}\right\} E_\theta(\phi_{m(i')}) \left\{\substack{\geq \\ = \\ \leq}\right\} E_\theta(\phi_{n(i'')}) \quad \text{as } \theta \begin{Bmatrix} \varepsilon\Theta_A \\ \varepsilon\Theta_0 \end{Bmatrix}. \tag{4.3}$$

Now $\{\phi_{n(i)}\}$ and its subsequence $\{\phi_{n(i'')}\}$ converge to ϕ_N, and $\{\phi_{m(i)'}\}$ converges to ϕ_M. Since $E_\theta(\cdot)$ is continuous with respect to weak* convergence (4.2), (4.3) can be used to show that $E_\theta\phi_N = E_\theta\phi_M$ for all $\theta \in \Theta_0 \cup \Theta_A$. The completeness assumption yields that $\phi_N = \phi_M$ a.e. $[\nu]$. Since the limit ϕ in (4.2) is unique only up to ν-null sets, both original subsequences converge to $\phi \equiv \phi_N$, say. Hence (4.2) holds.

The dominance result follows from Remark 2.3, Lemma 3.5, and the continuity of $E_\theta(\cdot)$. □

Can the sequence $\{\gamma_i\}$ be chosen in such a way that the limiting test ϕ is optimal, or at least no longer improvable using the $c-$ or $m-$ constructions? We do not have an answer in general. However, suppose $\mathcal{X}$ is countable, and $\nu(x) > 0$ for each $x \in \mathcal{X}$. If for a given $\gamma \in C$, every line parallel to γ passes through at most two points in $\mathcal{X}$, then by (3.3) we have that *every* test is γ-convex. If every such line passes through at most one point in $\mathcal{X}$, then by (3.4) *every* test is γ-monotone. Thus, since $\mathcal{X}$ is countable, there is only a countable number of vectors $\gamma \in C^{(1)} \cup C^{(2)}$ for which a given test ϕ_0 could possibly violate γ-convexity or γ-monotonicity. Let $\{\gamma_i\}$ be an enumeration of these vectors, and consider the sequence $\{\phi_{ij}\}_{i \le j \le i+1}$:

$$\begin{array}{ccccccccc} \phi_0 & \rightarrow & \phi_{11} & \rightarrow & \phi_{12} & & & & \\ & & \phi_{21} & \rightarrow & \phi_{22} & \rightarrow & \phi_{23} & & \\ & & \phi_{31} & \rightarrow & \phi_{32} & \rightarrow & \phi_{33} & \rightarrow & \phi_{34} \\ & & \cdot & & \cdot & & \cdot & & \cdot \quad \cdot \\ & & \cdot & & \cdot & & \cdot & & \cdot \quad \cdot \\ & & \cdot & & \cdot & & \cdot & & \cdot \quad \cdot \end{array} \tag{4.4}$$

given by

$$\phi_{11} = d_{\gamma_1}\phi_0, \phi_{ij} = d_{\gamma_j}\phi_{ij-1} \text{ if } j > 1, \text{ and } \phi_{i+11} = d_{\gamma_1}\phi_{ii+1}. \tag{4.5}$$

Now let ϕ be the limit of this sequence guaranteed to exist by Lemma 4.1. Since $\mathcal{X}$ is countable, the convergence (4.2) is actually pointwise. For each $J \ge 1$, the subsequence $\{\phi_{iJ}\}_{i \ge j-1}$ is a sequence of γ_J-convex (γ_J-monotone) tests if $\gamma_J \in C^{(2)}(\gamma_J \in C^{(1)})$. It can be shown using (3.3) ((3.4)) that a pointwise limit of such tests is also γ_J-convex (γ_J-monotone). Thus ϕ is γ-convex for all $\gamma \in C^{(2)}$, and γ-monotone for all $\gamma \in C^{(1)}$, hence can no longer be improved using these constructions.

Now suppose we have a continuous problem, where $\mathcal{X}$ is convex and ν and Lebesgue measure are mutually absolutely continuous on $\mathcal{X}$. Consider problem (1.5) with $p = q$ and $V = \mathsf{R}^p$. Now $C^{(2)} = C$. We conjecture that the sequence given in (4.4) and (4.5) with $\{\gamma_i\}$ being any countable dense subset of C converges to a test satisfying (1.6). However, the proof of this conjecture has so far eluded us. We note that a proof directly analogous to that in the previous paragraph will not work here since it is not true that a weak* limit of γ-convex tests is essentially γ-convex. It can be shown that the sequence of convex hulls of the acceptance regions of the tests is decreasing, hence has a convex limit. It is conjectured that this limit is in fact the acceptance region of a limiting test.

It is evident that the sequences as in (4.4) and (4.5) are motivated by theoretical considerations, and that in practice one would not use a preset sequence $\{\gamma_i\}$. Rather, at each stage, one would want to use the vector γ which is "best". If one possesses an objective function $U(\gamma, \phi)$, then the "best" γ_i could be defined to be the vector γ which maximizes $U(\gamma, \phi_{i-1})$, $i \geq 1$. For example, define

$$u_\theta(\phi) = \begin{cases} 1 - E_\theta\phi & \text{if } \theta \in \Theta_0 \\ E_\theta\phi & \text{if } \theta \in \Theta_A \end{cases},$$

and let $U(\gamma, \phi) = \int u_\theta(d_\gamma\phi)\pi(d\theta)$ for some measure π on $\Theta_0 \cup \Theta_A$, or let $U(\gamma, \phi) = \sup\{u_\theta(d_\gamma\phi) - u_\theta(\phi) | \theta \in \Theta_0 \cup \Theta_A\}$. A drawback is that, unless one is clever, maximizing such a function U entails calculating $d_\gamma\phi$ for several guesses at the maximizing γ. But after the first guess, one has already obtained a test which dominates the current one. Thus it seems reasonable to take this first-guess test, and try to improve it. For this reason, we recommend choosing γ_i by eye after finding ϕ_{i-1}. At worst, a poor choice of γ_i will yield a test ϕ_i identical a.e. $[\nu]$ to ϕ_{i-1}. We used this "method" in Example 1.1.

ACKNOWLEDGMENTS

We thank Professor Robert Wijsman for all the help he gave on this project, and a referee for careful reading and pertinent comments.

Research partially supported by National Science Foundation Grant #DMS 85-03321.

BIBLIOGRAPHY

Al-Rawwash, H. (1986). Improving inadmissible hypothesis testing procedures in exponential family statistical models. Ph.D. Thesis, U. of Illinois.

Birnbaum, A. (1955). Characterizations of complete classes of tests of some multiparametric hypotheses, with applications to likelihood ratio tests. *Ann. Math. Statist.* **26**, 21–36.

Brown, L., Cohen, A. and Strawderman, W. E. (1976). A complete class theorem for strict monotone likelihood ratio with applications. *Ann. Statist.* **4**, 712–722.

Eaton, M. L. (1970). A complete class theorem for multidimensional one-sided alternatives. *Ann. Math. Statist.* **41**, 1884–1888.

Ferguson, T. S. (1967). *Mathematical Statistics: A Decision-Theoretic Approach.* Academic Press, New York.

Lehmann, E. L. (1959). *Testing Statistical Hypotheses.* Wiley, New York.

Lehmann, E. L. and Scheffé, H. (1955). Completeness, similar regions, and unbiased estimation – Part II. *Sankhya* **15**, 219–236.

Matthes, T. K. and Truax, D. R. (1967). Tests of composite hypotheses for the multivariate exponential family. *Ann. Math. Statist.* **38**, 681–697.

Van Houwelingen, H. C. and Verbeek, A. (1985). On the construction of monotone symmetric decision rules for distributions with monotone likelihood ratio. *Scand. J. Statist.* **12**, 73–81.

ON ESTIMATING CHANGE POINT IN A FAILURE RATE

A. P. BASU

Department of Statistics
University of Missouri - Columbia
Columbia, Missouri 65211, U.S.A.

J. K. GHOSH

Stat./Math. Division
Indian Statistical Institute
Calcutta, India

S. N. JOSHI

Department of Statistics
Purdue University
West Lafayette, Indiana 47907, U.S.A.

1. INTRODUCTION

In reliability theory a widely accepted procedure is to apply "burn-in" techniques to screen out defective items and improve the lifetimes of remaining surviving units.

Formally, let $T_1, T_2, \ldots, T_n$ be a random sample from a lifetime distribution with d.f. $F(t)$ and density $f(t)$. The hazard rate $r(t)$ is defined as

$$r(t) = f(t)/\overline{F}(t) \text{ where } \overline{F}(t) = 1 - F(t).$$

We assume that $r(t)$ is a truncated "bath-tub" model i.e.

$$\begin{aligned} r(t) &= \lambda(t) \quad \text{if} \quad 0 \leq t \leq \tau \\ &= \lambda_0 \quad \text{if} \quad t > \tau \end{aligned} \tag{1.1}$$

where $\lambda(t)$ is nonincreasing and $\lambda(\tau) \geq \lambda_0$ with equality only if $\lambda(t)$ is strictly decreasing in $(\tau - \delta, \tau]$ for some $\delta > 0$. We wish to estimate the threshold τ. If one knew τ, items could be tested up to time τ and only survivors sold. This would be one way of screening. In our experience, screening in such situations is usually provided in a different way by subjecting items to a shock, thermal or electrical, and selling only survivors.

In Section 2 we propose two estimates for τ and prove their consistency.

It is of interest to study how our semiparametric estimates perform in specific parametric models. Nguyen et. al. [1984], hence forth abbreviated as NRW, have considered such a model, namely,

$$r(t) = a\ I(0 \le t \le \tau) + b\ I(t > \tau) \tag{1.2}$$

and proposed a consistent estimate for τ (when $a > b$, (1.2) is a special case of (1.1)). We have also introduced a restricted maximum likelihood estimate (m.l.e.) for purpose of comparison.

We carried out some simulations in Section 4 for model (1.2) (with $a > b$) for various values of the parameters a, b, and τ. When $F(\tau)$ is small i.e. when change takes place early in the lifetime our estimates of Section 2 perform well as compared to the NRW estimate or the m.l.e. .

In Section 3 we have obtained the asymptotic distribution of NRW estimate of τ. Our method for getting the asymptotic distribution would also apply to M-estimates with kernels that do not satisfy the usual conditions of differentiability or monotonicity but possess expectation having properties similar to those of the function $X(\cdot)$ of section 3. It should also be observed that the rescaling technique used in our method is due to Prakasa Rao ([1968], [1986]).

2. TWO NEW ESTIMATES OF τ

In our model it is reasonable to assume that

$$0 < F(\tau) < 1. \tag{2.1}$$

Moreover, an upper bound p_0 to $F(\tau)$ is assumed to be known, $p_0 < 1$; this would be a weak assumption in most practical situations.

Let $F_n(t)$ be the empirical d.f. of $T_1, T_2, \ldots, T_n$ and

$$y_n(t) = -\log \overline{F}_n(t),\ y(t) = -\log \overline{F}(t).$$

Let ξ_p and $\hat{\xi}_p$ denote p-th population and sample quantiles respectively.

Let p_1 be such that $p_0 < p_1 < 1$. Let k be the number of order statistics between $T_{([np_0])}$ and $T_{([np_1])}$ and let

$$\hat{\lambda}_0 = \frac{\sum T_{(i)} \log \overline{F}_n(T_{(i)})/(k+1) - (\sum T_{(i)}/(k+1))(\sum \log \overline{F}_n(T_{(i)})/(k+1))}{(\sum T_{(i)}^2/(k+1)) - (\sum T_{(i)}/(k+1))^2}$$

and the summations range over $i = [np_0] + 1$ to $i = [np_1]$.

Under (1.1) for $t > \tau$, $\log \overline{F}(t)$ is linear in t and $\hat{\lambda}_0$ is an ordinary least square estimate of the slope λ_0 treating $T_{(i)}$ as independent and $\log \overline{F}_n(T_{(i)})$ as dependent variables. It is well known (vide e.g. Serfling [1980] p. 59) that

$$\sqrt{n}\mathrm{Sup}_t |F_n(t) - F(t)| = Op(1), \tag{2.2}$$

and for $0 < p < 1$

$$\sqrt{n}(\hat{\xi}_p - \xi_p) = Op(1). \tag{2.3}$$

Using (2.2) it is easy to see that uniformly in $t \le \xi_p$, $p < 1$, we have

$$\sqrt{n}(y_n(t) - y(t)) = Op(1). \tag{2.4}$$

Now we claim that

$$\sqrt{n}(\hat{\lambda}_0 - \lambda_0) = O_p(1). \tag{2.5}$$

Note that $\hat{\lambda}_0$ can be expressed as a continuous function of quantities of the form

$$\frac{n}{(k+1)} \cdot \int_{T_{([np_0]+1)}}^{T_{([np_1])}} \phi(F_n(x), x) dF_n(x)$$

each of which can be handled using the following lemma and (2.5) can be proved.

Lemma. *Let* $T_{in} = \xi_i + O_p(n^{-1/2})$ $i = 1, 2$ *and let* $\phi(x, y)$ *be such that for some* $M > 0$ *and for some* $0 < \delta_1 < \delta_2 < 1$

(i) $|\phi(x_1, y) - \phi(x_2, y)| \le M|x_1 - x_2|$ *for all* $\xi_1 \le y \le \xi_2$ *and all* $\delta_1 \le x_1, x_2 \le \delta_2$,

(ii) $\phi(x, y)$ *is bounded in* $\xi_1 \le y \le \xi_2$, $\delta_1 \le x \le \delta_2$.

Then

$$\int_{T_{1n}}^{T_{2n}} \phi(F_n(x), x) dF_n(x) = \int_{\xi_1}^{\xi_2} \phi(F(x), x) dF(x) + O_p(n^{-1/2}).$$

The proof of the lemma is not hard and hence omitted.

Now we define our estimates of τ. Let $h_n = n^{-\frac{1}{4}}$ and $\varepsilon_n = c(\log n)n^{-\frac{1}{2}}$

$$\hat{\tau}_1 = \text{ Inf } \{t : y_n(t + h_n) - y_n(t) \le h_n \hat{\lambda}_0 + \varepsilon_n\}$$

and

$$\hat{\tau}_2 = \text{ Inf } \{t : \log \overline{F}_n(t) - \log(1 - p_0) \le \hat{\lambda}_0(\hat{\xi}_{p_0} - t) + \varepsilon_n\}.$$

To see the motivation for $\hat{\tau}_1$ note that $y_n(t + h_n) - y_n(t))/h_n$ is an estimate of the hazard rate $r(t)$ at t. For each fixed t, we test $H_{0t}:\ r(t) = \lambda_0$ vs $H_{1t}: r(t) > \lambda_0$, using the acceptance region $\{y_n(t + h_n) - y_n(t) \le h_n \hat{\lambda}_0 + \epsilon_n\}$. We then estimate τ as the smallest t for which H_{0t} is accepted. Formally $\hat{\tau}_1$ is as given above. Motivation of $\hat{\tau}_2$ is similar.

Theorem 1. *Let* (1.1) *and* (2.1) *hold. Then* $\hat{\tau}_1$ *and* $\hat{\tau}_2$ *are consistent for* τ.

PROOF. Note that for sufficiently small $\varepsilon > 0$ and for sufficiently large n, $\tau + h_n + \varepsilon < \xi_{p_0}$. Hence using (2.4)

$$\begin{aligned} y_n(\tau + \varepsilon + h_n) - y_n(\tau + \varepsilon) &= y(\tau + \varepsilon + h_n) - y(\tau + \varepsilon) + Op(n^{-\frac{1}{2}}) \\ &= \lambda_0 h_n + Op(n^{-\frac{1}{2}}) \qquad \text{by (1.1), and by (2.5)} \\ &= \hat{\lambda}_0 h_n + Op(n^{-\frac{1}{2}}). \end{aligned} \tag{2.6}$$

Now note that

$$(y_n(\tau + \varepsilon + h_n) - y_n(\tau + \varepsilon) \le \hat{\lambda}_0 h_n + \varepsilon_n) \Rightarrow (\hat{\tau}_1 \le \tau + \varepsilon) \tag{2.7}$$

Thus using (2.6) and (2.7) we have

$$P(\hat{\tau}_1 \leq \tau + \varepsilon) \to 1. \tag{2.8}$$

Now for sufficiently small $\varepsilon > 0$ we have $\tau - \varepsilon > 0$, hence using (2.4), we have

$$\begin{aligned} y_n(t+h_n) - y_n(t) &= \log \overline{F}(t) - \log \overline{F}(t+h_n) + Op(n^{-\frac{1}{2}}) \\ &\qquad \text{uniformly in } 0 \leq t \leq \tau - \varepsilon \\ &\geq h_n \lambda(x + h_n) + Op(n^{-\frac{1}{2}}) \\ &\qquad \text{uniformly in } 0 \leq t \leq \tau - \varepsilon \\ &> h_n \lambda_0 + h_n \delta_\varepsilon + Op(n^{-\frac{1}{2}}) \\ &\qquad \text{uniformly in } 0 \leq t \leq \tau - \varepsilon \\ &\qquad\quad \text{where } \delta_\varepsilon \text{ is such that } \lambda(\tau - \varepsilon/2) > \lambda_0 + \delta_\varepsilon \\ &> h_n \lambda_0 + h_n \delta_\varepsilon + O_p(n^{-1/2}) \\ &\qquad \text{uniformly in } 0 \leq t \leq \tau - \varepsilon \end{aligned} \tag{2.9}$$

Hence

$$P(\hat{\tau}_1 \geq \tau - \varepsilon) \to 1. \tag{2.10}$$

The relations (2.8) and (2.10) prove the consistency of $\hat{\tau}_1$.

Now for sufficiently small $\varepsilon > 0$, $\tau + \varepsilon < \xi_{p_0}$ hence using (2.4), we have

$$\begin{aligned} \log \overline{F}_n(\tau + \varepsilon) - \log(1 - p_0) &= \log \overline{F}(\tau + \varepsilon) - \log(1 - p_0) + Op(n^{-\frac{1}{2}}) \\ &= \hat{\lambda}_0(\hat{\xi} p_0 - \tau - \varepsilon) + Op(n^{-\frac{1}{2}}), \qquad \text{(by (2.5))}. \end{aligned}$$

Also,

$$(\log \overline{F}_n(\tau + \varepsilon) - \log(1 - p_0) \leq \hat{\lambda}_0(\hat{\xi}_{p_0} - \tau - \varepsilon) + \varepsilon_n) \Rightarrow (\hat{\tau}_2 \leq \tau + \varepsilon).$$

Hence

$$P(\hat{\tau}_2 \leq \tau + \varepsilon) \to 1. \tag{2.11}$$

Now, by (2.4), uniformly in $0 \leq x < \tau - \varepsilon$,

$$\begin{aligned} \log \overline{F}_n(x) - \log(1 - p_0) &= \log \overline{F}(x) - \log(1 - p_0) + Op(n^{-\frac{1}{2}}) \\ &= \int_x^{\xi_{p_0}} \lambda(t)\,dt + Op(n^{-\frac{1}{2}}) \\ &\geq (\hat{\xi}_{p_0} - x)\hat{\lambda}_0 + \delta + Op(n^{-\frac{1}{2}}) \qquad \text{for some } \delta > 0. \end{aligned}$$

Hence

$$P(\hat{\tau}_2 \geq \tau - \varepsilon) \to 1. \tag{2.12}$$

Consistency of $\hat{\tau}_2$ follows from (2.11) and (2.12). □

3. A PARAMETRIC EXAMPLE AND SOME PARAMETRIC ESTIMATES

The density specified by (1.2) has the form

$$f(t) = a\exp(-at)I(0 \le t \le \tau) + b\exp(-a\tau - b(t-\tau))I(t > \tau) \tag{3.1}$$

which is p.d.f. for (a, b, τ), $0 < a, 0 < b, 0 < \tau$. (If $a > b$, this is a special case of (1.1) but we will not assume this now.) If $a = b$ or $\tau = 0$ we have identifiability problems.

Note that τ is a change point not in the usual sense (e.g. Hinkley [1970]) where one has a sequence of parameters θ_t which change from one value θ_0 for $t < \tau$ to another value θ_1 for $t \ge \tau$.

Note that the density in (3.1) can be written as a mixture of a right truncated exponential and an untruncated exponential with the mixing proportion depending on the parameters a and τ.

Let $t_{(1)} \le t_{(2)} \le \ldots \le t_{(n)}$ be the ordered sample. If one sets any arbitrary value to the parameter "a" and sets $\hat{b} = 1/(t_{(n)} - \hat{\tau})$ where $t_{(n-1)} < \hat{\tau} < t_{(n)}$, then the likelihood at $\hat{a}$, $\hat{b}$, $\hat{\tau}$ may be made as large as we please making $\hat{\tau}$ as close to $t_{(n)}$ as needed. One may therefore say that in a sense a maximum likelihood estimate of τ is $\hat{\tau} = t_{(n)}$. Obviously $t_{(n)}$ is not consistent.

If one chooses a compact set of (a, b, τ)'s as the parameter space and imposes identifiability conditions like $\tau \ge \delta_1 > 0$, $|a - b| \ge \delta_2 > 0$ then Wald's general result (vide Wald [1949]) implies that m.l.e. $\hat{\tau}$ is consistent. Since τ is a point of discontinuity of the density, the general theory of Chernoff and Rubin [1956] ensures that $|\hat{\tau} - \tau|$ is in fact $Op(n^{-1})$ (better than usual $Op(n^{-\frac{1}{2}})$).

Either not being aware of the above result or because they do not want to impose any conditions on the parameter space, NRW proposed a new estimate $\hat{\tau}_3$ of τ. Using the n observations $t_1, \ldots, t_n$ they construct a kernel $X_n(t)$ such that the solution of $X_n(t) = 0$ provides a consistent estimate of τ. The construction of $X_n(t)$ is ingenious but apart from providing a consistent estimate the kernel does not seem to have any attractive properties. Existence of a consistent solution is a consequence of the fact that

$$X_n(t) \xrightarrow{\text{a.s.}} X(t)$$

where $X(t)$ is a non-stochastic function, $X(\tau) = 0$ and $X(t)$ is monotone in a neighborhood of τ. They show that $X_n(t) = 0$ has a "consistent" solution $\hat{\tau}_3$ which is their estimate. Since $X_n(t)$ is neither monotone nor sufficiently smooth (e.g., not differentiable at τ), it is hard to get the asymptotic distribution of $\sqrt{n}(\hat{\tau}_3 - \tau)$. In fact NRW fail to find it.

Now we derive the limiting distribution of $\hat{\tau}_3$. Let T be a r.v. having density $f(t)$ of (3.1).

Let

$$\begin{aligned} B_1(T,t) &= T, \\ B_2(T,t) &= I(T > t), \\ B_3(T,t) &= TI(T > t), \\ B_4(T,t) &= T^2 I(T > t). \end{aligned}$$

and

$$\underset{\sim}{B} = (B_1, B_2, B_3, B_4).$$

Let $T_1, \ldots, T_n$ be i.i.d. with density $f(t)$ and

$$\overline{B}_j(t) = n^{-1} \sum_{i=1}^{n} B_j(T_i, t) \quad j = 1, \ldots, 4,$$

$$\begin{aligned} H(\overline{\underset{\sim}{B}}(t)) = {} & (\overline{B}_4(t)/\overline{B}_2(t) - \overline{B}_3^2(t)/\overline{B}_2^2(t))^{\frac{1}{2}} (\overline{B}_2(t) - \overline{B}_2(t) \log \overline{B}_2(t) - 1) \\ & + (1 - \overline{B}_2(t))\overline{B}_3(t)/\overline{B}_2(t) + \overline{B}_1(t) \log \overline{B}_2(t). \end{aligned}$$

Let

$$\begin{aligned} X_n(0) &= 0, \\ X_n(T_{(i)}) &= H(\overline{\underset{\sim}{B}}(T_{(i)})), \quad i = 1, \ldots, n-1 \\ X_n(t) &= X_n(T_{(n-1)}), \qquad \text{for } t \geq T_{(n-1)} \end{aligned}$$

and $X_n(\cdot)$ at other points be defined by linear interpolation.

Let $\hat{\tau}_3$ be defined formally as follows. Choose a $\sqrt{n}$-consistent estimate $\hat{\tau}_c$ of τ and let $\hat{\tau}_3$ = zero of $X_n(t)$ nearest to $\hat{\tau}_c$. Fix a neighborhood $[\tau_1, \tau_2]$ of τ and consider

$$Y_n(t) = \sqrt{n}(X_n(t) - X(t)), \ \tau_1 \leq t \leq \tau_2$$

where

$$X(t) = H(\underset{\sim}{\mu}(t)), \ \underset{\sim}{\mu}(t) = E(\underset{\sim}{B}(\cdot, t)).$$

It can be checked from the appendix that

(i) $X(\tau) = 0$,

(ii) the derivatives $\dot{X}(t)$ and $\ddot{X}(t)$ exist for $t > \tau$ and $t < \tau$ and are continuous, (3.2)

(iii) $\dot{X}(\tau+)$ and $\dot{X}(\tau-)$ exist and are of the same sign (vide Appendix).

It can be checked using estimates like in (13.6) of Billingsley [1968, p. 104] and bounds on the derivatives of H, that

$$Y_n(t) - \sqrt{n}(H(\overline{\underset{\sim}{B}}(t)) - X(t)) = o_p(1).$$

From this one checks via the delta method that the finite dimensional distributions of $Y_n(\cdot)$ converge to a multivariate normal distribution. Tightness is proved by checking a condition analogous to (13.17) of Billingsley [1968, p.

106]. From these considerations it follows that

$$Y_n(\cdot)\xrightarrow{W}Y(\cdot)$$

where $Y(\cdot)$ is a zero mean Gaussian process.

Consider now a rescaled process in $C(-\infty,\infty)$ (with the topology of uniform convergence on compacts. For tightness in $C(-\infty,\infty)$, see Sen [1981]). Let

$$\begin{aligned} Z_n(h) &= Y_n(\tau+n^{-\frac{1}{2}}h) && \text{if } |h|\le \log n \\ &= Y_n(\tau+n^{-\frac{1}{2}}\log n) && \text{if } h\ge \log n \\ &= Y_n(\tau-n^{-\frac{1}{2}}\log n) && \text{if } h\le -\log n. \end{aligned}$$

Then

$$Z_n(\cdot)\xrightarrow{W}Z(\cdot),$$

where

$$Z(\cdot)=Y(\tau). \tag{3.3}$$

Now let

$$\begin{aligned} W_n(h) &= \sqrt{n}\ X_n(\tau+hn^{-\frac{1}{2}}) && \text{if } |h|\le \log n \\ &= \sqrt{n}\ X_n(\tau+n^{-\frac{1}{2}}\log n) && \text{if } h> \log n \\ &= \sqrt{n}\ X_n(\tau-n^{-\frac{1}{2}}\log n) && \text{if } h< -\log n. \end{aligned} \tag{3.4}$$

Then using (3.2), (3.3), and (3.4) it is easy to see that

$$W_n(\cdot)\xrightarrow{W}A(\cdot),$$

where $A(\cdot)$ is a Gaussian process on $C(-\infty,\infty)$ with the representation

$$\begin{aligned} A(h) &= Y(\tau)+h\dot{X}(\tau+) && \text{if } h>0 \\ &= Y(\tau)-h\dot{X}(\tau-), && \text{if } h<0 \end{aligned} \tag{3.5}$$

and

$$A(0)=Y(\tau).$$

Let, for $f\in C(-\infty,\infty)$,

$$A_1(f)=\text{Sup}\{t: f(t)=0\},$$
$$A_2(f)=\text{Inf}\{t: f(t)=0\},$$

if f has at least one zero and $A_1(f)$ and $A_2(f)$ equal to a constant, say, c otherwise. Note that A_1 and A_2 are measurable and continuous on a set which contains $A(\cdot)$ with probability one, hence

$$(A_1(W_n)-A_2(W_n))\xrightarrow{W}(A_1(A)-A_2(A)) \tag{3.6}$$

and $A_i(W_n)\xrightarrow{W}A_i(A)$ for $i=1,2$. Using (3.2) and (3.5) it is easy to see that

$(A_1(A) - A_2(A))$ is degenerate at zero, hence we have $w.p. \to 1$

$$A_1(W_n) \geq \sqrt{n}(\hat{\tau}_3 - \tau) \geq A_2(W_n). \tag{3.7}$$

Thus $\sqrt{n}(\hat{\tau}_3 - \tau)$ has split normal distribution of $A_1(A)$ (or of $A_2(A)$). It is easy to see that $w.p.1$, and $i = 1, 2$

$$\begin{aligned} A_i(A) &= \frac{-Y(\tau)}{\dot{X}(\tau+)} && \text{if} \quad Y(\tau) \leq 0 \text{ and } \dot{X}(\tau+) > 0 \\ & && \text{or if} \quad Y(\tau) \geq 0 \text{ and } \dot{X}(\tau+) < 0 \\ &= \frac{-Y(\tau)}{\dot{X}(\tau-)} && \text{if } Y(\tau) \geq 0 \text{ and } \dot{X}(\tau-) > 0 \\ & && \text{or if } Y(\tau) \leq 0 \text{ and } \dot{X}(\tau-) < 0. \end{aligned} \tag{3.8}$$

Hence d.f. of $\sqrt{n}(\hat{\tau}_3 - \tau)$ converges weakly to $G(t)$ where for $t > 0$

$$G(-t) = \Phi(-t|\dot{X}(\tau-)| \;/\; V(\tau)),$$

$$1 - G(t) = \Phi(-t|\dot{X}(\tau+)| \;/\; V(\tau)),$$

and

$$\begin{aligned} V^2(\tau) &= \mathrm{Var}(Y(\tau)) \\ &= \sum_{1 \leq i,\ j \leq 4} \sum \vartheta ij(\tau) \frac{\partial H}{\partial B_i}|_{\underset{\sim}{\mu}(\tau)} \frac{\partial H}{\partial B_i}|_{\underset{\sim}{\mu}(\tau)}. \end{aligned} \tag{3.9}$$

It can be shown that $V^2(\tau) > 0$. See appendix for further details. This completes the derivation of limiting distribution of $\hat{\tau}_3$.

For simulations, we also have compared $\hat{\tau}_4$ which is the m.l.e. with parameter space $\tau \leq \delta_1$, $|a - b| \leq \delta_2 > 0$.

Estimates of a and b and their limiting distributions: For each $\tau > 0$, formal differentiation of the likelihood function w.r.t. a and b yields $\hat{a}(\tau)$ and $\hat{b}(\tau)$, one can plug in an estimate of τ say $\hat{\tau}_3$ and get $\hat{a}$ and $\hat{b}$ the estimates of a and b respectively, it can be seen (vide NRW) that

$$\begin{aligned} \hat{a} &= (1 - \overline{B}_2(\hat{\tau}_3))/(\overline{B}_1(\hat{\tau}_3) - \overline{B}_3(\hat{\tau}_3) + \hat{\tau}_3\overline{B}_2(\hat{\tau}_3)) \\ &= H_1(\overline{\underset{\sim}{B}}, \hat{\tau}_3), \end{aligned}$$

say, and

$$\hat{b} = \overline{B}_2(\hat{\tau}_3)/(\overline{B}_3(\hat{\tau}_3) - \hat{\tau}_3\overline{B}_2(\hat{\tau}_3)).$$

The following is the sketch of the derivation of the limiting distribution of $\hat{a}$. Limiting distribution of $\hat{b}$ can be handled in a similar manner.

Using δ-method we have

$$n^{\frac{1}{2}}(\hat{a} - a) = n^{\frac{1}{2}}(H_1(\overline{\underset{\sim}{B}}(\hat{\tau}_3), \hat{\tau}_3) - H_1(\underset{\sim}{\mu}(\tau), \tau))$$

$$
\begin{aligned}
&= n^{\frac{1}{2}} \sum_{i=1}^{3} (\overline{B}_i(\hat{\tau}_3) - \mu_i(\tau)) \frac{\partial H_1}{\partial \mu_i(\tau)} + n^{\frac{1}{2}} (\hat{\tau}_3 - \tau) \frac{\partial H_1}{\partial \tau} + o_p(1) \\
&= n^{\frac{1}{2}} \sum_{i=1}^{3} (\overline{B}_i(\hat{\tau}_3) - \mu_i(\hat{\tau}_3)) \frac{\partial H_1}{\partial \mu_i(\tau)} \\
&\quad + n^{\frac{1}{2}} (\hat{\tau}_3 - \tau) \left[\frac{\partial H_1}{\partial \tau} + \sum_{i=1}^{3} \frac{\partial H_1}{\partial \mu_i(\tau)} \frac{\partial \mu_i(\tau)}{\partial \tau} \right] + o_p(1) \\
&= W_{3n}(\tau) + \mathcal{Q}_{1n}(h_n) - \mathcal{Q}_{1n}(0) + c(\tau) h_n + o_p(1),
\end{aligned} \tag{3.10}
$$

where

$$
\begin{aligned}
h_n &= n^{\frac{1}{2}} (\hat{\tau}_3 - \tau), \\
W_{3n}(t) &= n^{\frac{1}{2}} \sum_{i=1}^{3} (\overline{B}_i(t) - \mu_i(t)) \frac{\partial H_1}{\partial \mu_i(t)}, \\
\mathcal{Q}_{1n}(h) &= n^{\frac{1}{2}} \sum_{i=1}^{3} (\overline{B}_i(\tau + n^{-\frac{1}{2}} h) - \mu_i(\tau + n^{-\frac{1}{2}} h)) \frac{\partial H_1}{\partial \mu_i(\tau)}
\end{aligned}
$$

and

$$
c(\tau) = \frac{\partial H_1}{\partial \tau} + \sum_{i=1}^{3} \frac{\partial H_1}{\partial \mu_i(\tau)} \frac{\partial \mu_i(\tau)}{\partial \tau}.
$$

From (3.6), (3.7) and the remark following (3.6), it has been proved earlier that

$$
\sqrt{n}(\hat{\tau}_3 - \tau) - A_1(W_n(\cdot)) \xrightarrow{\text{W}} \delta,
$$

where δ is the measure degenerate at zero. This implies

$$
\sqrt{n}(\hat{\tau}_3 - \tau) - A_1(W_n(\cdot)) \xrightarrow{\text{P}} 0.
$$

It follows that

$$
c(\tau) h_n = A_1(W_n(\cdot)) c(\tau) + op(1),
$$

and using Theorem 8.2 of Billingsly [1968, p. 55] we have

$$
\operatorname{Sup}_{|h| < \log n} |\mathcal{Q}_{1n}(h) - \mathcal{Q}_{1n}(0)| \xrightarrow{\text{P}} 0. \tag{3.11}
$$

Thus

$$
n^{\frac{1}{2}} (\hat{a} - a) = W_{3n}(\tau) + c(\tau) A_1(W_n(\cdot)) + op(1) \tag{3.12}
$$

It is easy to check by the delta method that

$$
(W_{3n}(\tau), \sqrt{n}(H(\underset{\sim}{\overline{B}}(\tau)))) \xrightarrow{\text{w}} (X, Y)
$$

where (X, Y) has a bivariate normal distribution with mean zero and easily computable dispersion matrix. We may take Y to be equal to $Y(\tau) = A(0)$ without loss of generality; define $A(\cdot)$ as before. We first note that $(W_{3n}(\tau), W_n(\cdot))$ is a sequence of random variables taking values in $R \times C(-\infty, \infty)$, equipped with the product topology. Tightness follows from the tightness of the marginal distribution of $W_{3n}(\tau)$ and $W_n(\cdot)$. It is also easy to see that the joint distribution of $W_{3n}(\tau)$ and $W_n(T_1), \ldots, W_n(t_k)$ converge to that of $(X, A(t_1), \ldots, A(t_k))$. Since these finite dimensional distributions determine the distribution on $R \times C(-\infty, \infty)$, it follows from Prohorov's theorem that

$$(W_{3n}(\tau), W_n(\cdot)) \xrightarrow{\text{w}} (X, A(\cdot)). \tag{3.13}$$

It now follows from (3.8), (3.12) and (3.13) that $\sqrt{n}(\hat{a} - a)$ converges in distribution to $X + c(\tau)A_1(A(\cdot))$.

The limiting distribution can be calculated from bivariate normal tables. For example if $\dot{X}(\tau+) > 0 > \dot{X}(\tau-)$, then for any real "$d$",

$$\begin{aligned}\lim P\{\sqrt{n}(\hat{a} - a) \le d\} &= P\{X - c(\tau)Y(\tau)/\dot{X}(\tau+) \le d,\ Y(\tau) \ge 0\} \\ &+ P\{X - c(\tau)Y(\tau)/\dot{X}(\tau-) \le d,\ Y(\tau) < 0\}.\end{aligned}$$

In case $\hat{\tau}_3$ is replaced by $\hat{\tau}_4$ in the estimate for $\hat{a}$, then (3.10) and (3.11) continue to hold and $h_n \xrightarrow{\text{P}} 0$ since $(\hat{\tau}_4 - \tau) = 0_p(1/n)$. It follows that $\sqrt{n}\ (\hat{a} - a)$ has the same limiting normal distribution as $W_{3n}(\tau)$. This last fact has been noted by Nguyen and Pham [1987].

4. SIMULATION RESULTS

We obtained 100 samples each of size 100 and carried out simulations with $p_0 = .50$, $p_1 = .90$. We used the following "smoother" version of $\hat{\tau}_2$ for simulations (the summations below range over $i = [np_0] + 1$ to $i = [np_1]$)

$$\begin{aligned}\hat{\tau}_2 = \inf\{t\colon\ & \log \overline{F}_n(t) - \sum \log \overline{F}_n(T_{(i)})/(k+1) \\ & \le \hat{\lambda}_0\left(\sum T_{(i)}/(k+1) - t\right) + \varepsilon_n\} \\ & \text{if the infimum is less than or equal to } \hat{\xi}_{p_0} \\ = \hat{\xi}_{p_0} \quad & \text{otherwise.}\end{aligned}$$

For $\hat{\tau}_1$: $\varepsilon_n = .05$, $h_n = n^{-\frac{1}{4}}$.

For $\hat{\tau}_2$: $\varepsilon_n = .05$.

$\hat{\tau}_3$ is the solution of $X_n(\cdot)$ nearest to zero.

For $\hat{\tau}_4$: $\delta_1 = 3$, $\delta_2 = .01$.

The values of p_0, p_1, ε_n, h_n, δ_1 and δ_2 are chosen somewhat arbitrarily. The mi's and the Ri's are respectively means and mean square errors; R_{3a} is the mean square error using the limiting distribution of $\hat{\tau}_3$.

We like to make the following remarks on the simulations.

(1) Of the three estimates $\hat{\tau}_1$, $\hat{\tau}_2$, $\hat{\tau}_3$, $\hat{\tau}_1$ seems to do best in all the cases. (This

is a little surprising for we thought $\hat{\tau}_2$ should do better than $\hat{\tau}_1$).

(2) The (restricted) m.l.e. $\hat{\tau}_4$ is best when $a = 2$, $b = .5$ or $a = 3, b = 1$, i.e. when the amount of discontinuity in the density at τ is maximum in the cases simulated. A reason for this may be that the Chernoff-Rubin asymptotics (see Chernoff and Rubin [1956]) leading $0_p(1/n)$ errors for $\hat{\tau}_4$ is valid only when the extent of discontinuity is relatively large.

(3) $\sqrt{n}\ (\hat{\tau}_3 - \tau)$ has an asymptotic distribution with mean zero. So its asymptotic variance may be compared either with simulated variance ($= R3 - (m3)^2$) or with the mean square error ($= R3$). The asymptotic value provides good approximation to the simulated variance but not to the simulated mean square error, because the bias isn't negligible in all cases.

			$\hat{\tau}_1$	$\hat{\tau}_2$	$\hat{\tau}_3$	$\hat{\tau}_4$	$\hat{\tau}_3$
a	b	τ	m1(R1)	m2(R2)	m3(R3)	m4(R4)	R_{3a}
3	2	.15	.1061(.0115)	.1175(.0122)	.1743(.0131)	.9799(1.6256)	.0271
3	2	.10	.0792(.0094)	.1005(.0131)	.1890(.0255)	.9888(1.8319)	.0229
3	1	.15	.1371(.0076)	.1920(.0227)	.1957(.0194)	.1533(.0023)	.0199
3	1	.10	.0955(.0093)	.1705(.0401)	.1929(.0459)	.0994(.0019)	.0171
2	1	.20	.1459(.0159)	.2139(.0352)	.2728(.0441)	.3528(.3273)	.0385
2	1	.15	.1102(.0122)	.1923(.0447)	.2758(.0697)	.2726(.2075)	.0346
2	1	.10	.0883(.0136)	.1799(.0585)	.3322(.1323)	.4270(.6662)	.0310
2	.5	.20	.1846(.0109)	.3404(.1129)	.3092(.0793)	.1984(.0017)	.0568
2	.5	.15	.1246(.0084)	.3215(.1625)	.3289(.1649)	.1515(.0018)	.0513
2	.5	.10	.0839(.0059)	.3364(.2383)	.4036(.3270)	.1090(.0024)	.0463
1	.5	.40	.2548(.0454)	.4279(.1409)	.5456(.1766)	.4802(.2559)	.1538
1	.5	.30	.1736(.0285)	.3846(.1789)	.5517(.2788)	.4010(.2448)	.1382
1	.5	.20	.1120(.0201)	.3599(.2339)	.6643(.5294)	.4218(.4608)	.1241
1	.5	.15	.0826(.0172)	.3367(.2546)	.7568(.7140)	.4111(.4989)	.1176
1	.5	.10	.0613(.0125)	.3465(.3034)	.9064(.9721)	.5203(.7823)	.1114

APPENDIX

Here we prove some results mentioned in Section 3.

$$\mu_1(\tau) = a^{-1} + e^{-a\tau}(b^{-1} - a^{-1}),$$
$$\mu_2(\tau) = e^{-a\tau},$$
$$\mu_3(\tau) = b^{-1}e^{-a\tau}(1 + b\tau),$$
$$\mu_4(\tau) = b^{-2}e^{-a\tau}(b^2\tau^2 + 2b\tau + 2).$$

Let

$$\mu_{(3)}(\tau) = b^{-3}e^{-a\tau}(b^3\tau^3 + 3b^2\tau^2 + 6b\tau + 6),$$
$$\mu_{(4)}(\tau) = b^{-4}e^{-a\tau}(b^4\tau^4 + 4b^3\tau^3 + 12b^2\tau^2 + 24b\tau + 24).$$

Then

$$\vartheta_{11}(\tau) = 2e^{-a\tau}(b^{-1} - a^{-1}) + 2e^{-a\tau}(b^{-2} - a^{-2}) + 2a^{-2} - \mu_1^2(\tau),$$
$$\vartheta_{22}(\tau) = \mu_2(\tau) - \mu_2^2(\tau),$$
$$\vartheta_{33}(\tau) = \mu_4(\tau) - \mu_3^2(\tau),$$
$$\vartheta_{44}(\tau) = \mu_{(4)}(\tau) - \mu_4^2(\tau),$$
$$\vartheta_{12}(\tau) = \mu_3(\tau) - \mu_1(\tau)\mu_2(\tau),$$
$$\vartheta_{13}(\tau) = \mu_4(\tau) - \mu_1(\tau)\mu_3(\tau),$$
$$\vartheta_{14}(\tau) = \mu_{(3)}(\tau) - \mu_1(\tau)\mu_4(\tau),$$
$$\vartheta_{23}(\tau) = \mu_3(\tau) - \mu_2(\tau)\mu_3(\tau),$$
$$\vartheta_{24}(\tau) = \mu_4(\tau) - \mu_2(\tau)\mu_4(\tau),$$
$$\vartheta_{34}(\tau) = \mu_{(3)}(\tau) - \mu_3(\tau)\mu_4(\tau),$$

$$\frac{\partial H}{\partial B_1}|_{\mu(\tau)} = -a\tau,$$
$$\frac{\partial H}{\partial B_2}|_{\mu(\tau)} = e^{a\tau}(a^{-1} - b^{-1} - 2\tau - b\tau^2/2) + \tau(1 + ab^{-1})$$
$$+ \tau^2(a + b/2) + ab\tau^3/2 + b^{-1} - a^{-1},$$
$$\frac{\partial H}{\partial B_3}|_{\mu(\tau)} = 2e^{a\tau} - 2 - a\tau - b\tau - ab\tau^2 + b\tau e^{a\tau},$$
$$\frac{\partial H}{\partial B_4}|_{\mu(\tau)} = (b - be^{a\tau} + ab\tau)/2.$$

Let

$$X(t) = K_1(t) \quad \text{for } t \le \tau$$
$$= K_2(t) \quad \text{for } t > \tau,$$

then (vide NRW)

$$\begin{aligned} K_1(t) &= S(t)(\text{ at } \exp(-at) - 1 + \exp(-at)) \\ &\quad + (1 - \exp(-at))(t + a^{-1} + (b^{-1} - a^{-1})\exp(-a\tau + at)) \\ &\quad - (a^{-1} - a^{-1}\exp(-a\tau) + b^{-1}\exp(-a\tau))at, \\ S^2(t) &= a^{-2} + (b^{-1} - a^{-1})[2(\tau - t + b^{-1}) \\ &\quad - (b^{-1} - a^{-1})\exp(-a\tau + at)] \times \exp(-a\tau + at), \\ K_2(t) &= t + (a - b)\tau b^{-1}\exp(-a\tau - b(t - \tau)) \\ &\quad - (a\tau + b(t - \tau))(a^{-1} - a^{-1}\exp(-a\tau) + b^{-1}\exp(-a\tau)). \end{aligned}$$

Note that

$$\begin{aligned} \dot{S}(t) &= S^{-1}(t)a(b^{-1} - a^{-1})[\tau - t + b^{-1} - a^{-1} \\ &\quad - (b^{-1} - a^{-1})\exp(-a\tau + at)]\exp(-a\tau + at), \\ \dot{K}_1(t) &= \dot{S}(t)(at\exp(-at) - 1 + \exp(-at)) \\ &\quad - a^2 tS(t)\exp(-at) + at\exp(-at) \\ &\quad + a(b^{-1} - a^{-1})\exp(-a\tau + at), \\ \dot{K}_2(t) &= 1 - (a - b)\tau\exp(-a\tau - b(t - \tau)) \\ &\quad - b(a^{-1} - (b^{-1} - a^{-1})\exp(-a\tau)). \end{aligned}$$

Note that $\dot{K}_1(t)$ and $\dot{K}_2(t)$ are continuously differentiable in a neighborhood of τ hence $\dot{X}(t)$ and $\ddot{X}(t)$ exist for $\tau_1 \leq t < \tau$ and $\tau < t \leq \tau_2$ and are continuous. It is easy to see that

$$\begin{aligned} \dot{X}(\tau+) &= \dot{K}_2(\tau) = b^{-1}[a - b + (ab\tau - a^2\tau + b - a)\exp(-a\tau)] \\ \dot{X}(\tau-) &= \dot{K}_1(\tau) = a^{-1}b\dot{K}_2(\tau) \end{aligned}$$

(Thus $\dot{X}(\tau+)\dot{X}(\tau-) = a^{-1}b\dot{K}_2^2(\tau) > 0$ for $a \neq b$, $a > 0$, $b > 0$, $\tau > 0$). Asymptotic mean square error of $\hat{\tau}$ is

$$\frac{1}{2n}(\dot{X}^{-2}(\tau+) + \dot{X}^{-2}(\tau-))V^2(\tau).$$

Now note that $((\vartheta_{ij}(\tau)))$ the dispersion matrix of $\underset{\sim}{B}(\tau)$ is positive definite; if not then there exist, say, λ_1, λ_2 and λ_3 (not all zero) such that

$$\lambda_1 + \lambda_2 T + \lambda_3 T^2 = \text{ constant a.e. on } T > \tau$$

and

$$T = \text{ constant a.e. on } T < \tau$$

which is not possible. Also $\frac{\partial H}{\partial B_1}/_{\mu(\tau)} \neq 0$. Thus $V^2(\tau) > 0$. Now by continuity of $V^2(t)$ at τ, $V^2(t) > 0$ in a neighborhood of τ.

ACKNOWLEDGMENTS

The research of Asit P. Basu was partly carried out at the Indian Statistical Institute, Calcutta. This research was supported in part by a grant from the Graduate Research Council of the Graduate School, University of Missouri - Columbia, in part by the National Science Foundation under grant No. INT-8615383, and in part by UMC multipurpose Arthritis Center grant.

Research sponsored in part by the Air Force Office of Scientific Research, Air Force Systems Command, USAF, under grant number AFOSR-87-0139. The U.S. Government is authorized to reproduce and distribute reprints for Governmental purposes notwithstanding any copyright notation thereon.

BIBLIOGRAPHY

Billingsley, P. (1968) *Convergence of Probability Measures*, Wiley, New York.

Chernoff, H. and Rubin H. (1956) "The estimation of the location of a discontinuity in density", *Proceedings of the Third Berkeley Symposium on Mathematical Statistics and Probability* **1**, pp. 19–37, University of California Press.

Hinkley, D. V. (1970) "Inference about change-point in a sequence of random variables", *Biometrika* **57**, pp. 1–17.

Nguyen, H. T. and Pham, T. D. (1987) "Maximum-Likelihood estimation in the change-point hazard-rate model" [Preliminary Report], *The Institute of Mathematical Statistics Bulletin* **16**, No. 1 pp.36.

Nguyen, H. T., Rogers, G. S. and Walker, E. A. (1984) "Estimation in change-point hazard rate models", *Biometrika* **71**, pp. 299–304.

Prakasa Rao, B. L. S. (1968) "Estimation of the location of the cusp of a continuous density", *Ann. Math. Statist.* **39**, pp. 76–87.

Prakasa Rao, B. L. S. (1986) "Asymptotic theory of estimation in non-linear regression", to appear in *Proceedings of the Conference in honor of M. Joshi.*

Rubin, H. (1961) "The estimation of discontinuities in multivariate densities, and related problems in stochastic processes", *Proceedings of the Fourth Berkeley Symposium on Mathematical Statistics and Probability* **1**, University of California Press, pp. 563–574.

Sen, P. K. (1981) *Sequential Nonparametrics*, Wiley, New York.

Serfling, R. J. (1980) *Approximation Theorems of Mathematical Statistics*, Wiley, New York.

A NONPARAMETRIC, INTERSECTION-UNION TEST FOR STOCHASTIC ORDER

ROGER L. BERGER

Department of Statistics
North Carolina State University
Raleigh, North Carolina 27695-8203, U.S.A.

1. INTRODUCTION

A distribution function (cdf) $G(x)$ is said to be stochastically larger than another cdf $F(x)$ if $G(x) \leq F(x)$ for all x with strict inequality for some x. This relationship will be denoted by $G >^{st} F$. Stochastic ordering defines a situation in which, in a very strong sense, observations from a population with cdf G tend to be larger than observations from a population with cdf F.

To test if $G >^{st} F$, null and alternative hypotheses commonly considered are H_0': $G \equiv F$ and H_a': $G >^{st} F$. The Mann-Whitney-Wilcoxon rank sum test (RST) is often recommended as a nonparametric test for testing H_0' versus H_a'. See for example Lehmann (1975). Lee and Wolfe (1976) study a competitor to the RST, based on a restricted maximum likelihood estimate, for testing H_0' versus H_a'.

In this paper we consider the one sample problem in which F is a known standard and a random sample is available from G. The RST and Lee and Wolfe test are usually presented as two sample tests but they are applicable in the one sample problem. Both can be expressed in terms of a statistic of the form $\int_{-\infty}^{\infty} \hat{F}(x)d\hat{G}(x)$ where $\hat{F}$ and $\hat{G}$ are estimates of F and G respectively. For a one sample problem, $\hat{F}$ can be replaced by F. Moses (1964) has discussed this one sample application of the RST. Remarks similar to the following also apply to the one-sided one sample Kolmogorov-Smirnov test, another test that might be used to test H_0' versus H_a'. Now, consider the sense in which rejection by these tests confirms H_a'.

Using the asymptotic distribution of the RST statistic, it can be shown that the power of the RST converges to one for any pair (F, G) for which $\theta = \int_{-\infty}^{\infty} F(x)dG(x) > \frac{1}{2}$. The same is true of the Lee and Wolfe (1976) test. Here $\theta = P(X \leq Y)$ if X and Y are independent random variables with cdfs F and G respectively. If $G >^{st} F$ then $\theta > \frac{1}{2}$. But there are many other pairs (F, G) for which G is not stochastically larger than F and still $\theta > \frac{1}{2}$. If the model includes such pairs, then rejection of the null hypothesis by the RST or Lee and Wolfe test cannot be interpreted as convincing evidence that $G >^{st} F$ but only that $\theta > \frac{1}{2}$. Indeed, the power of the RST converges to 0 and α (the size of the test) according as $\theta < \frac{1}{2}$ or $\theta = \frac{1}{2}$. Thus the RST might be more properly thought of as a test of H_0'': $\theta \leq \frac{1}{2}$ versus H_a'': $\theta > \frac{1}{2}$ rather than as a test of H_0' versus H_a'.

To develop a nonparametric test for which "rejection of the null hypothesis" could be interpreted as convincing evidence that $G >^{st} F$, one might try to derive a size α test of H_0^*: G is not stochastically larger than F versus H_a^*: $G >^{st} F$. Unfortunately, as we shall show in Section 4, this is not practical. The uniformly most powerful level α test of H_0^* versus H_a^* is the "no data" test which rejects H_0^* with probability α, regardless of the data. The reason for this is that for any pair (F, G) satisfying H_a^*, there is a sequence of pairs (F, G_n), $n = 1, 2, \ldots$, satisfying H_0^* which converge to (F, G) in such a way that the power of any test at (F, G) is the limit of the powers at (F, G_n). Since (F, G_n) satisfies H_0^*, this limit must be at most α. The proof of this fact suggests that the difficulty may be that $G(x)$ may be less than $F(x)$ for most x's but $G(x)$ may be slightly greater than $F(x)$ for some x in the tail of the distributions. Thus, although $P(Y > x) = 1 - G(x) > 1 - F(x) = P(X > x)$ for most x's, G is not stochastically larger than F. In this paper we consider null and alternative hypotheses which are similar to H_0^* and H_a^* but which circumvent this troublesome tail behavior and allow the development of a reasonable level α test.

Specifically, in this paper we consider testing

$$\begin{aligned} H_0: \ & F(x) = G(x) \text{ for all } x \in [a, b] \\ & \text{or } F(x) < G(x) \text{ for at least one } x \in [a, b] \end{aligned}$$

versus

$$H_a: \ F(x) \geq G(x) \text{ for all } x \in [a, b]$$

with strict inequality for at least one $x \in [a, b]$. Here $-\infty < a \leq b < \infty$ are fixed constants which define the interval of x values on which $F(x)$ and $G(x)$ are to be compared. The constants a and b must satisfy certain constraints, specified in Section 2, but the constraints relax as the sample size becomes large. The relationship described in H_a is the same as that which defines stochastic ordering except that the inequality is required to hold only for $x \in [a, b]$ rather than for all x. If the interval $[a, b]$ is reasonably wide, H_0 and H_a provide practical approximations to H_0^* and H_a^*. Rejection of H_0 by a level α test (α small) will provide convincing evidence that $F(x) \geq G(x)$ for all $x \in [a, b]$. In many situations H_0 and H_a provide more meaningful and easily understood approximations to H_0^* and H_a^* than do the hypotheses H_0'' and H_a'' associated with RST.

In Section 2, a level α intersection-union test of H_0 versus H_a is described. In Section 3, the size of the test is shown to be the prescribed value. The test is shown to be the uniformly most powerful level α test among all monotone, nonrandomized, permutation invariant tests. The consistency class of the test is described. The problem of testing H_0^*: G is not stochastically larger than F versus H_a^*: $G >^{st} F$ is considered in Section 4.

2. DERIVATION OF THE TEST

Let F be a known cdf. Let $Y_1, \ldots Y_n$ be a random sample from an unknown cdf G. Note that neither F nor G are required to be continuous. We will note in later results when continuity simplifies something. Let $0 < \alpha < 1$ be the desired level of the test. Define the inverse of F in the usual way, namely, $F^{-1}(p) = \inf\{x\colon F(x) \geq p\}$. (If $F(x) < 1$ for all x, define $F^{-1}(1) = \infty$.) Let a and b be fixed values satisfying $F^{-1}(1-\alpha^{1/n}) \leq a \leq b < F^{-1}(1)$. Then a level α test of H_0 versus H_a is given by the following.

First define constants $p_0, \ldots, p_{n-1}$ by the relationship

$$P_{p_i}(B \leq i) = \alpha, \qquad i = 0, \ldots, n-1, \tag{2.1}$$

where B is a binomial random variable with parameters p_i and n. The constant p_i also depends on n and α but that dependence will be suppressed in the notation. Define $p_n = 1$. Next, for $i = 0, \ldots, n$, define critical values $c_i = F^{-1}(p_i)$. Since $p_0 = 1 - \alpha^{1/n}$, the restriction $F^{-1}(1-\alpha^{1/n}) \leq a$ is simply $c_0 \leq a$. Define integers I and J by the relationships $c_{I-1} \leq a < c_I$ and $c_{J-1} \leq b < c_J$. Note that $I \leq J$ since $a \leq b$. Let $Y_{(1)} \leq \ldots \leq Y_{(n)}$ denote the order statistics from the sample $Y_1, \ldots, Y_n$. Then the proposed test, that we will call test T, is defined as follows

$$\text{Reject } H_0 \text{ if and only if } Y_{(i)} \geq c_i,\ i = I, \ldots, J-1 \text{ and } Y_{(J)} > b. \tag{2.2}$$

Note that if $I = J$ then the set $i = I, \ldots, J-1$ is empty and test T simply rejects H_0 if and only if $Y_{(J)} > b$. But this corresponds to the atypical situation in which the interval $[a, b]$ is very short and $c_{J-1} \leq a \leq b < c_J$. Typically the interval is wide so that H_0 and H_a are similar to the stochastic ordering hypotheses H_0^* and H_a^* from Section 1. In all the remaining discussion we will assume that $I < J$. The arguments can be easily modified to handle the $I = J$ case.

Blyth (1986) discusses calculation and approximation of the values p_i. His notation for p_i is $p^{(1-\alpha)}(i)$. The values $p_0 = 1 - \alpha^{1/n}$ and $p_{n-1} = (1-\alpha)^{1/n}$ are easily calculated. Blyth's Table 3 contains the values of p_i for $n = 3, \ldots, 20$, and α=.05, .025, .01, and .005. Blyth gives very accurate approximations of p_i for $n > 20$.

The rejection region for test T can be related to the empirical cdf, $G_n(x) = n^{-1}\sum_{i=1}^{n} I(Y_i \leq x)$, of the sample $Y_1, \ldots, Y_n$. ($I(\cdot)$ is the indicator function.)

The condition $Y_{(i)} \geq c_i$ is equivalent to $G_n(x) \leq (i-1)/n$ for all $x < c_i$. Similarly $Y_{(J)} > b$ is equivalent to $G_n(x) \leq (J-1)/n$ for all $x \leq c_J$. These conditions define a step-shaped region as shown in Figures 1a and 1b. Test T rejects H_0 if and only if the graph of the empirical cdf lies entirely in the shaded, step-shaped region. The rejection region extends to $-\infty$, as in Figure 1a, if $a \geq c_1$ and, hence, $I \geq 2$. In this case the rejection region puts no restriction on $Y_{(1)}$. But if $c_0 \leq a < c_1$, then the rejection region has a finite lower endpoint, as in Figure 1b.

Test T was originally derived using the intersection-union (IU) method of test construction. This method was used as early as 1952 by Lehmann. More recently the method has been used by a variety of authors including Gleser (1973), Sasabuchi (1980), Cohen and Marden (1983), Berger (1984a) and Berger (1984b). The method has not always been identified as the IU method.

The IU method of test construction can be used if the null hypothesis can be conveniently expressed as a union of sets. In our problem, we express H_0 as $(U_{a \leq x \leq b} G_x) U G_=$ where, for each $x \in [a,b]$, $G_x = \{G\colon\ G(x) > F(x)\}$ and $G_= = \{G\colon\ G(x) = F(x) \text{ for all } x \in [a,b]\}$. The IU method prescribes that each of the hypotheses $H_{0x}\colon G \in G_x$ as well as $H_{0=}\colon G \in G_=$ be tested with a level α test and H_0 is rejected if and only if each of the individual tests rejects its hypothesis. That is the rejection region for the test of H_0 is the intersection of the rejection regions for each of the tests of H_{0x} and $H_{0=}$.

The usual test of H_{0x} is based on the statistic B_x = the number of sample values Y_i that are less than or equal to x. The hypothesis H_{0x} is rejected if B_x is sufficiently small. In particular, if $c_{i-1} \leq x < c_i$, then $p_{i-1} \leq F(x) < p_i$. Thus, from (2.1) we see that the level α test of H_{0x} rejects H_{0x} if and only if $B_x \leq i-1$. This is equivalent to $Y_{(i)} > x$. Now, if $a < c_i \leq b$, the

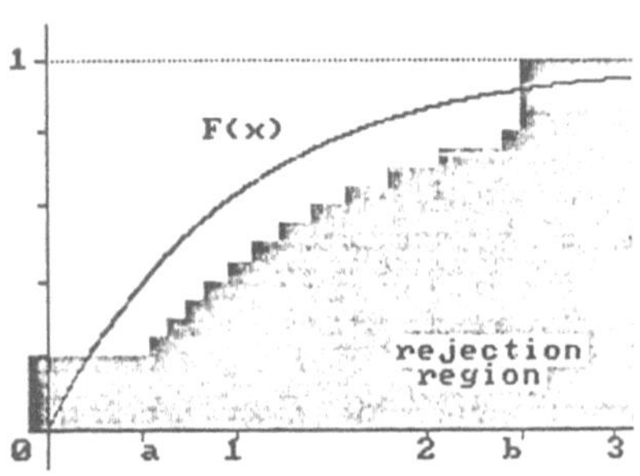

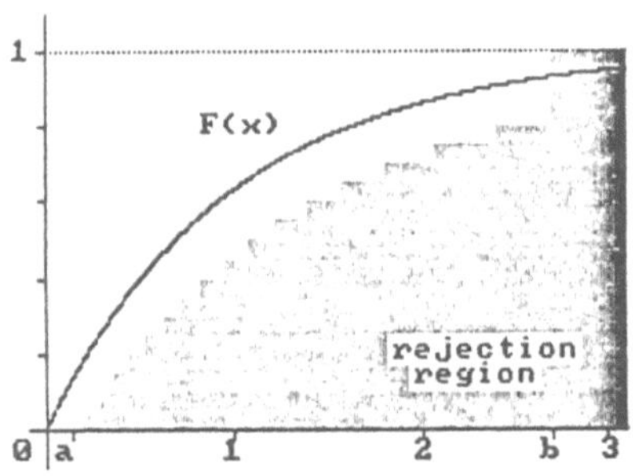

Figure 1. Rejection Region In Terms of Empirical CDF.
$F(x) = 1 - \exp(-x)$, $\alpha = .1$, $n = 20$.

requirement that $Y_{(i)} > x$ for all $x \in [c_{i-1}, c_i) \cap [a, b]$ is equivalent to $Y_{(i)} \geq c_i$. If $c_{i-1} \leq b < c_i$, the requirement that $Y_{(i)} > x$ for all $x \in [c_{i-1}, c_i) \cap [a, b]$ is equivalent to $Y_{(i)} > b$. By the definition of I and J, the only intervals of the form $[c_{i-1}, c_i)$ which intersect $[a, b]$ are $[c_{I-1}, c_I), \ldots [c_{J-1}, c_J)$. Thus the intersection over all $x \in [a, b]$ of the rejection regions of the tests of the hypotheses H_{0x} is the rejection region described in (2.2). According to the IU method, $H_{0=}$ must also be tested with a level α test. But note that for any $x \in [a, b]$, the test which rejects if $Y_{(i)} > x$ is also a level α test for $\overline{H}_{0x}$: $G \in \overline{G}_x = \{G \colon G(x) \geq F(x)\}$. Since $G_{0=} \subset \overline{G}_x$, this test which was already included as the test for H_{0x} is also a level α test of $H_{0=}$. Thus, no additional test for $H_{0=}$ need be included and, indeed, the inclusion of a different test for $H_{0=}$ would only serve to reduce the power of the resulting test.

Some properties of the test T that we have developed using the IU methodology will be discussed in the next section.

3. POWER PROPERTIES OF TEST

In this section we will discuss three properties of the power function of test T. First we will show that test T is a level α test and give conditions under which the size is exactly α. Next we will show that test T is a uniformly most powerful (UMP) level α test in the class of nonrandomized, monotone, permutation invariant tests. Finally we will describe the consistency class for test T.

3.1 Size of Test

Test T is a level α test of H_0 versus H_a. In Theorem 3.1, a condition is given under which the size of the test is exactly the specified value α. This property is important because sometimes the use of level α tests in constructing a test using the IU method will yield an overall test with size much less than α. Thus the constructed test will be overly conservative. Note, in particular, that the condition in Theorem 3.1 that ensures a size α test is satisfied if F is a continuous cdf and $[a, b]$ is wide enough that $I < J$.

Theorem 3.1. *Test T is a level α test of H_0. That is,*

$$\sup_{G \in H_0} P_G(\text{reject } H_0) \leq \alpha. \tag{3.1}$$

If in addition, for some j satisfying $I \leq j < J$, there exists an x such that $F(x) = p_j$ then test T has size exactly α, that is (3.1) is true with equality.

Proof. Fix $G \in H_0$. Either $G \in G_x$ for some $x \in [a, b]$ or $G \in G_=$. If $G \in G_x$ then, as explained in Section 2, the test will reject only if $Y_{(i)} > x$ where i satisfies $c_{i-1} \leq x < c_i$. Thus, $P_G(\text{reject } H_0) \leq P_G(Y_{(i)} > x) \leq \alpha$. If $G \in G_=$, then, as also explained in Section 2, for any $x \in [a, b]$ the test for H_{0x} is also a level α test for $H_{0=}$. Thus, $P_G(\text{reject } H_0) \leq P_G(Y_{(i)} > x) \leq \alpha$ and (3.1) is verified.

To prove the equality in (3.1) note that the definition of c_j as $F^{-1}(p_j)$, the right continuity of F, and the existence of an x with $F(x) = p_j$ together imply that $F(c_j) = p_j$ and $c_{j+1} > c_j$. The definitions of I and J imply that $a < c_j \leq b$. Fix an $x^* > b$ and for any ν satisfying $0 < \nu < 1 - p_j$ define $G_\nu(x)$ as the cdf of the probability measure which puts mass $p_j + \nu$ on the point c_j and mass $1 - p_j - \nu$ on the point x^*. For every ν, $0 < \nu < 1 - p_j$, $G_\nu \in H_0$ because $G_\nu(c_j) = p_j + \nu > p_j = F(c_j)$ and $c_j \in [a, b]$. Consider the inequalities which define the rejection region in (2.2). If G_ν is the true distribution, the inequalities corresponding to subscripts i satisfying $I \leq i \leq j$ are true with probability one. The inequalities for subscripts i satisfying $j + 1 \leq i \leq J$ are true if and only if $Y_{(j+1)} = x^*$ (recall $c_{j+1} > c_j$), that is, if at most j of the Y_i's equal c_j. But B, the random variable that counts the number of Y_i's equal to c_j, has a binomial distribution with success probability $p_j + \nu$. Combining (2.1), (3.1), and the fact that $G_\nu \in H_0$ we have

$$\alpha \geq \sup_{G \in H_0} P_G(\text{reject } H_0) \geq \overline{\lim}_{\nu \downarrow 0} P_{G_\nu}(\text{reject } H_0)$$

$$= \overline{\lim}_{\nu \downarrow 0} P_{p_j + \nu}(B \leq j) = P_{p_j}(B \leq j) = \alpha$$

Thus the equality in (3.1) is proved. □

The G_ν constructed in the proof of Theorem 3.1 is discontinuous. But a continuous G_ν could be used. For example, G_ν, rather than putting point masses at c_j and x^*, could spread mass $p_j + \nu$ uniformly on the interval $(c_j, c_j + \varepsilon_\nu)$ and spread mass $1 - p_j - \nu$ in any continuous fashion on (x^*, ∞). Here, $\varepsilon_\nu > 0$ must be chosen so small that $c_j + \varepsilon_\nu < \min\{c_{j+1}, b\}$ and $G_\nu(c_j + \varepsilon_\nu) = p_j + \nu > F(c_j + \varepsilon_\nu)$ so as to ensure that $G_\nu \in H_0$. This G_ν is continuous although the density of G_ν becomes more peaked as $\nu \to 0$. Hence Theorem 3.1, as well as other theorems in this paper, remain true if G is restricted to be a continuous cdf. In some problems it may be reasonable to assume that G is continuous. However, if a bound is put on the density of G, in addition to requiring continuity, then the equality part of Theorem 3.1 remains unproven.

It might be useful to express the power function of Test T, $\rho(G) = P_G(\text{reject } H_0)$, in terms of probabilities involving order statistics from a uniform sample. Gleser (1985) has explored this type of representation. However, care must be taken. Gleser's Theorem 1 is not immediately applicable since $\Gamma_1(x)$ (defined in (2.2') of Gleser) is not right-continuous at b. This is due to the fact that rejection by Test T requires $Y_{(J)} > b$ rather than $Y_{(J)} \geq b$.

3.2 Optimality Property

Test T is a UMP level α test of H_0 in the class of nonrandomized, monotone, permutation invariant tests. A nonrandomized test is called monotone if $(y_1, \ldots, y_n)$ is in the rejection region of the test and $y_i^* \geq y_i$, $i = 1, \ldots, n$, imply that $(y_1^*, \ldots, y_n^*)$ is in the rejection region of the test. Monotone tests have monotone power functions in that if $G >^{st} G^*$ then $P_G(\text{reject } H_0) \geq P_{G^*}(\text{reject } H_0)$ for any monotone test. Optimality properties such as this for IU tests in other problems have been discussed by Lehmann (1952), Berger (1982), and Cohen and Marden (1983). It is interesting to note that the following result does not require test T to have size exactly α. The distribution F may be discrete and the test may have size less than α and still the test will be UMP in this class of tests.

Theorem 2.2. *Test T is a UMP level α test of H_0 in the class of nonrandomized, monotone, permutation invariant tests.*

Proof. We will show that any monotone, permutation invariant, level α rejection region is a subset of the rejection region for test T defined in (2.2). Hence test T is UMP. We will show this by showing that any monotone, permutation invariant rejection region which is not a subset of the region in (2.2) has size greater than α.

Consider a monotone, permutation invariant rejection region which is not a subset of the region in (2.2). Let $(y_1, \ldots, y_n)$ be a point in this rejection region which does not satisfy all the inequalities in (2.2). Fix j, $I \leq j \leq J$ such that the inequality in (2.2) involving $y_{(j)}$ is not satisfied. That is $y_{(j)} < c_j$, if $j < J$, or $y_{(j)} \leq b$ if $j = J$. Set $y^* = \max(y_{(j)}, a)$. Recall that $a < c_I$ and $b < c_J$ so that, regardless of which subscript j we are considering, $F(y^*) < p_j$. Choose $\nu > 0$ so that $p_j - \nu > F(y^*)$. Fix $x^* > \max(y_{(n)}, b)$. Note that $x^* > y^*$. Let G_ν denote the cdf of the probability distribution that puts mass $p_j - \nu$ on y^* and $1 - p_j + \nu$ on x^*. Then $G_\nu \in H_0$ since $a \leq y^* \leq b$ and $G_\nu(y^*) = p_j - \nu > F(y^*)$. Let B be the random variable which counts the number of Y_i's equal to y^*. Under G_ν, B has a binomial distribution with success probability $p_j - \nu < p_j$.

Now consider sample points (n-vectors) consisting only of x^*'s and y^*'s. Since $x^* > \max(y_{(n)}, b) \geq y^* \geq y_{(j)}$, by the monotonicity and permutation invariance of the rejection region, all such sample points with at most j y^*'s are in this rejection region. Thus for this rejection region, by (2.1) we have

$$\sup_{G \in H_0} P_G(\text{reject } H_0) \geq P_{G_\nu}(\text{reject } H_0) \geq P_{G_\nu}(B \leq j)$$

$$= P_{p_j - \nu}(B \leq j) > P_{p_j}(B \leq j) = \alpha.$$

This rejection region has size greater than α as was to be shown. □

A nonparametric upper confidence bound for the cdf G might be used to construct a level α test of H_0. A test which rejected H_0 only if the $100(1-\alpha)\%$ upper confidence bound was less than or equal to $F(x)$ for all $x \in [a,b]$ would be a level α test of H_0. But the usual upper confidence bounds, such as those described by Sandford (1985), would produce a nonrandomized, monotone, permutation invariant test. Thus, by Theorem 3.2, test T is a more powerful test than a test constructed in this way.

Test T is not UMP among all level α tests. Test T is biased in that $P_G(\text{reject } H_0) < \alpha$ if $G(x) = F(x)$ for a nondegenerate interval of x values, even if $G \in H_a$. On the other hand, the randomized "no data" test that simply rejects H_0 with probability α, regardless of the data, has $P_G(\text{reject } H_0) = \alpha$ for all G. Thus the "no data" test has higher power than test T for some $G \in H_a$.

3.3 CONSISTENCY

In this section, consistency properties of test T are described. Let $\gamma = \inf_{a \le x \le b}[F(x) - G(x)]$. We show that the asymptotic value of $P_G(\text{reject } H_0)$ depends on the value of γ. An interpretation of the parameter γ is given at the end of this section. The following two theorems relate the power of the test to γ.

Theorem 3.3. *For test T, if $\gamma \le 0$ then $P_G(\text{reject } H_0) \le \alpha$, for every sample size n, and hence, $\lim_{n\to\infty} P_G(\text{reject } H_0) \le \alpha$.*

Theorem 3.4. *For test T, if $\gamma > 0$ then $\lim_{n\to\infty} P_G(\text{reject } H_0) = 1$.*

PROOF OF THEOREM 3.3. If $\gamma < 0$, $G(x) > F(x)$ for some $x \in [a,b]$. Thus $G \in H_0$, and by Theorem 3.1, $P_G(\text{reject } H_0) \le \alpha$. If $\gamma = 0$ and $G(x) = F(x)$ for some $x \in [a,b]$, then as explained in Section 2, the test which rejects if $Y_{(i)} > x$ (where $c_{i-1} \le x < c_i$) is a level α test of $\overline{H}_{0x}$: $G \in \overline{G}_x = \{G\colon G(x) \ge F(x)\}$. So also for this type of G we have $P_G(\text{reject } H_0) \le P_G(Y_{(i)} > x) \le \alpha$. The following argument handles the remaining case, $\gamma = 0$ and $G(x) < F(x)$ for all $x \in [a,b]$.

Since $[a,b]$ is compact, there exist $x_0 \in [a,b]$ and a sequence of values $x_m \in [a,b]$ such that $\lim_{m\to\infty} x_m = x_0$ and $\lim_{m\to\infty}[F(x_m) - G(x_m)] = 0$. Since $F(x_0) > G(x_0)$, the right continuity of F and G imply that x_m must approach x_0 from below and

$$G(x_0^-) = \lim_{x\uparrow x_0} G(x) = \lim_{x\uparrow x_0} F(x) = F(x_0^-). \tag{3.2}$$

Note that

$$0 < F(a) \le F(x_0^-) = G(x_0^-) \le G(x_0) < F(x_0) \le F(b) < 1. \tag{3.3}$$

Define a cdf $H(x)$ by

$$H(x) = \begin{cases} G(x)/G(x_0^-) & x < x_0. \\ 1 & x \ge x_0. \end{cases} \tag{3.4}$$

For any $\nu \in (0,1)$ define $G_\nu(x) = (1-\nu)G(x) + \nu H(x)$.

First we will show that for any $\nu \in (0,1)$, $G_\nu \in H_0$ and hence, by Theorem 3.1,

$$P_{G_\nu}(\text{reject } H_0) \leq \alpha. \tag{3.5}$$

Recall that $F(x_0^-) = G(x_0^-) < 1 = H(x_0^-)$. Thus

$$\begin{aligned}\lim_{m\to\infty}[F(x_m) - G_\nu(x_m)] &= F(x_0^-) - (1-\nu)G(x_0^-) - \nu H(x_0^-) \\ &= \nu(G(x_0^-) - 1) < 0.\end{aligned}$$

Thus $G_\nu(x_m) > F(x_m)$ for some x_m, verifying that $G_\nu \in H_0$ and (3.5).

Now we will show that

$$\lim_{\nu\downarrow 0} P_{G_\nu}(\text{reject } H_0) \geq P_G(\text{reject } H_0) \tag{3.6}$$

which with (3.5) will complete the proof that $P_G(\text{reject } H_0) \leq \alpha$. Let $Y = (Y_1, \ldots, Y_n)$ be a random sample from G. Let $Z_1, \ldots, Z_n$ be a random sample from H. Fix $\nu \in (0,1)$. Let $I_1, \ldots, I_n$ be independent Bernoulli random variables with $P(I_i = 1) = \nu$. All the Y's, Z's, and I's are mutually independent. Define $Y_i^* = Z_i I_i + Y_i(1 - I_i)$. That is $Y_i^* = Z_i$ if $I_i = 1$ and $Y_i^* = Y_i$ if $I_i = 0$. Then $Y^* = (Y_1^*, \ldots, Y_n^*)$ is a random sample from the cdf G_ν. Let A = the rejection region for test T and $B = I_1 + \ldots + I_n$. Note that B and Y are independent and $Y = Y^*$ if $B = 0$. Using Theorem 3.1 and the fact that $G_\nu \in H_0$ we have

$$\begin{aligned}\alpha \geq P_{G_\nu}(\text{reject } H_0) &= P(Y^* \in A) \geq P(Y^* \in A,\ B = 0) \\ &= P(Y \in A,\ B = 0) = P(Y \in A)P(B = 0) \\ &= P_G(\text{reject } H_0) \cdot (1-\nu)^n.\end{aligned}$$

Taking the limit as $\nu \downarrow 0$ yields (3.6). □

SKETCH OF PROOF OF THEOREM 3.4.

A proof of Theorem 3.4 can be based on the graphical representation of the rejection region described in Section 2.

Let $[a]$ denote the greatest integer less than or equal to a. Then the Weak Law of Large Numbers can be used to show that for any $v \in (0,1)$, $\lim_{n\to\infty} p_{[nv]} = v$. This convergence is uniform for $v \in [F(a), F(b)]$. These facts can be used to show that the step-shaped boundary, with vertices at the points $(F^{-1}(p_i), i/n)$, converges to $F(x)$ uniformly on the interval $[a,b]$. On the other hand, by the Glivenko-Cantelli Theorem, the empirical cdf converges to $G(x)$ uniformly on $[a,b]$ with probability one. Thus, if $\gamma = \inf_{a\leq x\leq b}[F(x) - G(x)] > 0$, for large n the boundary is greater than $F(x) - \gamma/2 \geq G(x) + \gamma/2$ for all $x \in [a,b]$. Test T will reject H_0 if $G_n(x) < G(x) + \gamma/2$ for all $x \in [a,b]$ and this event has probability approaching one as $n \to \infty$. □

Theorems 3.3 and 3.4 say that test T is a level α test of H_0^+: $\gamma \le 0$ versus H_a^+: $\gamma > 0$ and the test is consistent against all points in H_a^+. Letting X and Y denote independent random variables with cdfs F and G, respectively, the parameter γ plays the same role for test T that the parameter $\theta = P(X \le Y)$ plays for the RST. The parameter $\gamma = \inf_{a \le x \le b}[F(x) - G(x)]$ may be interpreted as follows:

$$\begin{aligned}\gamma > 0 \text{ if and only if } & P(X \le x) > P(Y \le x) \text{ for all } x \in [a,b] \\ & \text{and } P(X < x) > P(Y < x) \text{ for all } x \in [a,b] \\ \gamma \le 0 \text{ if and only if } & P(X \le x) \le P(Y \le x) \text{ for some } x \in [a,b] \\ & \text{or } P(X < x) \le P(Y < x) \text{ for some } x \in [a,b].\end{aligned}$$

If F is continuous, then each of the inequalities involving $<$ can be dropped in the above interpretation. As mentioned in Section 1, we believe that in many situations the hypothesis H_a^+: $\gamma > 0$ provides a more meaningful approximation to H_a^*: $G >^{st} F$ than does H_a'': $\theta > 1/2$, the alternative hypothesis associated with the RST.

4. UMP TEST OF H_0^* VERSUS H_A^*

There is no satisfactory level α test for testing H_0^*: G is not stochastically larger than F versus H_a^*: $G >^{st} F$. For any level α test of H_0^*, $P_G(\text{reject } H_0^*) \le \alpha$ for *every* cdf G including all $G \in H_a^*$. This means that the randomized "no data" test that rejects H_0^* with probability α, regardless of the data, and has $P_G(\text{reject } H_0^*) = \alpha$ for all G is the UMP level α test of H_0^* versus H_a^*. This fact is proved in the following theorem.

Theorem 4.1. *Consider testing H_0^* versus H_a^*. Let G be any cdf. Then there exists a sequence of cdf's G_m, $m = 1, 2, \ldots$, such that $G_m \in H_0^*$ for every m and, for any (possibly randomized) level α test of H_0^*,*

$$\alpha \ge \lim_{m \to \infty} P_{G_m} \text{ (reject } H_0^*) \ge P_G \text{ (reject } H_0^*). \tag{4.1}$$

PROOF. The proof is similar to the proof of Theorem 3.3. For each $m = 1, 2, \ldots$, fix a value x_m such that $F(x_m) < 1/m$ and let H_m be a cdf such that $H_m(x) = 1$ for $x \ge x_m$. Let $G_m = (1 - 1/m)G + (1/m)H_m$. Then $G_m(x_m) = (1 - 1/m)G(x_m) + 1/m \ge 1/m > F(x_m)$. So G_m is not stochastically larger than F. Let $y = (y_1, \ldots, y_n)$ denote a sample point and let $\phi(y)$ denote the test function for any level α test. That is $\phi(y)$ is the probability H_0^* is rejected given that y is observed. Let $Y = (Y_1, \ldots, Y_n)$ be a random sample from G, let $Z_m = (Z_{1m}, \ldots, Z_{nm})$ be a random sample from H_m and let $I_{1m}, \ldots, I_{nm}$ be Bernoulli random variables with success probability $1/m$. Assume all the

Y's, Z's and I's are mutually independent. Let $Y_m^* = (Y_{1m}^*, \ldots, Y_{nm}^*)$ where $Y_{im}^* = Z_{im}I_{im} + Y_i(1 - I_{im})$. Note that Y_m^* is a random sample from G_m. Let $B_m = I_{1m} + \ldots + I_{nm}$ and note that Y and B_m are independent and $Y = Y_m^*$ if $B_m = 0$. Thus the conditional distribution of Y_m^* given $B_m = 0$ is the same as the conditional distribution of Y given $B_m = 0$ and this is just the unconditional distribution of Y. Thus we have for every $m = 1, 2, \ldots,$

$$\begin{aligned} P_{G_m} \text{ (reject } H_0^*) &= E\phi(Y_m^*) = E(\phi(Y_m^*)|B_m = 0)P(B_m = 0) \\ &\qquad + E(\phi(Y_m^*)|B_m > 0)P(B_m > 0) \\ &\geq E(\phi(Y_m^*)|B_m = 0)P(B_m = 0) = E\phi(Y)(1 - 1/m)^n \\ &= P_G \text{ (reject } H_0^*)(1 - 1/m)^n. \end{aligned} \tag{4.2}$$

Since $G_m \in H_0^*$ for every $m = 1, 2, \ldots, P_{G_m}$(reject H_0^*) $\leq \alpha$ for every m. Thus taking limits as $m \to \infty$ in (4.2) yields (4.1). □

In the above proof, since $F(x_m) < 1/m$, we must have $\overline{\lim}_{m\to\infty} x_m \leq \inf\{x: F(x) > 0\}$. That is, the x_m are in the left tail of F. This suggests why we were able to develop a reasonable test of H_0 versus H_a. These hypotheses ignore the tails of F.

BIBLIOGRAPHY

Berger, R. L. (1982). Multiparameter hypothesis testing and acceptance sampling. *Technometrics* **24**, 295–300.

Berger, R. L. (1984a). Testing for the same ordering in several groups of means. *Design of Experiments: Ranking and Selection* 241–249. Ed. T. J. Santner and A. C. Tamhane. Marcel Dekker, New York.

Berger, R. L. (1984b). Testing whether one regression function is larger than another. *Comm. Statist. A – Theory Methods* **13**, 1793–1810.

Blyth, C. R. (1986). Approximate binomial confidence limits. *J. Amer. Statist. Assoc.* **81**, 843–855.

Cohen, A. and Marden, J. (1983). Hypothesis tests and optimality properties in discrete multivariate analysis. *Studies in Econometrics, Time Series, and Multivariate Statistics* 379–405. Academic Press, New York.

Gleser, L. J. (1973). On a theory of intersection-union tests. *IMS Bulletin* **2**, 233. Abstract.

Gleser, L. J. (1985). Exact power of goodness-of-fit tests of Kolmogorov type for discontinuous distributions. *J. Amer. Statist. Assoc.* **80**, 954–958.

Lee, Y. J. and Wolfe, D. A. (1976). A distribution-free test for stochastic ordering. *J. Amer. Statist. Assoc.* **71**, 722–727.

Lehmann, E. L. (1952). Testing multiparameter hypothesis. *Ann. Math. Statist.* **23**, 541–552.

Lehmann, E. L. (1975). *Nonparametrics: Statistical Methods Based on Ranks.* Holden-Day, Oakland.

Moses, L. E. (1964). One sample limits of some two-sample rank tests. *J. Amer. Statist. Assoc.* **59**, 645–651.

Sandford, M. D. (1985). Nonparametric one-sided confidence intervals for an unknown distribution function using censored data. *Technometrics* **27**, 41–48.

Sasabuchi, S. (1980). A test of a multivariate normal mean with composite hypotheses determined by linear inequalities. *Biometrika* **67**, 429–439.

ON ESTIMATING THE NUMBER OF UNSEEN SPECIES AND SYSTEM RELIABILITY

PETER J. BICKEL

Department of Statistics
University of California
Berkeley, California 94720, U.S.A.

JOSEPH A. YAHAV

Department of Statistics
The Hebrew University
Jerusalem, Israel

1. INTRODUCTION

The problem of estimating the number of unseen species and the problem of estimating system reliability that we want to discuss can be thought of in terms of a general proofreading problem.

A book contains k misprints, there are n "readers" who independently proofread the book and report the misprints they find. From this data we want to estimate k.

The system reliability problem we consider is to estimate the total number of errors k in a system on the basis of the errors found by n users. See for example Jewell (1983), Barlow and Singpurwalla (1984). The typical assumption in the system reliability literature is that errors are equally easy to detect and users may differ in their ability to detect errors.

The unseen species problem is to estimate the number of unseen species when a random sample of size n is drawn from an infinite population of animals of various species. See for example Good (1953), Efron and Thisted (1976) and Seber (1973). The underlying assumption in the species problem is that different species have different frequencies in the population and translating that to the proofreading example we have errors which differ in ease in detectability and readers who have the same ability to detect the errors.

It seems reasonable to assume both types of inhomogeneity. The ease with which animals are trapped may vary with time and location and errors in systems are certainly not equally likely to be found. It is the purpose of this paper to argue in particular that system reliability theory can benefit from the work done on estimating the number of unseen species. It turns out that asymptotically, while the inhomogeneity of the readers is not important the inhomogeneity of the errors is crucial.

2. FORMULATION OF THE GENERAL MODEL AND PROBLEM

Let $P_{ij} = Pr\{\text{Reader } i \text{ detects error "}j\text{"}\}$, where "$j$" is a label obtained by say a random permutation of $\{1, 2, \cdots, k\}$. Let X_{ij} be the indicator of the event that reader i detects error j. We assume the X_{ij} to be independent. Note that X_{ij} are observable if and only if $X_{ij} = 1$, so that the ith reader's list is $\{j\colon X_{ij} = 1\}$.

The following identities will lead our discussion.

$$k = T + D \tag{2.1}$$

where T denotes the number of distinct errors detected by the proofreading process, and D is the number of errors that were not detected. Of course,

$$k = E[T] + E[D]. \tag{2.2}$$

$E[T]$ can be estimated reasonably well by T so that our interest lies in the estimation of $E[D]$. We note that,

$$E[D] = \sum_{j=1}^{k} \prod_{i=1}^{n} (1 - P_{ij}). \tag{2.3}$$

There is no unbiased estimator of $E[D]$ and the maximum likelihood estimator of $E[D]$ is zero. The assumptions for asymptotic analysis in both the system reliability and species literatures that we make (compare Good (1953) and Barlow and Singpurwalla (1984)) are essentially that,

A: As $k \to \infty$ and $n(k) \to \infty$.

(i) $\max\limits_{i,j} P_{ij} \to 0$

(ii) $\dfrac{\#\left\{\sum_{i=1}^{n} P_{ij} \le x\right\}}{k} = G_k(x) \to G(x)$ weakly, where G is a probability distribution on $(0, \infty)$ with no mass at 0. That is, $\lim\limits_{x\to\infty} G(x) = 1$ and $G(0) = 0$.

(iii) $\sup\limits_{k} \int_{0}^{\infty} x dG_k(x) < \infty$.

A natural statistic for the problem is $(N_1, N_2, \cdots, N_n)$ where N_r denotes the number of errors that were detected by exactly r readers. This statistic behaves as follows:

Proposition 2.1. *Under A,*

$$\frac{1}{k} E[N_r] \to \int_0^\infty \frac{\lambda^r e^{-\lambda}}{r!} dG(\lambda) \qquad r = 0, 1, 2 \ldots \tag{2.4}$$

$$\frac{1}{k} E[D] \to \int_0^\infty e^{-\lambda} dG(\lambda) \tag{2.5}$$

$$\frac{1}{k}E[S] \to \int_0^\infty \lambda dG(\lambda) \tag{2.6}$$

where $S = \sum_{i,j} X_{ij}$ *is the total number of errors, multiplicities allowed, detected by all readers.*

PROOF. Let $n_j = \sum_{i=1}^n X_{ij}$. Now

$$\frac{1}{k}E[N_r] = \frac{1}{k}\sum_{j=1}^k P[n_j = r].$$

By A(i) and classical asymptotic theory if $\lambda_{jk} = \sum_{i=1}^n P_{ij}$ then

$$\max_r\{|P[n_j = r] - \frac{\lambda_{jk}^r e^{-\lambda_{jk}}}{r!}|: 0 \le \lambda_{jk} \le M\} \underset{k\to\infty}{\longrightarrow} 0.$$

$$\begin{aligned}\frac{1}{k}E[N_r] &= \int_0^M \frac{\lambda^r e^{-\lambda}}{r!}dG_k(\lambda) + O(1 - G_k(M)) \\ &= \int_0^M \frac{\lambda^r e^{-\lambda}}{r!}dG(\lambda) + O(\frac{1}{M}\sup_k \int_0^\infty \lambda dG_k(\lambda)) \\ &= \int_0^M \frac{\lambda^r e^{-\lambda}}{r!}dG(\lambda) + O(\frac{1}{M}\sup_k \int_0^\infty \lambda dG_k(\lambda))\end{aligned}$$

by (ii). The result now follows from (iii). Similarly $\frac{1}{k}\ E[D] = 1 - \frac{1}{k}\sum_{r=1}^n E(N_r)$ $\to 1 - \sum_{r=1}^\infty \int_0^\infty \frac{\lambda^r e^{-\lambda}}{r!}\ dG(\lambda) + o(1)$. To see this note that for each fixed R,

$$\frac{1}{k}\sum_{r=R}^n E[N_r] = \frac{1}{k}\sum_{j=1}^k P[n_j \ge R] = O\left(\sum_{r=R}^n \frac{\lambda^r e^{-\lambda}}{r!}\right) + O\left(\frac{1}{M}\sup_k \int_0^\infty \lambda dG_k(\lambda)\right).$$

and (2.5) follows. Claim (2.6) follows similarly. □

If we had a good estimator of $\int_0^\infty e^{-\lambda}dG(\lambda)$ say $\hat{\theta}$, we could use the solution of the equation $k = T + k\hat{\theta}$ as an estimator for k. If $P_{ij} \equiv Q_i$ i.e. the errors are homogeneous, then $\lambda_j \equiv \lambda$, G is degenerate, and we have

$$\int_0^\infty e^{-\lambda}dG(\lambda) = e^{-\int_0^\infty \lambda dG(\lambda)}.$$

In this case if $\hat{k}_2$ the unique solution of

$$k = T + ke^{-S/k} \tag{2.7}$$

then $\hat{k}_2$ is an asymptotically consistent estimator of k i.e. $\frac{\hat{k}_2}{k} \to 1$ in probability and $\frac{E[\hat{k}_2]}{k} \to 1$. If G is not degenerate then $\hat{k}_2$ underestimates k, as the following proposition shows.

Proposition 2.2. *Let r_0 be the unique solution of*

$$r = \int_0^\infty (1 - e^{-\lambda}) dG(\lambda) + r e^{-\frac{1}{r}\int_0^\infty \lambda dG(\lambda)}. \tag{2.8}$$

Then,

$$\frac{\hat{k}_2}{k} \xrightarrow{P} r_0, \qquad \frac{1}{k} E[\hat{k}_2] \to r_0 \text{ with } 0 < r_0 < 1,$$

unless G is degenerate.

PROOF. If G is degenerate then $r = 1$ is the unique solution of (2.8). Let $A(r)$ denote the R. H. S. of (2.8). Since $G(0) = 0$, $A(0) > 0$. For G nondegenerate, since $e^{-\lambda}$ is strictly convex $\int_0^\infty e^{-\lambda}\, dG(\lambda) > e^{-\int_0^\infty \lambda dG(\lambda)}$ and hence $A(1) < 1$. This proves that (2.8) has a solution for some r, $0 < r < 1$. Since,

$$A'(r) = e^{-\frac{1}{r}\int_0^\infty \lambda dG(\lambda)} \left(1 + \frac{1}{r}\int_0^\infty \lambda dG(\lambda)\right).$$

$A(r)$ is monotone increasing in r and $A'(r) < 1$, this yields the uniqueness of the solution of (2.8). Using proposition 2.1 we have $\frac{T}{k} \xrightarrow{P} \int_0^\infty (1 - e^{-\lambda}) dG(\lambda)$, $\frac{1}{k}E(T) \longrightarrow \int_0^\infty (1 - e^{-\lambda}) dG(\lambda)$, $\frac{S}{k} \xrightarrow{P} \int_0^\infty \lambda dG(\lambda)$ and $\frac{1}{k}E[S] \longrightarrow \int_0^\infty \lambda dG(\lambda)$. Dividing both sides of (2.7) by k we have $\frac{\hat{k}_2}{k} = \frac{T}{k} + e^{(-S/k)\cdot(k/\hat{k}_2)}$. Letting $k \to \infty$ we get (2.8) as the limit. If we put $\frac{T}{k} = a_k$ and $\frac{S}{k} = b_k$ we have then for (2.7) $r(a_k, b_k) = a_k + e^{-b_k/r(a_k, b_k)}$. Since $0 \le a_k \le 1$ and $0 \le b_k$ yield $0 \le r(a_k, b_k) \le 2$ the set $\{r(a_k, b_k)\}$ has limit points and these satisfy (2.8). By unicity of r_0 it must be the limit of $r(a_k, b_k)$. □

$\hat{k}_2$ is known as the Petersen-Chapman-Darroch estimator. See Jewell (1983).

3. ON ESTIMATING $E[D]$

Estimating $E[D]$ is a difficult statistical problem. Different researchers have suggested different approaches to this problem. B. M. Hill (1968) and (1979) approached this problem from the Bayesian point of view. Assuming a geometric prior on k, he suggested the mode of the posterior distribution as an estimator of k. Fisher, Corbet and Williams (1943) investigated the problem using a parametric empirical Bayes approach. They assumed G to have a gamma distribution and estimated the parameters of the gamma distribution using the method of moments. Good (1953) and Good and Toulmin (1956) and Efron and Thisted (1976) studied a lower bound $\Delta(t)$ for $E[D]$, where $\Delta(t)$ is the expected number of new species to be seen if an additional sample of size tn is taken. Clearly,

$$E[T] + \Delta(t) \leq k. \tag{3.1}$$

and

$$E[T] + \Delta(t) \uparrow k, \quad \text{as } t \to \infty.$$

Under A,

$$\frac{1}{k}\Delta(t) \to \int_0^\infty e^{-\lambda}(1 - e^{-\lambda t})dG(\lambda). \tag{3.2}$$

The limit in (3.2) can be written as

$$\int_0^\infty e^{-\lambda}(1 - e^{-\lambda t}))dG(\lambda) = \sum_{i=1}^{\infty}(-1)^i t^i \int_0^\infty \lambda^i e^{-\lambda}\frac{1}{i!}dG(\lambda). \tag{3.3}$$

Therefore an unbiased estimator for $\Delta(t)$ is:

$$\hat{\Delta}(t) = tN_1 - t^2 N_2 + t^3 N_3 \cdots \pm t^n N_n. \tag{3.4}$$

Unfortunately $\hat{\Delta}(t)$ is unstable for $t \geq 1$ and Good and Toulmin (1956) and Efron and Thisted (1976) proposed other estimators for $\Delta(t)$ of the form $\Sigma h_i N_i$, based on Euler's summation method. Bickel and Yahav (1985) suggested another approach, estimating best lower bounds for $E[D]$. These bounds are given by $B(r_1, \cdots, r_2) = \min\{\int_0^\infty e^{-\lambda}dG(\lambda)\colon \int_0^\infty \lambda dG(\lambda) = c_0, \int_0^\infty \lambda e^{-\frac{\lambda}{r_i}}dG(\lambda) = c_{ri}, r_1 = 1 < r_2 < \cdots < r_s < n\}$. Bickel and Yahav (1985) show

$$B(1) = \frac{\int_0^\infty \lambda e^{-\lambda}dG(\lambda)}{\log\left(\frac{\int_0^\infty \lambda dG(\lambda)}{\int_0^\infty \lambda e^{-\lambda}dG(\lambda)}\right)}. \tag{3.5}$$

Now $k \cdot B(1)$ can be estimated by $N_1\left(\log\frac{S}{N_1}\right)^{-1}$. Then under A, $\frac{N_1}{k}\left(\log\frac{S}{N_1}\right)^{-1} \xrightarrow{P} B(1)$ and $\frac{1}{k}E[N_1\left(\log\frac{S}{N}\right)^{-1}] \to B(1)$. Let $\hat{k}_1 = T + N_1\left(\log\frac{S}{N_1}\right)^{-1}$.

Proposition 3.1. *Under A*

$$\frac{\hat{k}_1}{k} \xrightarrow{P} \int_0^\infty (1 - e^{-\lambda})dG(\lambda) + B(1) \geq r_0 \quad \textit{given in proposition 2.2.}$$

Equality holds if and only if G is degenerate. In that case $\frac{\hat{k}_1}{k} \xrightarrow{P} 1$ *and* $\frac{\hat{k}_2}{k} \xrightarrow{P}$ 1.

PROOF. Using proposition 2.1 we have,

$$\frac{\hat{k}_1}{k} \xrightarrow{P} \int_0^\infty (1 - e^{-\lambda})dG(\lambda) + B(1) \leq 1. \tag{3.6}$$

From proposition 2.2 we have,

$$\frac{\hat{k}_2}{k} \to r_0 = \int_0^\infty (1 - e^{-\lambda})dG(\lambda) + r_0 e^{-\frac{1}{r_0}\int_0^\infty \lambda dG(\lambda)}. \tag{3.7}$$

Since $r_0 \leq 1$ we have,

$$r_0 \cdot e^{-\frac{1}{r_0}\int_0^\infty \lambda dG(\lambda)} \leq e^{-\int_0^\infty \lambda dG(\lambda)} \tag{3.8}$$

with equality if and only if G is degenerate. But,

$$e^{-\int_0^\infty \lambda dG(\lambda)} \leq B(1) \tag{3.9}$$

since $B(1) = \min_G\{\int_0^\infty e^{-\lambda}dG(\lambda)\colon \int_0^\infty \lambda dG(\lambda)$ and $\int_0^\infty \lambda e^{-\lambda}dG(\lambda)$ fixed$\}$ and $e^{-\int_0^\infty \lambda dG(\lambda)} = \min_G\{\int_0^\infty e^{-\lambda}dG(\lambda)\colon \int_0^\infty \lambda dG(\lambda)$ fixed$\}$. □

Proposition 3.1 asserts that unless G is degenerate $\underline{\hat{k}}_1$ is preferable to the Petersen-Chapman-Darroch estimator in the sense that its asymptotic bias is smaller, although both will be biased. Note that the assumption that the k errors are equally likely to be found is equivalent to the assumption that G is degenerate.

ACKNOWLEDGMENTS

This research was partially supported by Office of Naval Research contract N00014-80-C-0163.

BIBLIOGRAPHY

Bickel, P. J. and Yahav, J. A. (1985). "On estimating the number of unseen species: How many executions were there?" Department of Statistics, U. C. Berkeley Technical Report No. 43.

Barlow, R. S. and Singpurwalla, N. D. (1984). "Assessing the reliability of computer software and computing networks". ORC 84-7 University of California, Berkeley.

Efron, B. and Thisted, R. (1976). Estimating the number of unseen species. *Biometrika* **63**, 435–447.

Fisher, R. A., Corbet, A. S. and Williams, C. B. (1943). The relation between the number of species and the number of individuals in a random sample of an animal population. *J. Anim. Ecol.* **12**, 42–58.

Good, I. J. (1953). The population frequencies of species and the estimation of population parameters. *Biometrika* **40**, 237–264.

Good, I. J. and Toulmin, G. H. (1956). The number of new species, and the increase in population coverage, when a sample is increased. *Biometrika* **43**, 45–63.

Goodman, L. A. (1949). On the estimation of the number of classes in a population. *Ann. Math. Statist.* **20**, 572–579.

Hill, B. M. (1968). "Posterior distribution of percentiles: Bayes' theorem for sampling from a population." *Amer. Statist. Assoc.* **63**, 677–691.

Hill, B. M. (1979). "Posterior moments of the number of species in a finite population and the posterior probability of finding a new species." *Jour. of the Amer. Statist. Assoc.* **74**, 668–673.

Jewell, W. S. (1983). "Bayesian estimation of undetected errors." Technical Report ORC 83-11. University of California, Berkeley.

Olsen, D. E. (1977). "Estimating reliability growth ." IEEE Trans. on Reliability, R-26, 50–53.

Seber, G. A. F. (1973). *The Estimation of Animal Abundance.* New York, Hafner Press.

THE EFFECTS OF VARIANCE FUNCTION ESTIMATION ON PREDICTION AND CALIBRATION: AN EXAMPLE

RAYMOND J. CARROLL

Department of Statistics
The University of North Carolina
Chapel Hill, North Carolina 27514, U.S.A.

1. INTRODUCTION

Consider a heteroscedastic regression model, in which we observe N pairs (y, x) following the model

$$E(y|x) = f(x, \beta); \tag{1.1}$$

$$\text{Standard Deviation } (y|x) = \sigma g(x, \beta, \theta). \tag{1.2}$$

While our remarks hold generally, in what follows it suffices to consider the special case of linear regression for the mean and the power of the mean model for the standard deviation, i.e.,

$$f(x, \beta) = \beta_0 + \beta_1 x : \quad g(x, \beta, \theta) = f(x, \beta)^{\theta}. \tag{1.3}$$

When $\theta = 0$, we have the homoscedastic regression model, and ordinarily unweighted least squares will be used to estimate β. For other values of θ, generalized least squares can be used to estimate β, see Carroll & Ruppert (1987) for a discussion and a review of the literature. Generalized least squares is weighted least squares with estimated weights. The version of generalized least squares used here for each θ is fully iterated reweighted least squares, sometimes called quasi-likelihood, see McCullagh & Nelder (1983). In practice, θ is unknown and must be estimated. The theory of such estimation is given by Davidian & Carroll (1986).

The common folklore theory of generalized least squares states that as long as one's estimate $\hat{\theta}$ of θ is root-N consistent, the resulting generalized least squares estimate has the same asymptotic distribution as if θ were known. See Judge, et al (1985) and Carroll & Ruppert (1982, 1987) for references and proofs. Indeed, any generalized least squares estimate has the same limit distribution as weighted least squares based on the correct weights, i.e., the inverse of the square of (1.2).

The folklore theorem has an analogue in practice. In the linear regression model with a reasonably sized data set, since unweighted least squares is consistent its fitted values rarely differ much from the fitted values from a generalized least squares fit. Consequently, the usual practice is to treat the estimation of the variance function $g(x, \beta, \theta)$ fairly cavalierly, if at all. To quote Schwartz (1979), "there is one point of agreement among statistics texts and that is the minimal effect of weighting factors on fitted regression curves. Unless the variance nonuniformity is quite severe, the curve fitted to calibration data is likely to be nearly the same, whether or not the variance nonuniforming is included in the weighting factors". The narrow focus on estimating the mean is misplaced, as Schwartz later notes, see also Garden, et al. (1980). Sometimes the variance function is itself of importance. Box & Meyer (1985) state that "one distinctive feature of Japanese quality control improvement techniques is the use of statistical experimental design to study the effect of a number of factors on variance as well as the mean". Other times the variance function essentially determines the quantity of interest. This occurs, for example, in the estimation of the sensitivity of a chemical or biochemical assay, see Carroll, Davidian & Smith (1986). However, there are even more basic problems where the variance function is of considerable importance, namely prediction and calibration.

It is perhaps trite to state that how well one estimates the variance function has a large effect on how well one can do prediction and calibration. It is, however, a point that is rarely taken into account in practice, as any review of the techniques in the assay literature will show. There are two ways to see this point. The first is through an asymptotic theory outlined in section 3, where we show that the difference in the length of a prediction interval between θ known and unknown is asymptotically distributed with variance a monotone function of how well one estimates θ. The second and probably more useful way to see the effect of variance function estimation is through an example. The large costs involved in not weighting at all will be evident in this example, and will serve as an object lesson.

2. CALIBRATION AND PREDICTION

Calibration experiments start with a training or calibration sample $(y_1, x_1), \ldots, (y_N, x_N)$ and then fit models to the mean and variance structures. The real interest lies in an independent pair $(y_\square, x_\square)$. Sometimes $x_\square$ is known, and we wish to obtain confidence intervals for $y_\square$; this is prediction. Other times, $y_\square$ is easily measured but $x_\square$ is unknown and inference is to be made about it, see Rosenblatt & Spiegelman (1982) for a discussion of calibration.

For example, in an assay x might represent the concentration of a substance and y might represent a counted value or intensity level which varies with concentration. One will have a new value $y_\square$ of the count or intensity and wish to draw inference about the true concentration $x_\square$. The calibration sample is drawn so that we have a good understanding of how the response varies as a function of concentration. The regression equation relating the response to concentration is then inverted to predict the concentration from the observed response.

The remainder of this section we will assume that the responses are normally distributed, although this can be relaxed. Give a value $x_{\Box}$, the standard point estimate of the response $y_{\Box}$ is $f(x_{\Box}, \beta)$. Let $\hat{\beta}_G$ be a generalized least squares estimate, and define

$$S_G = S_G(\theta) = N^{-1} \sum_{i=1}^{N} f_\beta(x_i, \hat{\beta}_G) f_\beta(x_i, \hat{\beta}_G)^T / f(x_i, \hat{\beta}_G)^{2\theta},$$

$$\hat{\sigma}^2(\theta) = N^{-1} \sum_{i=1}^{N} \{y_i - f(x_i, \hat{\beta}_G)\}^2 / f(x_i, \hat{\beta}_G)^{2\theta},$$

where f_β is the derivative of f with respect to β. For large calibration data sets, the variance in the error made by prediction is

$$\text{Variance } \{y_{\Box} - f(x_{\Box}, \hat{\beta}_G)\} \simeq \sigma^2 q_N^2(x_{\Box}, \beta, \theta), \text{ where} \tag{2.1}$$

$$q_N^2(x_{\Box}, \beta, \theta) = g^2(x_{\Box}, \beta, \theta) + N^{-1} f_\beta(x_{\Box}, \beta)^T S_G^{-1} f_\beta(x_{\Box}, \beta).$$

Note that if the size N of the calibration data set is large, then the error in prediction is determined predominately by the variance function

$$\sigma^2 g^2(x_{\Box}, \beta, \theta),$$

and not by the calibration data set itself. An approximate $(1-\alpha)100\%$ confidence interval for the response $y_{\Box}$ is given by

$$I(x_{\Box}) = \{\text{all values } y \text{ in the interval} f(x_{\Box}, \hat{\beta}_G) \pm t_{1-\alpha/2}^{N-p} \hat{\sigma}_G q_N(x_{\Box}, \hat{\beta}_G, \theta)\}, \tag{2.2}$$

where $t_{1-\alpha/2}^{N-p}$ is the $(1-\alpha/2)$ percentage point of a t-distribution with $N-p$ degrees of freedom. For large sample sizes, this interval becomes

$$I(x_{\Box}) \simeq \{\text{all } y \text{ in the interval} f(x_{\Box}, \hat{\beta}_G) \pm t_{1-\alpha/2}^{N-p} \hat{\sigma}_G g(x_{\Box}, \hat{\beta}_G, \theta)\}. \tag{2.3}$$

The prediction interval (2.2) is only an approximate $(1-\alpha)100\%$ confidence interval because the function q_N is not known but rather must be estimated.

The effect of ignoring the heterogeneity can be seen through examination of (2.3). If $\hat{\sigma}_L^2$ is the unweighted mean squared error, then for large samples we have the approximation

$$\hat{\sigma}_L^2 \simeq \sigma^2 g_{\text{mean}}^2 = \sigma^2 N^{-1} \sum_{i=1}^{N} g^2(x_i, \beta, \theta).$$

Thus the unweighted prediction interval for large sample sizes is approximately

$$I_L(x_{\Box}) \simeq \{\text{all } y \text{ in the interval} f(x_{\Box}, \hat{\beta}_L) \pm t_{1-\alpha/2}^{N-p} \sigma g_{\text{mean}}\}. \tag{2.4}$$

Comparing (2.3) and (2.4) we see that where the variability is small, the unweighted prediction interval will be too long and hence pessimistic, and conversely where the variance is large.

Now suppose that we are given the value of the response $y_\square$ and wish to estimate and make inference about the unknown $x_\square$. The estimate of $x_\square$ is that value which satisfies $f(x_\square, \hat{\beta}_G) = y_\square$. The most common interval estimate of $x_\square$ is the set of all values x for which $y_\square$ falls in the prediction interval $I(x)$, i.e.,

$$\text{Calibration interval for } x_\square$$
$$= \{\text{all } x \text{ such that } y_\square \in I(x) \text{where } I(x) \text{ is given by (2.3)}\}.$$

The effect of not weighting is too long and pessimistic confidence intervals for $x_\square$ where the variance is small and the opposite where the variance is large. As far as we know, little work has been done to determine whether one can shorten the calibration confidence interval by making more direct use of the variance function.

3. ASYMPTOTICS

Assume throughout that the data are symmetrically distributed about their mean. Let $\hat{\beta}_G$ be any generalized least squares estimate of β based on an estimate of θ, call it $\hat{\theta}$. Davidian & Carroll (1986) introduce a class of estimators which depend on the data only through $\hat{\beta}_G$, the design $\{x_i\}$, and either sample variances from replicates at each design point or on transformations of the squared residuals

$$\{y_i - f(x_i, \hat{\beta}_G)\}^2.$$

This class of estimators includes most methods in the literature, see Judge, et al (1985). Davidian & Carroll (1986) show that all members of their class of estimators have the asymptotic expansion

$$N^{1/2}(\hat{\theta} - \theta) = W_N + a^T N^{1/2}(\hat{\beta}_G - \beta) + o_p(1). \tag{3.1}$$

In (3.1), a is a fixed vector and W_N is asymptotically normally distributed. Because the observations have symmetric distribution, W_N is asymptotically uncorrelated with $\hat{\beta}_G$.

Let $\hat{\beta}_G(\theta)$ and $\hat{\beta}_G(\hat{\theta})$ be generalized least squares estimates of β with θ known and unknown respectively, and let $\hat{\sigma}(\theta)$ and $\hat{\sigma}(\hat{\theta})$ be the corresponding estimates of σ. The length of the prediction intervals with θ known and unknown are proportional to $L(\theta)$ and $L(\hat{\theta})$ respectively, where

$$L(\theta) = \hat{\sigma}(\theta) q_N(x_\square, \hat{\beta}_G(\theta), \theta).$$

The random variable

$$\Delta L = N^{1/2}\{L(\hat{\theta}) - L(\theta)\}/\sigma$$

describes how well one approximates the length one would use if θ were known. Intuitively, we would like ΔL to have smallest possible variability.

Theorem. *Suppose that W_N in (3.1) is asymptotically normally distributed with mean zero and covariance $\mathrm{C} = \mathrm{C}(\hat{\theta})$ depending on the method of estimating θ. Then, under regularity conditions, ΔL is asymptotically normally distributed with variance an increasing function of $\mathrm{C}(\hat{\theta})$.*

NOTE. We have called C the covariance matrix since the result holds for vector valued θ in (1.2) and not just for the special case (1.3). The Theorem remains valid if ΔL is the normalized difference in length between the interval with θ unknown and the interval with completely specified variance function.

PROOF. (Sketch): It is easily seen that in the definition of L we may replace q_N by g. Further, $\Delta L = A_1 + A_2 + A_3$, where

$$A_1 = N^{1/2} g(x_\square, \hat{\beta}(\hat{\theta}), \hat{\theta})\{\hat{\sigma}(\hat{\theta}) - \hat{\sigma}(\theta)\}/\sigma$$

$$A_2 = N^{1/2}\{\hat{\sigma}(\theta)/\sigma\}\{g(x_\square, \hat{\beta}(\hat{\theta}), \hat{\theta}) - g(x_\square, \hat{\beta}(\hat{\theta}), \theta)\}$$

$$A_3 = N^{1/2}\{\hat{\sigma}(\theta)/\sigma\}\{g(x_\square;, \hat{\beta}(\hat{\theta}), \theta) - g(x_\square, \hat{\beta}(\theta), \theta)\}.$$

Now, $A_3 \xrightarrow{p} 0$ since, from Carroll & Ruppert (1987), we have that

$$N^{1/2}\{\hat{\beta}(\hat{\theta}) - \hat{\beta}(\theta)\}/\sigma \xrightarrow{p} 0,$$

By a Taylor series,

$$A_2 = N^{1/2} g_x(x_\square, \beta, \theta)(\hat{\theta} - \theta) + o_p(1).$$

Lemma A.3 of Carroll, Davidian & Smith shows that for some constant $b(x_\square)$,

$$A_1 = b(x_\square) N^{1/2}(\hat{\theta} - \theta\} + o_p(1).$$

This shows that for some constant $c(x_\square)$,

$$\Delta L = c(x_\square) N^{1/2}(\hat{\theta} - \theta) + o_p(1)$$

The proof is completed by applying (3.1). □

4. AN EXAMPLE

In Chapter 2, section 8, Carroll & Ruppert (1987) present the results of an assay for the concentration of an enzyme (esterase). There were 113 observations, of which 5 were deleted. The observed concentration of esterase was recorded and then a binding experiment was undertaken, so that the response is the count of the number of bindings. These data were given to us by another statistician and we are unable to give further detail into the background of the experiment. We do not know whether the recorded concentration of esterase has been accurately measured, although we will assume it has been and that there is little if any measurement error in this predictor. The lack of replicates in the response is rather unusual in our experience. Since the response is a count, one might expect Poisson variation, i.e., the power of the mean model holds with $\theta = 0.50$. In our experience with assays, such a model almost always underestimates θ, with values between 0.60 and 0.90 being much more common: see Finney (1976) and Raab (1981).

The eventual goal of the study is to take observed counts and infer the concentration of esterase, especially for smaller values of the latter. As is typical in these experiments, a calibration of training data set is taken for which the predictor variable esterase is known as is the counted response. Carroll & Ruppert (1982) plot the data, which appears reasonably although not perfectly linear. Actually, the logarithm of the response plotted against the logarithm of the predictor may appear more linear to some and less heteroscedastic. As is evident from that plot, the data exhibit rather severe heterogeneity of variance. The Spearman correlation between absolute studentized residuals and predicted values from an unweighted least squares fit is $\rho = 0.39$ with formal computed significance level ≤ 0.0001. Analysis as in Carroll & Ruppert (1982) indicate that the constant coefficient of variation model $\theta = 1.0$ is reasonable, although a value $\theta = 0.9$ might be even better. For $\theta = 1.0$, the Spearman correlation between absolute studentized residuals and predicted values if $\rho = -0.10$, with significance level 0.29. In Figure 1, we plot kernel regression estimates of the Anscombe studentized residuals, i.e., the absolute studentized residuals to the power 2/3, see McCullagh & Nelder (1983). Note that the plots indicate that $\theta = 1.0$ does a far better job of accounting for the heteroscedasticity.

In these data, the effect of not weighting should be to have prediction and calibration confidence intervals which are much too large for small amounts of esterase and conversely for large amounts. In Figure 2 we plot the 95% prediction intervals for the count response for unweighted versus weighted regression: the effect is clear. A similar plot for the calibration intervals shows the same effect: the unweighted analysis is much too conversative for small amounts of esterase, and much too liberal for larger amounts. As Oppenheimer, et al (1983) state, "Rather dramatic differences have been observed depending on whether a valid weighted or invalid unweighted analysis is used".

This example shows that the actual prediction intervals are sensitive to misspecification of the variance function. It should be clear by inference and the previous section that one should make efforts to estimate the structural variance parameter θ as well as possible.

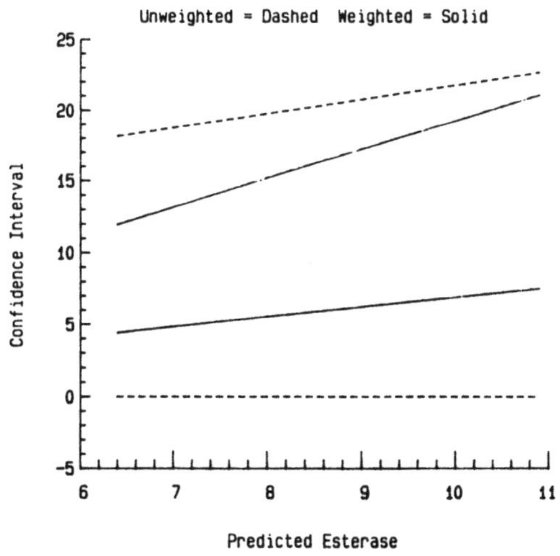

Figure 1. The Esterase Assay Data. This is a plot of the kernel regression fits to the Anscombe absolute residuals against the logarithms of the predicted values. The unweighted least squares fit is the solid line, while the generalized least squares fit for the constant coefficient of variation model is the dashed line. Endpoint effects have been adjusted for by selective deletion.

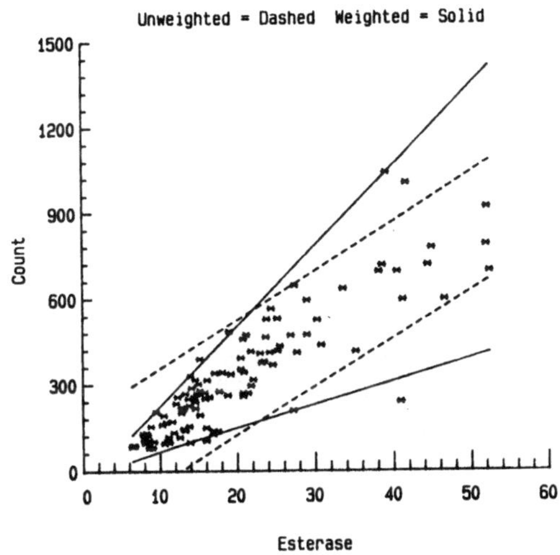

Figure 2. The Esterase Assay Data. These are the 95% prediction intervals for a new response. The dashed line is unweighted least squares, while the solid line is the constant coefficient of variation fit. The lower part of the least squares interval has been truncated at zero where necessary.

ACKNOWLEDGMENTS

Research supported by the Air Force Office of Scientific Research Contract AFOSR-F-49620-85-C-0144.

BIBLIOGRAPHY

Box, G. E. P. and Myer, R. D. (1985). Dispersion effects from fractional designs. *Technometrics* **28**, 19–28.

Carroll, R. J., Davidian, M. and Smith, W. (1986). Variance function estimation and the minimum detectable concentration in assays. Preprint.

Carroll, R. J., and Ruppert, D. (1982). Robust estimation in heteroscedastic linear models, *Annals of Statistics* **10**, 429–441.

Carroll, R. J., and Ruppert, D. (1987) *Transformations and Weighting in Regression.* Chapman and Hall, London.

Davidian, M. and Carroll, R. J. (1986). Variance function estimation in regression. Preprint.

Finney, D. J. (1976). Radiologand assay. *Biometrics* **32**, 721–740.

Garden, J. S., Mitchell, D. G. and Mills, W. N. (1980). Nonconstant variance regression techniques for calibration curve based analysis. *Analytical Chemistry* **52**, 2310–2315.

Judge, G. G., Griffiths, W. E., Hill, R. C., Lutkepohl, H. and Lee, T. C. (1985). *The Theory and Practice of Econometrics, Second Edition.* John Wiley and Sons, New York.

McCullagh, P. and Nelder, J. A. (1983). *Generalized Linear Models.* Chapman & Hall, New York.

Oppenheimer, L., Capizzi, T. P., Weppelman, R. M. and Mehto, H. (1983) Determining the lowest limit of reliable assay measurement. *Analytical Chemistry* **55**, 638–643.

Raab, G. M. (1981). Estimation of a variance function, with application to radioimmunoassay. *Applied Statistics* **30**, 32–40.

Rosenblatt, J. R. and Spiegelman, C. H. (1981). Discussion of the paper by Hunter and Lamboy. *Technometrics* **23**, 329–333.

Schwartz, L. M. (1979). Calibration curves with nonuniform variance. *Analytical Chemistry* **51**, 723–729.

ON ESTIMATING A PARAMETER AND ITS SCORE FUNCTION, II

C. A. J. KLAASSEN, A. W. VAN DER VAART AND W. R. VAN ZWET

Department of Mathematics
University of Leiden
Leiden, Netherlands

1. INTRODUCTION

A bound of Cramér-Rao type is provided for an estimator of a real-valued parameter θ in the presence of an abstract nuisance parameter η, such as an unknown distributional shape, on the basis of N i.i.d. observations. The bound consists of the reciprocal of the effective Fisher information in the sample, plus a term involving the integrated mean squared error of an estimator of a multiple of the so-called conditional score function for θ, for the case where θ is known. This implies that an estimator of θ can only perform well over a class of shapes η if it is possible to estimate the conditional score function for θ accurately over this class. For the special case where fully adaptive estimation may be possible, this result was given in a companion paper (Klaassen and van Zwet (1985)).

2. AN INEQUALITY OF CRAMÉR-RAO TYPE

Let $X_1, \ldots, X_N$ be independent and identically distributed (i.i.d.) random variables taking values in some measurable space $(\mathcal{X}, \mathcal{A})$, with a common density $f(\cdot;\eta,\theta)$ with respect to a σ-finite measure μ on $(\mathcal{X}, \mathcal{A})$. The parameter of interest θ belongs to an open subset Θ of $\mathbb{R}$ and the nuisance parameter η ranges over an arbitrary set H. For unknown η and θ, it is required to estimate θ and this is done by means of an estimator $T_N = T_N(X_1, \ldots, X_N)$ for some measurable function T_N: $\mathcal{X}^N \to \mathbb{R}$. We are interested in finding a lower bound for the variance of T_N under $f(\cdot;\eta,\theta)$. We shall write $P_{\eta\theta}$, $E_{\eta\theta}$ and $\sigma^2_{\eta\theta}$ for probabilities, expectations and variances under this model.

For every fixed $\theta \in \Theta$ and $j = 1, \ldots, N$, let $\psi(X_j;\theta)$ be a sufficient statistic for X_j with respect to $\eta \in H$. According to the factorization theorem this is equivalent to assuming that

$$f(x;\eta,\theta) = g(\psi(x;\theta);\eta,\theta)h(x;\theta) \quad \text{a.e. } [\mu], \tag{2.1}$$

where $g(\cdot;\eta,\theta)$ may be chosen to be the density of $\psi(X_1;\theta)$ with respect to a σ-finite measure ν_θ.

We shall assume that $f^{\frac{1}{2}}$ is differentiable in quadratic mean with respect to θ with a derivative which is not essentially zero, thus for every (η, θ)

$$\lim_{\theta' \to \theta} \int [(\theta' - \theta)^{-1}\{f^{\frac{1}{2}}(x;\eta,\theta') - f^{\frac{1}{2}}(x;\eta,\theta)\} - \frac{1}{2}\tau(x;\eta,\theta)f^{\frac{1}{2}}(x;\eta,\theta)]^2 d\mu(x) = 0, \tag{2.2}$$

$$I(\eta,\theta) = E_{\eta\theta}\tau^2(X_1;\eta,\theta) > 0. \tag{2.3}$$

Obviously (2.2) implies that $I(\eta,\theta) < \infty$. Note that $\tau(\cdot\,;\eta,\theta)$ is simply an L_2-version of the classical score function for θ, $\partial \log f(x;\eta,\theta)/\partial\theta$; $I(\eta,\theta)$ is the Fisher information concerning θ which is contained in a single observation X_1 and measures how well θ can be estimated when η is known. However, since η is unknown, one expects the information concerning θ to be smaller. As discussed in Begun, Hall, Huang and Wellner (1983), the information loss results from a reduction of the score function.

First we define score functions in the η-direction. We shall say that $\beta(\cdot\,;\eta,\theta)$ is an η-score function if there exists a sequence $\eta_k \in H$ such that

$$\lim_{k\to\infty} \int [k\{f^{\frac{1}{2}}(x;\eta_k,\theta) - f^{\frac{1}{2}}(x;\eta,\theta)\} - \frac{1}{2}\beta(x;\eta,\theta)f^{\frac{1}{2}}(x;\eta,\theta)]^2 d\mu(x) = 0. \tag{2.4}$$

It is easy to see that, in view of (2.1), (2.4) implies that

$$\beta(x;\eta,\theta) = b(\psi(x;\theta);\eta,\theta) \quad \text{a.e. } [P_{\eta\theta}], \tag{2.5}$$

where b satisfies

$$\lim_{k\to\infty} \int [k\{g^{\frac{1}{2}}(v;\eta_k,\theta) - g^{\frac{1}{2}}(v;\eta,\theta)\} - \frac{1}{2}b(v;\eta,\theta)g^{\frac{1}{2}}(v;\eta,\theta)]^2 d\nu_\theta(v) = 0, \tag{2.6}$$

so that b is an η-score function for the model $\{g(\cdot\,;\eta,\theta)\colon\ \eta \in H, \theta \in \Theta\}$. Let $B(\eta,\theta)$ denote the set of all η-score functions for the original model — i.e. functions β for which (2.4)–(2.5) hold for an appropriate sequence $\eta_k \in H$ — and let $\tilde{B}(\eta,\theta)$ be the closure in L_2 of the linear span of $B(\eta,\theta)$.

The effective score function τ_E for θ in the presence of the nuisance parameter η, is defined as

$$\tau_E(x;\eta,\theta) = \tau(x;\eta,\theta) - b_E(\psi(x;\theta);\eta,\theta), \tag{2.7}$$

where $b_E(\psi(x;\theta);\eta,\theta)$ is the L_2-projection of τ on $\tilde{B}(\eta,\theta)$, thus

$$\begin{aligned} I_E(\eta,\theta) &= E_{\eta\theta}\{\tau(X_1;\eta,\theta) - b_E(\psi(X_1;\theta);\eta,\theta)\}^2 \\ &= \min_{\beta\in\tilde{B}(\eta,\theta)} E_{\eta\theta}\{\tau(X_1;\eta,\theta) - \beta(X_1)\}^2. \end{aligned} \tag{2.8}$$

$I_E(\eta,\theta)$ is the effective Fisher information, which measures how well θ can be estimated when η is unknown (cf. Begun et al. (1983), but note that we do not assume that $B(\eta,\theta)$ itself is a linear space).

Let $C(\eta,\theta)$ denote the set of all square-integrable functions $b(\psi(x;\theta))$ with $E_{\eta\theta}\, b(\psi(X_1;\theta)) = 0$. In the special case where $\tilde{B}(\eta,\theta) = C(\eta,\theta)$, $b_E(v;\eta,\theta)$ equals the conditional expectation $E_{\eta\theta}(\tau(X_1;\eta,\theta)|\psi(X_1;\theta) = v)$ and τ_E and I_E

reduce to

$$\tau_C(x;\eta,\theta) = \tau(x;\eta,\theta) - E_{\eta\theta}(\tau(X_1;\eta,\theta)|\psi(X_1;\theta) = \psi(x;\theta)), \tag{2.9}$$

$$I_C(\eta,\theta) = E_{\eta\theta}\tau_C^2(X_1;\eta,\theta), \tag{2.10}$$

which are called the conditional score function and the conditional Fisher information for θ. In general, however, $\tilde{B}(\eta,\theta)$ may be a proper subset of $C(\eta,\theta)$, and hence

$$I_C(\eta,\theta) \leq I_E(\eta,\theta) \leq I(\eta,\theta) \tag{2.11}$$

as is clear from figure 1. Of course, we still have $\tau_C = \tau_E$ and $I_C = I_E$ if $E_{\eta\theta}(\tau(X_1;\eta,\theta)|\psi(X_1;\theta) = \psi(\cdot\,;\theta))$ happens to be in $\tilde{B}(\eta,\theta)$.

So far we have discussed various aspects of the model. Concerning the estimator T_N, we assume that, whenever $E_{\eta\theta}\, T_N^2 < \infty$ for a certain (η,θ), then

$$\sup_{(\eta',\theta')\in A_\varepsilon} E_{\eta'\theta'}T_N^2 < \infty \tag{2.12}$$

for some $\varepsilon > 0$, where

$$A_\varepsilon = \{(\eta',\theta') \colon \int |f(x;\eta',\theta') - f(x;\eta,\theta)|d\mu(x) < \varepsilon\} \tag{2.13}$$

consists of parameter values "close" to (η,θ). For simplicity we shall also assume that T_N is an unbiased estimator of θ, i.e. for all (η,θ),

$$E_{\eta\theta}T_N = \theta. \tag{2.14}$$

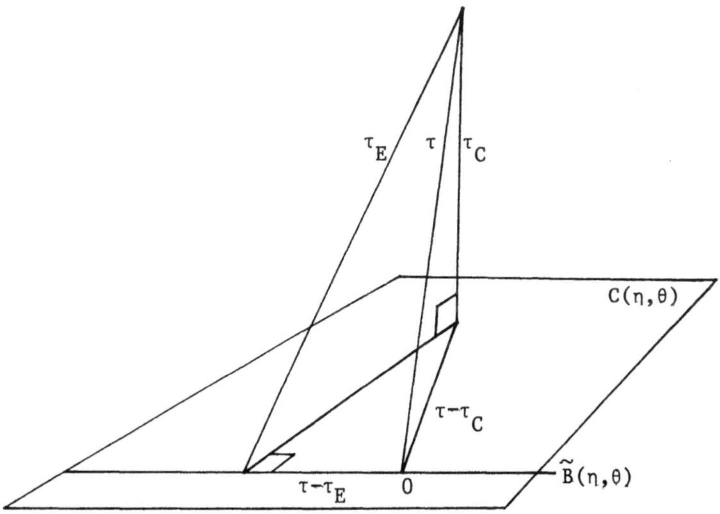

Figure 1.

The role of assumption (2.12)–(2.13) is to ensure that (2.14) implies that

$$E_{\eta\theta}T_N \sum_{j=1}^{N} \tau(X_j;\eta,\theta) = 1, \tag{2.15}$$

$$E_{\eta\theta}T_N \sum_{j=1}^{N} \beta(X_j) = 0 \qquad \text{for all } \beta \in \tilde{B}(\eta,\theta), \tag{2.16}$$

and hence in particular

$$E_{\eta\theta}T_N \sum_{j=1}^{N} \tau_E(X_j;\eta,\theta) = 1 \tag{2.17}$$

in view of (2.7)–(2.8). If we define

$$S_N(\eta,\theta) = \frac{1}{NI_E(\eta,\theta)} \sum_{j=1}^{N} \tau_E(X_j;\eta,\theta) \tag{2.18}$$

then (2.17) asserts that $S_N(\eta,\theta)$ and $T_N - S_N(\eta,\theta)$ are uncorrelated under $P_{\eta\theta}$. As a consequence

$$\begin{aligned} \sigma^2_{\eta\theta}(T_N) &= \sigma^2_{\eta\theta}(S_N(\eta,\theta)) + \sigma^2_{\eta\theta}(T_N - S_N(\eta,\theta)) \\ &= \{N\ I_E(\eta,\theta)\}^{-1} + \sigma^2_{\eta\theta}(T_N - S_N(\eta,\theta)), \end{aligned} \tag{2.19}$$

and hence

$$\sigma^2_{\eta\theta}(T_N) \geq \{N\ I_E(\eta,\theta)\}^{-1} \tag{2.20}$$

which is the form the Cramér-Rao inequality takes for unbiased estimation of θ in the presence of the nuisance parameter η. Note that this is the fixed sample size version of the corresponding asymptotic results in Begun et al. (1983).

However, (2.19) contains essentially more information than inequality (2.20). It implies that $\sigma^2_{\eta\theta}(T_N)$ can only come close to the Cramér-Rao bound (2.20) if $(T_N - \theta)$ is close to $S_N(\eta,\theta)$ under $P_{\eta\theta}$. It follows that if T_N performs well as an estimator of θ for all $\eta \in H$, then $(T_N - \theta)$ must resemble $S_N(\eta,\theta)$ under $P_{\eta\theta}$ for all $\eta \in H$ and $\theta \in \Theta$. It would seem therefore that $(T_N - \theta)$ contains information about the unknown function $\tau_E(\cdot\,;\eta,\theta)$. Let us try to extract this information.

For $S_N(\eta,\theta)$ as defined in (2.18) we have

$$\begin{aligned} &E_{\eta\theta}(S_N(\eta,\theta)|\psi(X_j;\theta) \quad \text{for } j \neq i; X_i = x) \\ &- E_{\eta\theta}(S_N(\eta\theta)|\psi(X_j;\theta) \quad \text{for } j \neq i; \psi(X_i;\theta) = \psi(x;\theta)) \\ &= \frac{1}{N\ I_E(\eta,\theta)}[\tau_E(x;\eta\theta) - E_{\eta,\theta}(\tau_E(X_1;\eta,\theta)|\psi(X_1;\theta) = \psi(x;\theta))] \end{aligned} \tag{2.21}$$

$$= \frac{1}{N\, I_E(\eta,\theta)} \tau_C(x;\eta,\theta)$$

in view of (2.7) and (2.9). If $(T_N - \theta)$ resembles $S_N(\eta,\theta)$ under $P_{\eta\theta}$, we may hope that

$$J_N(x;\theta) = \sum_{i=1}^{N} \{E_{\eta\theta}(T_N | \psi(X_j;\theta) \quad \text{for } j \neq i; X_i = x) \qquad (2.22)$$
$$- E_{\eta\theta}(T_N | \psi(X_j;\theta) \quad \text{for } j \neq i; \psi(X_i;\theta) = \psi(x;\theta))\}$$

can serve as an estimator of

$$J(x;\eta,\theta) = \frac{1}{I_E(\eta,\theta)} \tau_C(x;\eta,\theta). \qquad (2.23)$$

Note that since for each j, $\psi(X_j;\theta)$ is sufficient for X_j with respect to $\eta \in H$ for fixed $\theta \in \Theta$, J_N is indeed independent of η. For known θ, it is therefore a legitimate estimator of J.

We shall prove the following result.

Theorem 2.1. *Suppose that assumptions* (2.1)–(2.3) *and* (2.12)–(2.14) *are satisfied for every* (η,θ). *Then, for every* (η,θ),

$$\sigma^2_{\eta\theta}(T_N) \geq \frac{1}{N\, I_E(\eta\theta)} + \frac{1}{N} E_{\eta\theta} \int \{J_N(x;\theta) - J(x;\eta,\theta)\}^2 f(x;\eta,\theta)\, d\mu(x). \quad (2.24)$$

The theorem asserts that the Cramér-Rao bound (2.20) may be improved by adding N^{-1} times the integrated mean squared error (MSE) of the estimator J_N of the function J, which is an unknown multiple of the conditional score function τ_C. For practical purposes it is unsatisfactory that the right-hand side of (2.24) depends on the choice of T_N. However, one may obviously rephrase the theorem to assert only the existence of an estimator J_N such that (2.24) holds. The message of the theorem is then clear: the accuracy with which one can estimate θ for unknown η is delimited by the effective Fisher information on the one hand and by how well one can estimate $J(\cdot\,;\eta,\theta)$ for known θ on the other. Clearly the latter depends heavily on the class H. If $J(\cdot\,;\eta,\theta)$ runs through a large class of functions as η ranges over H, then the integrated MSE of any estimator of J may be quite large, especially for particularly irregular choices of J. If J is restricted to a smaller class of nicely behaved functions as $\eta \in H$, then the integrated MSE may be much smaller. Finally, if η is known so that H consists of a single element, then $J(\cdot\,;\eta,\theta)$ can serve as an estimator of itself and (2.24) reduces to the Cramér-Rao inequality (2.20).

In a sense, the result of the theorem is not surprising. Estimation of a parameter θ for an unknown distributional shape is based typically on a preliminary estimate of an unknown score function followed by a good estimate of θ for the distributional shape corresponding to the estimated score function.

For such estimators a result like (2.24) is to be expected. The interesting aspect of the theorem, however, is that it is not assumed that the estimator T_N is based on a preliminary estimate of a score function, but that an estimate of J for known θ is derived from T_N. In effect we are saying that a successful estimation procedure for θ must involve — either explicitly or implicitly — the estimation of J and that because of this, the accuracy of estimating J enters into the lower bound for the variance of the estimator of θ.

Although the theorem is purely a finite sample result, it obviously has asymptotic implications. An asymptotic analogue would imply that effective estimation of θ, i.e.

$$\{N\ I_E(\eta,\theta)\}^{\frac{1}{2}}(T_N-\theta) \xrightarrow{D} N(0,1), \tag{2.25}$$

is possible only if the function J can be estimated consistently with respect to integrated MSE for known θ. In this context it is unsatisfactory that J involves the conditional score function τ_C rather than the effective score function τ_E and, indeed, Klaassen (1987) has shown that a somewhat stronger version of (2.25) does entail consistent estimation of $\tau_E/I_E(\eta,\theta)$.

Of course this discrepancy disappears if $\tau_C=\tau_E$, i.e. if the function $E_{\eta\theta}(\tau(X_1;\eta,\theta)|\psi(X_1;\theta)=\psi(x;\theta))$ is an element of $\tilde{B}(\eta,\theta)$. This situation is rather common and examples, including non-i.i.d. models, are given by van der Vaart (1986), who also explicitly constructs an effective estimator of θ based on a preliminary consistent estimator of τ_C for such models.

An even more special case occurs if $\tau_C=\tau$, so that $I_E=I$ and $J=\tau/I$. Now (2.24) provides a finite sample analogue of the statement that fully adaptive estimation of θ is possible only if τ/I can be estimated consistently. This situation was discussed in the companion paper Klaassen and van Zwet (1985).

3. PROOF OF THE THEOREM

The proof resembles that of theorem 1.1 in Klaassen and van Zwet (1985). Let

$$f_N(x)=\prod_{j=1}^{N} f(x_j;\eta,\theta)$$

denote the density of $X=(X_1,\ldots,X_N)$ with respect to the N-fold product measure μ^N taken at the point $x=(x_1,\ldots,x_N)$ and write

$$\frac{f_N^{\frac{1}{2}}(x;\eta,\theta')-f_N^{\frac{1}{2}}(x;\eta,\theta)}{(\theta'-\theta)}=\frac{1}{2}\rho_N(x;\eta,\theta)+\Delta_N(x;\eta,\theta,\theta'), \tag{3.1}$$

with

$$\rho_N(x;\eta,\theta) = f_N^{\frac{1}{2}}(x;\eta,\theta)\sum_{i=1}^{N} \tau(x_i;\eta,\theta). \tag{3.2}$$

Since N is fixed, a standard argument shows that (2.2) implies

$$\lim_{\theta'\to\theta} \int \Delta_N^2(x;\eta,\theta,\theta')d\mu_N(x) = 0. \tag{3.3}$$

In view of (2.14) we have

$$1 = \int T_N(x)\frac{f_N^{\frac{1}{2}}(x;\eta,\theta') - f_N^{\frac{1}{2}}(x;\eta,\theta)}{(\theta'-\theta)}\{f_N^{\frac{1}{2}}(x;\eta,\theta') + f_N^{\frac{1}{2}}(x;\eta,\theta)\}d\mu_N(x) \tag{3.4}$$

$$= \int T_N(x)\{\frac{1}{2}\rho_N(x;\eta,\theta) + \Delta_N(x;\eta,\theta,\theta')\}\{f_N^{\frac{1}{2}}(x;\eta,\theta') + f_N^{\frac{1}{2}}(x;\eta,\theta)\}d\mu_N(x).$$

If $E_{\eta\theta}T_N^2 = \infty$, there is nothing to prove. Suppose therefore that $E_{\eta\theta}T_N^2 < \infty$. Since (2.2) ensures that

$$\lim_{\theta'\to\theta} \int |f(x;\eta,\theta') - f(x;\eta,\theta)|d\mu(x) = 0,$$

(2.12) and (2.13) yield

$$\limsup_{\theta'\to\theta} E_{\eta\theta'}T_N^2 < \infty. \tag{3.5}$$

Together, (3.3), (3.5) and the Cauchy-Schwarz inequality show that

$$\lim_{\theta'\to\theta} \int T_N(x)\Delta_N(x;\eta,\theta,\theta')\{f_N^{\frac{1}{2}}(x;\eta,\theta') + f_N^{\frac{1}{2}}(x;\eta,\theta)\}d\mu_N(x) = 0, \tag{3.6}$$

$$|\int T_N(x)\rho_N(x;\eta,\theta)\{f_N^{\frac{1}{2}}(x;\eta,\theta') - f_N^{\frac{1}{2}}(x;\eta,\theta)\}d\mu_N(x)|$$

$$\leq \{C^2 \int \{f_N^{\frac{1}{2}}(x;\eta,\theta') - f_N^{\frac{1}{2}}(x;\eta,\theta)\}2d\mu_N(x) \int \rho_N^2(x;\eta,\theta)d\mu_N(x)\}^{\frac{1}{2}} \tag{3.7}$$

$$+\{\int T_N^2(x)\{f_N^{\frac{1}{2}}(x;\eta,\theta') - f_N^{\frac{1}{2}}(x;\eta,\theta)\}^2d\mu_N(x) \int_{\{|T_N|>C\}} \rho_N^2(x;\eta,\theta)d\mu_N(x)\}^{\frac{1}{2}}$$

for every $C > 0$. As θ' tends to θ, the first term on the right tends to zero for every C in view of (2.2). Since $E_{\eta\theta}T_N^2 < \infty$ and $E_{\eta\theta}\tau^2(X_1;\eta,\theta) < \infty$, the second term converges to zero as $C \to \infty$. It follows that the left-hand side of (3.7) converges to zero, and together with (3.4) and (3.6) this proves (2.15). A similar argument produces (2.16) and (2.19) follows.

It remains to show that

$$\sigma_{\eta\theta}^2(T_N - S_N(\eta,\theta)) \geq \frac{1}{N}E_{\eta\theta}\int\{J_N(x;\theta) - J(x;\eta,\theta)\}^2 f(x;\eta,\theta)d\mu(x). \tag{3.8}$$

To see this, we copy the argument leading from (2.9) to (2.11) in Klaassen and van Zwet (1985), even though $S_N(\eta,\theta)$ is defined differently in that paper. We find

$$\begin{aligned}&\sigma^2_{\eta\theta}(T_N - S_N(\eta,\theta)) \qquad (3.9)\\ &\geq N^{-1}E_{\eta\theta}\int\{\sum_{i=1}^{N}[E_{\eta\theta}(T_N - S_N(\eta,\theta)|\psi(X_j;\theta) \text{ for } j\neq i; X_i = x)\\ &- E_{\eta\theta}(T_N - S_N(\eta,\theta)|\psi(X_j;\theta) \text{ for } j\neq i; \psi(X_i;\theta)=\psi(x;\theta))]\}^2 f(x;\eta,\theta)d\mu(x).\end{aligned}$$

In view of (2.21)–(2.23), (3.9) is identical to (3.8) and the proof complete.

ACKNOWLEDGMENTS

Research was supported in part by the Office of Naval Research Contract N00014-80-C-0163.

BIBLIOGRAPHY

Begun, J. M., Hall, W. J., Huang, W. M. and Wellner, J. A. (1983). Information and asymptotic efficiency in parametric-nonparametric models. *Ann. Statist.* **11**, 432–452.

Klaassen, C. A. J. (1987). Consistent estimation of the influence function of locally asymptotically linear estimators. *Ann. Statist.*, to appear.

Klaassen, C. A. J. and van Zwet, W. R. (1985). On estimating a parameter and its score function. *Proc. Berkeley Conference in Honor of Jerzy Neyman and Jack Kiefer*, L. Le Cam and R. A. Olshen, editors, Vol. II, 827–839.

van der Vaart, A. W. (1986). Estimating a real parameter in a class of semiparametric models. Submitted to *Ann. Statist.*

A SIMPLE TEST FOR THE EQUALITY OF CORRELATION MATRICES

KINLEY LARNTZ

School of Statistics
University of Minnesota
St. Paul, Minnesota 55108, U.S.A.

MICHAEL D. PERLMAN

Department of Statistics
University of Washington
Seattle, Washington 98195, U.S.A.

1. INTRODUCTION

1.1. THE TESTING PROBLEM

Suppose $R^{(1)}, \ldots, R^{(k)}$ are sample correlation matrices from k independent p-variate normal populations ($k \geqq 2$). Each $R^{(\beta)} \equiv (r_{ij}^{(\beta)})$ is based on a random sample of size n_β. Denote the βth population correlation matrix by $P^{(\beta)} \equiv (\rho_{ij}^{(\beta)})$. In this paper we consider the problem of testing the hypothesis

$$H_0\colon\ P^{(1)} = \ldots = P^{(k)} (\equiv P),$$

where $P \equiv (\rho_{ij})$ is an *unspecified* nonsingular correlation matrix, against the general alternative.

Seal (1964, p. 124) and Jennrich (1970, p. 910) have noted that this testing problem arises in biological applications and factor analysis studies. Seal stated that "unfortunately there is presently no criterion available to test the hypothesis" H_0. As pointed out by Jennrich (1970, p. 904), computation of the likelihood ratio test (LRT) statistic would require numerical maximization of a function of $p + p(p-1)/2 \doteq p^2/2$ variables under H_0, a difficult task unless the dimension p is very small. Kullback (1967) proposed the test which rejects H_0 for large values of the statistic

$$T_1 = \sum_{\beta=1}^{k} N_\beta \log \frac{|R|}{|R^{(\beta)}|},$$

where $N_\beta = n_\beta - 1$ and $R = \Sigma N_\beta R^{(\beta)} / \Sigma N_\beta$. Kullback asserted that under H_0, T_1 has an asymptotic χ^2 distribution with $(k-1)p(p-1)/2$ degrees of freedom. However, Aitken (1969, Section 5) and Jennrich (1970, Section 2) have shown that this assertion is incorrect, the asymptotic distribution of T_1 under H_0 in fact being a linear combination of independent χ^2 variates with the weights depending on the unknown value of the common correlation matrix P. One might attempt to estimate these weights from the data, as Aitken suggests, but this would entail difficult computations and would alter the significance level in an unknown way for finite sample size.

1.2. Jennrich's Test Procedure

Jennrich (1970) proposed a test statistic T_2 which suffers from neither of the above drawbacks — the computation of T_2 is only of moderate difficulty (involving the inversion of two $p \times p$ matrices), and T_2 does possess an asymptotic χ^2 distribution under H_0. For simplicity, we confine attention to the case $k = 2$ now, returning to the case $k \geqq 3$ in the final section. Let $\underline{r}^{(\beta)}$ denote the $p(p-1)/2$-dimensional column vector consisting of the elements $r_{ij}^{(\beta)}$ of $R^{(\beta)}(1 \leqq i < j \leqq p)$ arranged in lexicographic order, and let $c = n_1 n_2 (n_1 + n_2)^{-1}$. As $\min(n_1, n_2) \to \infty$, the random vector $c^{1/2}(\underline{r}^{(1)} - \underline{r}^{(2)})$ is asymptotically normal under H_0 with zero mean vector and nonsingular covariance matrix $\Gamma = \Gamma(P)$ which depends on P in a known way (see Jennrich (1970), Section 3). Also, the matrix $\overline{R} = \Sigma n_\beta R^{(\beta)} / \Sigma n_\beta$ is a consistent estimator of P under H_0, so $\hat{\Gamma} \equiv \Gamma(\overline{R})$ is a consistent estimator of Γ. Thus, Jennrich's test statistic

$$T_2 \equiv c(\underline{r}^{(1)} - \underline{r}^{(2)})' \hat{\Gamma}^{-1} (\underline{r}^{(1)} - \underline{r}^{(2)})$$

has an asymptotic χ^2 distribution with $p(p-1)/2$ degrees of freedom under H_0, and H_0 is rejected for large values of T_2. Fortunately, inversion of the $[p(p-1)/2] \times [p(p-1)/2]$ matrix $\hat{\Gamma}$ is not necessary, for Jennrich (1970, equation (4.6)) has provided a computationally simpler formula for T_2 which only requires inversion of two $p \times p$ matrices.

Although Jennrich's test has much better computational and distributional properties than the LRT and Kullback's test, it is basically a large sample test and can perform poorly for small samples, as we show in Section 3. The main purpose of this paper is to propose a test statistic T_3 which can be readily computed by hand, and which determines a test with reasonable small sample properties and with power comparable to that of Jennrich's test for large samples. Furthermore, our test remains valid even if one or more of the population correlation matrices $P^{(\beta)}$ are singular, a situation which may occasionally arise in application and under which Jennrich's statistic T_2 is undefined.

Section 3 contains a summary of our recommendations. In the final section we discuss the application of our procedures in the case $k \geqq 3$, and also discuss the one-sample problem. Also, we show how our procedure can be applied to obtain simultaneous confidence intervals for the population correlation coefficients in the one-sample case.

For additional work and references on tests for correlation matrices, see

Steiger and Browne (1984).

2. OTHER TEST PROCEDURES

2.1. A New Test Procedure

The basic idea is to apply the Fisher z-transformation to each sample correlation coefficient $r_{ij}^{(\beta)}$ obtaining

$$z_{ij}^{(\beta)} = \frac{1}{2}\ell n \left[\frac{(1 + r_{ij}^{(\beta)})}{(1 - r_{ij}^{(\beta)})} \right],$$

and to consider the $p(p-1)/2$-dimensional random column vector $\underline{z}^{(\beta)}$ consisting of the $z_{ij}^{(\beta)}$ arranged in lexicographic order. Let $d = (n_1 - 3)(n_2 - 3)(n_1 + n_2 - 6)^{-1}$. As $\min(n_1, n_2) \to \infty$, the random vector $d^{1/2}(\underline{z}^{(1)} - \underline{z}^{(2)})$ is asymptotically normal under H_0 with zero mean vector and covariance matrix $D\Gamma D$, where $D \equiv D(P)$ is the $[p(p-1)/2] \times [p(p-1)/2]$ diagonal matrix whose $(ij, ij)^{th}$ entry is $(1 - \rho_{ij}^2)^{-1}$. The significant aspect of the matrix $D\Gamma D$ is that only its off-diagonal elements involve the unknown matrix P, the diagonal elements of $D\Gamma D$ each being 1. By a result of Sidák (1967), therefore, one has that for any constant $b_{ij} > 0$,

$$\begin{aligned} &\text{Prob}\{d^{1/2}|z_{ij}^{(1)} - z_{ij}^{(2)}| \leqq b_{ij}, \quad 1 \leq i < j \leq p\} \\ &\geqq \prod_{1 \leqq i < j \leqq p} \text{Prob}\{d^{1/2}|z_{ij}^{(1)} - z_{ij}^{(2)}| \leqq b_{ij}\} \\ &= \prod_{1 \leqq i < j \leqq p} [2\Phi(b_{ij}) - 1] \end{aligned}$$

asymptotically under H_0 regardless of the value of P, where Φ is the c.d.f. of the standard normal distribution. Equality holds if $P = I$.

In particular, if we define the test statistic T_3 by

$$T_3 = d^{1/2} \max_{1 \leqq i < j \leqq p} |z_{ij}^{(1)} - z_{ij}^{(2)}|,$$

then for any $b > 0$

$$\text{Prob}\{T_3 \leqq b\} \geqq [2\Phi(b) - 1]^{p(p-1)/2}$$

asymptotically under H_0. Thus, we propose the (possibly conservative) level α test under which H_0 is rejected if $T_3 > b_\alpha$, where $b_\alpha > 0$ is chosen such that

$$[2\Phi(b_\alpha) - 1]^{p(p-1)/2} = 1 - \alpha. \tag{2.1}$$

In effect, this procedure simultaneously tests each of the $p(p-1)/2$ univariate hypothesis $\rho_{ij}^{(1)} = \rho_{ij}^{(2)}$, treating the differences $z_{ij}^{(1)} - z_{ij}^{(2)}$ as if they were independent, and relying on Sidák's result to provide an upper bound for the significance level.

Our procedure has several potential advantages over Jennrich's test especially for small sample size: (a) the value of T_3 is easily computable with only a hand calculator; (b) the z-transform of a sample correlation coefficient r converges in distribution to normality faster than r itself; (c) T_2 requires large sample sizes for $\hat{\Gamma} = \Gamma(\overline{R})$ to be an accurate estimate of $\Gamma = \Gamma(P)$ under H_0 (the same is true for T_4 and T_5). These facts, especially (c), may explain the breakdown of T_2, T_4, T_5 when the sample sizes are small relative to the dimension p (see Section 2.4). By contrast, the procedure T_3 does not require estimation of any unknown parameters and does not break down for small sample sizes. In fact, this procedure may be applied when $n_\beta < p+1$ for one or more of the samples (in which case $R^{(\beta)}$ is singular), and even when, under H_0, the common population correlation matrix P is singular, in which case the statistics T_2, T_4, and T_5 are not defined since $\overline{R}$ and hence $\hat{\Gamma}$ will be singular.

2.2. An Example

Jennrich (1970, Section 6) applied his test statistic T_2 to two 12×12 sample correlation matrices $R^{(1)}$ and $R^{(2)}$ computed from test scores of a group of 40 retarded children and a group of 89 normal children respectively ($p = 12$, $n_1 = 40$, $n_2 = 89$). The matrices $R^{(1)}$ and $R^{(2)}$ appear at the bottom of page 910 of Jennrich (1970). He found that $T_2 = 74.8$ which, when referred to the χ^2 distribution with $12(11)/2 = 66$ degrees of freedom, corresponds to an attained significance level of .22, hence is not significant at the 5 per cent level. Jennrich attributes this lack of significance to "the relatively small sample size, 40, of the sample of retarded children" and to "the similarity of the observed correlation matrices." Inspection shows, however, that $R^{(1)}$ and $R^{(2)}$ are not very similar. In fact, the average of the 66 absolute differences $|r_{ij}^{(1)} - r_{ij}^{(2)}|$ is .255, and only 4 of the 66 signed differences are negative. The maximum absolute difference of the z-transforms $|z_{ij}^{(1)} - z_{ij}^{(2)}|$ is .706, occurring for $(i,j) = (9,12)$, whence we find $T_3 = 3.59$. Setting $b_\alpha = 3.59$ in (2.1) and referring to a table of the normal distribution, one finds that this corresponds to an attained significance level of $\alpha = .02$. Thus, when applied to Jennrich's data, our procedure T_3 is far more sensitive than Jennrich's procedure T_2, another indication of the satisfactory power properties of T_3, especially for "relatively small" sample sizes.

2.3. Two Variants of Jennrich's Procedure

The test statistics T_4 and T_5 are similar to Jennrich's T_2, but are based on the vectors $\underline{z}^{(\beta)}$ rather than $\underline{r}^{(\beta)}$ in an attempt to exploit the more rapid convergence of the z-transforms to normality. Still, T_4 and T_5 share with T_2 the potential drawback (c) discussed earlier. The statistics T_4 and T_5 are defined by the formulae

$$T_4 = d(\underline{z}^{(1)} - \underline{z}^{(2)})'\hat{D}^{-1}\hat{\Gamma}^{-1}\hat{D}^{-1}(\underline{z}^{(1)} - \underline{z}^{(2)}),$$
$$T_5 = d(\underline{z}^{(1)} - \underline{z}^{(2)})'\tilde{D}^{-1}\tilde{\Gamma}^{-1}\tilde{D}^{-1}(\underline{z}^{(1)} - \underline{z}^{(2)}),$$

where $\hat{\Gamma} = \Gamma(\overline{R})$, $\hat{D} = D(\overline{R})$, $\tilde{\Gamma} = \Gamma(R^*)$, $\tilde{D} = D(R^*)$, and $R^* = (r^*_{ij})$ is an alternate estimator of P under H_0 defined in terms of $\underline{z}^{(1)}$ and $\underline{z}^{(2)}$ by the relation

$$\overline{z}_{ij} = \frac{1}{2}\ell n\left[\frac{(1+r^*_{ij})}{(1-r^*_{ij})}\right],$$

where $\overline{z}_{ij} = \Sigma_\beta(n_\beta - 3)z^{(\beta)}_{ij}/\Sigma_\beta(n_\beta - 3)$. The statistics T_4 and T_5 differ only in the way the asymptotic covariance matrix $D\Gamma D$ of $d^{1/2}(\underline{z}^{(1)} - \underline{z}^{(2)})$ is estimated. Under H_0, both T_4 and T_5 have asymptotic χ^2 distributions with $p(p-1)/2$ degrees of freedom. It should be kept in mind that as the sample sizes approach ∞, the tests based on T_2, T_4, and T_5 are asymptotically exact under H_0, whereas the test based on T_3 will be asymptotically conservative under H_0 unless $P = I$.

2.4. Summary of Monte Carlo Study

In Larntz and Perlman (1985) we conducted a Monte Carlo study of the actual significance levels and powers of the procedures based on T_2, T_3, T_4, and T_5. The values of p considered were 3, 5, and 10, while those of the sample sizes (n_1, n_2) were $n_1 = n_2 = 10, 20, 40, 100$, and 200. The results indicated that for small sample sizes, only T_3 has an actual significance level which can be trusted to lie in the vicinity of the nominal value α, the levels of T_2, T_4, and T_5 deviating greatly from α for some values of the matrix P under H_0. As n increases, the rejection rates for T_2, T_4, and T_5 appear to approach the nominal value, whereas the slightly conservative nature of T_3 becomes apparent. Thus, for small sample sizes T_3 comes closest to the nominal level, whereas any of T_2, T_4, or T_5 could be selected for the larger sample sizes. As expected, for a given sample size, the rejection rates tend to be less accurate and more variable as p increases.

As to power, no one of the four tests is uniformly better than the others. Preliminary investigations showed that the powers of T_2, T_4, and T_5 are comparable for moderate to large sample sizes, so in Larntz and Perlman (1985) emphasis was placed on comparing the powers of T_2 and T_3 under a variety of alternatives. When $p = 3$, we found that in those cases (75 out of 210 pairs of alternatives $(P^{(1)}, P^{(2)})$ where powers of T_2 and T_3 differed significantly, more often the advantage lay with T_2 (59 cases) than with T_3 (16 cases) (due to the arbitrary natures of the population correlation matrices selected for the study, the importance of these numbers should not be overemphasized). In addition, when the power of T_2 was significantly larger than that of T_3, the magnitude of the difference was occasionally quite substantial, whereas this is not the case when T_3 dominates T_2. Part of the reason for the often lower power of T_3 is, of course, its conservative nature, but this does not prevent the conclusion that when p is small and the sample sizes are moderate or large, T_2 is usually the preferable test (aside from considerations of ease of implementation, where T_3 has a considerable advantage).

When the dimension $p = 5$ or 10, however, the relative status of T_2 and T_3 changed markedly. Here the advantage in power lay with T_3 most often — the magnitude of the difference in power increasing as p increased. Surprisingly, when $p = 5$ and 10, we found eight pairs of alternatives $(P^{(1)}, P^{(2)})$ where the estimated power of T_3 was nearly 1, while that of T_2 was much lower, often nearly 0. Each of these eight pairs of alternatives involved some very large correlation coefficients ρ_{ij}, 0.9 or greater. There were no cases where the estimated power of T_2 was nearly 1 while that of T_3 was substantially lower. It should be mentioned that unlike T_2, the statistics T_4 and T_5 did not exhibit instances of extremely low power.

3. RECOMMENDATIONS AND COMMENTS

Because the nominal significance levels of the tests based on T_2, T_4, and T_5 have been found to be unreliable for small sample sizes, T_3 is the clear choice in this situation. Based on the cases studied, we propose the following rule-of-thumb: when the ratio of sample size to dimension does not exceed 4, i.e., when $\min(n_1, n_2)/p \leqq 4$, then T_3 is recommended. Thus, for the values of (p, n) studied in Larntz and Perlman (1985), T_3 should be used in the following cases: $(p, n) = (3,10), (5,10), (5,20), (10,10), (10,20), (10,40)$. (Note too that in the example of Section 2.2, $\min(n_1, n_2)/p = 40/12 \leqq 4$, and T_3 was found to be more sensitive than T_2.)

For moderate and large sample sizes the choice is less clear-cut. Certainly T_3 is the easiest procedure to implement, and it readily provides a set of simultaneous confidence intervals for all differences $\mu_{ij}^{(1)} - \mu_{ij}^{(2)}$, $1 \leqq i, j \leqq p$, where $\mu_{ij}^{(\beta)}$ denotes the z-transform of $\rho_{ij}^{(\beta)}$ — also see Section 4. For large sample sizes we suggest that any of T_2, T_4, T_5 should be used, since they are asymptotically exact (whereas T_3 is conservative) and more powerful for most alternatives. Of course, for moderate sample sizes ($4 \leqq \min(n_1, n_2)/p \leqq 6$, say), the convenience of T_3 may outweigh a relatively small loss in power. Finally, we remark that as the sample sizes $n_1, n_2 \to \infty$, all four tests are asymptotically consistent.

All four statistics depend upon approximate normality of the (normalized) sample correlations. The small sample size advantage of T_3 appears to be due to the fact that T_2, T_4, and T_5 additionally depend upon estimating the covariance structure of the sample correlations. T_3 does not "take advantage" of the actual covariance structure. In small sample sizes this structure is not well estimated by simply substituting R for P in the covariance formula.

It should also be noted (as was pointed out by the referee) that the advantage of the Sidák inequality over the Bonferroni inequality is often small in practical situations.

4. THE K-SAMPLE AND ONE-SAMPLE CASES

When the number k of populations is 3 or more, the hypothesis H_0 can be tested by means of Jennrich's extension of T_2, given in equation (7.2) of Jennrich (1970). Alternatively, our procedure can be extended to the k-sample case as described next, and should be expected to retain its advantageous small-sample properties. For fixed $i < j$ let $\underline{\mu}_{ij} = (\mu_{ij}^{(1)}, \ldots, \mu_{ij}^{(k)})$, $\underline{z}_{ij} = (z_{ij}^{(1)}, \ldots, z_{ij}^{(k)})$, and let H_{ij} denote the univariate hypothesis

$$H_{ij}\colon\ \mu_{ij}^{(1)} = \ldots = \mu_{ij}^{(k)}, \qquad 1 \leqq i < j \leqq p.$$

As $\min(n_1, \ldots, n_k) \to \infty$, the random vector $\underline{z}_{ij}$ has an asymptotic k-variate normal distribution with mean vector $\underline{\mu}_{ij}$ and covariance matrix $\Delta = \mathrm{Diag}[(n_1 - 3)^{-1}, \ldots, (n_k - 3)^{-1}]$. The hypothesis H_{ij} can be tested by means of the statistic

$$S_{ij} = \sum_{\beta=1}^{k} (n_\beta - 3)(z_{ij}^{(\beta)})^2 - \frac{\left[\sum_{\beta=1}^{k} (n_\beta - 3) z_{ij}^{(\beta)}\right]^2}{\sum_{\beta=1}^{k} (n_\beta - 3)}$$

(see Rao (1973), p. 434-5), which reduces to

$$S_{ij} = (n - 3) \sum_{\beta=1}^{k} (z_{ij}^{(\beta)} - \bar{z}_{ij})^2$$

when $n_1 = \ldots = n_k \equiv n$, where $\bar{z}_{ij} = k^{-1}\Sigma_\beta z_{ij}^{(\beta)}$. Under H_{ij}, S_{ij} has an asymptotic χ_{k-1}^2 distribution. Since H_0 is the intersection of the $p(p-1)/2$ univariate hypotheses H_{ij}, we consider the statistic

$$T_3^* = \max_{1 \leqq i < j \leqq p} S_{ij}$$

and propose the test which rejects H_0 if $T_3^* > \chi_{k-1,\varepsilon(\alpha)}^2$, where $\chi_{k-1,\varepsilon}$ is the upper ε point of the χ_{k-1}^2 distribution and $\varepsilon(\alpha) = (1-\alpha)^{2/p(p-1)}$. By Corollary 5′

of Khatri (1967),

$$\text{Prob}\{T_3^* \leqq \chi^2_{k-1,\varepsilon(\alpha)}\} \geqq \prod_{1 \leqq i < j \leqq p} \text{Prob}\{S_{ij} \leqq \chi^2_{k-1,\varepsilon(\alpha)}\} = 1 - \alpha$$

asymptotically under H_0, so this procedure yields a (possibly conservative) level α test.

Our proposed procedure for the k-sample p-variate problem may be regarded as a generalization of the Pearson-Wilks procedure for testing equality of the correlation coefficients from k independent bivariate normal populations (see Morrison (1976) or Pearson and Wilks (1933), pp. 434-5). For moderate sample sizes, the χ^2_{k-1} approximation can be improved if S_{ij} is modified as indicated in the formula for H on page 436 of Rao (1969).

Next, we consider briefly the one-sample case. Let $R = (r_{ij})$ be a sample correlation matrix based on a random sample of size n from a p-variate normal distribution with population correlation matrix $P = (\rho_{ij})$. Consider the problem of testing the hypothesis H_0: $P = P_0$, where $P_0 = (\rho_{ij,0})$ is a specified correlation matrix. If P_0 is the identity matrix, H_0 is the standard hypothesis of independence of a set of p-variates and may be tested by means of Bartlett's modified LRT (Morrison (1976)). See also M. Hills (1964) for a graphical method. For other values of P_0, Jennrich's method may be applied as follows. Let $\underline{r}$ denote the $p(p-1)/2$-dimensional column vector consisting of the elements $r_{ij} (i < j)$ of R, and similarly define $\underline{\rho}$ and $\underline{\rho}_0$ in terms of the elements of P and P_0. As $n \to \infty$, $\sqrt{n}(\underline{r} - \underline{\rho})$ is asymptotically normal with zero mean vector and covariance matrix $\psi \equiv \psi(P)$, where ψ depends on the elements of P in a known way. Thus the statistic

$$\tilde{T}_2 \equiv n(\underline{r} - \underline{\rho}_0)'\psi_0^{-1}(\underline{r} - \underline{\rho}_0)$$

has an asymptotic χ^2 distribution under H_0 with $p(p-1)/2$ degrees of freedom, where $\psi_0 = \psi(P_0)$, and Jennrich proposes rejection of H_0 for large values of $\tilde{T}_2$. He gives a computationally simpler formula for $\tilde{T}_2$ in equation (7.1) of (Jennrich (1970)).

Our method applied to the one-sample testing problem leads to the statistic

$$\tilde{T}_3 = (n-3)^{1/2} \max_{1 \leqq i < j \leqq p} |z_{ij} - \mu_{ij,0}|,$$

where z_{ij} is the z-transform of r_{ij} and $\mu_{ij,0}$ is the z-transform of $\rho_{ij,0}$. By Sidák's Theorem (Sidák (1967)) the test which rejects H_0 if $\tilde{T}_3 > b_\alpha$, where b_α is determined by (2.1), is a (possibly conservative) level α test of H_0. Our procedure may be expected to behave slightly better than Jennrich's for small samples since $\tilde{T}_3$ takes advantage of faster convergence to normality of the z_{ij} whereas $\tilde{T}_2$ depends upon normality of the elements of $\underline{r}$. However, the difference should not be expected to be as pronounced here as in the k-sample case ($k \geqq 2$), since the statistic $\tilde{T}_2$ does not require the estimation of a covariance matrix, unlike T_2. The procedure based on $\tilde{T}_3$ has also been discussed by Morrison (1976, pp. 119–120); he uses the somewhat weaker Bonferroni bound rather than the Sidák inequality.

On the other hand, our procedure does have the advantage that it readily yields a set of simultaneous confidence intervals for the $p(p-1)/2$ population correlation coefficients ρ_{ij}. If we let μ_{ij} denote the z-transform of ρ_{ij}, asymptotically the probability is at least $1-\alpha$ that the inequalities

$$z_{ij} - (n-3)^{1/2} b_\alpha \leqq \mu_{ij} \leqq z_{ij} + (n-3)^{1/2} b_\alpha \qquad (1 \leqq i < j \leqq p)$$

are simultaneously satisfied. These inequalities are then inverted to obtain simultaneous confidence bounds for the ρ_{ij}. Jennrich's procedure yields a (less convenient) ellipsoidal confidence region for the vector $\underline{\rho}$ but now requires the estimation of the asymptotic covariance matrix $\psi(P)$ by $\psi(R)$, and hence may be expected to perform badly (as in the k-sample case) if the sample size n is small relative to the dimension p.

Finally, we remark that in both the one-sample and k-sample case, our method can be applied much more readily than Jennrich's procedure if it is desired to restrict attention to some arbitrary subset L of the $p(p-1)/2$ indices (i,j) and test the hypothesis

$$\rho_{ij}^{(1)} = \ldots = \rho_{ij}^{(k)}, \qquad (i,j) \in L$$

(in the k-sample case) or

$$\rho_{ij} = \rho_{ij,0} \qquad (i,j) \in L$$

(in the one-sample case).

ACKNOWLEDGMENTS

This research report was supported in part by National Science Foundation Grant No. DMS-03489. We greatly appreciate the comments of the referee, whose suggestions led to substantial improvements in the paper.

BIBLIOGRAPHY

Aitkin, M. A. (1969). Some tests for correlation matrices. *Biometrika* **56**, 443–6.

Bartlett, M. S. (1933). On the theory of statistical regression. *Proceedings of the Royal Society, Edinburgh* **53**, 260–83.

Hills, M. (1964). On looking at large correlation matrices. *Biometrika* **56**, 249–253.

Jennrich, R. I. (1970). An asymptotic χ^2 test for the equality of two correlation matrices. *Journal of the American Statistical Association* **65**, 904–12.

Khatri, C. G. (1967). On certain inequalities for normal distributions and their applications to simultaneous confidence bounds. *Annals of Mathematical Statistics* **38**, 1853–67.

Knuth, D. E. (1969). *The Art of Computer Programming, Vol. 2: Seminumerical Algorithms.* Addison-Wesley, Reading, MA.

Kullback, S. (1967). On testing correlation matrices. *Applied Statistics* **16**, 80–5.

Larntz, K., and M. D. Perlman (1985). A simple test for the equality of correlation matrices. Unpublished report, School of Statistics, University of Minnesota, St. Paul, MN.

Morrison, D. F. (1976). *Multivariate Statistical Methods*, 2nd ed. McGraw-Hill, New York.

Pearson, E. S., and S. S. Wilks (1983). Methods of statistical analysis appropriate for k samples of two variables. *Biometrika* **25**, 353–78.

Rao, C. R. (1973). *Linear Statistical Inference and its Applications*, 2nd ed. John Wiley and Sons, Inc., New York, NY.

Seal, H. L. (1964). *Multivariate Statistical Analysis for Biologists.* John Wiley and Sons, Inc., New York, NY.

Sidák, Z. (1967). Rectangular confidence regions for the means of multivariate normal distributions. *Journal of the American Statistical Association* **62**, 626–33.

Steiger, J. H. and Browne, M. W. (1984). The comparison of interdependent correlations between optimal linear composites. *Psychometrika* **49**, 11–24.

CONDITIONS OF RAO'S COVARIANCE METHOD TYPE FOR SET-VALUED ESTIMATORS

H. MEISTER AND O. MOESCHLIN

Fachbereich Mathematik und Informatik
Lützowstr. 125
D-5800 Hagen 1, West Germany

1. INTRODUCTION

With the argument that the estimation of the values of a parameter function in the sense of the usual estimation theory leads to estimates which often with probability 1 do not coincide with the value that has to be estimated, we established in an earlier paper (see Meister and Moeschlin (1986a)) a concept of confidence estimation studying set-valued estimators (estimation correspondences) for set-valued parameter functions (determining those values, the statistician tends to accept) according to ideas of the usual point estimation theory, as they are: unbiasedness and minimal risk.

While we have treated in Meister and Moeschlin (1986a) estimation correspondences, their images being the convex compact subsets of a separable Banach space, basing on a theorem of Rådström, we make now use of an embedding theorem proved in Meister and Moeschlin (1986b). Introducing a certain jointly continuous bilinear function we are led to a quadratic loss function. The embedding theorem enables us not only to characterize unbiasedness of an estimation correspondence by the unbiasedness of some related real-valued functions, but also makes it possible to formulate (necessary and sufficient) conditions of the Rao-Covariance Method type for an unbiased estimation correspondence to be of minimal risk. In the case treated in Meister and Moeschlin (1986a) such results were only established for special subclasses of estimation correspondences, i.e. the polyhedral-valued ones.

Although we base in our approach on Banach space concepts one has to note that the presented set-up may not be confused with merely a set-up for a Banach space-valued statistical theory. This is founded on the fact that the space of estimates of our set-valued estimators may be seen as a subcone but not as the whole Banach space.

2. PRELIMINARIES

Let E be a separable Banach space; $\|.\|_E$ denoting its norm. By E^* we denote the space of continuous linear functionals on E (which is assumed to be separable). Let $D^* = \{d^i | i \in \mathbb{N}\}$ be a countable dense subset of the unit sphere in E^*.

$\mathcal{C}(E)$ is the space of nonempty convex and compact subsets of E. A convex structure on $\mathcal{C}(E)$ may be defined by

$$\sum_{i=1}^{r} t_i C_i = \left\{ \sum_{i=1}^{r} t_i c_i | c_1 \in C_1, \ldots, c_r \in C_r \right\}$$

for $t_1, \ldots, t_r \in \mathbb{R}_+$ and $C_1, \ldots, C_r \in \mathcal{C}(E)$.

With this convex structure in $\mathcal{C}(E)$ convexity can be introduced for real functions $f\colon \mathcal{C}(E) \to \mathbb{R}$ in a canonical way. The function $f\colon \mathcal{C}(E) \to \mathbb{R}$ is said to be convex, iff

$$f(tC_1 + (1-t)C_2) \leq tf(C_1) + (1-t)f(C_2)$$

holds for all $C_1, C_2 \in \mathcal{C}(E)$ and $t \in [0,1]$.

We first cite a result proved in Meister and Moeschlin (1986b).

Theorem 2.1. *The convex cone $\mathcal{C}(E)$ may be embedded as a closed convex cone into a Banach space $\mathcal{V}$, isomorphic to the subspace $\tilde{\mathcal{V}}$ of all bounded sequences in $\mathbb{R}^{\mathbb{N}}$ equipped with the maximum norm. On $\mathcal{C}(E)$ the isomorphism T is defined by:*

$$T(C) : = (\max_{x \in C} d^i(x) | i \in \mathbb{N}) \qquad (C \in \mathcal{C}(E)).$$

A jointly continuous bilinear function on $\tilde{\mathcal{V}}$ is given by

$$< x, y >_{\tilde{\mathcal{V}}} : = \sum_{i=1}^{\infty} \frac{1}{2^i} x_i y_i \qquad (x, y \in \tilde{\mathcal{V}}).$$

Let

$$< C, C' >_{\mathcal{C}(E)} : =< T(C), T(C') >_{\tilde{\mathcal{V}}} \qquad (C, C' \in \mathcal{C}(E)).$$

be the corresponding bilinear function on $\mathcal{C}(E)$.

$\mathcal{B}(E)$ means the Borel-σ-field on E, i.e. the σ-field generated by the system of open sets of E. The Borel-σ-field induced by the metric space structure of $\mathcal{C}(E)$ will be denoted by $\mathcal{B}(\mathcal{C}(E))$.

Furthermore, let $(\mathsf{H}, \mathcal{H}, P)$ be a probability space. By a correspondence Z from H to E we understand a set-valued mapping, which associates with every element $x \in \mathsf{H}$ a nonempty subset $Z(x)$ of E. On the other hand, the correspondence Z may be seen as a function from H to the set of all nonempty subsets of E. These two aspects of correspondences have led to different concepts of measurability. We base on the measurability concept of the later aspect of Z. The measurability of Z regarded as a function from H to $\mathcal{C}(\mathcal{E})$ is defined in the usual sense.

The notion of integral to be used in the sequel for mappings $Z\colon \mathsf{H} \to \mathcal{V}$ will be the Bochner-integral (see Bochner (1933) as introduced for instance, too, in Dunford and Schwartz (1958), chapter IV, 10.7, Def. 7. We denote it by $\int ZdP$. In addition to this notion of integral we refer to the integral introduced by R. Aumann (1965). Thereby let

$$\Phi(Z) : = \{f | f \text{ being } \mathcal{H} - \mathcal{B}(E) - \text{measurable, } f(x) \in Z(x)\} \qquad (x \in \mathsf{H})$$

be the set of measurable selectors of Z. (For the existence of a measurable selector within a correspondence Z compare Castaing and Valadier (1977), Theorem III.6, p.65). The Aumann-integral is now defined by:

$$\int^A ZdP : = \{z \in E | z = \int fdP \text{ for some } f \in \Phi(Z)\}.$$

Following F. Hiai and H. Umegaki (1977), Theorem 4.5, we have: If E has the Radon-Nikodym-Property (RNP), then

$$\int ZdP = \int^A ZdP. \tag{2.2}$$

(compare also G. Debreu, (1967).)

3. OPTIMALITY CONDITIONS IN CASE OF QUADRATIC LOSS FUNCTIONS

An estimation problem with set-valued parameter function is determined by an eight-tuple

$$(\mathsf{H}, \mathcal{H}, \Gamma, \mathcal{W}, E, \mathcal{S}, L, v),$$

what we call an *estimation experiment*; thereby $(\mathsf{H}, \mathcal{H})$ (sample space) is a measurable space. Γ (parameter space) is a nonempty set and $W\colon = \{P_\gamma(\cdot) | \gamma \in \Gamma\}$ (set of possible sample distributions) is a set of probability measures P_γ on $\mathcal{H}$, such that $\gamma \to P_\gamma(\cdot)$ is one-to-one. E denotes a real separable Banach space, $\mathcal{S}$ is a nonempty, closed and convex subcone of $\mathcal{C}(E)$ within $\mathcal{V}$ and $L\colon \Gamma \to \mathcal{S}$ (set-valued parameter function, parameter correspondence) is a mapping. $v\colon \Gamma \times \mathcal{S} \to \mathsf{R}_+$ ((convex) loss function) denotes a mapping, such that $v(\gamma, .)\colon \mathcal{S} \to \mathsf{R}_+$ is (convex) and $\mathcal{B}(\mathcal{C}(E)) - \mathcal{B}$-measurable for all $\gamma \in \Gamma$.

The given set-up follows the line of the usual (point-) estimation problem with the alteration that L is now set-valued – with values in the subcone $\mathcal{S}$ of $\mathcal{C}(E)$ – instead of being point-valued as in the case of a point-estimation problem. In the intuitive interpretation the convex and compact set $L(\gamma)$ determines those points in E, which the statistician tends to accept, provided that γ is the accurate parameter.

Definition 3.1. Let an estimation experiment $(\mathsf{H}, \mathcal{H}, \Gamma, \mathcal{W}, E, \mathcal{S}, L, v)$ be given.

Then any $\mathcal{H} - \mathcal{B}(\mathcal{C}(E))$-measurable mapping Z: $\mathrm{H} \to \mathcal{S}$ is called an *estimation correspondence* (to the given estimation experiment).

$\mathcal{Z}_\mathcal{S}$ denotes the set of all estimation correspondences.

By the concept laid down, the values of the estimation correspondences and the ones of the parameter correspondences lie in the same closed and convex subcone $\mathcal{S}$ of $\mathcal{C}(E)$. As Z has values in the closed, convex cone $\mathcal{S}$, it follows that $\int Z dP_\gamma \in \mathcal{S}$ for all $\gamma \in \Gamma$.

Definition 3.2. Let an estimation experiment be given.

1. An estimation correspondence Z is called *unbiased*, iff
$$\int Z dP_\gamma = L(\gamma) \qquad (\gamma \in \Gamma).$$
2. For $Z \in \mathcal{Z}_\mathcal{S}$ the mapping
$$\rho(.,Z): \quad \Gamma \to [0, \infty],$$
defined by
$$\rho(\gamma, Z): = \int v(\gamma, Z) dP_\gamma \qquad (\gamma \in \Gamma)$$
is called the *risk* of Z.
3. $Z \in \mathcal{Z}_\mathcal{S}$ is called *of finite risk*, iff
$$\rho(\gamma, Z) < \infty \qquad (\gamma \in \Gamma).$$
The set of all unbiased correspondences $Z \in \mathcal{Z}_\mathcal{S}$ with finite risk if denoted by $\mathcal{Z}_\mathcal{S}^{u\rho}$.
4. $\overline{Z} \in \mathcal{Z}'_\mathcal{S} \subset \mathcal{Z}_\mathcal{S}$ is called of *minimal risk within* $\mathcal{Z}'_\mathcal{S}$, iff
$$\rho(\gamma, \overline{Z}) \leq \rho(\gamma, Z) \qquad (\gamma \in \Gamma, Z \in \mathcal{Z}'_\mathcal{S}).$$

By the linearity of the Bochner-integral it is clear that the set of all unbiased estimation correspondences is convex.

If E has the RNP, (2.2) allows an intuitive interpretation of unbiasedness in connection with set-valued estimators: Z is unbiased, iff
$$L(\gamma) = \{z \in E | z = \int f dP_\gamma, \ f \in \Phi(Z)\} \qquad (\gamma \in \Gamma).$$

As an application of a Theorem of Hille (compare Diestel and Uhl (1977), Theorem 6, p.47) we get:

Theorem 3.3. *Let S: $\mathcal{V} \to \mathcal{V}'$ be a linear continuous mapping which is injective on $\mathcal{S}$, $\mathcal{V}'$ being a Banach-space. Z is an unbiased estimation correspondence for the experiment* $(\mathrm{H}, \mathcal{H}, \Gamma, \mathcal{W}, E, \mathcal{S}, L, v)$, *iff*
$$\int S \circ Z dP_\gamma = S(L(\gamma)) \qquad (\gamma \in \Gamma).$$

PROOF. Let Z be an estimation correspondence which is Bochner-integrable w.r.t. all $P_\gamma(\gamma \in \Gamma)$. Since both, $\int Z dP_\gamma$ and $L(\gamma)$ are elements of $\mathcal{S}$, and S is injective on $\mathcal{S}$, the unbiasedness of Z is equivalent to the validity of the equations

$$S\Big(\int Z dP_\gamma\Big) = S(L(\gamma)) \qquad (\gamma \in \Gamma).$$

By Hille's Theorem we have

$$\int S \circ Z dP_\gamma = S \int Z dP_\gamma \qquad (\gamma \in \Gamma).$$

This concludes the proof of the theorem. □

Let $(\mathsf{H}, \mathcal{H}, \Gamma, \mathcal{W}, \mathcal{S}, L, v)$ be an estimation experiment; and let Z: $\mathsf{H} \to \mathcal{S}$ be an estimation correspondence. Define the functions

$$z: \ \mathsf{H} \to \mathsf{R}^{\mathsf{N}} \text{ and } 1: \ \mathsf{H} \to \mathsf{R}^{\mathsf{N}} \text{ by}$$

$$z_i(x) := \sup_{y \in Z(x)} d^i(y), \ 1_i(\gamma) := \sup_{y \in L(\gamma)} d^i(y) \qquad (i \in \mathsf{N}). \qquad (3.4)$$

Corollary 3.5. *Z is unbiased, iff*

$$\int z_i dP_\gamma = 1_i(\gamma) \qquad (\gamma \in \Gamma, \ i \in \mathsf{N}).$$

PROOF. Use in 3.3 for S the mapping T defined in (2.1). □

Corollary 3.6. *Let $\mathcal{S} \subset \mathcal{C}(\mathsf{R}^n)$ be the cone of all balls with center $m \in \mathsf{R}^n$ and radius $r > 0$, i.e. the space $\{B(m,r) | m \in \mathsf{R}^n, \ r \in \mathsf{R}_+\} =: \mathcal{C}_0(\mathsf{R}^n)$. Let further m_L and r_L be the center and radius functions of L. Z: $\mathsf{H} \to \mathcal{S}$ with $Z(x) =: B(m(x), r(x))(x \in \mathsf{H})$ is unbiased, iff*

$$\int m dP_\gamma = m_L(\gamma), \ \int r dP_\gamma = r_L(\gamma) \qquad (\gamma \in \Gamma).$$

A generalization of Rao's Covariance Method, compare Rao (1952), allows a characterization of unbiased ($\mathcal{S}$-valued) estimators with minimal variance.

Theorem 3.7. *Let be given the estimation experiment* $(\mathsf{H}, \mathcal{H}, \Gamma, \mathcal{W}, E, \mathcal{S}, L, v)$ *with*

$$v(\gamma, Z) := < L(\gamma) - Z, L(\gamma) - Z > \qquad (\gamma \in \Gamma, Z \in Z_S).$$

Then, an estimation correspondence $\overline{Z} \in Z_S^{up}$ is of minimal risk (variance) within Z_S^{up}, iff

$$\int < Z' - \overline{Z}, \overline{Z} > dP_\gamma \geq 0 \qquad (\gamma \in \Gamma) \qquad (3.8)$$

holds for all estimation correspondences $Z' \in Z_S^{up}$.

PROOF. The proof is an application of the method of directional derivatives (cp. Meister and Moeschlin (1986a)). □

Note that in (3.8) the inequality sign may not be replaced by an equality sign. The reason that our result does not fully coincide with the one of Rao lies in the nature of the estimation problem considered here. Note that the space $\mathcal{C}(E)$ is not the whole space $\mathcal{V}$, but only a subcone of it. Though, the condition

$$\mathrm{cov}_\gamma(z_i' - \overline{z}_i, \overline{z}_i) = 0 \qquad (\gamma \in \Gamma, i \in \mathbb{N})$$

given now in the terminology of (3.4) is sufficient but not necessary for $\overline{Z}$ being an unbiased estimation correspondence with minimal variance.

Theorem 3.8. *Let be given the estimation experiment* $(\mathsf{H}, \mathcal{H}, \Gamma, \mathcal{W}, E, \mathcal{S}, L, v)$ *with the specifications of Corollary 3.6 and with*

$$v(\gamma, B(m,r)) = \|m - m_L(\gamma)\|^2 + \|r - r_L(\gamma)\|^2.$$

Then an estimation correspondence $\overline{Z} = B(\overline{m}, \overline{r}) \in Z_S^{up}$ *is of minimal risk (variance) within* Z_S^{up}, *iff*

$$\mathrm{cov}_\gamma(m^o, \overline{m}) = 0$$
$$\mathrm{cov}_\gamma(r^o, \overline{r}) \geq 0$$

holds for all zero estimators m^o *and* r^o *with* $r^o \geq -\overline{r}$.

BIBLIOGRAPHY

Aumann, R. J. (1965). Integrals of set-valued functions, *J. Math. Anal. Appl.* **12**, 1–12.

Bochner, S. (1933). Integration von Funktionen, deren Werte die Elemente eines Vektorraumes sind, *Fund. Math.* **20**, 262–276.

Castaing, C. and M. Valadier (1977). Convex analysis and measurable multifunctions, *Lecture Notes in Mathematics*, **Vol. 580**, Berlin.

Debreu, G. (1967). Integration of correspondences, *Proc. Fifth Berkeley Sympos. on Math. Statist. and Probability*, **Vol. II**, Part I, 351–372.

Diestel, J. and J. J. Uhl (1977). Vector Measures, *Math. Surveys* **15**, AMS.

Dunford, N. and J. T. Schwartz: *Linear Operators, Part I; General Theory*, New York, 1958.

Hiai, F. and H. Umegaki (1977). Integrals, Conditional Expectations and Martingales of Multivalued Functions, *J. Multivariate Anal.* **7**, 149–182.

Meister, H. and O. Moeschlin (1986). *Unbiased set-valued estimators with minimal risk*, preprint, FernUniversität Hagen.

Meister, H. and O. Moeschlin: *An embedding problem for the cone of convex sets*, preprint, FernUniversität Hagen.

Rao, C. R. (1952). Some theorems on minimum variance estimation, *Sankhya* **12**, 27–42.

From a private communication with R. Vitale we learnt that an embedding result similar to the one used by us, compare Theorem 2.1, was already obtained by L. Hörmander, *Arkiv. f. Math.* **3**, 1954.

CONSERVATION OF PROPERTIES OF OPTIMALITY OF SOME STATISTICAL TESTS AND POINT ESTIMATORS UNDER EXTENSIONS OF DISTRIBUTIONS

D. PLACHKY

Inst. f. Math. Statistik
Einsteinstr. 62
4400 Münster, West Germany

1. INTRODUCTION

It is shown that the property of a (point) estimator to be MVU is preserved under extensions of distributions in the dominated case. Furthermore, starting from the fact that the generalized Pitman estimator in the non-parametric model concerning extensions of distributions is locally optimal, a simple measurability condition is given for the property of a generalized Pitman estimator to be MVU.

Based on the fact that analogous to the result that the property of a (point) estimator to be MVU is preserved under extensions of distributions (in the dominated case) every optimal test at level α for testing two simple hypotheses is UMP in the non-parametric model concerning extensions of distributions, a simple sufficient condition for restricted hypotheses is given, in order that these tests are maximin-tests at level α for the corresponding restricted hypotheses.

2. TERMINOLOGY

Let $(\Omega, \mathcal{A})$ denote a measurable space and $\mathcal{A}_o$ a sub-σ-algebra of $\mathcal{A}$. The symbol $\mathcal{P}$ stands for a family of probability measures defined on $\mathcal{A}$ and $E(\mathcal{P})$ is introduced as the set consisting of all probability measures Q defined on $\mathcal{A}$, such that there exists a probability measure $P \in \mathcal{P}$ with the property $P|\mathcal{A}_o = Q|\mathcal{A}_o$. For the special case $\mathcal{P} = \{P_o\}$ the notation $E(P_o)$ is used instead of $E(\{P_o\})$ and $\hat{E}(P_o)$ denotes the set of $P \in E(P_o)$ with the property $P << P_o$.

By $D_o(\mathcal{P})$ is the set of all estimators $d_o \in \cap_{P\in\mathcal{P}} L_2(P)$ of zero denoted. An estimator $d^* \in \cap_{P\in\mathcal{P}} L_2(P)$ is called locally MVU for $P = P_o$ if the variance of d^* with respect to P_o is minimal among all unbiased estimators $d \in \cap_{P\in\mathcal{P}} L_2(P)$ (i.e. $E_p(d) = E_P(d^*)$ holds for all $P \in \mathcal{P}$, where E_p denotes the expectation with respect to P). Furthermore, $\hat{d} = d - E_{P_o}(d|\mathcal{A}_o)$ is called a *generalized Pitman estimator* for the non-parametric model $E(P_o)$, where $d \in \cap_{P\in E(P_o)} L_2(P)$ is fixed.

If $Q : \mathcal{A} \to \mathrm{R}$ is non-negative, normed and finitely additive (in the sequel Q

will be called *finitely additive probability measure*), $E_Q(f)$ denotes $\int f dQ$ in the sense of Dunford and Schwartz (1964), where f belongs to $B(\Omega, \mathcal{A})$ consisting of all bounded and $\mathcal{A}$-measurable functions (i.e. $\int f dQ$ can be approximated by Darboux type sums, c.f. Hewitt and Stromberg (1969), p.164). The dual space of $B(\Omega, \mathcal{A})$ consisting of all bounded, finitely additive set functions defined on $\mathcal{A}$ is denoted by $ba(\Omega, \mathcal{A})$ (c.f. Dunford and Schwartz (1964), IV.5.1.).

3. MAIN RESULTS

The first part of the following theorem is a slight generalization of a result of Plachky and Rüschendorf (1987) and a result of Rüschendorff (1987), whereas the second part shows that the property of an estimator to be MVU is preserved with respect to extensions of probability measures.

Theorem 1.

a) $d_o \in D_o(\hat{E}(P_o))$ *implies* $d_o = E_{P_o}(d|\mathcal{A}_o)P_o$*-a.e.*

b) *Let* $d^* \in \cap_{P\in\mathcal{P}} L_2(P|\mathcal{A}_o)$ *be MVU with respect to* $\mathcal{P}|\mathcal{A}_o = \{P|\mathcal{A}_o : P \in \mathcal{P}\}$. *Then* d^* *is MVU with respect to* $\cup_{P\in\mathcal{P}}\hat{E}(P)$, *if* $\mathcal{P}$ *is dominated by a* σ*-finite measure. Furthermore, for any MVU estimator* d^* *with respect to* $\cup_{P\in\mathcal{P}}\hat{E}(P)$ *and any* $P \in \mathcal{P}$ *the equation* $d^* = E_P(d^*|\mathcal{A}_o)P$*-a.e. is valid.*

PROOF. The proof of a) is based on the observation that by

$$\int_A \left[\frac{d_o^\pm}{E_{P_o}(d_o^\pm|\mathcal{A}_o)} I_{\{E_{P_o}(d_o^\pm|\mathcal{A}_o)>0\}} + I_{\{E_{P_o}(d_o^\pm|\mathcal{A}_o)=0\}}\right] dP_o$$

for $A \in \mathcal{A}$, elements of $E_o(P_o)$ are defined ($d_o^\pm$ denotes the positive resp. negative part of d_o). Therefore, $E_P(d_o) = 0$ for all $P \in \hat{E}(P_o)$ implies

$$\int \frac{d_o^+}{E_{P_o}(d_o^+|\mathcal{A}_o)} I_{\{E_{P_o}(d_o^+|\mathcal{A}_o)>0\}} dP_o \leq \int d_o^- dP_o.$$

Jensen's inequality yields $\int d_o^+ dP_o$ as a lower bound for the left hand side of the last equation, which implies together with $\int d_o^+ dP_o = \int d_o^- dP_o$ that $E^2_{P_o}(d_o^+|\mathcal{A}_o)$ $= E_{P_o}(d^{+2}|\mathcal{A}_o)$ holds P_o-a.e., from which the assertion a) follows.

For the proof of part b) let μ denote the probability measure defined by $\sum_{i=1}^\infty c_i P_i$, $c_i > 0$, $i \in \mathbb{N}$, $\sum_{i=1}^\infty c_i = 1$, where $\{P_i\colon i \in \mathbb{N}\}$ is a subset of $\mathcal{P}$ with the same system of zero sets as $\mathcal{P}$. Then

$$E_\mu(d_o|\mathcal{A}_o) = \sum_{i=1}^\infty c_i E_{P_i}(d_o|\mathcal{A}_o)\frac{d(P_i|\mathcal{A}_o)}{d(\mu|\mathcal{A}_o)}\mu - \text{a.e.}$$

for any $d_o \in D_o(\cup_{P\in\mathcal{P}}\hat{E}(P))$ together with $E_{P_i}(d_o|\mathcal{A}_o) = d_o\mu$-a.e. on $\{\frac{d(P_i|\mathcal{A}_o)}{d(\mu|\mathcal{A}_o)} > 0\}$, $i \in \mathbb{N}$, according to part a) of Theorem 1, implies $E_\mu(d_o|\mathcal{A}_o) = d_o P$-a.e. for all $P \in \mathcal{P}$, from which $E_\mu(d_o|\mathcal{A}_o) \in D_o(\mathcal{P}|\mathcal{A}_o)$ follows. Hence for any

$d^* \in \cap_{P\in\mathcal{P}} L_2(P|\mathcal{A}_o)$ with the property to be a MVU estimator relative to $P|\mathcal{A}_o$ the equation $E_Q(d^* d_o) = 0$ is implied for all $Q \in \cup_{P\in\mathcal{P}} \hat{E}(P)$ because of $E_{Q|\mathcal{A}_o}(d^* E_\mu(d_o|\mathcal{A}_o)) = 0$, $E_\mu(d_o|\mathcal{A}_o) = d_o\mu$-a.e., and $Q << \mu$. If now d^* is MVU with respect to $\cup_{P\in\mathcal{P}} \hat{E}(P)$, then d^* is locally MVU relative to μ, which implies together with $E_\mu(d_o|\mathcal{A}_o) \in D_o(\mathcal{P}|\mathcal{A}_o)$ for any $d_o \in D_o(\cup_{P\in\mathcal{P}} \hat{E}(P))$ the equation $E_\mu(E_\mu(d^*|\mathcal{A}_o)d_o) = 0$, from which $d^* = E_\mu(d^*|\mathcal{A}_o)\mu$-a.e. follows. Furthermore,

$$\begin{aligned} E_P(d^*|\mathcal{A}_o) &= E_\mu(d^* \frac{dP}{d\mu}|\mathcal{A}_o) / \frac{d(P|\mathcal{A}_o)}{d(\mu|\mathcal{A}_o)} \\ &= d^* E_\mu(\frac{dP}{d\mu}|\mathcal{A}_o) / \frac{d(P|\mathcal{A}_o)}{d(\mu|\mathcal{A}_o)} \;=\; d^*\mu \;-\; \text{a.e.} \end{aligned}$$

holds for any $P \in \mathcal{P}$. □

Theorem 1 yields as an immediate application a result due to L. Rüschendorf concerning the locally MVU property for P_o of the generalized Pitman estimator $\hat{d} = d - E_{P_o}(d|\mathcal{A}_o)$ in the non-parametric model $E(P_o)$ (cf. Rüschendofr (1987)) and, as a slight extension, also relative to $\hat{E}(P_o)$. Furthermore, it is not difficult to characterize the family consisting of all $d \in \cap_{P\in\hat{E}(P_o)} L_2(P)$, such that the corresponding generalized Pitman estimator $\hat{d}$ is uniformly MVU in the non-parametric model $\hat{E}(P_o)$.

Theorem 2.

a) *For any $d \in \cap_{P\in E(P_o)} L_2(P)$ the generalized Pitman estimator $\hat{d} = d - E_{P_o}(d|\mathcal{A}_o)$ is locally MVU for P_o with respect to $\hat{E}(P_o)$.*

b) *Let $P_o|\mathcal{A}_o$ be not $\{0,1\}$-valued. Then the generalized Pitman estimator $\hat{d}$ is uniformly MVU relative to $\hat{E}(P_o)$ if and only if d is P_o-a.e. $\mathcal{A}_o$-measurable.*

PROOF. Part a) follows immediately from part a) of Theorem 1. For the proof of part b) one starts from the observation that $\hat{d}$ is uniformly MVU relative to $\hat{E}(P_o)$ if and only if $E_P(\hat{d}d_o) = 0$ holds for all $d_o \in D_o(\hat{E}(P_o))$ and any $P \in \hat{E}(P_o)$ (cf. Lehmann (1983)), which is equivalent to $E_P(dd_o) = E_{P_0}(dd_o)$ for all $d_o \in D_o(\hat{E}(P_o))$ and any $P \in \hat{E}(P_o)$ for any $d_o \in D_o(\hat{E}(P_o))$ because of

$$\begin{aligned} E_P(dd_o) &= E_P(d_o E_{P_o}(d|\mathcal{A}_o)) \\ &= E_P(E_{P_o}(d_o d|\mathcal{A}_o)) \\ &= E_{P_o}(\frac{dP}{dP_o} E_{P_o}(dd_o|\mathcal{A}_o)) \\ &= E_{P_o}[E_{P_o}(\frac{dP}{dP_o}|\mathcal{A}_o) E_{P_o}(dd_o|\mathcal{A}_o)] \\ &= E_{P_o}(dd_o) \end{aligned}$$

for every $P \in \hat{E}(P_o)$ and all $d_o \in D_o(E(P_o))$ (observe that $E_{P_o}(d_o|\mathcal{A}_o) = d_o P_o$-

a.e. resp.

$$E_{P_o}(\frac{dP}{dP_o}|\mathcal{A}_o) = \frac{d(P|\mathcal{A}_o)}{d(P_o|\mathcal{A}_o)} P_o - \text{a.e.}$$

for all $P \in \hat{E}(P_o)$ holds). According to Theorem 1 the condition $dd_o - E_{P_o}(dd_o|\mathcal{A}_o) \in D_o(\hat{E}(P_o))$ is equivalent to the property of dd_o to be P_o-a.e. $\mathcal{A}_o$-measurable if dd_o belongs to $\cap_{P\in\hat{E}(P_o)} L_2(P)$. The last part of b) follows from Lemma 1 of the appendix. □

REMARK 1. Let for a given $\mathcal{A}$-measurable function d_1: $\Omega \to \mathbb{R}$ denote the set of all $d \in \cap_{P\in\hat{E}(P_o)} L_2(P)$ with the property $\sup |d|/(1+|d_1|^k) < \infty$ by $B_k(d_1)$ ($k \in \mathbb{N}$ fixed). Then $\hat{d}$ for a given $d \in \cap_{P\in\hat{E}(P_o)} L_{4k}(P)$ is MVU relative to $\hat{E}(P_o)$ in $B_k(d)$ if and only if $dd_o - E_{P_o}(dd_o) \in D_o(\hat{E}(P_o))$ holds for any $d_o \in D_o(\hat{E}(P_o)) \cap B_k(d)$. If in addition $P_o|\mathcal{A}_o$ is not $\{0,1\}$-valued, the last condition is equivalent to the property of d_o to be P_o-a.e. $\mathcal{A}_o$-measurable. This follows directly from the technique of proof of Theorem 2.

The property of a test to be most powerful is also preserved with respect to extensions of distributions:

REMARK 2. In Plachky and Rüschendorff (1987), the following analogous result to part b) of Theorem 1 has been shown: Let P_i denote probability measures on $\mathcal{A}$, $i = 1, 2$, and $\mathcal{A}_o$ a sub-σ-algebra of $\mathcal{A}$. Then any optimal test ϕ^* at level α for testing $\{P_1|\mathcal{A}_o\}$ against $\{P_2|\mathcal{A}_o\}$ is UMP at level α for testing $E(P_1)$ against $E(P_2)$ and any UMP test ψ^* at level α for testing $E(P_1)$ against $E(P_2)$ is $(Q_1 + Q_2)/2$-a.e. measurable with respect to $\mathcal{A}_o$ for any $Q_1 \in E(P_1)$ and every $Q_2 \in E(P_2)$, such that $E_{Q_2}(\psi^*) < 1$ is valid. Here $E(P_i)$ can be replaced by $\hat{E}(P_i)$, $i = 1, 2$, since the main tool for the proof of this result is assertion a) of Theorem 1. It is natural to ask which properties of optimality of ψ^* are preserved, if $E(P_i)$ is replaced by subsets $\mathcal{Q}_i$ of $E(P_i)$, $i = 1, 2$. The following result is a natural answer to this question:

Theorem 3. *Let $\mathcal{A}_o$ denote a sub-σ-algebra of the σ-algebra $\mathcal{A}$ of subsets of a set Ω and let $Q_i^{(o)}$ defined by $\int_A f_i dQ_o$, $A \in \mathcal{A}$, for given $f_i \in B(\Omega, \mathcal{A}_o)$, $f_i \geq 0$, $i = 1, 2$, finitely additive probability measures on $\mathcal{A}$, such that $Q_i^{(o)}|\mathcal{A}_o$ are σ-additive, $i = 1, 2$, where Q_o is a given finitely additive probability measure defined on $\mathcal{A}$ such that $Q_o|\mathcal{A}_o$ is σ-additive. Then any optimal test ψ^* at level α for testing $\{P_1\}$ against $\{P_2\}$ is a maximin-test at level α for testing $\mathcal{Q}_1$ against $\mathcal{Q}_2$ if $\mathcal{Q}_i \subset E(P_i)$ has the property that $Q_i^{(o)}$ belongs to the closet, convex hull $\overline{\text{con}}(\mathcal{Q}_i)$ of $\mathcal{Q}_i$, $i = 1, 2$, with respect to the weak*-topology of $ba(\Omega, \mathcal{A})$.*

PROOF. Let $\Phi_\alpha(\mathcal{Q}_1)$ denote the family of all tests at level α relative to $\mathcal{Q}_1$

and $\Phi_\alpha(Q_1^{(o)})$ the set of all tests ϕ satisfying $E_{Q_1^{(o)}}(\phi) \le \alpha$. Then

$$Q_1^{(o)} \in \overline{\text{con}}\,\mathcal{Q}_1$$

implies

$$\Phi_\alpha(\mathcal{Q}_1) \subset \Phi_\alpha(Q_1^{(o)}),$$

from which

$$E_{Q_2^{(o)}}(\phi^*) \ge \sup\{E_{Q_2^{(o)}}(\phi): \quad \phi \in \Phi_\alpha(\mathcal{Q}_1)\}$$

follows, because

$$\begin{aligned} E_{Q_2}(\phi^*) &= E_{Q_2^{(o)}}(\phi^*) \\ &= \sup\{E_{Q_2^{(o)}}(\phi): \quad \phi \in \Phi_\alpha(Q_1^{(o)})\} \end{aligned}$$

is valid for any $Q_2 \in \mathcal{Q}_2$, since according to Lemma 5 of the appendix for every $\phi \in \Phi_\alpha(Q_1^{(o)})$ there exists an $\mathcal{A}_o$-measurable test ψ, such that $E_{Q_i^{(o)}}(\phi) = E_{Q_i^{(o)}}(\psi)$, $i = 1, 2$, holds. This implies

$$\begin{aligned} &\inf\{E_{Q_2}(\phi^*): \quad Q_2 \in \mathcal{Q}_2\} \\ &\ge \sup\{E_{Q_2^{(o)}}(\phi): \quad \phi \in \Phi_\alpha(\mathcal{Q}_1)\} \\ &\ge \sup\{\inf\{E_{Q_2(\phi)}: \quad Q_2 \in \mathcal{Q}_2\}: \quad \phi \in \Phi_\alpha(Q_1)\}, \end{aligned}$$

since from $Q_2^{(o)} \in \overline{\text{con}}(\mathcal{Q}_2)$ the inequality

$$E_{Q_2^{(o)}}(\phi) \ge \inf\{E_{Q_2(\phi)}: \quad Q_2 \in \mathcal{Q}_2\}$$

for all $\phi \in \Phi_\alpha(\mathcal{Q}_1)$ follows. This leads to the property of ϕ^* to be a maximin-test, i.e.

$$\inf\{E_{Q_2}(\phi^*): \quad Q_2 \in \mathcal{Q}_2\} = \sup\{\inf\{E_{Q_2}(\phi): \quad Q_2 \in \mathcal{Q}_2\}: \quad \phi \in \Phi_\alpha(\mathcal{Q}_1)\}$$

holds, because ϕ^* belongs $\Phi_\alpha(\mathcal{Q}_1)$. □

Theorem 4. *Let $(\Omega, \mathcal{A})$ denote a measurable space, G a family of mappings g: $\Omega \to \Omega$, such the mapping defined by $(\omega, g) \to g(\omega)$ is $(\mathcal{A} \otimes \mathcal{A}(G), \mathcal{A})$-measurable, where $\mathcal{A}(G)$ is a σ-algebra of subsets of G. Furthermore, $\mathcal{A}_o$ denotes the sub-σ-algebra of $\mathcal{A}$ consisting of all $A \in \mathcal{A}$, such that $(P_1 + P_2)(A \Delta g^{-1}(A)) = 0$ holds for any $g \in G$, where P_i are (σ-additive) probability measures $\mathcal{A}$, $i = 1, 2$. Furthermore, the existence of a G-invariant, finitely additive probability measure ν on $\mathcal{A}(G)$ is assumed. Then any optimal test at level α for $\{P_1|\mathcal{A}_o\}$ against $\{P_2|\mathcal{A}_o\}$ is a maximin-test at level α for testing $\mathcal{Q}_1$ against $\mathcal{Q}_2$, where $\mathcal{Q}_i = \{P_i^g: g \in G\}$, $i = 1, 2$.*

PROOF. Theorem 3 together with Lemma 7 and Lemma 8 of the appendix imply the assertion of Theorem 4. □

APPENDIX

Lemma 1. *There exists a bounded $d_o \in D_o(\hat{E}(P_o))$ satisfying $d_o \neq 0$ P_o-a.e. if and only if $P_o|\mathcal{A}_o$ is not $\{0,1\}$-valued.*

PROOF. Let a bounded $d_o \neq 0$ P_o-a.e. belong to $D_o(\hat{E}(P_o))$ and let $P_o|\mathcal{A}_o$ be $\{0,1\}$-valued. Then

$$\int d_o dP_o = 0 = \int d_o d(P_o|\mathcal{A}_{oo}),$$

where $\mathcal{A}_{oo}$ denotes the σ-algebra $d_o^{-1}(\mathcal{B})$ ($\mathcal{B}$ σ-algebra of Borel subsets of $\mathbb{R}$) induced by d_o, implies the existence of an atom A_o of $\mathcal{A}_{oo}$, such that $P_o = \delta_{\omega_o}$ (for any fixed $\omega_o \in A_o$) is valid and, therefore, $\int d_o dP_o = d_o(\omega_o)$ for all $\omega_o \in A_o$ holds, if P is $\{0,1\}$-valued. This leads to the contradiction $d_o = 0$ P_o-a.e. The converse direction follows from the fact that d_o defined by $P_o(A_o^c)I_{A_o^c} - P_o(A_o)I_{A_o}$ for some $A_o \in \mathcal{A}_o$ with the property $0 < P_o(A_o) < 1$ belongs to $D_o(\hat{E}(P_o))$ and satisfies $d_o \neq 0$. □

Lemma 2. *Let $(\Omega, \mathcal{A})$ be a measurable space, $\mathcal{A}_o$ a sub-σ-algebra of $\mathcal{A}$ and Q a finitely additive probability measure defined on $\mathcal{A}$ such that $P = Q|\mathcal{A}_o$ is σ-additive. Then $E_Q(f|\mathcal{A}_o))$ defined for $f \in B(\Omega, \mathcal{A})$ by $\frac{d(\int f dQ)}{dP}$ ($\int f dQ$ denotes the set function defined by $\int_{A_o} f dQ$, $A_o \in \mathcal{A}_o$) belongs P-a.e. to $B(\Omega, \mathcal{A}_o)$ and is P-a.e. uniquely determined by $\int_{A_o} E_Q(f|\mathcal{A}_o)dP = \int_{A_o} f dQ$ for all $A_o \in \mathcal{A}_o$.*

PROOF. Existence and uniqueness follows immediately from the fact that P is σ-additive. □

Lemma 3. *For every $f \in B(\Omega, \mathcal{A})$ and $g \in B(\Omega, \mathcal{A}_o)$ the equation $E_Q(fg|\mathcal{A}_o) = gE_Q(f|\mathcal{A}_o)$ holds $Q|\mathcal{A}_o$-a.e.*

PROOF. The assertion is obvious for non-negative functions $g \in B(\Omega, \mathcal{A}_o)$ with finite range. For $g_n \in B(\Omega, \mathcal{A}_o)$, $g_n \geq 0$, $g_n(\Omega)$ finite, let $g_n \uparrow g$ (uniform convergence!) hold. Then $Q(\{|g_n - g| \geq \varepsilon\}) \to 0$ implies $Q(\{|fg_n - fg| \geq \varepsilon\}) \to 0$ for any $\varepsilon > 0$, since f is bounded (cf. Dunford and Schwartz (1964), III.3.7). This implies the assertion. □

Lemma 4. *For any non-negative $f \in B(\Omega, \mathcal{A}_o)$ and $h \in B(\Omega, \mathcal{A})$ the equation $\int hf dQ = \int h dQ_f$ is valid, where Q is a finitely additive probability measure defined on $\mathcal{A}$ and Q_f denotes the finitely additive probability measure on $\mathcal{A}$ defined by $Q_f(A) = \int_A f dQ$, $A \in \mathcal{A}$.*

PROOF. The assertion is obvious for non-negative $h \in B(\Omega, \mathcal{A})$, $h(\Omega)$ finite, and $h_n \uparrow h$ (uniform convergence), $h_n \geq 0$, $h_n(\Omega)$ finite, implies

$$Q(\{|h_n - h| > \varepsilon\}) \to 0, \; Q_f(\{|h_n - h| > \varepsilon) \to 0$$

and, hence,

$$Q(\{|fh_n - fh| > \varepsilon\}) \leq Q(\{|h_n - h| \geq \frac{\varepsilon}{M}\}) \to 0 \quad (M = \sup|f|),$$

for every $\varepsilon > 0$, from which $\int h_n f dQ \to \int hf dQ$ and $\int h_n f dQ = \int h_n dQ_f \to \int h dQ_f$ follows. □

Lemma 5. *Let $(\Omega, \mathcal{A})$ be a measurable space, $\mathcal{A}_o$ a sub-σ-algebra of $\mathcal{A}$, Q_o a finitely additive probability measure defined on $\mathcal{A}$, such that $Q_o|\mathcal{A}_o$ is a probability measure. Furthermore, for given $f_i \in B(\Omega, \mathcal{A}_o)$ the following finitely additive probability measures Q_i on $\mathcal{A}$ defined by $Q_i(A) = \int_A f_i dQ_o$, $A \in \mathcal{A}$, $i = 1,2$, are introduced. Then for every $\phi \in B(\Omega, \mathcal{A})$, $0 \leq \phi \leq 1$, there exists some $\psi \in B(\Omega, \mathcal{A}_o)$, $0 \leq \psi \leq 1$, such that $\int \phi dQ_i = \int \psi dQ_i$, $i = 1,2$, holds.*

PROOF. It will be shown that ψ defined by $E_{Q_o}(\phi|\mathcal{A}_o)$ (cf. Lemma 2), which can be chosen such that $0 \leq \psi \leq 1$ holds, is of the type described in Lemma 5. From Lemma 3 the equation $E_{Q_o}(\phi f_i|\mathcal{A}_o) = f_i E_{Q_o}(\phi|\mathcal{A}_o) Q_o|\mathcal{A}_o$-a.e. follows. Furthermore,

$$\begin{aligned} E_{Q_o}(\phi f_i|\mathcal{A}_o) &= \frac{d(\int \phi f_i dQ_o)}{d(Q_o|\mathcal{A}_o)} \\ &= \frac{d(\int f_i dQ_o)}{d(\int f_i dQ_o)} \frac{d(\int f_i dQ_o)}{d(Q_o|\mathcal{A}_o)} \\ &= E_{Q_i}(\phi|\mathcal{A}_o) \cdot f_i \end{aligned}$$

holds $Q_o|\mathcal{A}_o$-a.e., $i = 1,2$ (the integrals occurring in connection with Radon-Nikodym derivatives have to be interpreted as set functions defined on $\mathcal{A}_o$). Therefore, the equation

$$f_i E_{Q_i}(\phi|\mathcal{A}_o) = f_i E_{Q_o}(\phi|\mathcal{A}_o) Q_o|\mathcal{A}_o - \text{a.e.}, \ i = 1,2,$$

is valid. The last equation together with

$$I_{\{f_i=0\}} E_{Q_i}(\phi|\mathcal{A}_o) = E_{Q_i}(\phi I_{\{f_i=0\}}|\mathcal{A}_o) = 0 \quad Q_o|\mathcal{A}_o - \text{a.e.}$$

imply

$$E_{Q_i}(\phi|\mathcal{A}_o) = E_{Q_o}(\phi|\mathcal{A}_o) \quad Q_i|\mathcal{A}_o - \text{a.e.},$$

i.e. $E_{Q_i}(\phi) = E_{Q_i}(\psi)$, $i = 1,2$, holds. □

Lemma 6. *Let $(\Omega, \mathcal{A})$ be a measurable space. Then for any finitely additive probability measure Q defined on $\mathcal{A}$ there exists a net P_α of (σ-additive) probability measures, such that $P_\alpha(A) \to Q(A)$ for every $A \in \mathcal{A}$ is valid. P_α can be chosen as a convex combination of Dirac measures.*

PROOF. Let M denote the closed, convex hull $\overline{\text{con}}\{\delta_\omega \colon \omega \in \Omega\}$ of the set consisting of all Dirac measure δ_ω, $\omega \in \Omega$, defined on $\mathcal{A}$ with respect to the weak*-topology of $ba(\Omega, \mathcal{A})$. Then every probability measure $\mu_o \in \text{con}\{\delta_\omega \colon \omega \in$

$\Omega\}$ is an internal point of M, since for any $\mu_1 \in \text{con}\{\delta_\omega\colon \omega \in \Omega\}$ and $\eta \in \mathbb{R}$ the equation $\mu_o + \eta\mu_1 = \sum_{i=1}^m (\alpha_i^{(o)} + \alpha_i^{(1)}\eta)\delta_{\omega_i}$ holds, $(\mu_o + \eta\mu_1)\,(\{\omega_1, \ldots, \omega_m\}^c) = 0$, $\sum_{i=1}^m \alpha_i^{(o)} = \sum_{i=1}^m \alpha_i^{(1)} = 1$, $0 \leq \alpha_i^{(o)}, \alpha_i^{(1)} \leq 1$, $i = 1, \ldots, m$, i.e. for η sufficiently small $\alpha_i^{(o)} + \alpha_i^{(1)}\eta \geq 0$, $i = 1, \ldots, m$, is valid. It will now be proved that M is equal to the set of all finitely additive probability measures μ defined on $\mathcal{A}$. Otherwise, there would exist a finitely additive measure μ_o on $\mathcal{A}$ and a $h \in B(\Omega, \mathcal{A})$, $h \neq 0$, such that $\int h d\delta_\omega \leq \varepsilon_1 < \varepsilon_2 \leq \int h d\mu_o$ holds for all $\omega \in \Omega$ and some $\varepsilon_i \in \mathbb{R}$, $i = 1, 2$ (cf. Dunford and Schwartz (1964), V.2.10 and V.3.9). Without loss of generality $0 \leq h \leq 1$ can be assumed, since h can be replaced by $\hat{h} = (h + m)/2m$, $m = \sup|h|(> 0)$. Therefore, $m \leq \int h d\mu_o \leq m$ and $m \leq \varepsilon_1 < \varepsilon_2 \leq \int h d\mu_o$ must be valid. □

REMARK. Let $(\Omega, \mathcal{A})$ denote a measurable space and G a family of $(\mathcal{A}, \mathcal{A})$-measurable mappings $g\colon \Omega \to \Omega$. Then the existence of a G-invariant, finitely additive probability measure Q on $\mathcal{A}$ is equivalent to the existence of a net P_α of (σ-additive) probability measures defined on $\mathcal{A}$, such that $P_\alpha(A) \to Q(A)$ and $P_\alpha(A)$ - $P_\alpha(g^{-1}(A)) \to 0$ is valid for all $A \in \mathcal{A}$ and every $g \in G$.

PROOF. Let Q be any G-invariant, finitely additive probability measure on $\mathcal{A}$. Because of $M = \overline{\text{con}}\{\delta_\omega\colon \omega \in \Omega\}$ there exists a net P_α of (σ-additive!) probability measures, such that $P_\alpha(A) \to Q(A)$, $A \in \mathcal{A}$, is valid, from which

$$P_\alpha(A) - P_\alpha(g^{-1}(A)) \to Q(A) - Q(g^{-1}(A)) = 0, \quad A \in \mathcal{A}, g \in G,$$

follows. The converse direction follows directly from

$$P_\alpha(A) - P_\alpha(g^{-1}(A)) \to 0, \quad A \in \mathcal{A}, g \in G,$$

together with

$$P_\alpha(A) \to Q(A), \quad A \in \mathcal{A},$$

i.e.

$$Q(A) = Q(g^{-1}(A)), \quad A \in \mathcal{A}, g \in G,$$

is valid. □

Lemma 7. *Let $(\Omega, \mathcal{A})$ denote a measurable space, $\mathcal{A}_o$ a sub-σ-algebra, and Q_i finitely additive probability measures on $\mathcal{A}$, such that $P_i = Q_i|\mathcal{A}_o$ is σ-additive, $i = 1, 2$. Then*

$$\int_A \frac{dP_i}{d\frac{P_1+P_2}{2}}\, d\left(\frac{Q_1 + Q_2}{2}\right) = \int E_{(Q_1+Q_2)/2}(I_A|\mathcal{A}_o)dP_i$$

holds for all $A \in \mathcal{A}$.

PROOF. From Lemma 2 of the appendix

$$\int E_{(Q_1+Q_2)/2}(I_A|\mathcal{A}_o)dP_i = \int \frac{dP_1}{d(P_1+P_2)/2} d(\rho|\mathcal{A}_o), \quad A \in \mathcal{A},$$

where ρ is defined by $\rho(B) = \int_B I_A d(Q_1+Q_2)/2$, $B \in \mathcal{A}$, follows. Furthermore,

$$\int f d(\rho|\mathcal{A}_o) = \int_A f d((Q_1+Q_2)/2|\mathcal{A}_o)$$

is valid for any $f \in B(\Omega, \mathcal{A}_o)$, which yields the assertion of Lemma 7. A second proof of Lemma 7 can be based on the observation that there is a net $Q_i^{(\alpha)}$ of (σ-additive) probability measures on $\mathcal{A}$, such that $Q_i^{(\alpha)}(A) \to Q_i(A)$, $A \in \mathcal{A}$, holds. Furthermore, there exists a subnet α_β, $\beta \in D(A)$ (the directed set depends on $A \in \mathcal{A}$), such that

$$\lim_{\alpha_\beta} \int_{A_o} E_{(Q_1^{(\alpha_\beta)}+(Q_2^{(\alpha_\beta)})/2}(I_A|\mathcal{A}_o)dP_i$$

exists for all $A_o \in \mathcal{A}_o$ (cf. Dunford and Schartz (1964), V.4.2. and IV.8.5). Since

$$\int_{A_o} E_{(Q_1^{(\alpha_\beta)}+(Q_2^{(\alpha_\beta)})/2}(I_A|\mathcal{A}_o)d(P_1+P_2)/2 = \int_{A_o} I_A d(Q_1^{(\alpha_\beta)} + (Q_2^{(\alpha_\beta)})/2$$

holds for all $A_o \in \mathcal{A}_o$, the weak*-limit

$$\lim_{\alpha_\beta} E_{(Q_1^{(\alpha_\beta)}+(Q_2^{(\alpha_\beta)})/2}(I_A|\mathcal{A}_o)$$

is a version of

$$E_{(Q_1+Q_2)/2}(I_A|\mathcal{A}_o),$$

from which the assertion of Lemma 7 follows, since Lemma 7 is obvious if Q_i is σ-additive on $\mathcal{A}$, $i = 1, 2$. □

The technique of proof applied in the second proof of Lemma 7 is used to show that the following result is true.

Lemma 8. *Let $(\Omega, \mathcal{A})$ denote a measurable space, and G a family of mappings $g\colon \Omega \to \Omega$, such that the mapping defined by $(\omega, g) \to g(\omega)$, $\omega \in \Omega$, $g \in G$, is $(A \otimes \mathcal{A}(G), \mathcal{A})$-measurable, where $\mathcal{A}(G)$ denotes a σ-algebra of subsets of G. Furthermore, $\mathcal{A}_o$ denotes the sub-σ-algebra of $\mathcal{A}$ consisting of all $A \in \mathcal{A}$, such that $(P_1 + P_2)(A \Delta g^{-1}(A)) = 0$ holds for any $g \in G$, where P_i are (σ-additive) probability measures on $\mathcal{A}$, $i = 1, 2$. Furthermore, Q_i denotes the finitely additive probability measures on $\mathcal{A}$ defined by $Q_i(A) = \int P_i(g^{-1}(A)\nu(dg)$, $A \in \mathcal{A}$, $i = 1, 2$, where ν is a G-invariant, finitely additive probability measure on $\mathcal{A}(G)$. Then*

$$\int E_{(Q_1+Q_2)/2}(I_A|\mathcal{A}_o)dP_i = \int P_i(g^{-1}(A))\nu(dg)$$

holds for any $A \in \mathcal{A}$, $i = 1, 2$.

PROOF. Let ν_α denote a net of (σ-additive) probability measures on $\mathcal{A}(G)$, such that $\nu_\alpha(C) \to \nu(C)$ holds for all $C \in \mathcal{A}(G)$. Then there exists a subnet α_β, $\beta \in D(A)$, such that the weak*-limit

$$\lim_{\alpha_\beta} \int I_{g^{-1}(A)} \nu_{\alpha_\beta}(dg)$$

exists, i.e.

$$\lim_{\alpha_\beta} \int_{A_o} (\int I_{g^{-1}(A)} \nu_{\alpha_\beta}(dg)) dP_i = \int_{A_o} (\lim_{\alpha_\beta} \int I_{g^{-1}(A)} \nu_{\alpha_\beta}(dg)) dP_i$$

is valid for all $A_o \in \mathcal{A}_o$, $i - 1, 2$ (cf. Dunford and Schwartz (1964), V.4.2 and IV.8.5). This implies together with Fubini's theorem

$$\begin{aligned}\int_{A_o} E_{(Q_1+Q_2)/2}(I_A|\mathcal{A}_o) d(P_1+P_2)/2 &= \lim_{\alpha_\beta} \int (P_1 + P_2)/2(g^{-1}(A) A_o) \nu_{\alpha_\beta}(dg) \\ &= \lim_{\alpha_\beta} \int_{A_o} (\int I_{g^{-1}(A)} \nu_{\alpha_\beta}(dg)) d(P_1 + P_2)/2 \\ &= \int_{A_o} (\lim_{\alpha_\beta} \int I_{g^{-1}(A)} \nu_{\alpha_\beta}(dg)) d(P_1 + P_2)/2\end{aligned}$$

for all $A_o \in \mathcal{A}_o$. Since, furthermore, for any $g_o \in G$ the equation

$$\lim_{\alpha_\beta} \int I_{g^{-1}(A)} o g_o \nu_{\alpha_\beta}(dg) = \lim_{\alpha_\beta} \int I_{g^{-1}(A)} \nu_{\alpha_\beta}(dg)$$

holds $(P_1 + P_2)/2$-a.e., since ν is G-invariant,

$$\lim_{\alpha_\beta} \int I_{g^{-1}(A)} \nu_{\alpha_\beta}(dg)$$

is a version of $E_{(Q_1+Q_2)/2}(I_A|\mathcal{A}_o)$. This implies

$$\begin{aligned}\int_{A_o} E_{(Q_1+Q_2)/2}(I_A|\mathcal{A}_o) dP_i &= \lim_{\alpha_\beta} \int_{A_o} (\int I_{g^{-1}(A)} \nu_{\alpha_\beta}(dg)) dP_i \\ &= \lim_{\alpha_\beta} \int P_i(g^{-1}(A) \cap A_o) \nu_{\alpha_\beta}(dg),\end{aligned}$$

for any $A \in \mathcal{A}$ and $A_o \in \mathcal{A}_o$, $i = 1, 2$, where for the last equality sign Fubini's theorem has been applied. Choosing $A_o = \Omega$ yields the assertion of Lemma 8. □

REMARK. It seems to be interesting to ask, whether it is possible to derive the main results of the article by a sufficiency argument. This is only possible for uninteresting special cases, because of the following characterizations of sufficiency:

a) $\mathcal{A}_o$ is sufficient for $E(P_o)$ if and only if $E(P_o) = \{P_o\}$ holds, i.e. the extension of $P_o|\mathcal{A}_o$ to $\mathcal{A}$ as a probability measure is uniquely determined, namely equal to P.

b) $\mathcal{A}_o$ is sufficient for $\hat{E}(P_o)$ if and only if P is an extremal point of $E(P)$. The last assertion follows from the fact that an extremal point Q of $E(P_o)$ is characterized by one of the following two properties:

(i) $Q' << Q$ for $Q' \in E(P_o)$ implies $Q' = Q$;

(ii) For any $A_1 \in \mathcal{A}$ there exists an $A_o \in \mathcal{A}_o$ such that $Q(A_1 \Delta A_o) = 0$ holds.

Clearly (ii) implies (i) and (ii) follows from the proof of part a) of Theorem 1 if one replaces d_o^+ by the indicator function of A_1.

ACKNOWLEDGMENTS

I would like to thank my colleague L. Rüschendorf for some valuable remarks.

BIBLIOGRAPHY

Dunford, N. and J. Schwartz (1964). *Linear Operators, Part I.* Interscience, New York.

Hewitt, E. and K. Stromberg (1969). *Real and Abstract Analysis.* Springer, Heidelberg.

Lehmann, E. L. (1983). *Theory of point estimation.* Wiley, New York.

Plachky, D. and L. Rüschendorf (1987). Conservation of the UMP- resp. Maximin-property of statistical tests under extensions of probability measures. Goodness-of-Fit, *Colloquia Mathematica, Societas János Bolyai*, **45**, North-Holland, 439–458.

Rüschendorf, L. (1987). Unbiased estimation in nonparametric classes of distributions. *Statist. and Decis.* 1/2, Vol. 5, 89–104.

SOME RECENT RESULTS IN SIGNAL DETECTION

C. RADHAKRISHNA RAO

Center for Multivariate Analysis
University of Pittsburgh
Pittsburgh, Pennsylvania 15260, U.S.A.

1. INTRODUCTION

Some of the problems of signal detection considered by electrical engineers can be described in terms of a linear model

$$Y = \xi + \eta \tag{1.1}$$

where ξ is an n-vector signal, η is an n-vector noise variable perturbing the signal during transmission and Y is the received message. The general problem associated with the model (1.1) is that of detecting the presence of a signal in a received message and estimating it when present. Different types of problems arise depending on the nature of ξ and η.

We consider two types of problems, one where ξ is considered as a specified vector or a specified function of unknown parameters, and another where ξ is considered as a stochastic vector distributed independently of the noise vector η.

The following notations are used. A' denotes the transpose of a matrix A when its elements are real and A^* the conjugate transpose of A when its elements are complex.

(i) $X \sim N_p(\mu, \Sigma)$, i.e., a real p-vector X has a p-variate real normal distribution with the probability density function (p.d.f.)

$$(2\pi)^{-p/2}|\Sigma|^{-1/2}\exp[-\frac{1}{2}(x-\mu)'\Sigma^{-1}(x-\mu)]. \tag{1.3}$$

(ii) $X \sim \tilde{N}_p(\mu, \Sigma)$, i.e., a complex vector X has a p-variate complex normal distribution with the p.d.f.

$$(\pi)^{-p}|\Sigma|^{-1}\exp[-(x-\mu)^*\Sigma^{-1}(x-\mu)]. \tag{1.4}$$

(iii) $S \sim W_p(f, \Sigma)$, i.e., a real $p \times p$ positive definite matrix S has the Wishart distribution on f degrees of freedom with the p.d.f.

$$2^{-pf/2}[\Gamma_p(f/2)]^{-1}|\Sigma|^{-f/2}|S|^{(f-p-1)/2}\exp(-\frac{1}{2}tr\Sigma^{-1}S) \tag{1.5}$$

where

$$\Gamma_p(a) = \pi^{p(p-1)/4}\prod_{i=1}^{p}(a - \frac{i-1}{2}).$$

(iv) $S \sim \tilde{W}_p(f, \Sigma)$, i.e., a complex $p \times p$ positive definite matrix S has the complex Wishart distribution with the p.d.f.

$$[\tilde{\Gamma}_p(f)]^{-1}|\Sigma|^{-f}|V|^{f-p}\exp(-tr\Sigma^{-1}S) \tag{1.6}$$

where

$$\tilde{\Gamma}_p(a) = \pi^{p(p-1)/2}\prod_{i=1}^{p}(a-i+1).$$

2. PROBLEMS INVOLVING A FIXED SIGNAL

2.1 Test for a Specified Signal

Consider the model

$$X = \xi + \eta \tag{2.1.1}$$

where ξ is a fixed vector and $\eta \sim N_p(0, a^{-1}\Sigma)$ with a known scalar a and an unknown covariance matrix Σ. Let S be a $p \times p$ positive definite matrix variable such that $S \sim W_p(f, \Sigma)$. We have independent observations on X and S, on the basis of which we wish to test the hypothesis

$$H_0: \ \xi = 0 \qquad \text{versus} \qquad H_1: \ \xi = \delta \text{ (specified)}. \tag{2.1.2}$$

Rejection of H_0 at a chosen level of significance would indicate that the received message X contains the signal δ, i.e., it is not pure noise.

It may be noted that when $p = 1$, the appropriate test of H_0 is the one-sided t test

$$t = \frac{a^{1/2}X}{(s/f)^{1/2}} > c \text{ if } \delta > 0 \text{ (or } < c \text{ if } \delta < 0) \tag{2.1.3}$$

on f degrees of freedom. When $p > 1$, one generally uses Hotelling's

$$T^2 = \frac{a(f-p+1)}{p}X'S^{-1}X \tag{2.1.4}$$

which is distributed as F on p and $f - p + 1$ degrees of freedom. The test (2.1.4), however, does not involve the specified δ. A more powerful test than (2.1.4) is recently suggested by Khatri and Rao (1985b) based on the following considerations.

Let C be a $p \times (p-1)$ matrix of rank $p-1$ such that $\delta'C = 0$ and consider the transformation

$$Y = \begin{pmatrix} Y_1 \\ Y_2 \end{pmatrix} = \begin{pmatrix} \delta' \\ C' \end{pmatrix} X, \ V = \begin{pmatrix} \delta' \\ C' \end{pmatrix} S \ (\delta : C). \tag{2.1.5}$$

Then, we have

$$Y \sim \begin{cases} N_p\left(\binom{0}{0}, \quad a^{-1}\Sigma_*\right) \text{ under } H_0 \\ N_p\left(\binom{\delta'\delta}{0}, \quad a^{-1}\Sigma_*\right) \text{ under } H_1 \end{cases}$$

$$V \sim W_p(f, \Sigma_*), \ \Sigma_* = \binom{\delta'}{C'}\Sigma\ (\delta : C) \tag{2.1.6}$$

and the problem (2.1.2) reduces to testing

$$H_0 : \ E(Y_1) = 0 \qquad \text{versus} \qquad H_1 : \ E(Y_1) = \delta'\delta > 0 \text{ given } E(Y_2) = 0. \tag{2.1.7}$$

Such a problem involving a conditional test was considered in Rao (1946). The appropriate test is $t > t_\alpha$ where

$$t = \frac{a^{1/2}(f-p+1)^{1/2}\delta' S^{-1}X}{[(1+aX'S^{-1}X)\delta'S^{-1}\delta - a(\delta'S^{-1}X)^2]^{1/2}} \tag{2.1.8}$$

has t distribution on $(f-p+1)$ degrees of freedom and t_α is the upper tail $\alpha\%$ of the t distribution.

In testing (2.1.7), we used the condition $E(Y_2) = 0$. But in practice it may be necessary to check whether this holds. For this, we use Hotelling's

$$T^2 = \frac{a(f-p+2)}{p-1}\left(X'S^{-1}X - \frac{(\delta'S^{-1}X)^2}{\delta'S^{-1}\delta}\right) \tag{2.1.9}$$

which has F distribution on $(p-1)$ and $(f-p+2)$ degrees of freedom.

The tests (2.1.8) and (2.1.9) can be extended to the case where η and S have complex normal and Wishart distributions respectively. For example, the test for H_0: $\mu = 0$ versus H_1: $\mu = \delta$ is $t \geq t_\alpha$ where

$$t = \text{ real part of } \frac{[2(f-p+1)]^{1/2}\delta^* S^{-1}X}{[(a^{-1}+X^*S^{-1}X)\delta^*S^{-1}\delta - \delta^*S^{-1}XX^*S^{-1}\delta]^{1/2}} \tag{2.1.10}$$

has t distribution on $2(f-p+1)$ degrees of freedom. Further the hypothesis that $E(Y_2) = 0$ is tested by Hotelling's

$$T^2 = \frac{a(f-p+2)}{p-1}\left(X^*S^{-1}X - \frac{\delta^*S^{-1}XX^*S^{-1}\delta}{\delta^*S^{-1}\delta}\right) \tag{2.1.11}$$

which has F distribution on $2(p-1)$ and $2(f-p+2)$ degrees of freedom.

2.2 Discrimination Between Noise and a Specified Signal

In section 2.1, we considered the problem of testing for pure noise against a specified signal on the basis of an observed message X and an independent estimate $f^{-1}S$ of Σ. The interpretation of such a test at a chosen level of significance is not simple, specially when the same estimate $f^{-1}S$ of Σ is used repeatedly to test for signals in a number of incoming future messages. In order to provide a satisfactory solution, we consider the problem of signal detection as one of discrimination between alternative populations $N_p(0, \Sigma)$, (or $\tilde{N}_p(0, \Sigma)$), and $N_p(\delta, \Sigma)$, (or $\tilde{N}_p(\delta, \Sigma)$), when Σ is unknown but an estimate $f^{-1}S$ of Σ is available. In such a case, the estimated linear discriminant function (LDF) is proportional to

$$y = \delta' S^{-1} X, \text{ (or } \delta^* S^{-1} X). \tag{2.2.1}$$

[In the sequel we follow the practice of giving the results for the real case first and then for the complex case within brackets as in (2.2.1)]. The discriminatory power of (2.2.1) when applied to future observations is a monotone function of the discrimination index (DI)

$$\begin{aligned} \rho(S, \Sigma) &= \frac{[E(y|S, \xi = \delta) - E(y|S, \xi = 0)]^2}{V(y|S)} \\ &= \frac{(\delta' S^{-1} \delta)^2}{\delta' S^{-1} \Sigma S^{-1} \delta}, \text{ (or } \tilde{\rho}(S, \Sigma) = \frac{(\delta^* S^{-1} \delta)^2}{\delta^* S^{-1} \Sigma S^{-1} \delta}). \end{aligned} \tag{2.2.2}$$

If Σ is known, the true LDF is $\delta' \Sigma^{-1} X$, (or $\delta^* \Sigma^{-1} X$), and the optimal DI is

$$\rho(\Sigma, \Sigma) = \delta' \Sigma^{-1} \delta, \text{ (or } \tilde{\rho}(\Sigma, \Sigma) = \delta^* \Sigma^{-1} \delta) \tag{2.2.3}$$

which is greater than or equal to (2.2.2) by the Cauchy-Schwartz inequality, so that there is, in general, loss of information in using an estimated Σ. The expression (2.2.2) involves the unknown covariance matrix Σ, and therefore, the realized discriminatory power by using a particular estimate of Σ in classifying future observations remains unknown. We raise the question as to whether an estimate of (2.2.2) can be obtained in terms of known (observed) values to provide some idea of the performance of the estimated LDF. One such is the plug-in estimate $\rho(S, f^{-1}S)$, or $\tilde{\rho}(S, f^{-1}S)$), but it is known to be a highly biased estimator of (2.2.2). In a recent paper, Khatri and Rao (1985a) provided a satisfactory solution, which is as follows.

It is shown that in the real case

$$B = \frac{(\delta' S^{-1} \delta)^2}{(\delta' \Sigma^{-1} \delta)(\delta' S^{-1} \Sigma S^{-1} d)} \quad \text{and} \quad G = \frac{\delta' \Sigma^{-1} \delta}{\delta' S^{-1} \delta} \tag{2.2.4}$$

are independently distributed with the p.d.f. of B as

$$\frac{\Gamma(\frac{f+1}{2})}{\Gamma(\frac{f-p+2}{2})\Gamma(\frac{p-1}{2})} b^{(f-p)/2}(1-b)^{(p-3)/2} \tag{2.2.5}$$

and that of G as

$$\frac{1}{2^{(f-p+1)/2}\Gamma(\frac{f-p+1}{2})} e^{-g/2} g^{(f-p+1)/2}. \tag{2.2.6}$$

In the complex case, defining $\tilde{B}$ and $\tilde{G}$ with δ' replaced by δ^* in (2.2.4), it is shown that $\tilde{B}$ and $\tilde{G}$ are independently distributed with the p.d.f. of $\tilde{B}$ as

$$\frac{\Gamma(f+1)}{\Gamma(f-p+2)\Gamma(p-1)} \tilde{b}^{(f-p+1)}(1-\tilde{b})^{p-2} \tag{2.2.7}$$

and that of $\tilde{G}$ as

$$\frac{1}{\Gamma(f-p+1)} e^{-\tilde{g}} \tilde{g}^{(f-p)}. \tag{2.2.8}$$

The distribution (2.2.7) was earlier obtained by Reed, Mallet and Brennan (1974). The distributions (2.2.3) - (2.2.8) are independent of the unknown parameters which enables us to draw inferences on $\rho(S,\Sigma)$, (or $\tilde{\rho}(S,\Sigma)$), through the pivotal quantities B and G, (or $\tilde{B}$ and $\tilde{G}$) as discussed below.

1. Using the expressions for the moments of the beta distribution (Rao (1973), p. 168)

$$E(B \text{ or } \tilde{B}) = \frac{f-p+2}{f+1}$$

we have

$$E[\rho(S,\Sigma), \text{ (or } \tilde{\rho}(S,\Sigma))] = \frac{f-p+2}{f+1}[\delta'\Sigma^{-1}\delta, \text{ (or } \delta^*\Sigma^{-1}\delta)]. \tag{2.2.9}$$

If the average efficiency is to be maintained at about half the optimal efficiency, then from (2.2.9) we have

$$\frac{f-p+2}{f+1} \simeq \frac{1}{2} \text{ or } f \simeq 2p \tag{2.2.10}$$

for both real and complex cases, i.e., the degrees of freedom on which Σ is estimated should be at least twice the number of components of the signal. This result for the complex case is mentioned in Reed, Mallet and Brennan (1974). Similarly we can equate the ratio in (2.2.9) to any desired ratio other than (1/2) and find the degrees of freedom f needed for the estimation of Σ.

2. Perhaps, a more satisfactory way of using the distributions (2.2.5) and (2.2.7) is as follows. Let b_α, (or $\tilde{b}_\alpha$), be the lower $\alpha\%$ point of the distribution (2.2.5), (or (2.2.7)). Then we can make the confidence statement that

$$\rho(S,\Sigma) \geq b_\alpha \delta'\Sigma^{-1}\delta, \text{ (or } \tilde{\rho}(S,\Sigma) \geq \tilde{b}_\alpha \delta^*\Sigma^{-1}\delta) \tag{2.2.11}$$

with a confidence coefficient of $(1-\alpha)\%$.

3. The results (2.2.10) and (2.2.11) still involve the unknown quantity Σ. We raise the question, whether the actual magnitude of $\rho(S,\Sigma)$ for given S can be assessed through known values. Using the joint distribution of B and G as in (2.2.5) and (2.2.6) it is shown in Khatri and Rao (1985a) that $E[\rho(S,\Sigma) - gD^2]^2$ attains its minimum at $g = (f-p+2)(f-p-3)/f(f+1)$ so that

$$\hat{\rho}(S,\Sigma) = \frac{(f-p+2)(f-p-3)}{f(f+1)} D^2 \tag{2.2.12}$$

which does not involve unknown parameters, is a close approximation to $\rho(S,\Sigma)$. Similarly

$$\hat{\tilde{\rho}}(S,\Sigma) = \frac{(f-p+2)(f-p-1)}{f(f+1)} D^2 \tag{2.2.13}$$

is a close approximation to $\tilde{\rho}(S, \Sigma)$.

4. We can also obtain the exact lower confidence bound to $\rho(S, \Sigma)$, which provides a satisfactory answer to the problem raised. We define the random variable

$$Z = \frac{1}{2} BG = \frac{f}{2} \frac{\rho(S, \Sigma)}{D^2} \qquad (\text{or } \tilde{Z} = \frac{f \tilde{\rho}(S, \Sigma)}{D^2}) \tag{2.2.14}$$

which has the confluent hypergeometric distribution with the p.d.f.

$$\frac{e^{-z} z^{m-1}}{\Gamma(m)} \frac{\Gamma(a+b)}{\Gamma(a)} \Psi(b, m-a+1; z) \tag{2.2.15}$$

where $m = (f-p+1)/2$, $a = (f-p+2)/2$, $b = (p-1)/2$, (or $m = f-p+1$, $a = f-p+2$, $b = p-1$), and

$$\Psi(b, c; z) = \frac{1}{\Gamma(b)} \int_0^\infty t^{b-1}(1+t)^{c-b-1} \exp(-zt)\, dt. \tag{2.2.16}$$

The percentage points of this distribution at various levels are tabulated in Khatri, Rao and Sun (1986). If z_α (or $\tilde{z}_\alpha$) is the lower $\alpha\%$ point of this distribution, then

$$\rho(S, \Sigma) \geq \frac{2z_\alpha}{f} D^2 \qquad (\text{or } \tilde{\rho}(S, \Sigma) \geq \frac{\tilde{z}_\alpha}{f} D^2) \tag{2.2.17}$$

provides the lower confidence bound to $\rho(S, \Sigma)$, (or $\tilde{\rho}(S, \Sigma)$), with a confidence coefficient of $(1-\alpha)\%$.

Several approximations to the distribution (2.2.15) are also obtained in Khatri, Rao and Sun (1986) when f is large compared to p, from which fairly accurate percentage points can be easily obtained.

3. PROBLEMS INVOLVING A RANDOM SIGNAL

Following the recent papers by Wax and Kailath (1984) and Bai, Krishnaiah and Zhao (1986), let us consider the model

$$X(t) = As(t) + n(t),\ t = t_1, \ldots, t_n, \tag{3.1}$$

where $X(t)$ is a p-vector message received at time t, $s(t)$ is a q-vector random signal, $n(t)$ is a p-vector noise component and A is an unknown $p \times q$ matrix with elements independent of t. The following assumptions are made regarding the model (3.1).

(i) $s(t_i)$, $i = 1, \ldots, n$, are i.i.d. with the common distribution $N_q(0, \Psi)$, (or $\tilde{N}_q(0, \Psi)$), $q < p$.

(ii) $n(t_i)$, $i = 1, \ldots, n$, are i.i.d. with the common distribution $N_p(0, \sigma^2 \Sigma_1)$, (or $\tilde{N}_p(0, \sigma^2 \Sigma_1)$).

(iii) $s(t_i)$ and $n(t_j)$ are independent for all i and j.

Under these assumptions

$$X(t) \sim N_p(0, \Sigma_2 = \Gamma + \sigma^2\Sigma_1), \text{ (or } \tilde{N}_p(0, \Sigma_2 = \Gamma + \sigma^2\Sigma_1))$$

where

$$\Gamma = A\Psi A', \text{ (or } A\Psi A^*) \tag{3.2}$$

is of rank $q < p$. Further, if

$$S_2 = \sum_1^n X(t_i)X(t_i)', \text{ (or } \sum_1^n X(t_i)X(t_i)^*) \tag{3.3}$$

then

$$S_2 \sim W_p(n, \Sigma_2), \text{ (or } S_2 \sim \tilde{W}_p(n, \Sigma_2)). \tag{3.4}$$

We consider the problem of testing

$$H_0: \text{ Rank } \Gamma = q < p \qquad \text{versus} \qquad \text{rank } \Gamma \text{ is arbitrary} \tag{3.5}$$

and also of estimating q, the rank of Γ.

3.1 CASE 1: $\Sigma_1 = I$

In this case $\Sigma_2 = \Gamma + \sigma^2 I$, and it is well-known (see Anderson (1963)) that the likelihood criterion for testing (3.5) is

$$L_q = \frac{(\ell_{q+1} \cdots \ell_p)^s}{[(\ell_{q+1} + \ldots + \ell_p)/(p-q)]^{s(p-q)}} \tag{3.1.1}$$

where $\ell_1 \geq \ldots \geq \ell_p$ are the eigen values of $n^{-1}S_2$ and $s = n/2$ in the real case and $s = n$ in the complex case. In large samples, i.e., as $n \to \infty$

$$-2\log L_q \sim \chi^2([(p-q)(p-q+1)/2] - 1), \text{ (or } \chi^2[(p-q)^2 - q - 1]). \tag{3.1.2}$$

As for the estimation of q, Zhao, Krishnaiah and Bai (1986a) suggested a new information theoretic criterion which is more general than that proposed by Akaike (1972), Schwartz (1978) and Rissanen (1978). Their method consists in choosing as an estimate of q the number $\hat{q}$ such that

$$I(\hat{q}, C_n) = \min\{I(0, C_n), \ldots, I(p-1, C_n)\} \tag{3.1.3}$$

where

$$I(k, C_n) = -\log L_k + C_n\nu(k, p)$$
$$\nu(k, p) = [k(2p - k + 1)/2], \text{ (or } k(2p - k + 1))$$

which is the number of free parameters when rank $\Gamma = k$, and C_n are such that

$$\lim_{n\to\infty} (C_n/n) = 0,$$
$$\lim_{n\to\infty} (C_n/\log\log n) = \infty. \tag{3.1.4}$$

Zhao, Krishnaiah and Bai (1986a) proved that $\hat{q}$ determined as in (3.1.3) is a strongly consistent estimate of q.

3.2 CASE 2: $\Sigma_1 = I$, $\sigma^2 = 1$

In this case, the likelihood ratio criterion for testing the hypothesis (3.5) is derived by Zhao, Krishnaiah and Bai (1986) as

$$\log L_q = s \sum_{i=1+\min(\tau,q)} (\log \ell_i + 1 - \ell_i) \tag{3.2.1}$$

where $s = n/2$ in the real case and $s = n$ in the complex case and τ is the number of eigen values ℓ_i which are greater than unity. The large sample distribution of $-2\log L_q$ is no longer χ^2. But as suggested by Rao (1983) in a slightly different situation, the test of the hypothesis in this case can be carried out in two stages, first as in Case 3.1 taking σ^2 as unknown, and then examining whether $\sigma^2 = 1$. However, the criterion of Zhao, Krishnaiah and Bai can be used for the estimation of q, i.e., by minimizing

$$-\log L_k + C_n \nu(k,p) \tag{3.2.2}$$

where C_n is as in (3.1.4) and

$$\nu(k,p) = [k(2p-k+1)/2], \text{ (or } k(2p-k+1)). \tag{3.2.3}$$

CASE 3: Σ_1 ARBITRARY AND σ^2 UNKNOWN

In addition to S_2 defined in (3.3), we suppose that another independent variable S_1 is observable and has the distribution

$$S_1 \sim W_p(f, \Sigma_1), \text{ (or } S_1 \sim \tilde{W}_p(f, \Sigma_1)). \tag{3.3.1}$$

Let $\ell_1 \geq \ldots \geq \ell_p$ be the roots of the equation $|n^{-1}S_2 - f^{-1}S_1| = 0$. The likelihood ratio criterion for the hypothesis (3.5) is derived by Rao (1983) in the form

$$-2\log L_q = -s \log \prod_{i=q+1}^{p} \left[\left(\frac{n\ell_i + f\hat{\sigma}^2}{n+f} \right)^{n+f} \frac{1}{\ell_i^n \hat{\sigma}^{2f}} \right] \tag{3.3.2}$$

where $s = 1$ in the real case and $s = 2$ in the complex case, and $\hat{\sigma}^2$ is the root of the equation

$$\frac{(p-q)n}{f+n} = \sum_{i=q+1}^{p} \frac{nm_i}{nm_i + f\sigma^2}. \tag{3.3.3}$$

As n and $f \to \infty$ such that $[f/(n+f)] \to \beta \in (0,1)$, the statistic (3.3.2) is distributed as

$$\chi^2\left(\frac{(p-q)(p-q+1)}{2} - 1 \right), \text{ (or } \chi^2((p-q)^2 - q - 1)). \tag{3.3.4}$$

For estimating q, the method of Zhao, Krishnaiah and Bai is to minimize

$$-\log L_k + C_n \nu(k,p) \tag{3.3.5}$$

where C_n is as in (3.1.4) and

$$\nu(k,p) = \frac{k(2p-k+1)}{2}, \text{ (or } k(2p-k+1)). \tag{3.3.6}$$

CASE 4: Σ_1 IS ARBITRARY AND $\sigma^2 = 1$

In this case the likelihood ratio criterion is derived by Zhao, Krishnaiah and Bai (1986b) as

$$-2\log L_q = s\log \prod_{i=1+\min(q,\tau)} \left[\left(\frac{n\ell_i + f}{n+f}\right)^{n+f} \ell_i^{-n}\right] \tag{3.4.1}$$

where $s = 1$ for the real case and $s = 2$ for the complex case, and τ is the number of ℓ_i which are greater than unity. The statistic does not have an asymptotic χ^2 distribution but is useful in the estimation of q. As in other cases we choose q to minimize

$$-\log L_k + C_n\nu(k,p) \tag{3.4.2}$$

where C_n is as in (3.1.4) and

$$\nu(k,p) = \frac{k(2p-k+1)}{2}, \text{ (or } k(2p-k+1)) \tag{3.4.3}$$

The estimates of q obtained in (3.2.2), (3.3.5) and (3.4.2) are strongly consistent as n and $f \to \infty$ such that $[f/(n+f)] \to \beta \in (0,1)$. Detailed proofs are given in two papers by Zhao, Krishnaiah and Bai (1986a, 1986b).

4. EXPONENTIAL SIGNAL MODELS

Let $y_t = \xi_t + n_t$, $t = 1,\ldots,n$, be observations taken at equal intervals of time on signals corrupted by noise. The signal ξ_t is considered to be of the form

$$\xi_t = a_1 e^{s_1 t} + \ldots + a_m e^{s_m t} \tag{4.1}$$

where $a' = (a_1,\ldots,a_m)$ and $s' = (s_1,\ldots,s_m)$ are unknown complex vector parameters. Often the value of m, the number of signals (exponential terms in (4.1)), is itself unknown and has to be considered as an essential parameter under estimation. The noise components are taken to be independently distributed with mean zero and common variance σ^2. The problem has a long history starting with the pioneering work of Prony (1795) two hundred years ago. In a series of papers Tufts and Kumaresan (see Kumaresan (1982), Tufts and Kumaresan (1982) and Kumaresan and Tufts (1982) and the numerous references therein) suggested some new approaches to the problem based on Prony's parametrization of the signal process described below.

Prony (1795) observed that ξ_i as defined in (4.1) satisfy the recurrence relations

$$\xi_{i+m} + g_1\xi_{i+m-1} + \ldots + g_m\xi_i = 0, \; i = 1, \ldots, n-m, \tag{4.2}$$

where $g' = (g_m, \ldots, g_1, 1)$ is a function of s, which may be regarded as an alternative parameter to s. The equations (4.2) lead to the observational equations

$$-y_{i+m} = g_1 y_{i+m-1} + \ldots g_m y_i, \; i = 1, \ldots, n-m \tag{4.3}$$

which can be written in the matrix form

$$GY = 0, \text{ with } Y' = (y_1, \ldots, y_n) \tag{4.4}$$

choosing the $(n-m) \times n$ matrix G appropriately with each row containing the row vector g' and a number of zeros, the position of g' being shifted by one element when we go from one row to the next row. Much of the previous work is centered on the equations (4.3, 4.4) and the estimation of g by minimizing

$$Y^*G^*GY \qquad \text{or} \qquad Y^*G^*GY/g^*g \tag{4.5}$$

fixing an appropriate value for m. Once g is estimated, $\exp(s_i)$ are obtained as the roots of the polynomial equation

$$0 = 1 + g_1 z^{-1} + \ldots + g_m z^{-m} = \prod_{r=1}^{m}(1 - z^{-r}e^{\hat{s}_r}) \tag{4.6}$$

and a is estimated by minimizing

$$(Y - Xa)^*(Y - Xa) \tag{4.7}$$

taking

$$X \triangleq \begin{pmatrix} \exp(\hat{s}_1) & \ldots & \exp(\hat{s}_m) \\ . & \ldots & . \\ \exp(\hat{s}_1 n) & \ldots & \exp(\hat{s}_m n) \end{pmatrix} \tag{4.8}$$

as fixed. Tufts and Kumaresan in the papers cited above make some refinements by starting with a larger value, say $\ell > m$, and use the extra terms to reduce the noise part of the model. However, it is not clear whether minimizing the first expression in (4.5) as Tufts and Kumaresan do ignoring the correlations between the components of GY lead to consistent estimators of the unknown parameters. It may be noted that the covariance matrix of GY as defined in (4.4) is $\sigma^2 GG^*$, in which case the appropriate quadratic form to be minimized is

$$Y^*G^*(GG^*)^{-1}GY \tag{4.9}$$

and not (4.5), although the minimization problem associated with (4.9) is far more complicated.

It is shown (see Smyth (1985) and references therein) that the estimates of a and s obtained from (4.6) and (4.7) using the g estimated from (4.9) are, indeed, maximum likelihood estimates, i.e., those obtained by minimizing

$$\sum_{t=1}^{n} |y_t - a_1 e^{s_1 t} - \dots - a_m e^{s_m t}|^2 \tag{4.10}$$

with respect to a_i and s_j, under the assumption that the error terms are normally distributed. We can also look upon the estimates of a_i and s_i obtained by minimizing (4.10) as non-linear least squares estimates without any distributional assumptions. Smyth (1985) has developed an efficient algorithm for obtaining g which minimizes (4.9), and then estimating s and a through the steps in (4.6) and (4.7), which solves the non-linear least squares problem (4.10) for given m. Osborne (1975) provides the necessary theory and algorithms for solving problems such as (4.10).

The choice of m can be made by a suitable information theoretic model selection criterion such as the one used in the previous section. If the minimum value of (4.10) is denoted by R_m, then under the assumption of normality of η_t, the error components, the information theoretic criterion takes the form

$$n \log R_m + C_n(8m) \tag{4.11}$$

where C_n are chosen to satisfy the condition (3.1.4). The AIC criterion (Akaike (1972)) corresponds to the choice $C_n = 1$, which may be sufficient to provide a fairly accurate estimate of m. Note that R_m is also the minimum value of (4.9) when minimized with respect to the vector g.

However, a more relevant and satisfactory method in finite samples is provided by the cross validation (CV) approach, although the computations may be extremely heavy. (See for instance papers by Rao (1984) and Rao and Boudreau (1985) for such an approach in a prediction problem.)

In the CV method we leave one of the values, say y_i, but replace it by a variable Y_i. For any choices of Y_i and m, using Smyth's algorithm, we compute

$$R(Y_i, m) = \min_{a,s} \sum_{t=1}^{n} |y_t - a_1 e^{s_1 t} - \dots - a_m e^{s_m t}|^2 \tag{4.12}$$

where $a' = (a_1, \dots, a_m)$ and $s' = (s_1, \dots, s_m)$. Then for given m, by a suitable computer program, we find $\hat{y}_{im}$ such that

$$R(\hat{y}_{im}, m) = \min_{Y_i} R(Y_i, m) \tag{4.13}$$

which provides $\hat{y}_{im}$ as an estimate of Y_i for given m. Then comparing $\hat{y}_{im}$ with the observed y_i, the cross validation error (CVE) is obtained as

$$R_*(m) = \sum_{i=1}^{n} (y_i - \hat{y}_{im})^2, \tag{4.14}$$

and finally m is chosen as that value for which $R_*(m)$ is a minimum.

As observed earlier, the computations involved in the above procedure are extremely heavy. However, simplication may be effected in some ways.

1. If n is large, we may choose every alternative or every third value among the components of $(y_1, \ldots, y_n)$ for cross validation. This cuts down on the number of terms in (4.14) and reduces the computing time considerably.

2. It may be noted that $\hat{y}_{im}$ can be computed by an alternative method as

$$\hat{y}_{im} = \sum_{j=1}^{m} \hat{a}_j e^{\hat{s}_j i} \tag{4.15}$$

where $\hat{a}_j$, $\hat{s}_j$, $j = 1, \ldots, m$, are the values minimizing the expression,

$$\sum_{t=1}^{i-1} |y_t - \sum_{1}^{m} a_j e^{s_j t}|^2 + \sum_{t=i+1}^{n} |y_t - \sum_{1}^{m} a_j e^{s_j t}|^2. \tag{4.16}$$

Due to the absence of the term y_i in (4.16), one cannot take full advantage of Prony's reparametrization. But if the optimum values $\hat{a}_j$ and $\hat{s}_j$ can be found directly through some other algorithm, then $\hat{y}_{im}$ can be obtained as in (4.15).

3. We defined $R(Y_i, m)$ as the minimum of the expression on the right hand side of (4.12). For purposes of estimating m, one could use an approximation

$$R(Y_i, m) = \sum_{t=1}^{m} |y_t - \hat{a}_1 e^{\hat{s}_1 t} - \ldots - a_m e^{\hat{s}_m t}|^2 \tag{4.17}$$

where a_i and s_j are estimates obtained by minimizing the second expression in (4.5) to estimate g and then applying the formulae (4.6) - (4.8).

ACKNOWLEDGMENTS

Research sponsored by the Air Force Office of Scientific Research (AFSC) under Contract F49620-85-C-0008. The United States Government is authorized to reproduce and distribute reprints for governmental purposes notwithstanding any copyright notation hereon.

BIBLIOGRAPHY

Akaike, H. (1972). Information theory and an extension of the maximum likelihood principle. In *Proc. Second International Symposium on Information Theory.* Supp. to *Problems of Information Theory and Control*, 267–281.

Anderson, T. W. (1963). Asymptotic theory for principal component analysis, *Ann. Math. Statist.* **34**, 122–148.

Bai, Z. D., Krishnaiah, P. R. and Zhao, L. C. (1986). Signal processing using model selection methods. Tech. Rept. 86–03, Center for Multivariate Analysis, Pittsburgh.

Khatri, C. G. and Rao, C. Radhakrishna (1985a). Effects of estimated noise covariance matrix in optimal signal detection. Tech. Rept. 85-38, Center for Multivariate Analysis, Pittsburgh.

Khatri, C. G. and Rao, C. Radhakrishna (1985b). Test for a specified signal when the noise covariance matrix is unknown. Tech. Rept. 85-47, Center for Multivariate Analysis, Pittsburgh.

Khatri, C. G., Rao, C. R. and Sun, Y. N. (1986). Tables for obtaining confidence bounds for realized signal to noise ratio with an estimated discriminant function. *Communications in Statistics, Simulation and Computation* **15**, 1–14.

Kumaresan, Ramdas (1982). *Estimating the parameters of exponentially damped or undamped sinusoidal signals in noise.* Ph.D. thesis, University of Rhode Island.

Kumaresan, R. and Tufts, D. W. (1982). Estimating the parameters of exponentially damped sinusoids and pole-zero modeling in noise. *IEEE Trans. Acoust., Speech, Signal Processing* **ASSP-30**, 833–840.

Osborne, M. R. (1975). Some special non-linear least squares problem. *Siam J. Numer. Anal.* **12**, 571–592.

Prony, R. de (1795). Essai experimentale et analytique. *J. Ecole Polytechnique* (Paris), 24–76.

Rao, C. Radhakrishna (1946). Tests with discriminant functions in multivariate analysis. *Sankhya* **7**, 407–414.

Rao, C. Radhakrishna (1948). The utilization of multiple measurements in problems of biological classification. *J. Roy. Statist. Soc.* **B10**, 159–203.

Rao, C. Radhakrishna (1949). On some problems arising out of discrimination with multiple characters. *Sankhya* **9**, 343–364.

Rao, C. Radhakrishna (1973). *Linear Statistical Inference and its Applications* (Second Edition), Wiley, New York.

Rao, C. Radhakrishna (1983). Likelihood ratio tests for relationships between two covariance matrices. In *Studies in Econometrics, Time Series, and Multivariate Statistics* (Eds. S. Karlin, T. Amemiya and L. A. Goodman), Academic Press, 529–544.

Rao, C. Radhakrishna (1984). Prediction of future observations in polynomial growth curve models. Proc. Indian Statistical Institute Golden Jubilee International Conference on *Statistics: Applications and New Directions*, 512–520.

Rao, C. Radhakrishna and Boudreau, R. (1985). Prediction of future observations in a factor analytic type growth model. In *Multivariate Analysis* **VI** (ed. P. R. Krishnaiah), Elsevier Science Publishers, 449–466.

Reed, I. S., Mallet, J. D. and Brennan, L. E. (1974). Rapid convergence rates in adaptive arrays. *IEEE Transactions on Aerospace and Electronic Systems.* **10**, 853–863.

Rissanen, J. (1978). Modeling by shortest data description. *Automatica* **14**, 465–471.

Schwartz, G. (1978). Estimating the dimension of a model. *Ann. Statist.* **6**, 461–464.

Smyth, G. K. (1985). Coupled and separable iterations in non-linear estimation. Ph.D. thesis, Australian National University.

Tufts, D. W. and Kumaresan, R. (1982). Estimation of frequencies of multiple sinusoids: Making linear prediction perform like maximum likelihood. *Proc. IEEE, Special Issue on Spectral Estimation* **70**, 975–989.

Wax, M. and Kailath, T. (1985). Detection of signals by information theoretic criteria. *IEEE Trans. Accoustics, Speech, and Signal Processing* **ASSP-33**, 387–392.

Zhao, L. C., Krishnaiah, P. R. and Bai, Z. D. (1986a). On the detection of the number of signals in the presence of white noise. *J. Multivariate Analysis* **20**.

Zhao, L. C., Krishnaiah, P. R. and Bai, Z. D. (1986b) On the detection of number of signals when the noise covariance matrix is arbitrary. *J. Multivariate Analysis* **20**.

PART 4 - Design, and Comparison of Experiments and Distributions

COMPARISON OF EXPERIMENTS AND INFORMATION IN CENSORED DATA

PREM K. GOEL

Department of Statistics
The Ohio State University
Columbus, Ohio 43210-1247, U.S.A.

1. INTRODUCTION

Let $\mathcal{E} = \{(X, S_X, \mathcal{A}); P_\theta : \theta \in \Theta\}$ denote a statistical experiment in which a random variable X, defined on a measurable space $(S_X, \mathcal{A})$, is to be observed. The distribution of X, denoted by P_θ, involves an unknown parameter θ which belongs to a fixed, but arbitrary, collection of states of nature Θ. Let $\mathcal{F} = \{(Y, S_Y, \mathcal{B}); Q_\theta\text{: } \theta \in \Theta\}$ denote another experiment with the same parameter space Θ. Starting with the pioneering work of Blackwell (1951, 1953), the literature on "Comparison of Experiments" has grown steadily. The maturity of literature on "Comparison of Experiments" is evidenced by the publication of a book by Heyer (1982), and a forthcoming book by Torgersen (1985). In this paper, we shall not make any attempt to exhaustively review this topic. After a brief discussion of some concepts, we shall concentrate our effort on the information theoretic connections and the comparison of various experiments with censored data.

A decision theoretic notion of the experiment $\mathcal{E}$ being "more informative than" the experiment $\mathcal{F}$, is as follows:

Definition 1.1. If for every decision problem and any decision procedure $\delta(Y)$, there exists a decision procedure $\delta^*(X)$ based on $\mathcal{E}$ such that the risk functions $R(\theta, \delta(Y))$ and $R(\theta, \delta^*(X))$ satisfy

$$R(\theta, \delta^*(X)) \leq R(\theta, \delta(Y)) \text{ for all } \theta \in \Theta, \tag{1.1}$$

then the experiment $\mathcal{E}$ is said to be *more informative* than the experiment $\mathcal{F}$.

The following Bayesian version of this definition was suggested by Bohnenblust, Shapley and Sherman (see Blackwell (1951)) in an unpublished memorandum.

Definition 1.2. The experiment $\mathcal{E}$ is *more informative* than the experiment $\mathcal{F}$ if for every decision problem involving θ and every prior distribution π on Θ, the Bayes risk $r(\pi, \delta(Y)) = E^\pi[R(\theta, \delta(Y))]$ of any decision procedure $\delta(Y)$ based on the experiment $\mathcal{F}$ satisfies

$$r(\pi, \delta(Y)) \geq r(\pi, \delta^\pi(X)), \tag{1.2}$$

where $\delta^{\pi}(X)$ denotes the Bayes procedure with respect to the prior distribution π, and $r(\pi, \delta^{\pi}(X))$ denotes the minimum Bayes risk for this decision problem, obtainable from the experiment $\mathcal{E}$.

Yet another concept "the experiment $\mathcal{E}$ being sufficient for the experiment $\mathcal{F}$", developed by Blackwell (1951), is based on the idea of *randomization* or *the existence of Markov kernel.*

Definition 1.3. The experiment $\mathcal{E}$ is sufficient for the experiment $\mathcal{F}$, if there exists a *stochastic transformation* of X on $(S_X, \mathcal{A})$ to a random variable $Z(X)$ on $(S_Y, \mathcal{B})$ such that, for each $\theta \in \Theta$, the random variables $Z(X)$ and Y have identical distributions.

Lehmann (1986a) (see §3.4) states the above definition in the following equivalent form.

Definition 1.3*. The experiment $\mathcal{E}$ is sufficient for the experiment $\mathcal{F}$ if $\exists$ a function $h(x, U)$, where U is a random variable, distributed independently of X, with a known distribution, such that the distributions of $h(X, U)$ and Y are identical for each $\theta \in \Theta$.

A more precise measure-theoretic definition of this concept in terms of the existence of a Markov kernel is stated in an excellent review article on this topic by Torgersen (1976). For families $\mathcal{P} = \{P_\theta\colon \theta \in \Theta\}$ and $\mathcal{Q} = \{Q_\theta\colon \theta \in \Theta\}$ of probability measures dominated by σ-finite measures μ and ν, with pdf f and g respectively, this definition is equivalent to (see DeGroot (1970), §14.17) the existence of a non-negative function $h(x, y)$ on $S_X \times S_Y$ such that,

$$g(y|\theta) = \int h(x,y) f(x|\theta) d\mu(x), \text{ for } y \in S_Y \text{ and } \theta \in \Theta \tag{1.3}$$

and

$$\int h(x,y) d\nu(y) = 1 \text{ for each } x \in S_X, \tag{1.4a}$$

$$0 < \int h(x,y) d\mu(x) < \infty \text{ for each } y \in S_Y. \tag{1.4b}$$

The equivalence of these definitions under various conditions on Θ has been established by Blackwell (1953), Boll (1955), Le Cam (1964), Richter and Schafer (1983) and Feldman and Ramamoorthi (1986).

The requirement that one experiment have everywhere smaller risk than the other for all decision problems involving θ is clearly quite restrictive for the experimental comparison purposes. It is well known that this definition produces only a partial ordering on the set of all possible experiments involving Θ. Usually, it will be the case that the experiment $\mathcal{E}$ has smaller risk than the experiment $\mathcal{F}$ for some decision problems and/or prior distributions whereas the experiment $\mathcal{F}$ has smaller risk than the experiment $\mathcal{E}$ for other decision problems and/or prior distributions. Thus these two experiments are not comparable from this point of view.

Various attempts have been made to relax these requirements in order to obtain meaningful comparisons. Le Cam (1964) has generalized the "sufficiency" notion to "approximate sufficiency" and "ε-deficiency" of $\mathcal{E}$ relative to $\mathcal{F}$ and developed an elegant theory for these concepts and their connections to asymptotic decision theory (see e.g., Le Cam (1986), Strasser (1985)). These concepts have been applied to various experiments and decision problems (see e.g., Hansen and Torgersen (1974), Swenson (1980), and Torgersen (1970, 1972, 1981).

Another direction in which these strict requirements have been relaxed is the so called "Nonequivalent Comparison of Experiments". Stone (1961) suggested the requirement that only for a fixed decision problem the experiment $\mathcal{E}$ have a smaller risk than the experiment $\mathcal{F}$. Furthermore, Goel and DeGroot (1979) suggested the requirement that the experiment $\mathcal{E}$ have smaller Bayes risk than the experiment $\mathcal{F}$ only for a smaller class of prior distributions π on Θ, but for every decision problem. Both of these articles provide examples of the usefulness of these relaxed requirements. A recent article by Lehmann (1986b), which was presented at the Purdue Symposium, also shows the usefulness of these ideas. He first shows that for some very reasonable problems the sufficiency definition produces noncomparability. This leads him to define the concept, '*the experiment $\mathcal{E}$ is more effective than the experiment $\mathcal{F}$*' if the experiment $\mathcal{E}$ provides decision procedures with smaller risk than that based on the experiment $\mathcal{F}$ only for an interesting subclass of decision problems. He then considers the class of all monotone decision problems (Karlin and Rubin (1956)) for location experiments to obtain quite meaningful comparisons.

Principal methods used to determine comparability/non-comparability of experiments are as follows: If $\mathcal{E}$ is more informative than $\mathcal{F}$, then one needs to determine the stochastic transformation $h(X, U)$ required by Definition 1.3*. However, in order to prove that $\mathcal{E}$ is not sufficient for $\mathcal{F}$, one can either construct a decision problem in which the minimum Bayes risk obtainable from the experiment $\mathcal{E}$ is more than that from the experiment $\mathcal{F}$ or find a measure of information for which the experiment $\mathcal{E}$ has more information than the experiment $\mathcal{F}$. Usually the construction of a decision problem for showing non-comparability of $\mathcal{E}$ and $\mathcal{F}$ is quite a difficult task. Therefore, Lehmann (1986b) interprets this approach to mean that "we attempt to construct a statistical task which can be performed on the basis of $\mathcal{F}$, but which either cannot be performed, or at least not as well on the basis of $\mathcal{E}$. For this purpose, it is most convenient to find a function $\alpha(\theta)$ which has an unbiased estimator based on $\mathcal{F}$ but not based on $\mathcal{E}$". He justifies this interpretation partially through a statement in DeGroot (1966), which is a restatement of the stochastic transformation definition in terms of the existence of *nonnegative unbiased estimators* of $Q_\theta(Y \in B)$, for every set $B \in \mathcal{B}$, based on the random variable X.

Since the loss function $L(\theta, \delta)$ is a function of θ and $\delta(X)$ only (and not of $E_\theta[\delta(X)]$), we do not know of any statistical decision problem which requires the construction of an unbiased estimator (incorporating bias as a component in the loss function is not reasonable, because it is a function of $E_\theta[\delta(X)]$). Furthermore, Bayesian statisticians do not necessarily believe in the unbiasedness requirement. Thus the fact that the non-existence of an unbiased estimator of $\alpha(\theta)$ based on $\mathcal{E}$, when one exists based on $\mathcal{F}$, implies that $\mathcal{E}$ is not sufficient for $\mathcal{F}$ is not so obvious. We state the following proposition, with a simple proof, to establish the validity of Lehmann's statement. This will also convince the Bayesian statisticians that unbiasedness may be a useful tool in decision analysis for comparing experiments.

Proposition 1.1. *Suppose that the experiment $\mathcal{E}$ is sufficient for the experiment $\mathcal{F}$. Let $\alpha(\theta)$ be a function of θ for which $\exists$ an unbiased estimator $T(Y)$ based on $\mathcal{F}$. Then there must exist an unbiased estimator of $\alpha(\theta)$ based on $\mathcal{E}$.*

PROOF. Given that $T(Y)$ is an unbiased estimator of $\alpha(\theta)$, we have

$$\alpha(\theta) = \int T(y) g(y|\theta) d\nu(y), \qquad \forall\, \theta \in \Theta.$$

However, since $\mathcal{E}$ is sufficient for $\mathcal{F}$, it follows from (1.3) that

$$\alpha(\theta) = \int T(y) \int h(x,y) f(x|\theta) d\mu(x) d\nu(y), \qquad \forall\, \theta \in \Theta.$$

Now, a change in the order of integration provides

$$\alpha(\theta) = \int \left[\int T(y) h(x,y) d\nu(y)\right] f(x|\theta) d\mu(x) \qquad \forall\, \theta \in \Theta. \tag{1.5}$$

If we let $U(x) = \int T(y) h(x,y) d\nu(y)$, then

$$\alpha(\theta) = \int U(x) f(x|\theta) d\mu(x) \qquad \forall\, \theta \in \Theta, \tag{1.6}$$

i.e., $U(X)$ is an unbiased estimator of $\alpha(\theta)$ based on $\mathcal{E}$. Hence, if one can construct an unbiased estimator of some parametric function based on the random variable Y for which there is no unbiased estimator based on X, then $\mathcal{E}$ cannot be sufficient for $\mathcal{F}$. □

2. INFORMATION MEASURES

Usually, mathematical statisticians are concerned with finding optimal inferential or decision procedures for problems of interest and studying their operating characteristics after the experiment to be conducted has been chosen. Therefore, from their perspective, a ***statistical experiment*** could be considered as a synonym for ***statistical information contained in the experiment.*** Thus, given a choice between two experiments, the process of choosing an appropriate one is equivalent to comparing information about θ contained in these experiments and choosing the one with larger information. However, as is well known, there is no unique measure of 'statistical information' contained in an experiment. Nevertheless, the statement that 'the experiment $\mathcal{E}$ is more informative than the experiment $\mathcal{F}$' in the sense of Definition 1.1 must imply that 'the experiment $\mathcal{E}$ has more information that the experiment $\mathcal{F}$' regardless of the measure of information. In this section, we review these concepts and in addition provide a Bayesian analog of Hellinger transform, which naturally suggests an extension of this transform to arbitrary prior distributions π on Θ.

In the classical statistics framework, this notion has been formalized for Fisher information as follows. Let $\mathcal{I}_X(\theta)$ and $\mathcal{I}_Y(\theta)$ denote the Fisher information matrices. If the experiment $\mathcal{E}$ is sufficient for the experiment $\mathcal{F}$, then $\mathcal{I}_X(\theta) - \mathcal{I}_Y(\theta)$ is a non-negative definite matrix (see Stone (1961), Goel and DeGroot (1979)). Using this result, Ferentino and Papaioannou (1982) prove that the information measures defined by the trace, the determinant, and the ith largest eigen value of the Fisher information matrix also satisfy this notion of more information. They also prove that for one dimensional θ, this result is valid for three other measures of information, suggested by Mathai, Vajda and Boekee, which are monotone functions of $E_\theta(|\partial \log f(x|\theta)/\partial\theta|^\alpha)$ for all $\alpha \geq 1$.

In the Bayesian statistics formulation, measures of information in an experiment are defined either as the difference of the expected utility of the posterior Bayes action and the utility of the prior Bayes action or as a 'divergence' between the posterior and the prior distributions or between the marginal (prior-predictive) distributions corresponding to two prior distributions.

For any prior distribution π and any decision d in a given decision problem, let

$$\rho(\pi, d) = E^\pi[L(\theta, d)] = \int_\Theta L(\theta, d)d\pi(\theta). \tag{2.0a}$$

Also, let d^π denote the (unique) Bayes decision with respect to the prior distribution π, i.e.,

$$\rho(\pi, d^\pi) = \inf_{d \in D} \rho(\pi, d) = -V[\pi] \quad \text{(say)} \tag{2.0b}$$

The function $V[\pi]$ can be interpreted as the expected utility of the Bayes action for some decision problem when the Decision Maker's uncertainty about θ is reflected by π. Furthermore, every convex functional $V[.]$ on the space Ξ of all probability measures on Θ can be shown to have arisen via (2.0a) and (2.0b) for some decision problem. If the observation x updates the prior distribution

π to $\pi(x)$, then the observed value of the sample information $I(x)$ is defined by (see e.g., DeGroot (1984))

$$I(x) = \rho(\pi(x), d^{\pi}) + V[\pi(x)]. \tag{2.1a}$$

In words, $I(x)$ is the expectation, with respect to the posterior distribution, of the difference between the loss from the Bayes decision and the loss which would have been attained if the observation x were not available. Apriori, the *Expected Value of Sample Information* $E[I(X)]$, where the expectation is taken with respect to the prior-predictive (marginal) distribution of X, is equivalent to the expected gain in utility (see e.g., DeGroot (1984), i.e.,

$$E[I(X)] = E_X\{V[\pi(X)]\} - V[\pi]. \tag{2.1b}$$

It is well known that if the experiment $\mathcal{E}$ is sufficient for the experiment $\mathcal{F}$, then

$$E[I(X)] \geq E[I(Y)] \tag{2.2}$$

for every decision problem and every prior distribution π on Θ (see e.g., DeGroot (1962, 1970)).

For the information measures defined through 'divergence measures' let π_1 and $\pi_2 \in \Xi$ be two prior distributions, and let

$$p_i(x) = \int f(x|\theta) d\pi_i(\theta), \qquad i = 1, 2 \tag{2.3}$$

be the corresponding marginal distributions of X. Furthermore, let

$$I_X(\pi_1, \pi_2) = \int p_1(x) \log[p_1(x)/p_2(x)] d\mu(x), \tag{2.4}$$

denote the Kullback-Leibler divergence between π_1 and π_2. If the experiment $\mathcal{E}$ is sufficient for the experiment $\mathcal{F}$, then

$$I_X(\pi_1, \pi_2) \geq I_Y(\pi_1, \pi_2) \tag{2.5}$$

for all priors π_1 and $\pi_2 \in \Xi$ (see Goel and DeGroot (1979)). A most general result of this type can be stated in terms of the so called ϕ-divergence of π_1 and π_2, defined below.

Definition 2.1. Let p_1 and p_2 be the marginal distributions of X corresponding to the prior distributions π_1 and π_2 on Θ and let $\phi(u): R^+ \to R$ be a convex function. Then the ϕ-divergence of π_1 and π_2, based on the experiment $\mathcal{E}$, is defined by

$$I_{\phi,X}(\pi_1, \pi_2) = \int p_2(x) \phi[p_1(x)/p_2(x)] d\mu(x). \tag{2.6}$$

Several well-known divergence measures are special cases of ϕ-divergence measures for various convex functions. Table 1 provides a list of these functions

Table 1. ϕ-functions and divergence measures.

$\phi(u)$	information/divergence measure
$u \log u$	$I(\pi_1, \pi_2)$ Kullback-Leibler.
$-\log u$	$I(\pi_2, \pi_1)$ Directed Divergence.
$(u-1)\log u$	$J(\pi_1, \pi_2)$ J-Divergence.
$0.5(\sqrt{u}-1)^2$	Kolmogorov's measure of distance (Hellinger and Matusita distances are monotone functions).
$\|u-1\|/2$	Variation (City-Block) distance.
$1-u^\alpha, 0<\alpha<1$	Basis for Chernoff's measure of discriminatory information and generalized Bhattacharya measure.
$u^\alpha \mathrm{Sgn}\ (\alpha-1), 0<\alpha(\neq 1)$	Renyi's entropy of order α is a monotone function.
$[u^\alpha + u^{(1-\alpha)}]/(\alpha-1), \alpha \neq 1$	Divergence of degree α
$(u-1)^2$	Kagan's measure or Chi-squared divergence.
$\|(1-u^{1/m})\|^m, m>0$	Invariants of Jeffreys.
$1-\min(u,1)$	Distance function used in Pattern Recognition problems.

for some widely used information measures. Some of these measures and their applications in statistical contexts are discussed in Rao (1980).

The ϕ-divergence measure has been used to define *Expected Amount of Sample Information* in the context of hierarchical models by Goel (1983). Ferentino and Papaioannou (1982) prove that for the information measures defined by (2.6), the result corresponding to (2.2) is also valid. We state their result here.

Theorem 2.1. *If the experiment $\mathcal{E}$ is sufficient for the experiment $\mathcal{F}$, then*

$$I_{\phi,X}(\pi_1,\pi_2) \geq I_{\phi,Y}(\pi_1,\pi_2) \tag{2.7}$$

for all priors π_1 and $\pi_2 \in \Xi$, and all convex functions ϕ.

Taneja (1986) proves the corresponding result for some more divergence measures which are essentially monotone functions of (2.6) for some convex functions ϕ.

Another important functional which has been used in studying the implication of the sufficiency of the experiment $\mathcal{E}$ for the experiment $\mathcal{F}$ is the well known Hellinger transform. For a prior π, with finite support, which assigns probability λ_i to θ_i, $i = 1, 2, \ldots, k$, the Hellinger transform for the experiment $\mathcal{E}$ is defined by

$$H_X(\pi) = \int \Pi[f(x|\theta_k)]^{\lambda_i} d\mu(x). \tag{2.8}$$

It is well known that if the experiment $\mathcal{E}$ is sufficient for the experiment $\mathcal{F}$,

then

$$H_Y(\pi) \geq H_X(\pi) \tag{2.9}$$

holds for all priors $\pi \in \Xi_0$ with only a finite number of points in its support (see Torgersen (1976). Thus the Hellinger transform acts as negative of an information measure. Furthermore, in (2.9), if the equality holds for all priors π in Ξ_0, then the experiments $\mathcal{E}$ and $\mathcal{F}$ are equivalent. However, validity of (2.9) for all priors π in Ξ_0 does not imply that the experiment $\mathcal{E}$ is sufficient for the experiment $\mathcal{F}$.

We shall now rewrite (2.8) in a form so as to provide an interesting Bayesian interpretation of this function. The integrand in (2.8) can be written as

$$\exp[\Sigma\lambda_i \log(f(x|\theta_i))] \tag{2.10}$$

or, equivalently as

$$\exp[\Sigma\lambda_i \log\{\lambda_i(x)/\lambda_i\}]p_\pi(x), \tag{2.11}$$

where $\lambda_i(x)$ is the posterior probability of θ_i and $p_\pi(x)$ is the marginal distribution of X with respect to the prior distribution π. Now let $I_\pi(x)$ be the Kullback-Leibler information measure for discriminating between the prior and the posterior distribution, i.e.,

$$I_\pi(x) = \int \log[\pi(\theta)/\pi(\theta|x)]d\pi(\theta). \tag{2.12}$$

Then (2.11) can be written as $\{\exp[-I_\pi(x)]p_\pi(x)\}$ and the Hellinger transform can be expressed as

$$H_X(\pi) = \int \exp[-I_\pi(x)]p_\pi(x)d\mu(x). \tag{2.13}$$

The representation of the Hellinger transform suggests a generalization to all priors for which the integral (2.13) exists. In view of (2.9), the general version may be more useful for showing that an experiment is not sufficient for the other. We shall pursue this topic further in a subsequent paper.

3. CENSORING EXPERIMENTS

Let $(X_i, Y_i), i = 1, 2, \ldots, n$ be independent and identically distributed random variables, where X_i corresponds to the true life time of the ith item on test and Y_i, distributed independently of X_i, is the censoring variable for this item. Thus the observations $(Z_i, \delta_i), i = 1, 2, \ldots, n$ consist of independent and identically distributed random variables (Z, δ) defined by

$$Z = \min(X, Y) \quad \text{and} \quad \delta = I_{[X \leq Y]}, \tag{3.0}$$

where $I_{[A]}$ denotes the indicator function of the set A. Here, Z denotes the censored *or* uncensored life time of the item under test and $\delta = 1$ indicates that the observation was uncensored. This is the well known (non-informative) randomly right-censored model.

In general, the experimenter is interested in making inference about the unknown parameters when the distribution of X is modeled as a parametric family or else about some functional of the distribution of X, when it is modeled in a nonparametric framework. The variable Y is thus considered as a confounding variable which inhibits the experimenter's ability to learn about the parameters involved in the distribution of X (see Hollander, Proschan and Sconing (1985a)). Suppose that an experimenter has the choice of several censoring variables. Then, it is desirable to compare various censored experiments which observe (Z, δ) based on different censoring variables. Hollander, Proschan and Sconing (1985a, 1985b) consider this problem. A natural and simple method for comparing two experiments is to choose an appropriate measure of information and then compare the expected amount of information contained in them. In Hollander, Proschan and Sconing (1985a), the authors choose the Shannon entropy when X and Y are both discrete random variables. Given $p_i = P(X = i), q_i = P[Y = i]$, they define the expected information in the discrete censored experiment based on (X, Y) as

$$H(\mathbf{p}, \mathbf{q}) = E_Y[E_Z[-\log p(Z)|Y = y]]. \tag{3.1}$$

Thus this information is defined as the average of Shannon entropies for the experiments with degenerate censoring points $Y = y$. With $H(\mathbf{p}, \mathbf{q})$ as the expected information, they show that

$$E[\text{Information } (X)] \le E[\text{Information } (Z, \delta)] \quad \text{for every } X, Y. \tag{3.2}$$

Let (Z_i, δ_i) be the censored variable associated with Y_i, $i = 1, 2$ such that Y_2 is stochastically larger than Y_1 (to be denoted by $Y_2 \ge_{\text{st}} Y_1$) then it is shown that

$$E[\text{Informaion } (Z_1, \delta_1)] \le E[\text{Information } (Z_2, \delta_2)] \quad \text{for every } X. \tag{3.3}$$

Under regularity conditions on the distribution of X, it is also shown that for sufficiently large i,

$$E[\text{Information } (Z|Y = i)] \text{ is a concave increasing function of } i. \tag{3.4}$$

In Hollander, Proschan and Sconing (1985b), the authors state that the results in (3.3) and (3.4) "should represent a sort of 'acid-test' for the applicability of any measure of information". Thus one will need to check on the validity of these results for any measure to be called an information measure in the context of censored experiments. However, the proofs of these results require substantial amount of algebraic manipulations and do not throw any light on the underlying phenomena since they are specific to this information measure. Similar effort is also spent on proving these results for a generalization of Shannon's entropy information measure in which $\{-\log p(z)\}$ is replaced by $\phi(p(z))$, where $\phi(x)$ is decreasing and $\phi(x)/x$ is concave.

The problem of defining information measure satisfying these properties for continuous random variables X and Y is also discussed in Hollander, Proschan and Sconing (1985a). It is pointed out that the Shannon entropy is not satisfactory, because in some cases it is not defined while in others it may be negative. For example, in the exponential distribution case with $p(x) = \lambda \exp(-\lambda x), x > 0$, it is shown that $H(p(x)) \leq 0$ if $\lambda \leq e$, and it is stated that "an observation will decrease our knowledge if $H(.)$ is used as a measure of information". Therefore the authors reject Shannon entropy based information measure in the continuous case, and suggest the following measure of spread of the censored observations (Z, δ) as a measure of information

$$H^{(1)}(X,Y) = E_Y[E_X\{(X-\mu)^2 I_{[X \leq y]} + (y-\mu)^2 I_{[X>y>\mu]}\}]. \tag{3.5}$$

For an uncensored variable, it reduces to the variance of X. Again, the proof of 'acid-test' is unique to this measure. In Hollander, Proschan and Sconing (1985b), 'measures of information in censored models are developed by adapting measures of dependence' between X and Z. The 'acid-test' for most of these measures requires different tools again. Hence an alternative analysis which avoids going through this tedius checking process for every measure of information will certainly be useful. We shall now present the result *which establishes the fact that the results* (3.2) *and* (3.3) *must hold for every information measure.*

Theorem 3.1. *Let $\mathcal{E}$ and $\mathcal{F}$ be two randomly right-censored experiments based on the random variables (X, Y_1) and (X, Y_2) respectively such that $X \sim F_X(x|\theta), \theta \in \Theta$ and $Y_1 \geq_{st} Y_2$. Then the experiment $\mathcal{E}$ is sufficient for the experiment $\mathcal{F}$.*

The following Lemma (see Lehmann (1986a), Lemma 1, page 84) will be used in the proof of the above result.

Lemma 3.1. *Let G_1 and G_2 be two cdf on the real line. Then $G_1(y) \leq G_2(y)$ for all y, iff $\exists$ two non-decreasing functions g and h and a random variable V such that*

(i) *$g(u) \geq h(u)$ for all u,*

(ii) *The distribution functions of $g(V)$ and $h(V)$ are given by G_1 and G_2 respectively.*

In order to simplify the exposition, we will assume that the random variables X, Y_1 and Y_2 are absolutely continuous. In this case g and h can be assumed to be strictly increasing.

PROOF OF THEOREM 3.1. The joint distribution $p(z, \delta)$ of the random variables (Z, δ), defined by (3.1), in the censored-data experiment based on the random variable (X, Y), is given by

$$\begin{aligned} p(z, \delta = 0|\theta) &= f_Y(z)\overline{F}_X(z|\theta), \\ p(z, \delta = 1|\theta) &= \overline{F}_Y(z)f_X(z|\theta) \end{aligned} \tag{3.6}$$

where $\overline{F}_X$ and $\overline{F}_Y$ denote the survival functions of X and Y and f_X, f_Y respectively denote the Radon-Nikodym derivatives of F_X and F_Y with respect to σ-finite measures μ and ν. Since the censoring is non-informative, the distributions of Y_i, $i = 1, 2$ do not depend on θ. Note that both the experiments $\mathcal{E}$ and $\mathcal{F}$ are defined on the same parameter space Θ. Furthermore,

$$\text{Prob } (\delta = 1) = P_\theta(X > Y) = \int \overline{F}(y|\theta)g(y)d\nu(y). \tag{3.7}$$

In order to prove that the experiment $\mathcal{E}$ is sufficient for the experiment $\mathcal{F}$, we will explicitly mention a stochastic transformation from (Z_1, δ_1) to (Z^*, δ^*) which will be shown to satisfy the requirements given in (1.3) and (1.4). Consider the following *randomization mechanism*:

(i) If $\delta_1 = 0$ and $Z_1 = z$, then define $\delta^* = 0$ and $Z^* = h(g^{-1}(z))$.

(ii) If $\delta_1 = 1$ and $Z_1 = z$, then generate a random sample w with pdf

$$f_W^*(w) = f_V(w)/\overline{F}_V(g^{-1}(z)), \qquad w \geq g^{-1}(z), \tag{3.8}$$

where f_V and F_V denote the pdf and the cdf of the random variable V. Now define (Z^*, δ^*) as follows:

$$\begin{aligned} &\text{If } h(w) < z \text{ then define } \delta^* = 0 \text{ and } Z^* = h(w), \text{ and} \\ &\text{if } h(w) \geq z \text{ then define } \delta^* = 1 \text{ and } Z^* = z. \end{aligned} \tag{3.9}$$

Note that, if Z_1 is a censored observation, then Z^* is defined to be an appropriate censored observation. However, if Z_1 is an non-censored observation then Z^* is defined to be a censored observation with probability

$$p(z) = \text{ Prob } [g^{-1}(z) \leq V < h^{-1}(z)]/\text{Prob } [V \geq g^{-1}(z)]. \tag{3.10}$$

It is clear from the definition that for each fixed (Z_1, δ_1), the above randomization mechanism produces a stochastic transformation in that equations (1.4a, 1.4b) are satisfied. Furthermore, it follows from (3.6)–(3.10) that

$$p(z^*, \delta^* = 1|\theta) = \overline{F}_{Y1}(z^*)f_X(z^*|\theta)\{\text{Prob } [V \geq h^{-1}(z^*)]/\text{Prob } [V \geq g^{-1}(z^*)]\}.$$

It follows from Lemma 3.1 that $g(V) \sim F_{Y1}$ and $h(V) \sim F_{Y2}$. This is equivalent to

$$\overline{F}_{Y1}(z^*) = \text{ Prob } [V \geq g^{-1}(z^*)] \text{ and } \overline{F}_{Y2}(z^*) = \text{ Prob } [V \geq h^{-1}(z^*)], \tag{3.11}$$

and therefore,

$$p(z^*, \delta^* = 1|\theta) = \overline{F}_{Y2}(z^*)f_X(z^*|\theta). \tag{3.12}$$

Now it easily follows from (3.8) and (3.9) that

$$p(z^*, \delta^* = 0|\theta) = p(g^{-1}(h(z^*)), \delta = 0|\theta) + \int_0^\infty p(z, \delta = 1|\theta) \cdot p(z) d\mu(z). \quad (3.13)$$

The first term on the right hand side of (3.12) is equal to

$$f_{Y1}(g(h^{-1}(z^*))) \cdot \overline{F}_X((g(h^{-1}(z^*))|\theta), \quad (3.14a)$$

and the second term on the right hand side of (3.12) is equal to

$$\int_{A(z^*)} \overline{F}_{Y1}(z) f_X(z|\theta) \cdot \frac{[f_V(h^{-1}(z^*))]}{\Pr[V \geq g^{-1}(z)]} \; d\mu(z), \quad (3.14b)$$

where $A(z^*) = \{z\colon\ z^* < z \leq g(h^{-1}(z^*))\}$. Using (3.11), one can write (3.14b) as

$$f_{Y1}[g(h^{-1}(z^*))]\{F_X(g(h^{-1}(z^*))|\theta) - F_X(z^*|\theta)\}. \quad (3.15)$$

It now follows from (3.13), (3.14a,b) and (3.15) that

$$p(z^*, \delta^* = 0|\theta) = f_{Y1}[g(h^{-1}(z^*))]\overline{F}_X(z^*|\theta).$$

However, this expression is same as

$$p(z^*, \delta^* = 0|\theta) = f_{Y2}(z^*)\overline{F}_X(z^*|\theta), \quad (3.16)$$

which along with (3.12) is equivalent to the statement $(Z^*, \delta^*) \sim (Z_2, \delta_2)$, i.e., this stochastic transformation from (Z_1, δ_1) generates the distribution of (Z_2, δ_2) for every $\theta \in \Theta$. Hence the experiment $\mathcal{E}$ is sufficient for the experiment $\mathcal{F}$. □

Corollary 3.2. *An uncensored experiment based on the random variable X is sufficient for a censored experiment (Z_2, δ_2) based on (X, Y) for any censoring variable Y.*

PROOF. An uncensored experiment is equivalent to a censoring experiment with the censoring variable being equal to infinity. Thus it is stochastically larger than every other censoring variable. The result now follows from Theorem 3.1. □

REMARK. Hollander, Proschan and Sconing (1985b), show that the ϕ-divergence between the distributions of X and Z as a measure of information does not pass their 'acid-test' in that (3.3) fails to hold for some distributions of X and Z. However, it is well known that ϕ-divergence is a reasonable measure of information in the decision theoretic contexts and, as mentioned in Section 2, the fact that the experiment $\mathcal{E}$ is sufficient for the experiment $\mathcal{F}$ implies that appropriately defined ϕ-divergence based on experiment $\mathcal{E}$ are larger than those based on the experiment $\mathcal{F}$.

We believe that these seemingly contradictory statements arise because they use dependence between X and Z, i.e., the distance between the distributions of X and Z for a fixed $\theta \in \Theta$, as a measure of information in the censoring experiment based on (X, Y), rather than the distance between the distributions of the observables (Z, δ) for two different states of nature. The second concept is much more relevant for defining measures of information since the experiment (Z, δ) is meant for learning about θ, rather than to measure how close the distribution of Z is to that of X for a fixed θ.

ACKNOWLEDGMENTS

This research is supported in part by the Air Force Office of Scientific Research, Air Force Systems Command, USAF under contract number AFOSR-84-0162 and by the National Science Foundation under grant number DMS-8400683. The U.S. Government is authorized to reproduce and distribute reprints for Governmental purposes notwithstanding any copyright notation thereon. I would like to thank Myles Hollander for arousing my interest in Censored Data problem and to Morrie DeGroot and C. Srinivasan for stimulating discussion.

BIBLIOGRAPHY

Blackwell, D. (1951). Comparison of experiments. *Proc. 2nd Berkeley Symp. Math. Statist. Probab.* 93–102, Berkeley: Univ. of Calif. Press.

Blackwell, D. (1953). Equivalent comparison of experiments. *Annals Math. Statist.* **24** 265–272.

Boll, C. H. (1955). Comparison of Experiments in the infinite Case and the use of invariance in establishing sufficiency. *Unpublished Ph.D. Thesis* Palo Alto: Stanford Univ.

DeGroot, M. H. (1962). Uncertainty, information and sequential experiments. *Annals Math. Statist.* **33** 404–419.

DeGroot, M. H. (1966). Optimal allocation of observations. *Annals Inst. Statist. Math.* **18** 13–28.

DeGroot, M. H. (1970). *Optimal Statistical Decisions.* New York: McGraw-Hill.

DeGroot, M. H. (1984). Changes in utility as information. *Theory and Decision* **17** 283–303.

Ferentino, K. and Papaioannou, T. (1982). Information in experiments and sufficiency. *Jour. Statist. Planning Inference* **6** 309–317.

Feldman, D. and Ramamoorthi, R. V. (1986). Equivalent comparison of experiments — another proof of Blackwell's theorem. *To Appear.*

Goel, P. K. and DeGroot, M. H. (1979). Comparison of experiments and information measures. *Annals Statist.* **7** 1066–1077.

Goel, P. K. (1983). Information measures and Bayesian hierarchical models.

Jour. Amer. Statist. Assoc. **78** 408–410.

Hansen, O. H. and Torgersen, E. N. (1974). Comparison of linear normal experiments. *Annals Statist.* **2** 367–373.

Heyer, H. (1982). *Theory of Statistical Experiments.* New York: Springer-Verlag.

Hollander M., Proschan, F. and Sconing, J. (1985a). Information in censored models, *Tech. Report M701, Department of Statistics*, Tallahassee: Florida State University.

Hollander, M, Proschan, F. and Sconing, J. (1985b). Measures of dependence for evaluating information in censored models. *Tech. Report M706, Department of Statistics*, Tallahassee: Florida State University.

Karlin, S. and Rubin, H. (1956). The theory of decision procedures for distributions with monotone likelihood ratio. *Annals Math. Statist.* **27** 272–299.

Le Cam L. (1964). Sufficiency and approximate sufficiency. *Annals Math. Statist.* **35** 1419–1455.

Le Cam L. (1986). *Asymptotic Methods in Statistical Decision Theory.* New York: Springer-Verlag.

Lehmann, E. L. (1986a). *Testing Statistical Hypotheses.* 2nd Ed. New York: John Wiley.

Lehmann, E. L. (1986b). Comparing location experiments. *Tech. Report No. 75, Department of Statistics*, Berkeley, CA: University of California.

Rao, C. R. (1980). Diversity and dissimilarity coefficients: a unified approach. *Tech. Report No. 80-10, Department of Mathematics and Statistics*, Pittsburgh: University of Pittsburgh.

Richter, K. and Schafer, W. (1983). On comparison of experiments for an arbitrary set of states of nature. *Technical Report, Department of Mathematics*, GDR: Freiberg Mining Acad.

Stone, M. (1961). Non-equivalent comparison of experiments and their use for experiments involving location parameters. *Annals Math. Statist.* **32** 326–332.

Strasser, H. (1985). *Mathematical Theory of Statistics: Statistical Experiments and Asymptotic Decision Theory.* New York: Walter de Gruyter.

Swenson, A. R. (1980). Deficiencies between linear normal experiments. *Annals Statist.* **8** 1142–1155.

Taneja, I. J. (1986). Statistical aspects of divergence measures. *Jour. Statist. Planning Inference.* To appear.

Torgersen, E. N. (1970). Comparison of experiments when the parameter space is finite. *Z. Wahrs. Verw. Gebiete.* **16** 219–249.

Torgersen, E. N. (1972). Comparison of translation experiments. *Annals Math. Statist.* **43** 1381–1399.

Torgersen, E. N. (1976). Comparison of statistical experiments. *Scand. Jour. Statist.* **3** 186–208.

Torgersen, E. N. (1981). Measure of information based on comparison with total information and with total ignorance. *Annals Statist.* **9** 638–657.

Torgersen, E. N. (1985). *Comparison of Statistical experiments.* To be published.

A NOTE ON APPROXIMATE D-OPTIMAL DESIGNS FOR $G \times 2^M$

YONG B. LIM

Department of Experimental Statistics
Louisiana State University
Baton Rouge, Louisiana 70803, U.S.A.

W. J. STUDDEN

Department of Statistics
Purdue University
West Lafayette, Indiana 47907-2005, U.S.A.

H. P. WYNN

University Statistical Laboratory
The City University
Northampton Square
London, England

1. INTRODUCTION

In experimental designs, we frequently have a response variable y which depends on both qualitative and quantitative factors. The qualitative factors may be, for example, type of fertilizer, method of treatment, sex of patients, type of drug, curing condition, etc. The quantitative factors could be temperatures, or the amount of concentration. In addition, the number of levels of some of the quantitative factors may be reduced to two, thus making them, in effect, qualitative factors. The reason for this may be cost considerations, ease of experimentation or to conduct preliminary studies. In the following all of the qualitative factors will have only two levels.

The purpose of this note is to show that one can leave part of the model more general and still analyze the situation from a D-optimal viewpoint; and that the analysis is equivalent to a type of weighted model selection which has a Kiefer-Wolfowitz equivalence theorem.

The general part of the model may consist of those factors of more interest to the experimenter.

2. DEFINITIONS AND FORMULATION

The basic criterion of design optimality which we shall use here is that of D-optimality developed largely by Kiefer (1959, 1961) and Kiefer and Wolfowitz

(1959, 1960). It is assumed that for each point z in some multidimension factor space a random variable or response $Y(z)$ can be observed. The variable $Y(z)$ has expected value

$$EY(z) = \sum \theta_i f_i = \theta' f(z)$$

and

$$\text{var } Y(z) = \sigma^2$$

where f(z) is a $K \times 1$ column vector of known functions $f_i(z)$, $i = 1, \ldots, K$ and θ is a $K \times 1$ vector of unknown parameters. Uncorrelated observations are taken at $z_1, z_2, \ldots, z_N$ (not necessarily distinct). If the least squares estimates of θ are used, they have covariance matrix given by $(\sigma^2/N)\ M^{-1}(\xi)$ where M(ξ) has elements

$$m_{ij}(\xi) = \int_z f_i(z) f_j(z) d\xi(z) \tag{1}$$

and $\xi(z)$ has mass $1/N$ at each point z_i. We shall assume that our design measure ξ is an arbitrary probability measure. As usual, some approximation to exact designs would be necessary in practice.

The D-optimal criterion is to choose ξ to maximize the determinant $|M(\xi)|$. This is known, by the celebrated Kiefer-Wolfowitz theorem, to be equivalent to minimizing the supremum over z of

$$d(z, \xi) = f'(z) M^{-1}(\xi) f(z). \tag{2}$$

In this general setting we let $z = (x, y_1, \ldots, y_m)$. Here x, which may be multidimensional or more general, consists of the more important factor(s) while $y_1, \ldots, y_m$ are each set at two levels ± 1. For the x factor we consider a general model, denoted here by G, which involves $g+1$ linearly independent functions $h_0(x), h_1(x), \ldots, h_g(x)$. The model G corresponds to or incorporates the variables of main interest. For example if the experiments are categorized by height, weight and sex, and height $= x$ is the important variable, one might use $h_i(x) = x^i$, $i = 0, \ldots, g$. Then take weight at two levels, light and heavy and sex at two levels, male and female. If height and weight are important then we may use $x = (x_1, x_2)$ and $h_i(x)$, $i = 0, 1, \ldots, g$, could consist of a general polynomial model in two variables, leaving sex at two levels. The functions h_i should really be considered as function on $z = (x, y_1, \ldots, y_m)$ as they will constitute part of our more general model involving $f_1, \ldots, f_K$.

In addition to these $h_i(x)$ we allow m linear variables $y_1, \ldots, y_m$. To the functions h_i we will eventually add $m + \binom{m}{2}$ terms $y_1, y_2, \ldots, y_m$ and $y_i y_j$, $1 \le i < j \le m$ where each $y_i \in [-1, 1]$. These will be eventually restricted, without loss of generality, to take on only the values ± 1.

To describe the interaction between the two parts of the model it is convenient to introduce a number of submodels of G. This will be used later to describe the equivalence with the weighted model selection formulation in section 4. Thus let G_j, $j = 1, 2, \ldots, r$ denote distinct submodels of G, i.e. each G_j consists of a distinct subset of the functions $h_0, h_1, \ldots, h_g$. The interactions will be of the form $y_i \times G_j$ *or* $y_i y_l G_j$, consisting of y_i or $y_i y_l$ times all the functions in G_j. For convenience we assume that $h_0 \equiv 1$ and that h_0 is contained in each submodel, so that y_i and $y_i y_j$ for $i \neq j$ will be included in the model. Then the totality of our functions f_i, $i = 1, \ldots, K$ consists of

$$G$$

$$y_1 G_{i_1}, y_2 G_{i_2}, \ldots, y_m G_{i_m} \tag{3}$$

$$y_1 y_2 G_{i_{1,2}}, y_1 y_3 G_{i_{1,3}}, \ldots, y_{m-1} y_m G_{i_{m-1,m}}.$$

For later analysis it will be sufficient to note how many y_i *or* $y_i y_l$ occur with each specific G_j. This we denote by s_j, $j = 1, \ldots, r$. That is, there are s_1 terms like $y_i G_1$ *or* $y_i y_l G_1$, s_2 of the type $y_i G_2$ *or* $y_i y_l G_2$, *etc.* So (3) is reduced to

$$G$$

$$G_1, G_2, \ldots, G_r$$

$$s_1, s_2, \ldots, s_r$$

To illustrate, (3) might consist of 1, x, x^2, x^3, y_1, $y_1 x$, y_2, $y_2 x$, y_3, $y_3 x$, $y_3 x^2$, y_4, $y_4 x^2$, and $y_i y_j$, $1 \leq i < j \leq 4$. Here, $G = \{1, x, x^2, x^3\}$, $G_1 = \{1, x\}$, $G_2 = \{1, x, x^2\}$, $G_3 = \{1, x^2\}$, $s_1 = 2$, $s_2 = 1$ and $s_3 = 1$.

3. SYMMETRY REDUCTION

In analysing the above situation, all of the functions listed in (3) are viewed as functions on $z = (x, y_1, \ldots, y_m)$. An arbitrary design is a probability measure on z. A symmetry argument will show the following.

Lemma 3.1. *A solution ξ to the D-optimal design problem has each y_i on ± 1 with marginal probability $1/2$ and a measure μ on x; the full ξ being a product of these measures. The D-optimal design problem reduces to maximizing*

$$\Delta = |M_G(\mu)| \prod_{i=1}^{r} |M_i(\mu)|^{s_i}, \tag{4}$$

where M_G denotes the information matrix corresponding to G and M_i to G_i, $i = 1, 2, \ldots, r$.

PROOF. We consider the functions in (3) in blocks as indicated. Thus G is one block. Each $y_j G_{i_j}$, $j = 1, \ldots, m$, is a block and all the terms $y_j y_l G_{i_{j,l}}$, $j < l$, are in a separate block. A symmetry argument on each y_i shows that the D-optimal design must have these blocks orthogonal. That is, integrals of products of functions from different blocks are zero. To show that y_i must be ± 1 we observe that $d(z, \xi)$ will split according to the blocks. The terms with $y_1 G_{i_1}$ separate out and y_1^2 factors out with a nonnegative coefficient. Terms from $y_1 y_l G_{i_{1,l}}$ also will enter with y_1^2. Thus the supremum over y_1 must be at $y_1 = \pm 1$. Similarly for y_2, etc. The symmetry on y_i then gives weight 1/2 to $y_i = \pm 1$. We may then assume without loss of generality that ξ is of the form $\xi = \mu \times \rho$, where μ is on x and ρ has equal mass on the 2^m points $(y_1, \ldots, y_m)$ with each $y_i = \pm 1$. (The measure ξ is only constrained by certain moment conditions and may differ from the one described).

If the design ξ is as described above then the determinant of the overall information matrix is seen to equal to Δ in (4). □

4. WEIGHTED MODEL SELECTION

The situation resulting in Δ given in (4) is exactly the same as certain analysis of Atkinson and Cox (1974) and Läuter (1974). In these situations a number of models (in our case G and G_i, $i = 1, 2, \ldots, r$) are tentatively being considered. (Here $y_1, \ldots, y_m$ are not present.) We have assumed that the models G_i are all distinct in our formulation. One of the G_i may, however, be equal to G, i.e., some y_i may interact with a general model G. In this case we could omit G, if desired, and replace the corresponding s_i by $s_i + 1$. For convenience of notation we shall set $G = G_0$ and assume it is distinct from the other G_i. To design an experiment to distinguish between the models consideration is given directly to the weighted D-optimal criterion Δ. At least two methods for choosing the weights seem to present themselves. The first is to make them proportional to some prior likelihood of the different models. The other is to make them proportional to the inverse of the dimension of the model.

For the criterion Δ, a Kiefer-Wolfowitz equivalence theorem has been described in Laüter (1974) and Atkinson and Cox (1974). Thus the following are equivalent:

$$
\begin{aligned}
&\text{(i)} && \mu^* \text{ maximizes } \Delta(\mu) \\
&\text{(ii)} && \mu^* \text{ minimizes } \sup_x \sum_{j=0}^{r} s_j d_j(x, \mu) \qquad (5)\\
&\text{(iii)} && \sum s_i d_i(x, \mu^*) = \sum s_i k_i,
\end{aligned}
$$

where k_i is the dimension of G_i.

The analysis of Δ in previous sections gives another interpretation of the conditions in (5).

5. EXAMPLES

EXAMPLE 5.1. Consider for G the simple case $G = \{1, x, \ldots, x^n\}$ with $x \in [-1, 1]$ and let $G_i = \{1, x, \ldots, x^i\}$, $i = 0, 1, \ldots, n$. Thus in our model we have $x \in [-1, 1]$ and, in addition, m variables $y_1, \ldots, y_m$ on ± 1, with 1st order interactions between the y_i. The subset G_i appears as interactions with s_i of the y terms and

$$\Delta(\mu) = |M_n(\mu)| \times \prod_{i=1}^{n} |M_i(\mu)|^{s_i} \tag{6}$$

The maximization of (6) is given in Studden and Lau (1985) using canonical moments. We use the notation from that paper to describe the solution here. One can show that, except for some constants independent of μ,

$$|M_l| = \prod_{i=1}^{l} (\zeta_{2i-1}\zeta_{2i})^{l+1-i}, \quad l = 1, \ldots, n, \tag{7}$$

where $\zeta_i = q_{i-1}, p_i, i \geq 1, q_0 = 1, 0 \leq p_i \leq 1$ and $q_i = 1 - p_i$. The quantities $p_1, p_2, \ldots$ give a convenient parameterization to the design part μ. Substituting (7) into (6) the general model $G \times 2^m$ can be solved, at least in terms of the p_i. For $n = 2$ the quantity Δ reduces to

$$\Delta = (p_1 q_1 p_2)^{s_1} [(p_1 q_1 p_2)^2 (q_2 p_3 q_3 p_4)]^{s_2+1}$$

Maximizing Δ in terms of p_1, p_2, p_3, p_4 we find $p_1 = p_3 = 1/2, p_4 = 1$ and

$$p_2 = \frac{2(s_2 + 1) + s_1}{3(s_2 + 1) + s_1}. \tag{8}$$

The resulting measure μ^* can be shown to give weight $p_2/2$ to ± 1 and $1 - p_2$ to zero.

For $n = 3$ it can be shown that the measure μ^* has weight $w/2$ on ± 1 and $(1 - w)/2$ on $\pm t$ where $t = (p_2 q_4)^{1/2}$, $w = p_2 p_4/(q_2 + p_2 p_4)$ and

$$p_2 = \frac{3(s_3 + 1) + 2s_2 + s_1}{5(s_3 + 1) + 3s_2 + s_1}, \quad p_4 = \frac{2(s_3 + 1) + s_2}{3(s_3 + 1) + s_2}. \tag{9}$$

We remark that the usual D-optimal design for polynomial regression corresponds to $s_i = 0$, $i = 1, 2, \ldots, n$. This same solution results if these are zero only for i up to $n - 1$. In this case we have a 'full product' model for some of factors.

EXAMPLE 5.2. Consider for G the quadratic polynomials on the k-cube,

$$G = \begin{pmatrix} & x_1, & x_2, & \dots, & x_k \\ 1, & x_1^2, & x_2^2, & \dots, & x_k^2 \\ & x_1x_2, & x_1x_3, & \dots, & x_{k-1}x_k \end{pmatrix}, \quad x_i \in [-1,1], i = 1, \dots, k, k \geq 2 \tag{10}$$

and

$$G_1 = \{1, x_1, \dots, x_k\}.$$

Thus we have the quadratic polynomials on the k-cube, m variables $y_1, \dots, y_m$, all the 1st order interactions between x_i and y_j and interactions between y_j in our model. The subset G_1 interacts with each of the y terms and

$$\Delta(\mu) = |M_G(\mu)| \cdot |M_1(\mu)|^m.$$

Note that the model is invariant under the group H of sign changes and permutations of x_i's, $i = 1, \dots, k$. By the invariance theorem (Kiefer(1959)) there exists a symmetric D-optimal design μ^* under H. By the same argument as one used in the proof of Lemma 3.1, $d(z, \xi^*)$ is a quartic function of x_i with the positive coefficient of x_i^4 and symmetric w.r.t. x_i. So $d(z, \xi^*)$ can be maximized at $x_i = \pm 1$ or 0. Thus we restrict μ to a symmetric design on E, $E = \{x : |x_i| = 0 \text{ or } 1\}$. For $i = 1, 2, \dots, k$, let E_i be the subset of E consisting of those $\binom{k}{i} \cdot 2^i$ elements with $(k-i)$ components of x being equal to zero. Then a symmetric design μ^* on $E_{r_1} \cup E_{r_2} \cup E_{r_3}$ is D-optimal iff

$$0 \leq r_1 \leq (k-1)u^* \frac{(1-t^*)}{(1-u^*)} \leq r_2 \leq k-1, r_3 = k,$$

$$\mu^*(E_{r_1}) = \frac{k}{(k-r_1)(r_2-r_1)}[r_2 - (k+r_2-1)u^* + (k-1)v^*], \tag{11}$$

$$\mu^*(E_{r_2}) = \frac{k}{(k-r_2)(r_2-r_1)}[-r_1 + (k+r_1-1)u^* - (k-1)v^*]$$

and

$$\mu^*(E_k) = 1 - \mu^*(E_{r_1}) - \mu^*(E_{r_2}),$$

$$\text{where} \quad t^* = \frac{(2k+2m+1) + \sqrt{4(k+m)^2 + 12(k+m) + 17}}{4(k+m+2)}$$

$$\text{and} \quad u^* = \frac{k+2m+3}{k^2 + k(2m+3) + 2}[(k-1)t^* + 1].$$

A proof of this fact is given in an Appendix.

The weights for a symmetric D-optimal design with $r_1 = 0$ and $r_2 = k-1$ are listed in Table 1 for some k and m. That would be beneficial if fewer points in the design are desired.

We remark that a symmetric D-optimal design for quadratic regression on

Table 1. Weights for a symmetric D-optimal design on E with a minimal support.

	m	0	1	2	3	4		m	0	1	2	3	4
	$\mu^*(E_2)$	.583	.655	.706	.743	.772		$\mu^*(E_3)$	.510	.578	.629	.670	.702
k=2	$\mu^*(E_1)$	.321	.284	.252	.226	.204	k=3	$\mu^*(E_2)$	.424	.378	.339	.306	.279
	$\mu^*(E_0)$	.096	.061	.042	.031	.024		$\mu^*(E_0)$	.066	.044	.032	.024	.019
	$\mu^*(E_4)$	.451	.516	.567	.609	.643		$\mu^*(E_5)$	.402	.465	.516	.558	.593
k=4	$\mu^*(E_3)$	.502	.451	.408	.371	.341	k=5	$\mu^*(E_4)$	.562	.509	.464	.426	.393
	$\mu^*(E_0)$	.047	.034	.025	.020	.016		$\mu^*(E_0)$	.036	.026	.020	.016	.013
	$\mu^*(E_6)$	.362	.423	.472	.515	.551		$\mu^*(E_7)$	.330	.387	.436	.477	.513
k=6	$\mu^*(E_5)$	.610	.556	.511	.471	.438	k=7	$\mu^*(E_6)$	.648	.595	.550	.511	.477
	$\mu^*(E_0)$	.028	.021	.017	.014	.011		$\mu^*(E_0)$	.022	.018	.014	.012	.010
	$\mu^*(E_8)$	.302	.357	.404	.445	.481		$\mu^*(E_9)$	.279	.331	.377	.417	.452
k=8	$\mu^*(E_7)$	.680	.628	.584	.545	.510	k=9	$\mu^*(E_8)$	.706	.656	.613	.574	.540
	$\mu^*(E_0)$	.018	.015	.012	.010	.009		$\mu^*(E_0)$	.015	.013	.010	.009	.008

the k-cube corresponds to $m = 0$, and Farrell, et al (1967) have shown that $E_{r_1} \cup E_{r_2} \cup E_k$ supports a symmetric D-optimal design iff

$$r_2 = k-1, r_1 < k-1 \text{ for } k \leq 5$$

$$r_2 = k-1 \text{ or } k-2, r_1 < k-2 \text{ for } k \geq 6.$$

But we have to choose r_2 to be $k-1$ for any k and sufficiently large m since $(k-1) \cdot u^* \frac{1-t^*}{1-u^*}$ converges to $k-1$ as m goes to infinity.

6. APPLICATIONS

1. A civil engineer want to predict unconfined compressive strength of 100 % cement for two controlled variables, x_1 = freeze/thaw cycles and y_1 = water cement ratio. Suppose his model consists of G, y_1G_1 over $\{(x_1, y_1) : 0 \leq x_1 \leq 300, .5 \leq y_1 \leq .75\}$, where $G = \{1, x_1, x_1^2, x_1^3\}$ and $G_1 = \{1, x_1, x_1^2\}$, and $N = 18$ observations are to be taken. First let us find an approximate design μ^* over $[-1, 1]$ from Example 5.1. The model corresponds to $n = 3$, $s_1 = 0$, $s_2 = 1$, $s_3 = 0$ in Example 5.1. So from (9), $\mu^*(\pm 1) = 5/18$ $\mu^*(\pm\sqrt{5/32}) = 4/18$. Since

the D-optimal criterion is invariant under linear transformations of controlled variables, an approximate D-optimal design ξ^* is $\mu_x^* \times \rho$, where μ_x^* is the measure on $\{x : 0 \leq x \leq 300\}$ induced by μ^*, i.e. $\mu_x^*(0) = \mu_x^*(300) = 5/18$ and $\mu_x^*(150(1 \pm \sqrt{5/32})) = 4/18$, and $\rho(.5) = \rho(.75) = 1/2$. Simple approximation of ξ^* to an exact design with $N = 18$ observations would be to take 3 observations at (0,.5) and (300,.75) and 2 observations at all the other points on the support of ξ^*.

2. Suppose a civil engineer want to predict the unconfined compressive strength of mixed type cement for three controlled variables, x_1 = freeze/thaw cycles, x_2 = cement bentonite ratio and y_1 = water cement ratio, and his model is given by (10) with $k = 2$ and $m = 1$ over $\{(x_1, x_2, y_1) : 0 \leq x_1 \leq 300, .96 \leq x_2 \leq 1, .5 \leq y_1 \leq .75\}$. $N = 60$ observations are to be taken. From Table 1, an approximate design μ^* puts mass .655/4 at $(\pm 1, \pm 1)$, .284/4 at $(\pm 1, 0)$ and $(0, \pm 1)$ and mass .061 at (0,0). Using the invariance property of the D-optimal criterion, an approximate D-optimal design ξ^* is $\mu_x^* \times \rho$, where μ_x^* is the measure on $\{(x_1, x_2) : 0 \leq x_1 \leq 300, .96 \leq x_2 \leq 1\}$ induced by μ^* and $\rho(.5) = \rho(.75) = 1/2$. An approximation of ξ^* to an exact design with $N = 60$ observations would be to take 5 observations at 8 corners and 2 observations at all the other points on the support of ξ^*.

3. Suppose an experimenter is only interested in main effects and all the 1st order interactions in $3^k \times 2^m$ factorial experiments. Then a symmetric design given by (11) in Example 5.2 determines an approximate D-optimal design for this model. In the case of $k = 2$, $m = 1$ and $N = 13$, an approximate D-optimal design in Table 1 suggests to take 2 observations at 4 corners and one observation at 4 middles of edges and the center of the 3^2 factorial space. Simple evaluation of determinants of information matrices of such designs over possible allocations of levels of the 2 level factor shows that an exact D-optimal design seems to be a design which takes +1 at the center and -1 at 4 middles or vice versa. But as m gets larger, approximations to exact designs would be more difficult.

ACKNOWLEDGMENTS

The idea of this paper arose from conversations at the Workshop on Efficient Data Collection and Related Inference, held at Univ. of California at Berkely in summer 1984.

This research was supported by a David Ross Grant from Purdue University, National Science Foundation Grant No. MCS-8200631 and MCS-8410183.

APPENDIX: DERIVATION OF μ^* IN EXAMPLE 5.2

Farrell, et al(1967) have shown algebraically that there exists a symmetric D-optimal design for quadratic regression on the k-cube. Kono (1962) has used geometric arguments for the moment space to get a symmetric D-optimal design. In the following we generalize the geometric ideas contained in Kono (1962).

Let

$$u = \int x_1^2 \mu(dx)$$

and

$$v = \int x_1^2 x_2^2 \mu(dx).$$

Noting that μ is a symmetric design on E, we get

$$M_G(\mu) = \begin{bmatrix} uI_k & 0 & 0 \\ 0 & (u-v)I_k + v11' & 0 \\ 0 & 0 & vI_{\frac{k(k-1)}{2}} \end{bmatrix}$$

and

$$M_1(\mu) = \begin{bmatrix} 1 & 0 \\ 0 & uI_k \end{bmatrix}$$

where 1 is the $k \times 1$ vector of ones and I is the identity matrix. So

$$\begin{aligned}\Delta(\mu) &= \left[u^k \cdot (u-v)^{k-1} \cdot (u+(k-1)v - ku^2) \cdot v^{\frac{k(k-1)}{2}}\right] \cdot \left(u^k\right)^m \\ &= u^{k(m+1)} \cdot v^{\frac{k(k-1)}{2}} \cdot (u-v)^{k-1} \cdot (u+(k-1)v-ku^2). \end{aligned} \tag{12}$$

Simple algebra shows that $\Delta(\mu)$ is maximized at

$$\begin{aligned} u^* &= \frac{k+2m+3}{k^2+k(2m+3)+2}[(k-1)t^*+1] \\ \text{and } v^* &= t^* \cdot u^*, \\ \text{where } t^* &= \frac{(2k+2m+1)+\sqrt{4(k+m)^2+12(k+m)+17}}{4(k+m+2)}. \end{aligned} \tag{13}$$

The space of (u,v) is the convex hull of $\left\{\left(\frac{i}{k}, \frac{i(i-1)}{k(k-1)}\right), i = 0, \ldots, k\right\}$ since

$$u = \sum_{i=0}^{k} \frac{i}{k} \cdot \mu(E_i)$$

$$v = \sum_{i=2}^{k} \frac{\binom{i}{2}}{\binom{k}{2}} \mu(E_i) = \sum_{i=2}^{k} \frac{i(i-1)}{k(k-1)} \cdot \mu(E_i).$$

So we need to show that (u^*, v^*) is in that convex hull for the existence of a symmetric D-optimal design. Now we try to find r_1, r_2, r_3, $\mu^*(E_{r_1})$ and $\mu^*(E_{r_2})$ such that

$$u^* = \sum_{i=1}^{3} \frac{r_i}{k} \cdot \mu^*(E_{r_i})$$

$$\text{and} \quad v^* = \sum_{i=1}^{3} \frac{r_i(r_i-1)}{k(k-1)} \cdot \mu^*(E_{r_i}) \quad \text{where} \quad 0 \le r_1 < r_2 < r_3 \le k.$$

Since $t^* > \frac{k+m+1}{k+m+2}$ and u^* is a linear increasing function of t^*,

$$\begin{aligned} u^* &> \frac{k+2m+3}{k^2+k(2m+3)+2}\left[(k-1)\cdot\frac{k+m+1}{k+m+2}+1\right] \\ &> \frac{k-1}{k}. \end{aligned}$$

So we have to choose r_3 to be k. Let L_1 be the line which passes through $\left(\frac{r_1}{k}, \frac{r_1(r_1-1)}{k(k-1)}\right)$ and (u^*, v^*) and L_2 be the line which passes through $(1,1)$ and $\left(\frac{r_2}{k}, \frac{r_2-1}{k(k-1)}\right)$. Suppose u_1 is the abscissa of the intersection point of L_1 and L_2. Then $u^* \le u_1 \le 1$ if and only if there exists a symmetric D-optimal design μ^* whose support is $E_{r_1} \cup E_{r_2} \cup E_k$. It can be easily checked that

$$u_1 = \frac{r_2(k\cdot u^* - r_1) + r_1\left[(r_1-1)u^* - (k-1)v^*\right]}{r_2(k\cdot u^* - r_1) - r_1(k-r_1) + k(k-1)(u^*-v^*)}$$

and

$$u^* \le u_1 \le 1 \quad \text{iff} \quad r_1 \le (k-1)u^* \frac{(1-t^*)}{(1-u^*)} \le r_2. \tag{15}$$

By the direct substitution of (13) into $u^* \cdot \frac{(1-t^*)}{(1-u^*)}$ we get

$$u^* \cdot \frac{(1-t^*)}{(1-u^*)} < 1,$$

which assures the existence of a symmetric D-optimal design on $E_{r_1} \cup E_{k-1} \cup E_k$. We get (11) from (14) and (15).

BIBLIOGRAPHY

Atkinson, A. C. and Cox, D. R. (1974). Planning experiments for discriminating between models. *J. Roy. Statist. Soc. Ser. B.* **36**, 321–348.

Farrell, R. H., Kiefer, J. and Walbran, J. (1967). Optimum multivariate designs. Proc. Fifth Berkeley Symp. Vol. I, Univ. of California Press, 113–138.

Kiefer, J. (1959). Optimum experimental designs. *J. Roy. Statist. Soc. Ser. B.* **21**, 272–319.

Kiefer, J. (1961). Optimum designs in regression problems, II. *Ann. Math.*

Statist. **32**, 298–325.

Kiefer, J. (1961). Optimum experimental designs V, with applications to systematic and rotatable designs. Proc. Fourth Berkeley Symp. Vol. I, Univ. of California Press, 381–405.

Kiefer, J. and Wolfowitz, J. (1959). Optimum designs in regression problems. *Ann. Math. Statist.* **30**, 271–294.

Kiefer, J. and Wolfowitz, J. (1961). The equivalence of two extremum problems. *Canad. J. Math.* **12**, 363–366.

Kono, K. (1962). Optimal design for quadratic regression on the k-cube. *Mem. Fac. Sci. Kynshu Univ. Ser. A.* **16**, 114–122.

Kurotschka, V. (1981). A general approach to optimum design of experiments with qualitative and quantitative factors. Proc. of I.S.I. Golden Jubilee Intern. Conf. on Statistics: Applications and New Directions, 353–368

Lau, T. S. and Studden, W. J. (1985). Optimal designs for trigonometric and polynomial regression using canonical moments. *Ann. Statist.* **13**, 383–394.

Läuter, E. (1974). Experimental planning in a class of models. *Math. Operationsforsch. U. Statist.* **5** 379–398.

SOME STATISTICAL DESIGN ASPECTS OF ESTIMATING AUTOMOTIVE EMISSION DETERIORATION FACTORS

GARY C. MCDONALD

Mathematics Department
General Motors Research Laboratories
Warren, Michigan 48090-9055, U.S.A.

1. INTRODUCTION

To determine an automobile manufacturer's compliance with the vehicle emissions standards for hydrocarbon (HC), carbon monoxide (CO), and the oxides of nitrogen (NO_x), the Environmental Protection Agency (EPA) adjusts emission results from low mileage test vehicles with deterioration factors (see Title 40 Code of Federal Regulations 1978a). With this approach, the emission level of a prototype vehicle with 50,000 miles is used as the basis for determining compliance with the standards and the deterioration factors account for changes in emission control efficiency resulting from mileage accumulation. Lorenzen (1980) has developed various statistical characteristics of an EPA audit procedure in which the emissions of a tested production vehicle are compared to these "deteriorated standards". The importance of these deterioration factors in the interplay between compliance probabilities and emission design targets is clearly indicated in that article. The method of determining vehicle deterioration factors suffers from at least two serious statistical shortcomings. First, no account is taken of between car variability and, second, the single car process itself yields estimates with substantial uncertainty. McDonald (1981) addressed the latter point and provided a method to calculate an uncertainty measure. In a sense the first aspect of between cars variability is addressed by end-of-line audit tests of production vehicles.

The construction of a deterioration factor for a specific engine-system combination and emission involves the testing of a prototype vehicle at various mileage points, the fitting of a least squares line, and the computation of the ratio of two points on this line (Title 40 Code of Federal Regulations 1978b). This process is given by the following:

1. The vehicle is tested at 5,000 miles (approximately) and continues to be tested up to 50,000 miles in increments of 5,000 (approximately).
2. The vehicle is tested before and after major maintenance when major maintenance is required.
3. A least squares line, say $\hat{y}(x)$, is fitted to the resulting emission-mileage data.

4. The deterioration factor (DF) is defined as DF = $\hat{y}$ (50,000 miles)/$\hat{y}$ (4,000 miles), and is an estimate of the ratio of the true *mean* response at 50,000 miles to the true *mean* response at 4,000 miles.

A method for assessing the uncertainty (resulting from test variability) in the resulting deterioration factor was given by McDonald (1981). In that publication it was also noted, that for the particular example given, a more reliable estimate of the deterioration factor as quantified by the length of a confidence interval could be obtained by altering the experimental design. In this context the experimental design is the specification of mileages at which emission measurements are to be taken along with the frequency (i.e., repeats). The particular alteration suggested was motivated and specified by designs known to provide 'best' estimates of the slope parameter in a simpler linear regression model using least squares estimation criteria. This idea will be exploited in this article.

By focusing on the reliability of the slope parameter estimate it is possible to construct designs which require total mileage accumulation *less* than the 50 thousand miles and concurrently improve the statistical reliability of the DF estimate. This reliability improvement with less mileage is achieved through repeat tests at the design endpoints.

This article provides the statistical assumptions and derivations which quantify the tradeoff between reduced mileage accumulation and increased number of tests. Although the formulation here is quite general and does permit an evaluation of any design (i.e., mileage and testing frequency specification), one particular class of designs is considered in detail. This class is generated by a systematic relaxation of a 50,000 mileage accumulation requirement. Recently (see Title 40 Code of Federal Regulations 1985a) the rigidity of the previously described testing schedule has been relaxed somewhat. The testing mileage increments now are not necessarily confined to 5,000 miles, and the 50,000 testing schedule *may* be reduced by the EPA Administrator if it is deemed that such a change continues to meet the objective of this procedure. Once the emission data is collected, the procedure for establishing the DF generally remains as described earlier (see Title 40 Code of Federal Regulations 1985b); however, there are provisions for outlier screening and repeated measurement averaging. These latter aspects will not be considered here. The recent provisos for altering the test plan, and possibly setting aside the 50,000 mile accumulation requirement, is very much in the spirit of this article. Thus, we focus on the procedure described earlier and how modifications can affect statistical properties of the DF.

2. BASIC MODEL AND FORMULATION

Assume a linear relationship between emissions and mileage; i.e.,

$$y = \beta_0 + \beta_1 X + e,$$

where y is emissions in grams per mile (HC, CO, or NO_x), X is the mileage (expressed in thousands) at which the emissions are measured, and e is the error term assumed normally distributed with mean zero and variance σ^2. As a collection the e's are assumed independent.

An exact $100(1-\alpha)\%$ confidence interval for the deterioration factor (DF) is given by (r_1, r_2), where $r_1 < r_2$ denote the two roots of a quadratic equation $A\theta^2 + B\theta + C = 0$. The coefficients A, B and C are functions of the least squares estimates of β_0 and β_1 as well as other quantities. The precise form of these coefficients is given by McDonald (1981) and is discussed in Section 4. Due to the complexity of the effect of the experimental design on the roots of the quadratic equation (see Section 4), a simpler criteria for evaluating designs is desired. One such criteria is provided by the reliability of the least squares estimate of the slope. We first exploit this simple approach in the remainder of this section and in Section 3.

The variance of the least squares estimate of the slope, say $\hat{b}_1$, is given by

$$V(\hat{b}_1) = \sigma^2/\Sigma x_i^2,$$

where $x_i = X_i - \bar{X}$, with $\bar{X}$ being the average of all the X values at which emissions are measured. If measurements are repeat then the mileage is repeated in the average and summation.

A $100(1-\alpha)\%$ confidence interval for β_1 is given by

$$\hat{b}_1 \pm t_{\alpha/2;n-2}\, s/\left[\Sigma x^2\right]^{1/2},$$

where s^2 is the usual sample estimate of σ^2 and $t_{\alpha/2;n-2}$ is the upper $(\alpha/2)$ percentage point of a t-variate with $n-2$ degrees-of-freedom.

Suppressing subscripts, the square of the length of the confidence interval is given by

$$L^2 = 4t^2 s^2/\Sigma x^2.$$

Taking an expectation and square root gives

$$\left(EL^2\right)^{1/2} = 2t\sigma/\left[\Sigma x^2\right]^{1/2}.$$

Design 1 is said to be "better than" design 2 if and only if $(EL^2)^{1/2}$ for design 1 is less than the corresponding quantity for design 2. This condition can be stated as:

$$\left(\Sigma x^2\right)_1^{1/2} > \left(t_{df1}/t_{df2}\right)\left(\Sigma x^2\right)_2^{1/2}$$

or

$$\left(\Sigma x^2\right)_1 > \left(t_{df1}/t_{df2}\right)^2 \left(\Sigma x^2\right)_2 . \tag{2.1}$$

The quantities $df1$ and $df2$ are the degrees-of-freedom associated with design 1 and design 2 respectively. This is essentially the analytical justification for the criteria adopted by the California Air Resources Board (1982) to permit alterations in the durability-data test interval. However, the Board did retain the requirement "at a minimum, tests must be performed at 5,000 and 50,000 miles". As we shall see, this requirement is not necessary to insure the condition (2.1) and hence a better design in the sense described.

3. ALTERNATIVES TO A STANDARD PROCEDURE

We define as the "standard procedure" the design specified by one emission measurement at 5,000 miles and up to 50,000 miles at increments of 5,000 miles. (As noted in Section 1, this may not necessarily be the current procedure, but it will suffice to illustrate the concepts). At 25,000 miles one additional measurement is taken after vehicle maintenance. This design yields eleven points (with one repeat) and

$$\Sigma x^2 = 2068.18.$$

Using the relation (2.1) with design 2 being the "standard procedure," and specifying a 95% confidence requirement, we now have that an alternative design (say design 1) is better than the standard procedure if

$$\begin{aligned}\left(\Sigma x^2\right) &> \left(t_{df1}/2.262\right)^2 (2068.18) \\ &= t^2_{df1}(404.21).\end{aligned} \tag{3.1}$$

The df1 is the degrees-of-freedom associated with the alternative design, i.e., the number of specified measurements less two.

We now consider a particular class of alternative designs with the goal of trading mileage for increased number of measurements. The alternatives can be stated as follows:

(M) Take n measurements at 5,000 miles and at M miles. At mileages between 5,000 and M follow the current procedure. Let M be 30, 35, 40, 45, or 50 thousand miles.

The standard procedure is a special case; namely $M = 50{,}000$ and $n = 1$. Our goal now is to find, for each value of M, the smallest n such that the alternative design is better than the current, i.e., the smallest n for which (3.1) is satisfied.

To illustrate, let's take $M = 45$. Then the frequency-mileage (in thousands) design is given by:

$$n - 5\text{'s};\ 1 - 10;\ 1 - 15;\ 1 - 20;\ 2 - 25;\ 1 - 30;\ 1 - 35;\ 1 - 40;\ n - 45\text{'s},$$

so

$$\Sigma X^2 = 2050n + 5700$$
$$\Sigma X = 50n + 200$$
$$\Sigma x^2 = 2050n + 5700 - [(50n + 200)^2/(2n + 8)].$$

And now,

n	$df1$	$(\Sigma x^2)_1$	$t^2_{df1}(404.21)$
1	8	1500	2149.44
2	10	2300	2006.49

Thus, $n = 2$ is the smallest value of n for which (3.1) is satisfied and this design, double measurement at 5 and 45 thousand miles, is better than the standard design which runs to 50 thousand miles. For two repeat measurements, mileage requirement is reduced 10 percent — five thousand miles.

In a similar fashion the best designs within the alternative class are given by:

Terminal mileage (1000s)		# of measurements repeated on ends	Total number of measurements	
50	(0)	1	11	(0)
45	(10)	2	12	(9)
40	(20)	3	13	(18)
35	(30)	4	14	(27)
30	(40)	6	17	(55)

The percent reduction in mileage and percent increase in measurements are noted in parentheses.

4. OTHER DESIGN CONSIDERATIONS

The exact confidence limits on the DF are the roots of $A\theta^2 + B\theta + C = 0$, and the design goal would be to minimize the difference of the roots, $r_2 - r_1$, for a fixed confidence level. We have (see McDonald (1981))

$$r_2 - r_1 = 2(m_1 - m_2)[\ell_{12}^2 - \ell_{11}\ell_{22}]^{1/2}/(\ell_{11} + 2m_2\ell_{12} + m_2^2\ell_{22}), \qquad (4.1)$$

where $m_1 = 50, m_2 = 4$, and

$$A = \ell_{11} + 2m_2\ell_{12} + m_2^2\ell_{22},$$
$$B = -2[\ell_{11} + (m_1 + m_2)\ell_{12} + m_1 m_2 \ell_{22}],$$
$$C = \ell_{11} + 2m_1\ell_{12} + m_1^2\ell_{22},$$

and ℓ_{ij} is the (i, j)-element of the symmetric matrix

$$L = [\underline{b}\,\underline{b}' - s^2(X'X)^{-1}F]. \tag{4.2}$$

The vector $\underline{b}'$ is the least squares estimate of (β_0, β_1); and $F \equiv F_{1,n-2;\alpha}$ is the upper α^{th} percentage point of a central F variate with 1 and $n-2$ df; and X is an $n \times 2$ matrix with the first column having ones and the second column giving the mileage at which the emission is measured.

To minimize $r_2 - r_1$ it suffices to maximize the quantity

$$\begin{aligned}\Delta \equiv \Delta(L, m_2) &= (\ell_{11} + 2m_2\ell_{12} + m_2^2\ell_{22})/[\ell_{12}^2 - \ell_{11}\ell_{22}]^{1/2} \\ &= (1, m_2)L\begin{pmatrix}1\\ m_2\end{pmatrix}/[-\det L]^{1/2}.\end{aligned} \tag{4.3}$$

While the difference of the roots as given in (4.1) does depend on both m_1 and m_2, the criterion given in (4.3) depends on m_2 but not m_1. In a sense this may indicate the sensitivity of the ratio estimate to the precision of the denominator. Insofar as Δ is a stochastic quantity (L depends on $\underline{b}$ which in turn depends on the realized emission y), it would be desirable to choose an experimental design to maximize the expected value of Δ, say $E(\Delta)$. However, this appears to be intractable due to the analytic form of Δ. Since

$$E(L) = \underline{\beta}\,\underline{\beta}' + \sigma^2(X'X)^{-1}(1 - F), \tag{4.4}$$

we turn to construction of designs to maximize $\Delta(E(L), m_2)$ rather than $E(\Delta)$.

Using equations (4.4) and (4.3) the quantity $\Delta(E(L), m_2)$, a design criterion to be maximized, can be written as

$$\Delta(EL, m_2) = K_1(cK_2)^{-1/2} - K_3(c/K_2)^{1/2}, \tag{4.5}$$

where

$$K_1 = (1, m_2)\underline{\beta}\,\underline{\beta}'\begin{pmatrix}1\\ m_2\end{pmatrix},$$
$$K_2 = \underline{\beta}'X'X\underline{\beta} + \sigma^2(1 - F),$$
$$K_3 = (m_2, -1)X'X\begin{pmatrix}m_2\\ -1\end{pmatrix},$$

and

$$c = \sigma^2(F - 1)[\det(X'X)]^{-1}.$$

In many applications the value of c will be very close to zero and the second term on the right hand side of equation (4.5) will be negligible compared to the first term. Dropping the second term, K_1 which is design independent and positive, and the $\sigma^2(F-1)$ portion of c yields a somewhat simplified objective function (independent of m_2) to maximize for a fixed value of n:

$$\tilde{\Delta}(EL) = [\det(X'X)]/K_2. \tag{4.6}$$

The numerator in $\tilde{\Delta}$ is the denominator in the expression for $V(\hat{b}_1)$ given in Section 2 and provided the basis for constructing the designs given earlier in the text. Thus the K_2 inclusion can be viewed as the modification required to link the design results applicable to a regression slope to that applicable to the DF.

EXAMPLE. $\beta_0 = .3$ $\beta_1 = .002$ $s = .018$ $m_2 = 4$

- *Standard Design*
  $X =$ (5, 10, 15, 20, 25, 30, 35, 40, 45, 50)
  $f =$ (1,1,1,1,2,1,1,1,1,1), $n = 11$, $F_{1,9;.05} = 5.12$.
- *Alternate 1*
  $X =$ (5, 10, 15, 20, 25, 30, 35, 40, 45)
  $f =$ (1,1,1,1,2,1,1,1,1), $n = 10$, $F_{1,8;.05} = 5.32$.
- *Alternate 2*
  $X =$ (5, 10, 15, 20, 25, 30, 35, 40, 45)
  $f =$ (2,1,1,1,2,1,1,1,2), $n = 12$, $F_{1,10;.05} = 4.96$.

	Standard Design	*Alternate* 1	*Alternate* 2	
ΣX	300	250	300	
ΣX^2	10,250	7,750	9,800	
K_1	.095	.095	.095	(design indept.)
K_2	1.390	1.230	1.478	
K_3	8,026	5,910	7,592	
c	$(2.422 \cdot 10^{-4})^2$	$(3.055 \cdot 10^{-4})^2$	$(2.156 \cdot 10^{-4})^2$	
$K_1(cK_2)^{-1/2}$	332.21	280.06	361.92	
$K_3(c/K_2)^{1/2}$	1.649	1.628	1.346	
$\Delta(EL, m_2)$	330.56	278.43	360.57	
$\tilde{\Delta}(EL, m_2)$	16,370.85	12,199.09	18,674.93	

Thus, among the three designs, Alternate 2 maximizes $\Delta(EL, m_2)$ as well as the approximate criterion $\tilde{\Delta}(EL)$. The latter criterion is applicable for fixed n since an F-value, which is dependent on n, was dropped in deriving (4.6).

Without this simplification the modified criterion would be $(cK_2)^{-1/2}$ which for this example is ordered as $\tilde{\Delta}$.

5. SUMMARY

A test procedure design is described by the mileages (X) at which emission measurements are taken and the frequency (f) of the measurements at these mileages. Thus, a standard procedure, with maintenance at 25 thousand miles, is specified by

$$X = (5, 10, 15, 20, 25, 30, 35, 40, 45, 50)$$

and

$$f = (1, 1, 1, 1, 2, 1, 1, 1, 1, 1)$$

for a total of 11 measurements. A class of alternative designs described by

$$X = (5, 10, 15, 20, 25, \ldots, M)$$
$$f = (n, 1, 1, 1, 2, 1, \ldots, n)$$
$$M = 30, 35, 40, 45, \text{ or } 50$$

has been evaluated with respect to improving the reliability of the estimated slope in the assumed linear relationship between mileage and vehicle emission. Allowing repeated measurements at the design end-points can permit significant reduction in required accumulated mileage for estimating DF's with no loss in statistical accuracy (compared to procedures requiring 50,000 miles).

While focusing on the slope of the regression function provides a simple method to derive improved designs, assessing the direct effect on the DF is somewhat complicated. A criterion has been derived which facilitates this comparison.

BIBLIOGRAPHY

California Air Resources Board: Manufacturer's Advisory Correspondence 82-01, January 7, 1982.

Lorenzen, T. J. (1980). "Determining Statistical Characteristics of a Vehicle Emissions Audit Procedure", *Technometrics* **22**, 483–493.

McDonald, G. C. (1981). Confidence intervals for vehicle emission deterioration factors, *Technometrics* **23**, 239–242.

Title 40 Code of Federal Regulations (1978a), Section 86.079-28 (a) (4) (4) (ii-iii), The National Archives of the United States, July 1, 1978.

Title 40 Code of Federal Regulations (1978b), Section 86.080-26 (a) (4) and

Section 86.079-28 (a) (4) (i), The National Archives of the United States, July 1, 1978.

Title 40 Code of Federal Regulations (1985a), Section 86.084-26 (a) (4) (i), The National Archives of the United States, July 1, 1985.

Title 40 Code of Federal Regulations (1985b), Section 86.085-28 (a) (4) (i) (B), The National Archives of the United States, July 1, 1985.

PEAKEDNESS IN MULTIVARIATE DISTRIBUTIONS

I. OLKIN

Department of Statistics
Stanford University
Stanford, California 94305-2195, U.S.A.

Y. L. TONG

School of Mathematics
Georgia Institute of Technology
Atlanta, Georgia, U.S.A.

1. INTRODUCTION

Peakedness is a descriptive index of a distribution that provides an indication of concentration. Although the notion is intuitive, a formal definition for the univariate case was first provided by Birnbaum (1948).

Definition 1.1. A (univariate) random variable Y is said to be more peaked about μ than a random variable X and ν if

$$P\{|Y-\mu|<t\} \geq P\{|X-\nu|<t\} \text{ for all } t. \tag{1.1}$$

When $\mu=\nu=0$ we say that Y is more peaked than X, and write

$$Y \overset{p}{\geq} X. \tag{1.2}$$

Even in the univariate case, various questions arise almost immediately. Does $Y_1 \overset{p}{\geq} X_1$ and $Y_2 \overset{p}{\geq} X_2$ imply $Y_1+Y_2 \overset{p}{\geq} X_1+X_2$? If $\{Y_m\}$ and $\{X_m\}$ are sequences for which $Y_m \overset{p}{\geq} X_m$, $m=1,2,\ldots$ and $\lim \mathcal{L}(Y_m)=Y$, $\lim \mathcal{L}(X_m)=X$, does it follow that $Y \overset{p}{\geq} X$? The answer to both questions is in the affirmative, as we shall see.

These two questions deal with univariate peakedness. But an extension to multivariate peakedness can also be considered. This was first discussed by Sherman (1955).

Definition 1.2. An n-dimensional random variable Y is said to be more peaked than X if they have densities and if

$$P\{Y \epsilon A\} \geq P\{X \epsilon A\} \tag{1.3}$$

holds for all $A \,\epsilon\, \mathcal{A}_n$, the class of compact, convex, symmetric (about the origin) sets in R^n.

The studies of peakedness, for certain specific multivariate distributions and geometric regions, play an important role in the development of theory of inequalities in statistics (see e.g., Tong (1980, chapters 2-4)). For example, Šidak's result (1971) concerns the probability contents of n- dimensional rectangles for the multivariate normal and t distributions. That result was generalized by Das Gupta, Eaton, Olkin, Perlman, Savage and Sobel (1972) to a larger class of geometric regions and for all elliptically contoured distributions. A recent paper of Jensen (1985) deals with partial ordering of probability contents of all compact, convex, and symmetric regions in R^n via the ordering of the scale matrices of elliptically contoured distributions, thus it represents a generalization of the main theorem in Das Gupta-Eaton-Olkin-Perlman-Savage-Sobel (1972).

In the present paper we study peakedness for a larger class of distributions, obtain some results on closure properties (Section 2), further study the multivariate case (Section 3), and discuss some applications (Section 4).

2. CLOSURE PROPERTIES

A closure property for the sum of random vectors was obtained by Sherman (1955) as follows. Let $\mathcal{G}$ denote the closed convex cone generated by the indicator functions of all compact, convex, symmetric (about the origin) sets in the norm that is the maximum of the uniform and L_1 norms.

Proposition 2.1. *(Sherman, 1955). Let X_1, X_2, Y_1, Y_2 be independent continuous random vectors with densities f_1, f_2, g_1, g_2, respectively. If*

$$f_1, f_2, g_1, g_2 \in \mathcal{G}, \tag{a}$$

$$Y_1 \overset{p}{\geq} X_1, \quad Y_2 \overset{p}{\geq} X_2, \tag{b}$$

then

$$(Y_1 + Y_2) \overset{p}{\geq} (X_1 + X_2). \tag{2.1}$$

This proposition provides a sufficient condition for (2.1) to hold. Condition (a) is not always easy to verify. However, the following sufficient condition for (a) to hold is quite straightforward to verify, and indeed, it is well-known that many densities have this property.

Proposition 2.2. *If $f : R^n \to [0, \infty)$ is a continuous density function that is bounded, symmetric (about the origin) and log-concave, then f is in $\mathcal{G}$.*

PROOF. It suffices to show that there exists an array $\{A_{mr}\}$ for $r = 1, 2, \ldots, m = 1, 2, \ldots$, of compact, convex, and symmetric sets and a sequence of functions $f_m(x)$ of the form $f_m(x) = \sum_{r=1}^{\infty} I_{A_{mr}}(x)$, where I_A is an indicator function, such that

$$\lim_{m\to\infty} \sup_x |f_m(x) - f(x)| = 0, \tag{2.2}$$

$$\lim_{m\to\infty} \int |f_m(x) - f(x)|dx = 0. \tag{2.3}$$

Define

$$A_{mr} = \{x \mid x \in R^n, f(x) \geq r/m\},$$

then $A_{mr} \in \mathcal{A}_n$ for every m and r (convexity follows because all log-concave density functions are unimodal), and $A_{m,r+1} \subset A_{mr}$ for every r and m. Now denote $B_{mr} = A_{mr} \bigcap \bar{A}_{m,r+1}$ and define $f_m(x) = r/m$ for $x \in B_{mr}$, that is

$$f_m(x) = \begin{cases} \sum_{r=1}^{\infty} I_{A_{mr}}(x), & \text{for } x \in A_{m1}, \\ 0, & \text{otherwise.} \end{cases}$$

Then $|f_m(x) - f(x)| \leq 1/m$ holds for all $x \in R^n$. Consequently, $f_m(x)$ converges to $f(x)$ in the uniform norm and (2.2) holds.

To show that (2.3) also holds, note that there exists a convex function $Q(x)$ of x with $Q(x) = Q(-x)$ for which $f(x)$ is of the form $\exp(-Q(x))$. It follows that there exists a large M_0 and $K > 0$ such that

$$\|x\| > \frac{1}{K}\log m \Rightarrow Q(x) \geq K\|x\|$$

for all $m \geq M_0$. Thus for $m \geq M_0$ the volume (vol) of A_{m1} satisfies

$$\begin{aligned} &\text{vol}\{x|f(x) > \frac{1}{m}\} = \text{vol}\{x|Q(x) < \log m\} \\ &\leq \text{vol}\{x \mid K\|x\| \leq \log m\} \leq \text{vol}\{x \mid \max_{1\leq i\leq n} |x_i| \leq K^{-1}\log m\} \\ &= (2\log m)^n/K^n. \end{aligned}$$

Consequently

$$\begin{aligned} \int |f_m(x) - f(x)|dx &\leq \int_{A_{m1}} \frac{1}{m}dx + \int_{\bar{A}_{m1}} |f_m(x) - f(x)|dx \\ &\leq \frac{1}{m}(\frac{2}{K}|\log\frac{1}{m}|)^n + \int_{\bar{A}_{m1}} |f_m(x) - f(x)|dx \equiv I_1 + I_2, \end{aligned}$$

say. Clearly $I_2 \to 0$ as $m \to \infty$. Also, since (by l'Hospital's rule)

$$\lim_{m\to\infty} (1/m)|\log(1/m)|^n = 0,$$

$I_1 \to 0$, which completes the proof. □

An application of Propositions 2.1 and 2.2 is the fact that: If the densities of $X_1, \ldots, X_n$, $Y_1, \ldots, Y_n$ are bounded, symmetric (about the origin), and log-concave, $X_1, \ldots, X_n$ are independent, $Y_1, \ldots, Y_n$ are independent, then

$Y_m \overset{p}{\geq} X_m (m = 1, \ldots, n)$ implies that

$$\sum_1^n c_m Y_m \overset{p}{\geq} \sum_1^n c_m X_m \tag{2.4}$$

for all real numbers $c_1, \ldots, c_n$.

Two special cases follow: (i) If $n = 2, c_1 = c_2 = 1$ and $X_2 = Y_2 \equiv U$, then (2.4) reduces to $Y + U \overset{p}{\geq} X + U$; (ii) when $X_1, \ldots, X_n$ are independent identically distributed, $Y_1, \ldots, Y_n$ are independent identically distributed, and $Y_i \overset{p}{\geq} X_i$, then (2.4) yields $\frac{1}{n} \sum_1^n Y_m \overset{p}{\geq} \frac{1}{n} \sum_1^n X_m$.

Proposition 2.3. *If $\{X_m\}_{m=1}^{\infty}$ and $\{Y_m\}_{m=1}^{\infty}$ are two sequences of n-dimensional random vectors, with $Y_m \overset{p}{\geq} X_m$, $m = 1, 2, \ldots$, $\mathcal{L}(X_m) \overset{d}{\to} \mathcal{L}(X)$, $\mathcal{L}(Y_m) \overset{d}{\to} \mathcal{L}(Y)$, then $Y \overset{p}{\geq} X$.*

PROOF. Let $I_A(x)$ be the indicator function of $A \in \mathcal{A}_n$, and let X_m, X, Y_m, and Y have distribution functions F_m, F, G_m, G, respectively. Then

$$\begin{aligned} P\{X \epsilon A\} &= \int I_A(x) dF(x) = \lim_{m \to \infty} \int I_A(x) dF_m(x) \\ &\leq \lim_{m \to \infty} \int I_A(x) dG_m(x) = P\{Y \epsilon A\}. \end{aligned}$$

The equalities are a consequence of the Helly-Bray Lemma, and the inequality follows by hypothesis. □

Two other properties that are related to the above deal with marginal distributions and linearly transformed variables.

Proposition 2.4. *If $X = (\dot{X}, \ddot{X})$ and $Y = (\dot{Y}, \ddot{Y})$ are n-dimensional random variables partitioned comformably with $\dot{X}$ and $\dot{Y}$ q-dimensional, and if $Y \overset{p}{\geq} X$, then $\dot{Y} \overset{p}{\geq} \dot{X}$.*

PROOF. For any given $A \epsilon$ in $\mathcal{A}_q \subset R^q$ define

$$A_m = A \bigcap \{(x_{q+1}, \ldots, x_n) \Big| \, |x_i| \leq m, \; q+1 \leq i \leq n\}.$$

Then $A_m \epsilon \mathcal{A}_n$ for all m and

$$P\{\dot{X} \epsilon A\} = \lim_{m \to \infty} P\{X \epsilon A_m\} \leq \lim_{m \to \infty} P\{Y \epsilon A_m\} = P\{\dot{Y} \epsilon A\}.$$

The equalities follow from the monotone convergence theorem, the inequality follows by hypothesis. □

Proposition 2.5. *$Y \overset{p}{\geq} X$ holds if and only if $CY \overset{p}{\geq} CX$ holds for all $k \times n$ matrices $C, k \leq n$.*

PROOF. The "if" part follows immediately by choosing C to be the identity matrix of order n. The "only if" part is proved in three steps:

(i) First consider the case that $k = n$ and C is nonsingular. For any given $A \in \mathcal{A}_n$, then $Cx \in A$ if and only if $x \in C^{-1}A$ and $C^{-1}A \in \mathcal{A}_n$. Thus

$$P\{CX \in A\} = P\{X \in C^{-1}A\} \leq P\{Y \in C^{-1}A\} = P\{CY \in A\}.$$

(ii) If C is any $n \times n$ matrix of rank $q \leq n$, then C can be decomposed as

$$C = E_1 \cdots E_r D F_1 \cdots F_s,$$

where $D = \text{diag}(I_q, 0)$, the E_i's and F_j's are elementary transformations. Each elementary transformation is nonsingular, so that by (i), $E_i Y \overset{p}{\geq} E_i X$ and $F_j Y \overset{p}{\geq} F_j X$ for all i, j.

Furthermore, since DX and $\dot{X}$ DY and $\dot{Y}$ are identically distributed, by Proposition 2.4 we have $DY \overset{p}{\geq} DX$. Consequently, $CY \overset{p}{\geq} CX$ holds for all $n \times n$ matrices C.

(iii) If C is $k \times n$, with $k \leq n$, define the $n \times n$ matrix $C^* = \text{diag}(C, I_{n-k})$, and apply (ii) to yield $C^* Y \overset{p}{\geq} C^* X$, then apply Proposition 2.4 to obtain $CY \overset{p}{\geq} CX$. □

REMARK 2.6. After obtaining Proposition 2.5 we now see that Proposition 2.4 is a special case (by choosing $C = \text{diag}(I_q, 0)$ in Proposition 2.5). Of course, the proof of Proposition 2.5 depends on Proposition 2.4.

Choosing $C = \begin{pmatrix} I_q & I_q \\ 0 & 0 \end{pmatrix}$ in Proposition 2.5 yields the following fact:

FACT 2.7. For $n = 2q$ let $X = (\dot{X}, \ddot{X}), Y = (\dot{Y}, \ddot{Y})$ be such that $\dot{X}, \ddot{X}, \dot{Y}, \ddot{Y}$ are q- dimensional random variables. If $Y \overset{p}{\geq} X$, then $\dot{Y} + \ddot{Y} \overset{p}{\geq} \dot{X} + \ddot{X}$.

REMARK 2.8. Assume that $\dot{X}$ and $\ddot{X}$ are independent, $\dot{Y}$ and $\ddot{Y}$ are independent, and that $\dot{Y} \overset{p}{\geq} \dot{X}$, $\ddot{Y} \overset{p}{\geq} \ddot{X}$, . Does this imply $(\dot{Y}, \ddot{Y}) \overset{p}{\geq} (\dot{X}, \ddot{X})$? Without any additional conditions on the densities the answer to this question is negative. To see this, consider $n = 2, q = 1$, and let the densities of $\dot{X}, \ddot{X}, \dot{Y}, \ddot{Y}$ be as given in the example in Sherman (1955). If $(\dot{Y}, \ddot{Y}) \overset{p}{\geq} (\dot{X}, \ddot{X})$ were true, then Fact 2.7 would imply $(\dot{Y} + \ddot{Y}) \overset{p}{\geq} (\dot{X} + \ddot{X})$ which is known to be false.

3. SCHUR PROPERTIES

The questions posed in this section deal with comparisons of linear combinations for which the weights are ordered by majorization. (For definitions and properties of majorization see Marshall and Olkin, 1979.)

The following result of Proschan (1965) deals with univariate random variables.

Theorem 3.1 *If $Z_1, \ldots, Z_N$ are independent, identically distributed univariate random variables with common log-concave density function $f(z)$ symmetric about the origin, and if $a = (a_1, \ldots, a_N)$ and $b = (b_1, \ldots, b_N)$ are two real nonnegative vectors for which $a \succ b$ (a majorizes b), then*

$$\sum_1^N b_i Z_i \overset{p}{\geq} \sum_1^N a_i Z_i.$$

Univariate densities that are log-concave are said to be strongly unimodal (Ibragimov, 1956) or Pólya frequency functions of order 2 (Karlin, 1968, p. 12 or Barlow and Proschan, 1975, p. 76). It was shown by Proschan (1965) that Theorem 3.1 fails if the strong unimodality condition is replaced by the weaker condition: $f(z)$ is unimodal in the sense that $\{z : f(z) \geq u\}$ is an interval for all u. The Cauchy distribution provides a counterexample.

This section primarily concerns two generalizations of Theorem 3.1 along different directions. The first generalization (Theorem 3.2) deals with the peakedness of linear combinations of i.i.d. random vectors (according to Definition 1.2), and the second generalization (Theorem 3.2) involves peakedness of univariate random variables which are exchangeable but not necessarily independent (according to Defintion 1.1). Our proofs of the theorems are not an adaptation of Proschan's proof for the univariate case, but instead depend on an application of the result of Anderson (1955) and Sherman (1955). It has the virtue of removing the nonnegativity condition of the weights.

Theorem 3.2. *If $Z_1, \ldots, Z_n$ are independent, identically distributed k-dimensional random vectors with common log-concave density function $f(z)$ symmetric (about the origin), and if $(a_1, \ldots, a_N) \succ (b_1, \ldots, b_N)$, then*

$$\sum_1^N b_i Z_i \overset{p}{\geq} \sum_1^N a_i Z_i.$$

Theorem 3.3. *Suppose that $Z_1, \ldots, Z_n$ are exchangeable univariate random variables with joint density function $\phi(z)$, and $(a_1, \ldots, a_n)$ and $(b_1, \ldots, b_n)$ are two real n-dimensional vectors. If the conditional density g of*

$$\{c(Z_1 + Z_2) + \sum_3^N a_i Z_i\} \mid (Z_1 - Z_2) = v$$

is symmetric (about the origin), and is unimodal for all fixed c, v, and $a_3, \ldots, a_n$, then $a \succ b$ implies that

$$\sum_1^n b_i Z_i \overset{p}{\geq} \sum_1^n a_i Z_i.$$

To prove Theorems 3.2 and 3.3 we first observe two lemmas.

Lemma 3.4. *Let Z_1 and Z_2 be independent, identically distributed k-dimensional random variables with common density $f(z)$, which is log-concave and symmetric about the origin, and for nonvanishing c, let $g(v_1|v_2)$ denote the conditional density of $V_1 = c(Z_1 + Z_2)$, given $V_2 = (Z_1 - Z_2) = v_2$. Then, for every fixed v_2 such that g exists, g is symmetric (about the origin) and is a log-concave function of v_1.*

PROOF. The joint density of V_1 and V_2 is

$$h(v_1, v_2) = f(\frac{v_1 + cv_2}{2c})f(\frac{v_1 - cv_2}{2c})|J|$$

where $|J|$ is the Jacobian of the linear transformation ($|J| > 0$), and the conditional density of V_1 given $V_2 = v_2$ is of the form

$$g(v_1|v_2) = h(v_1, v_2)/q(v_2),$$

where $q(v_2) = \int h(v_1, v_2)dv_1$. The fact that $h(-v_1, v_2) = h(v_1, v_2)$ for all fixed v_2 implies that $g(v_1|v_2) = g(-v_1|v_2)$. To show that $g(v_1|v_2)$ is log-concave in v_1, first note that for all v_1, v_1^*, and for all $\alpha \in [0,1]$,

$$\begin{aligned}\log h(\alpha v_1 + (1-\alpha)v_1^*,\ v_2) &= \log f[\frac{\alpha(v_1 + cv_2) + (1-\alpha)(v_1^* + cv_2)}{2c}])\\ &\quad + \log f[\frac{\alpha(v_1 - cv_2) + (1-\alpha)(v_1^* - cv_2)}{2c}])\\ &\quad + \alpha \log|J| + (1-\alpha)\log|J|\\ &\geq \alpha \log h(v_1, v_2) + (1-\alpha)\log h(v_1^*, v_2).\end{aligned}$$

Thus

$$\begin{aligned}&\log g(\alpha v_1 + (1-\alpha)v_1^*|v_2)\\ &\quad \geq \alpha[\log h(v_1, v_2) - \log q(v_2)] + (1-\alpha)[\log(hv_1^*, v_2) - \log q(v_2)]\\ &\quad = \alpha \log g(v_1|v_2) + (1-\alpha)\log g(v_1^*|v_2),\end{aligned}$$

which completes the proof. □

Note that, by choosing $n = 1$ in Lemma 3.4, the conditions in Theorem 3.3 are weaker than that in Theorem 3.1.

Lemma 3.5. *Under the assumptions stated in Theorem 3.2 the conditional density g of $c(Z_1 + Z_2) + \sum_3^N a_i Z_i$ given $Z_1 - Z_2 = v_2$ is symmetric (about the origin) and log-concave for all c, v_2 and $a_i (i = 3, \ldots, N)$ such that g exists.*

PROOF. If $a_3 = \cdots = a_N = 0$ then the lemma follows immediately from Lemma 3.4; thus we assume that at least one of the a_i's is not zero. It is known that the convolution of two symmetric (about the origin) and log- concave densities is also symmetric and log-concave (the symmetry property can be easily checked, the log-concavity property can be found in Das Gupta (1973, Theorem 4.2)). Thus the marginal density, or equivalently the conditional density of $\sum_3^n a_i Z_i$ given $Z_1 - Z_2 = v_2$ (by independence), is symmetric and log-concave. Consequently, if $c = 0$ then $\sum_3^n a_i Z_i$ and $Z_1 - Z_2$ are independent, so that the Lemma follows immediately. If $c \neq 0$ then the conditional density g is the convolution of the conditional densities of $c(Z_1 + Z_2)$ and that of $\sum_3^n a_i Z_i$, given $Z_1 - Z_2$, which is also symmetric and log- concave. □

PROOF OF THEOREM 3.2 Let $(a_1, \ldots, a_n) \succ (b_1, \ldots, b_n)$. Without loss of generality (see Marshall and Olkin (1979)) it can be assumed that

$$a_1 > b_1 \geq b_2 > a_2, \quad a_1 + a_2 = b_1 + b_2 = 2c, \tag{3.1}$$

and $b_i = a_i$ for $i = 3, \ldots, n$. For any given $A \in \mathcal{A}_n$ write

$$P\{\sum_1^n a_i Z_i \in A\} = \int P\{[c(Z_1+Z_2)+\sum_3^n a_i Z_i] \in A - \lambda v_2 | Z_1 - Z_2 = v_2\} dH(v_2), \tag{3.2}$$

where $\lambda = (a_1 - a_2)/2$ and H is the distribution function of $Z_1 - Z_2$. Since g is symmetric and log- concave, it is symmetric and unimodal. By the main theorem in Anderson (1955) the conditional probability on the right-hand side of (3.2) is a non-increasing function of λ for every fixed v_2, and consequently, the unconditional probability on the left-hand side of (3.2) is nonincreasing in λ. □

PROOF OF THEOREM 3.3 Again assume that (3.1) holds. For $c = (a_1 + a_2)/2$, $\lambda = (a_1 - a_2)/2 > 0$,

$$a_1 Z_1 + a_2 Z_2 = c(Z_1 + Z_2) + \lambda(Z_1 - Z_2),$$

so that

$$P\{|\sum_1^n a_i Z_i| \leq t\}$$

$$= \int P\{-t - \lambda v_2 \leq c(Z_1 - Z_2) + \sum_3^n a_i Z_i \leq t - \lambda v_2 | Z_1 - Z_2 = v_2\}\, dH(v_2),$$

where H is the distribution function of $Z_1 - Z_2$. Since the conditional density g is symmetric and unimodal, the conditional probability is a nonincreasing function of λ for every fixed v_2, and the unconditional probability is a nonincreasing function of λ. □

4. APPLICATIONS

A number of examples of these results can be given. The present choices are illustrative of a variety of applications, and obviously are not exhaustive.

Application 4.1. The first application deals with elliptically contoured distributions. If the density function of the random vector $Z = (Z_1, \ldots, Z_n)$ is permutation invariant and elliptically contoured, that is, it is of the form $\phi(z) = h(zTz')$, where h is nonincreasing, the positive definite matrix $T = (\tau_{ij})$ has elements

$$\tau_{ij} = \begin{cases} \tau_1 & \text{for } i = j \\ \tau_1\tau_2 & \text{for } i \neq j \end{cases}$$

then Theorem 3.3 applies.

To see this, for $a_1 > b_1 \geq b_2 > a_2$ let $c = (a_1 + a_2)/2, \lambda = (a_1 - a_2)/2$,

$$V_1 = c(Z_1 + Z_2) + \sum_3^n a_i Z_i, \quad V_2 = Z_1 - Z_2, \quad V_i = Z_i \quad \text{for } i = 3, \ldots, n.$$

Then the joint density of $V = (V_1, \ldots, V_n)$ is $h(v(\Lambda T \Lambda')v')|\Lambda|$, where Λ satisfies $V = Z\Lambda^{-1}$. Partition the matrix $\Lambda T \Lambda'$ as

$$\Lambda T \Lambda' = \begin{pmatrix} C_{11} & C_{12} \\ C_{21} & C_{22} \end{pmatrix},$$

where C_{11} is a 2×2 matrix. An elementary calculation shows that

$$C_{11} = \text{diag}(\tau_1(1 + \tau_2)/2c^2, \tau_1(1 - \tau_2)/2).$$

Since the marginal density of a subset of elliptically contoured distributed random variables is again elliptically contoured, and depends on $\Lambda T \Lambda'$ only through the submatrix C_{ii} (e.g. see Kelker (1970)), the density of (V_1, V_2) is a nonincreasing function of

$$(v_1, v_2)C_{11}(v_1, v_2)' = \frac{1}{2}\tau_1[(1 + \tau_2)v_1^2/c^2 + (1 - \tau_2)v_2^2].$$

Consequently the conditional density of V_1 given $V_2 = v_2$ is a nonincreasing function of v_1^2. Thus it is symmetric and unimodal, and the conditions of Theorem 3.3 are satisfied.

Note that, as a special case, if the distribution of $(Z_1, \ldots, Z_n)$ is multivariate normal with means 0, variances τ_1 and correlations $\tau_2, -1/(n - 1) \leq \tau_2 \leq 1$, then Theorem 3.3 applies.

Application 4.2 This application pertains to independent identically distributed multivariate normal vectors. Let $Z_1, \ldots, Z_n$ be independent identically distributed multivariate normal vectors with mean vector 0 and common positive definite covariance matrix Σ. The common density $f(z)$ is symmetric and log-concave, so that Theorem 3.2 applies.

Application 4.3. In this application we obtain a monotonicity property for a confidence probability. As in Proschan (1965), Theorems 3.2 and 3.3 yield the following monotonicity property: Let $\{W_i\}$ be a sequence of independent identically distributed n-dimensional random variables with mean vector μ and density function $f(w)$. Let $A \subset R^n$ be convex and symmetric (about the origin), $\bar{W}_n = \frac{1}{n}\sum_1^n W_m (n = 1, 2, \ldots)$, and let $A + \bar{W}_n$ denote the confidence region for μ on the basis of $W_1, \ldots, W_n$. If $f(w)$ is symmetric about μ (i.e., $f(\mu - v) = f(\mu + v)$ for every v) and is a log-concave function of w, then the confidence probability $P[\mu \epsilon A + \bar{W}_n]$ is a nondecreasing function of n.

If A is compact, the monotonicity follows immediately from Theorem 3.2 and from the choice

$$a_1 = \cdots = a_n = \frac{1}{n}, \quad a_{n+1} = 0,$$

$$b_1 = \cdots = b_n = b_{n+1} = \frac{1}{n+1}.$$

If A is not compact, then we require a sequence of compact sets $\{A_m\}$ which converge to A; the result follows from an application of Theorem 3.2 and the monotone convergence theorem.

Application 4.4. Here we obtain a monotonicity of the type I error in certain multivariate hypothesis-testing problems. Consider the sequence $\{W_i\}$ of n-dimensional random variables described in Application 4.3. For testing the hypothesis

$$H_0 : \mu = \mu_0 \quad \text{vs} \quad H_1 : \mu \neq \mu_0,$$

consider the procedure: reject H_0 if $(\bar{W}_n - \mu_0) \notin A$ for a set A which is symmetric (about the origin) and convex. Then by Theorem 3.2 the type I error is a non-increasing function of the sample size n.

ACKNOWLEDGMENTS

This research was partially supported by the NSF Grants NSF MCS81-04262, MCS81-00775, A01 and DMS-8502346.

Abstract: Peakedness provides one of the principal descriptive indices of a distribution. A definition for the univariate case was developed by Birnbaum (1948) and generalized to the multivariate case by Sherman (1955). Of course, other multivariate definitions can be developed. In this paper we study and extend some closure properties of peakedness. In particular, two multivariate generalizations are given of a majorization inequality of Proschan (1965). The results are shown to apply to the family of elliptically contoured distributions, and to have implications in estimation and for tests of hypotheses.

Key Words and Phrases: Peakedness in multivariate distributions, inequalities, majorization, log-concave densities.

AMS 1980 Subject Classifications. 62H99, 60D15.

BIBLIOGRAPHY

Anderson, T. W. (1955). The integral of a symmetric unimodel function over a symmetric convex set and some probability inequalities. *Proc. Amer. Math. Soc.*, **6**, 170–176.

Barlow, R. E. and Proschan, F. (1975). *Statistical Theory of Reliability and Life Testing.* New York: Holt, Rinehart and Winston.

Birnbaum, Z. W. (1948). On random variables with comparable peakedness. *Ann. Math. Statist.*, **19**, 76–81.

Das Gupta, S. (1973). S-unimodal function: Related inequalities and statistical applications. *Sankhyā, Ser. B.*, **38**, 301–314.

Das Gupta, S., Eaton, M. L., Olkin, I., Perlman, M. D., Savage, L. J., and Sobel, M. (1972). Inequalities on the probability content of convex regions for elliptically contoured distributions. *Proc. Sixth Berkeley Symp. Math. Statist. Probab.*, 2, (L. M. Le Cam, J. Neyman and E. L. Scott, eds.), 241–265, University of California Press: Berkeley, CA.

Ibragimov, I. A. (1956). On the composition of unimodal distributions. *Theor. Probability Appl.*, **1**, 255–260.

Jensen, D. R. (1985). Ordering ellipsoidal measures: Scale and Peakedness orderings. *SIAM J. Appl. Math.*, **44**, 1226–1231.

Karlin, S. (1968). *Total Positivity*, Vol 1. Stanford: Stanford University Press.

Kelker, D. (1970). Distribution theory of spherical distributions and a location-scale parameter generalization. *Sankhyā, Ser. A*, **32**, 419–430.

Marshall, A. W., and Olkin, I. (1979). *Inequalities: Theory of Majorization and Its Applications.* New York: Academic Press.

Proschan, F. (1965). Peakedness of distributions of convex combinations. *Ann. Math. Statist.*, **36**, 1703–1706.

Sherman, S. (1955). A theorem on convex sets with applications. *Ann. Math. Statist.*, **26**, 763–767.

Šidak, Z. (1971). On probabilities of rectangles in multivariate student distribution: Their dependent on correlations. *Ann. Math. Statist.*, **42**, 169–175.

Tong, Y. L. (1980). *Probability Inequalities in Multivariate Distributions.* New York: Academic Press.

SPATIAL DESIGNS

JEROME SACKS AND SUSANNAH SCHILLER

Department of Statistics
University of Illinois
Urbana, Illinois 61801, U.S.A.

1. INTRODUCTION

Suppose observation are to be taken at sites t in a space T and a model is hypothesized connecting the observations to the sites by

$$y(t) = f(t) + \varepsilon(t).$$

Inference about f is usually the ultimate goal but a question that has first to be answered is: how should the sites be selected? The focus here is on a specific approach to this design question in a context where T is a subset of $R^d, d \geq 2$. A treatment in one-dimension was given in Sacks and Ylvisaker (1984), (1985) – the approach here stems from the (1985) considerations. A chief concern of our analysis lies in computational problems and a large part of this report relates to such issues.

The following specific assumptions underlie the questions to be addressed.

A. T, the background space of possible sites, is a finite subset of R^d having N points.
B. Observations are taken at a subset S of T; the cardinality of S, $\#(S)$, is k.
C. The measurement errors $\{\varepsilon(s), s \in S\}$ are uncorrelated with mean 0 and variance σ^2, σ^2 specified. We will usually assume $\{\varepsilon(s)\}$ are Gaussian.
D. Let X be a Gaussian process on T with mean 0 and Cov $(X(t), X(u)) = R(t,u)$. X will act as a prior on the class of allowable f's. We will write

$$Y(s) = X(s) + \varepsilon(s), \ s \in S \tag{1.1}$$

and regard $Y(s)$ as the observation at site s.

The *best linear predictor* (BLP) $\hat{X}(t)$ of $X(t)$ based on the observations $\{Y(s), s \in S\}$ is of the form

$$\hat{X}(t) = c'(t)y$$

where $c(t)$ and y are k-vectors given by

$$\begin{aligned} c(t) &= (R_S + \sigma^2 I)^{-1} R(\cdot, t) \\ y &= (y(s_1), \dots, y(s_k))' \\ R(\cdot, t) &= (R(s_1, t), \dots, R(s_k, t))' \\ R_S &= k \times k \text{ matrix } [R(s_i, s_{i'})]. \end{aligned} \tag{1.2}$$

The prediction error of $\hat{X}(t)$ is

$$E(\hat{X}(t) - X(t))^2 = R(t,t) - R'(\cdot,t)(R_S + \sigma^2 I)^{-1} R(\cdot,t). \tag{1.3}$$

We denote the prediction covariances by

$$\begin{aligned} G(t,u) &= \text{Cov } (\hat{X}(t) - X(t), \hat{X}(u) - X(u)) \\ &= R(t,u) - R'(\cdot,t)(R_S + \sigma^2 I)^{-1} R(\cdot,u). \end{aligned} \tag{1.4}$$

Consider the objective function

$$J_{\text{Max}}(S) = \max_{t \in T} E(\hat{X}(t) - X(t))^2.$$

The primary problem we address here and below is that of finding the best design for the criterion J_{Max} that is, to find the set of cardinality k which minimizes the maximum prediction error:

$$\min_{\#(S)=k} J_{\text{Max}}(S).$$

There are many variations to this problem of finding a "best" design. Different objective functions J can be treated, alternative formulations allowing replicate observations at sites can be stated and the form of model (1.1) can be altered. For example, we will also report results about the criteria

$$J_{\overline{X}}(S) = E[\tilde{X} - \overline{X}]^2 \tag{1.7}$$

where $\overline{X} = \frac{1}{N} \Sigma X(t)$, $\tilde{X}$ its BLP, and

$$J_{DET}(S) = \det G_S, \tag{1.8}$$

where G_S is the $N - k \times N - k$ matrix with entries $G(t,u)$ for $t, u \in T - S$ (see (1.4)). The criterion (1.8) is considered by Shewry and Wynn (1986).

A formulation, related to that of Sacks and Ylvisaker (1985), which allows replicate observations, poses the model

$$y(s) = X(s) + n^{-1/2}(s)\varepsilon(s) \tag{1.9}$$

where $n(s)$ = the number of observations at site $s \in S$. The cardinality restriction on S becomes $\sum_{s \in S} n(s) = k$. A deterministic trend can be incorporated through

$$y(s) = g(s) + X(s) + n^{-1/2}(s)\varepsilon(s)$$

where g follows a linear model.

Since (1.6) is a natural global criterion we deal mainly with that objective. We defer consideration of the replicated case: the computational challenge in the unreplicated case is enough.

Prediction of linear functionals of the form $\int X(t)\Phi(dt)$ (e.g., $\overline{X}$) is of central concern in Sacks and Ylvisaker (1985). Some of that work relates to earlier work in numerical integration as described by Sacks and Ylvisaker (1970), with even earlier roots in the work of Suldin (1960). It has 'newly' found currency in the computational complexity literature (see Wozniakowski, 1985, Smale, 1985).

The specific assumption of finite T is most easily recognized as natural in combinatorial design settings: an example of a $3^2 2^3$ factorial will be discussed in Section 3. From a 'practical' viewpoint the assumption of finite T is a minor restriction: its use makes computational issues easier to confront. Other justifications can be given through considerations like that in Li (1984).

Adequate structure theorems are not obviously accessible, though there is some work by Ylvisaker (1985). Some of the difficulties may be seen in the problem of finding reasonable asymptotic solutions (Ylvisaker, 1975; Micchelli and Wahba, 1979) in dimension 2 or more. For such reasons, the contents of this article relate largely to the computational issues involved in finding best designs. While in Section 2 we treat algorithmic questions connected with finding solutions to (1.6), the information obtained by applying the algorithm gives some perspective on the effects of different criteria and alternate covariance structures on the nature of good designs. Our principal 'findings' are:

(1) An annealing algorithm (for a definition see Section 2) is useful, but by no means fool-proof, for finding good sub-optimal solutions.
(2) The designs do not appear to be robust to changes in the covariance structure (see Section 3).
(3) Designs for (1.6) tend to 'fill in the middle' (for example, see Section 3, Figure 3.3), in conformity with intuitive needs expressed by practitioners. This is in contrast with designs which minimize J_{DET} in (1.8) and traditional designs (D-optimum, etc.) in standard settings.

The fact that useful algorithms, like (2.7) of Section 2, are available, at least for modest sized (n, k), will hopefully lead to a capability for computing good designs in contexts where solutions are not easy to conjecture.

Our computing efforts were greatly assisted by the use of the Cray X-MP at the National Center for Supercomputing Applications at the University of Illinois at Urbana-Champaign. Conversations with Toby Mitchell, William Welch, Henry Wynn, and Donald Ylvisaker were of great value in the course of this work.

2. ALGORITHMS

A natural and relatively simple way to proceed to minimize (1.6) or any of the objective functions J, is by a 'single excursion' algorithm, a routine which plays a key role in the DETMAX algorithm of Mitchell (1974). Such an algorithm is defined by:

(a) Pick a starting set S_0 at random with $\#(S_0) = k$. (2.1)

(b) At stage $j+1$, choose $t^* \in T - S_j$ such that

$$J(S_j \cup t^*) = \min_{t \in T - S_j} J(S_j \cup t)$$

(find the best single point addition to S_j).

(c) Choose $s^* \in S_j \cup t^*$ such that

$$J(S_j \cup t^* - s^*) = \min_{s \in S_j \cup t^*} J(S_j \cup t^* - s)$$

(find the best single point deletion from $S_j \cup t^*$).

(d) $S_{j+1} = S_j \cup t^* - s^*$.

This algorithm was applied to several examples like those in Section 3. All too often a set S_j was reached with $s^* = t^*$ and the algorithm stopped, but S_j was not close to the global optimum. Nevertheless, this idea is a reasonable starting point for devising other algorithms, as will be seen below.

The inherent difficulty in finding a solution to (1.6) is the fact that it is hard to recognize when a particular set S is only locally optimum (in some sense, like a one point exchange) rather than globally optimum. Annealing algorithms, a class of 'scheduled' random search algorithms, attempt to surmount such obstacles. The basic ingredients of annealing algorithms are:

(a) At stage j pick a set S_{j+1} *at random.* (2.2)

(b) Go from S_j to S_{j+1} with *prob.* π_j.

(c) Continue *until* ....

In order to implement the algorithm the phrases in italics in (2.2) must be made specific. The specification of (b) has received the most attention in the literature (see for example Geman and Geman, 1984, Hajek, 1985, Bohachevsky, Johnson, and Stein, 1986). There is little difficulty surrounding the choice in (a) but, as expected, (c) remains the most vexing and obstinate part of the specification and accordingly has been largely ignored in the literature.

One example of π_j is:

$$\pi_j = \min(1, \exp(-\gamma_j[J(S_{j+1}) - J(S_j)])) \tag{2.3}$$

where

$$\gamma_j = \log j/\gamma \ (j \geq 2)$$

(γ has to be specified). Another is

$$\pi_j = 1 \text{ if } J(S_{j+1}) - J(S_j) \leq 0$$
$$\pi_j = 1/2 \text{ otherwise.} \tag{2.4}$$

The specification in (2.4) is related to one used in the Metropolis algorithm (Metropolis et. al. (1953), Bohachevsky, et. al. (1986)). For (2.3) it has been shown that, under reasonable conditions and for γ large enough, the S_j's converge to a global optimum (Geman and Geman, 1984, Hajek, 1985).

The interpretation of the algorithm is simple: If S_{j+1} improves on S_j then with probability 1 go to S_{j+1}; if S_{j+1} is worse than S_j then go to S_{j+1} with probability $\pi_j < 1$. The latter aspect is supposed to permit enough movement to avoid being trapped at a local optimum.

Consider the problem, related to a 2^7 factorial experiment, where

$$T = \text{ the } 2^7 \text{ vertices of the unit cube in 7-space}, k = 8, \sigma^2 = 0,$$
$$R(s,t) = e^{-.5\|s-t\|^2}, \text{ and objective function}, J_{\overline{X}}. \tag{2.5}$$

Figure 2.1 shows the behavior of the annealing algorithm using (2.3). Other problems will induce similar pictures but not always with a successful end, at least in a reasonable amount of time. Use of a fixed and large γ rarely led to convergence in an acceptable length of time. One reason that (2.3) induces a process which takes a very long time to settle to a minimum is that π_j is nearly one when $J(S_{j+1}) - J(S_j)$ is positive but small, a situation typical of a number of problems.

An example where the annealing process failed to reach the optimum in less than an astronomical number of tries was

$$T = 100 \text{ points randomly picked from the square } [0,5] \times [0,5],$$
$$k = 10, \sigma^2 = 0, R(s,t) = e^{-.5\|s-t\|^2}, J = J_{\overline{X}}. \tag{2.6}$$

The presumed optimum had been found by repeating the 'single-excursion' algorithm several times. Despite the failure of the annealing algorithm to converge in the example of (2.6) it did reach a good sub-optimal solution quite early in the run. This is shown in Figure 2.2.

Because exact optimal solutions can rarely be found in reasonable time, we sought algorithms which find good sub-optimal solutions quickly. Such algorithms were also not readily found. By only requiring a 'successful' solution 'most of the time' we were able to obtain the useful algorithm we now present. In so doing we also incorporated consideration of (2.2c), the stopping rule.

We now specify the algorithm; the parameters n_0, δ', α, δ, and M on which it depends are discussed following (2.7).

(a) Pick S_0 at random from T ($\#(S_0) = k$), let $\pi_0 = .7$. (2.7)

(b) At stage $j+1$ pick t at random from $T - S_j$. Find $s^* \in S_j$ such that

$$J(S_j \cup t - s^*) = \min_{s \in S_j} J(S_j \cup t - s).$$

If $J(S_j \cup t - s^*) \leq J(S_j)$ then $S_{j+1} = S_j \cup t - s^*$. Otherwise, $S_{j+1} = S_j \cup t - s$ with probability π_j, $S_{j+1} = S_j$ with probability $1 - \pi_j$.

(c) If $S_{j+1} = S_j$ in (b) do not change stage number but pick t' at random from $T - S_j - t$ and repeat (b) using t' instead of t.

(d) Repeat (c), if necessary, n_0 times, $n_0 \leq N - k$. If, after n_0 tries, no

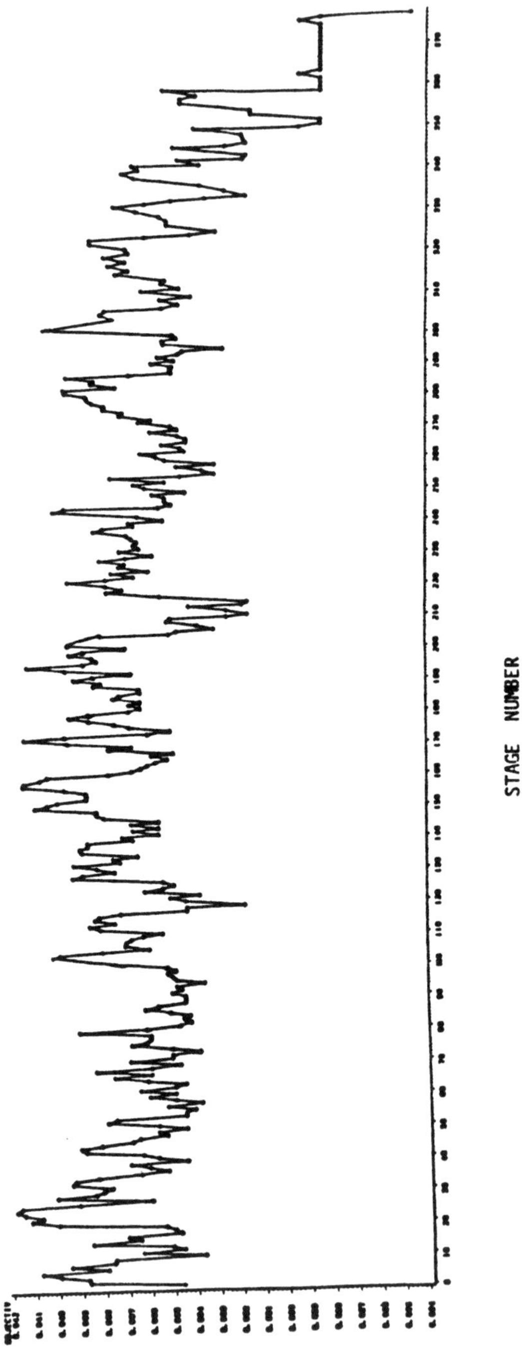

Figure 2.1. Annealing Process: π Depends on Fixed γ

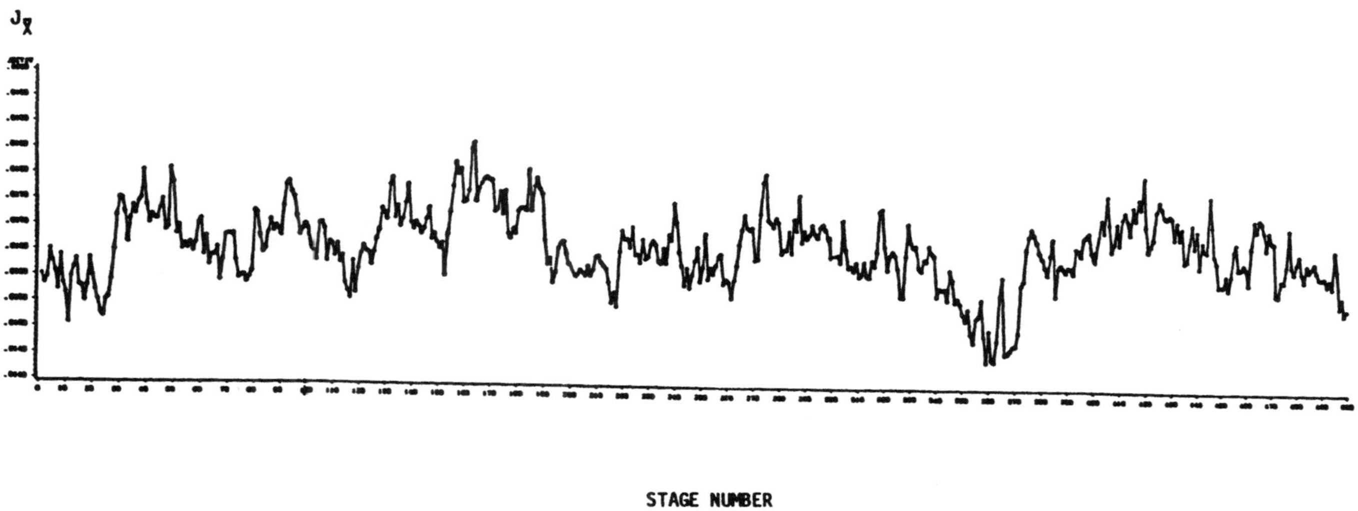

Figure 2.2. Annealing Process: π Depends on Sequential γ

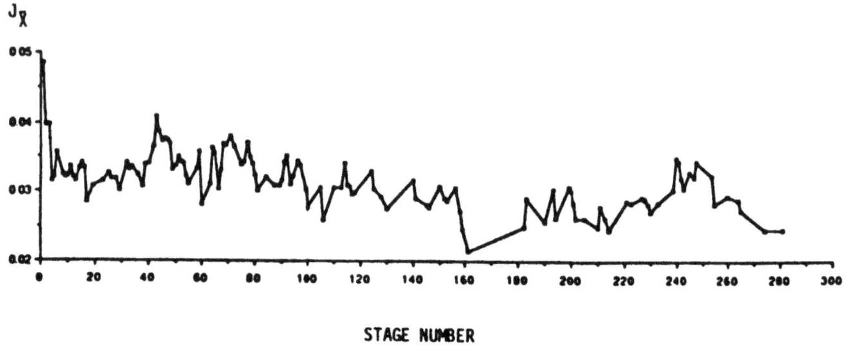

Figure 2.3. Annealing Process: π Depends on α and δ.

movement has occurred take $S_{j+1} = S_j$, $\pi_{j+1} = \min(1, \pi_j/(1-\delta'))$ change the stage to $j+2$ and return to (b).

(e) If $J(S_{j+1}) \leq (1-\alpha) \min_{i\leq j} J(S_i)$ let $\pi_{j+1} = (1-\delta)\pi_j$. Otherwise $\pi_{j+1} = \pi_j$.

(f) Stop if there have been M tries since a change in Π_j.

For the problems tried the behavior of this algorithm did not appear to be affected by α. The choice of δ played a role – large δ's made it difficult to move around after a few downward movements. We chose $n_0 = N - k$ and $\delta' = \delta$. This appeared to cause unnecessary delays; δ''s larger than δ and n_0 less than $N-k$ will reduce the time spent at stage (c), a stage which is mostly 'wasted'. The specification of M is clearly critical for stopping the process. The limited data collected thus far does not point as yet to clear recommendations or conclusions about the choice of M. For many of the examples in Section 3 α=.01, $.3 \leq \delta \leq .5$, and $100 \leq M \leq 250$ worked well.

The algorithm of (2.7) exhibits markedly different behavior than that of (2.3). Figure 2.3, a counterpart to Figures 2.1 and 2.2 for a slightly different but closely related problem, shows the difference: there is a tendency to steadily decline and then to oscillate with little additional improvement.

More investigation is needed to refine the algorithm and to employ it in more challenging situations. Values of $n = 400$ appear manageable and we expect problems with $n = 1000$ and $k = 50$ to be amenable to (2.7).

3. EXAMPLES

The algorithm (2.7) was used in several instances in order to obtain insight into its behavior and to explore the impact of the assumptions about the underlying model. The features we concentrated on were the effects on design behavior of different covariance models, presence of measurement error, alternate criteria, and different types of background spaces. While only a few possibilities were examined, indications of general behavior emerged, as will be seen below.

The covariance structures we selected were two of the traditional spatial covariances: the Gaussian and the exponential. The Gaussian covariance is given by:

$$R_G(s,t) = \exp(-\theta_G\|s-t\|^2), \tag{3.1}$$

and

$$R_E(s,t) = \exp(-\theta_E\|s-t\|) \tag{3.2}$$

defines the exponential covariance. In one dimension these covariance structures produce different asymptotic behavior (see Sacks and Ylvisaker, 1970) due to structural differences: R_G is analytic, but R_E lacks a derivative at the origin. In higher dimensions we also expect differences of behavior but, as indicated in Section 1, there is little in the way of theoretical results. In some examples, however, there are differences in the designs which appear in numerical comparisons but are not readily evident in visual inspection of the designs.

In one class of examples, we defined T to be the regular grid in the unit square with $n = 121$ points and we let $k = 15$ (recall k is the number of points in the design). This choice of k introduces an element of asymmetry. We set $\theta_G = 20$, a scaling which produces a nearest neighbor correlation of $e^{-.2} = 0.8187$. The parameter θ_E was set to be 20/11=1.82 producing a nearest neighbor correlation of $e^{-.182} = 0.8336$. Perhaps a better choice would have been $\theta_E = 2$ which would match the nearest neighbor correlation in the Gaussian case – such a choice should produce equivalent results.

An important difference between R_G and R_E in this example is that R_G is smaller than R_E, especially as the distance between points increases. This explains some of the differences in Figure 3.1 and Table 3.1 where the effect on the objective function J_{Max} (see (1.6)) is examined in the context of no measurement error i.e., $\sigma^2 = 0$. Visual examination of the designs in Figure 3.1a and 3.1b does not detect serious differences but Table 3.1 indicates a strong effect of using the wrong design.

The introduction of measurement error has some small effect on the designs as can be seen by comparing Figure 3.2 (covariance = R_E, objective = J_{Max})

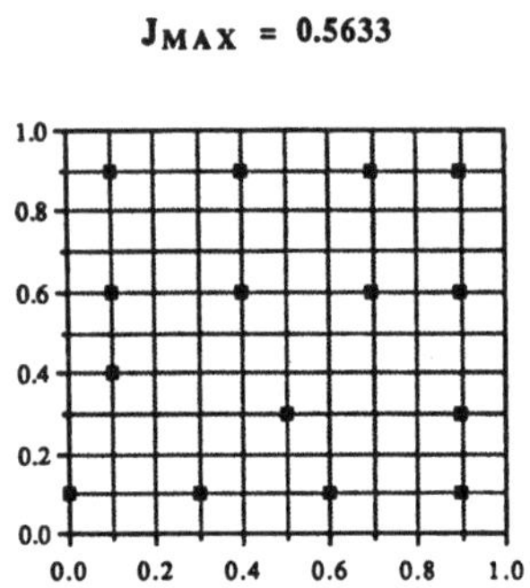

Figure 3.1a. Gaussian Covariance Structure, $\sigma^2 = 0.0$.

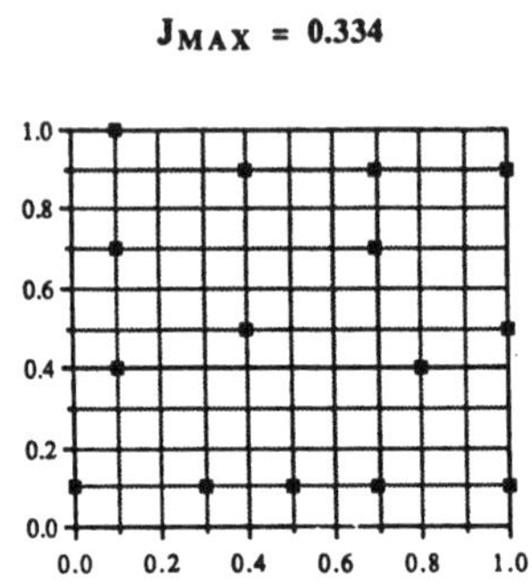

Figure 3.1b. Exponential Covariance Structure, $\sigma^2 = 0.0$.

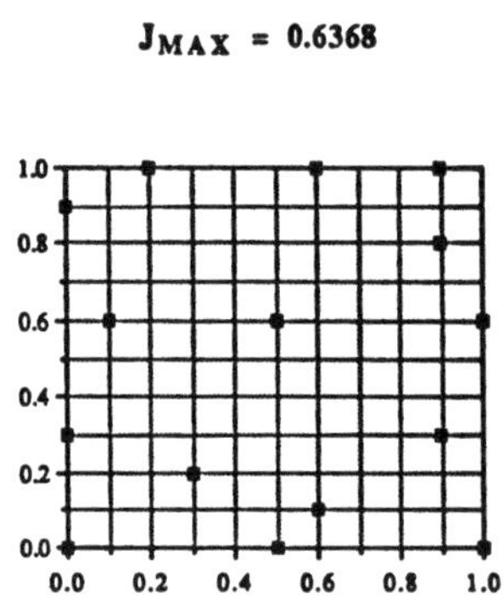

Figure 3.2. Exponential Covariance Structure, $\sigma^2 = 2.0$.

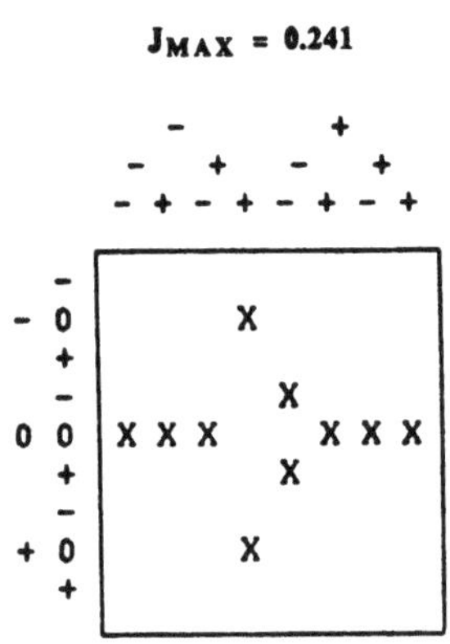

Figure 3.3a. Gaussian Covariance Structure, $\sigma^2 = 0.0$.

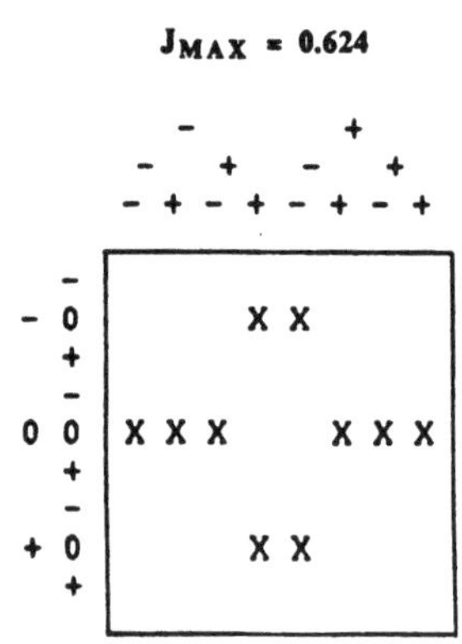

Fig 3.3b. Gaussian Covariance Structure, $\sigma^2 = 2.0$.

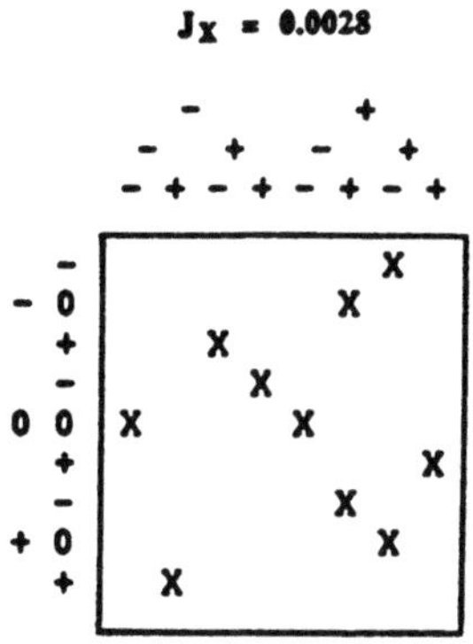

Figure 3.4a. Gaussian Covariance Structure, $\sigma^2 = 0.0$.

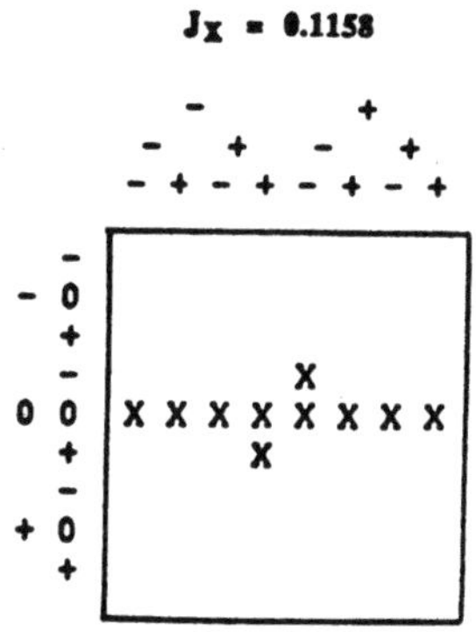

Figure 3.4b. Gaussian Covariance Structure, $\sigma^2 = 2.0$.

Table 3.1

	DESIGNS ($J_{\text{Max}}, \sigma^2 = 0$)	
COVARIANCE	Best for Exponential	Best for Gaussian
Exponential	0.3341	0.5633
Gaussian	0.3996	0.6574

Table 3.2

	DESIGNS (Covariance R_G, J_{Max})	
MEASUREMENT ERROR	Best for $\sigma^2 = 0.0$	Best for $\sigma^2 = 2.0$
$\sigma^2 = 0.0$	0.5633	0.5989
$\sigma^2 = 2.0$	0.8502	0.8502

Table 3.3

	DESIGNS (Covariance R_E, $\sigma^2 = 0$)	
CRITERIA	Best for $J_{\overline{X}}$	Best for J_{Max}
$J_{\overline{X}}$	.09	.11
J_{Max}	.80	.64

with Figure 3.1b. There is an apparent movement toward the boundary in Figure 3.2.

The introduction of the measurement error, and $\sigma^2 = 2$ is a substantial amount of such error, has only a small effect on the prediction error, at least as measured in Table 3.2. A similar pattern holds when R_G is the covariance. Smaller values of σ^2 have correspondingly smaller effects.

The criterion $J_{\overline{X}}$ of (1.7) is a global criterion very different from J_{Max}. It is therefore of some interest that designs constructed for J_{Max} perform reasonably well against the criterion $J_{\overline{X}}$ and vice versa, suggesting a degree of robustness. For example, Table 3.3 shows that the J_{Max}-design used in the $J_{\overline{X}}$-criterion does about as well as the $J_{\overline{X}}$-design does in the J_{Max}-criterion for the covariance R_E.

The setting of factorial designs brings forward a different type of background space T. We took an example where T correspond to a $3^2 2^3$ factorial experiment (2 factors each at three levels, 3 factors each at 2 levels). This T can be identified with the set of 72 points in 5-dimensions with coordinates $(u_1, u_2, u_3, u_4, u_5)$ where $u_i = -1, 0$, or 1 if i=1, 2, $u_i = -1$ or 1 if i=3,4,5. We took $k = 10$ (i.e., there are 10 experimental runs allowed.) We examined the effects of the J_{Max} and the $J_{\overline{X}}$ criteria and the presence of measurement error. The covariance structure was R_G, $\theta_G = 0.1$.

Under the criterion J_{Max}, the introduction of measurement error had little effect on the optimal design structure that had been found for the case of no measurement error. In both cases the designs are highly centralized in that there are few points in the design without a 0 level in at least one of the first two factors.

Figure 3.3 lists the designs found for J_{Max}. The $J_{\overline{X}}$ designs which appear in Figure 3.4 are more strongly affected by σ^2 but are also highly centralized.

4. DETERMINANT OPTIMALITY

We conclude with some comments about (1.8), an objective function in the spirit of the D-optimality criterion of 'standard' optimal design problems. The comments and results reported stem from conversations with Henry Wynn and are connected with the contents of Shewry and Wynn (1986).

The usual calculation of the determinant of a block matrix produces, in the case $\sigma^2 = 0$, the identity

$$\det R = \det R_s \cdot \det G_S \tag{4.1}$$

where G_S is defined following (1.8) and $R_S = R(s, s')$ for $s, s' \in S$. Therefore, in order to minimize $J_{\text{Det}} = \det G_S$ it is enough to maximize $\det R_S$. Since R_S depends only on elements of S this optimization is easier to carry out than the

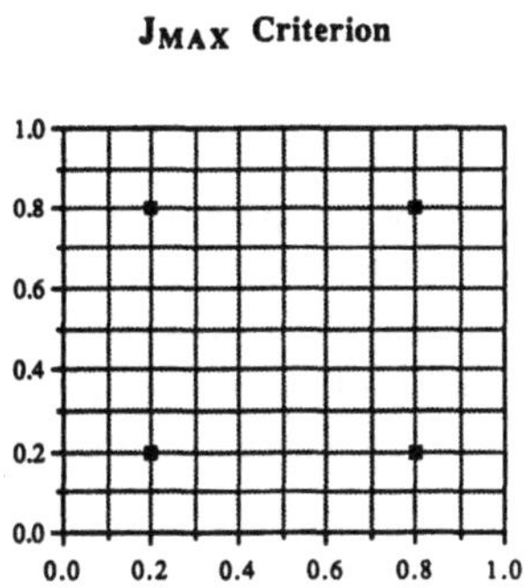

Figure 4.1a. Gaussian Covariance Structure, $\sigma^2 = 0.0$.

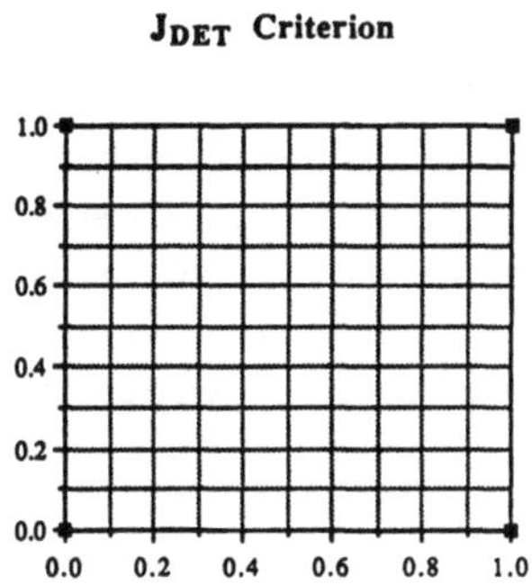

Fig 4.1b. Gaussian Covariance Structure, $\sigma^2 = 0.0$.

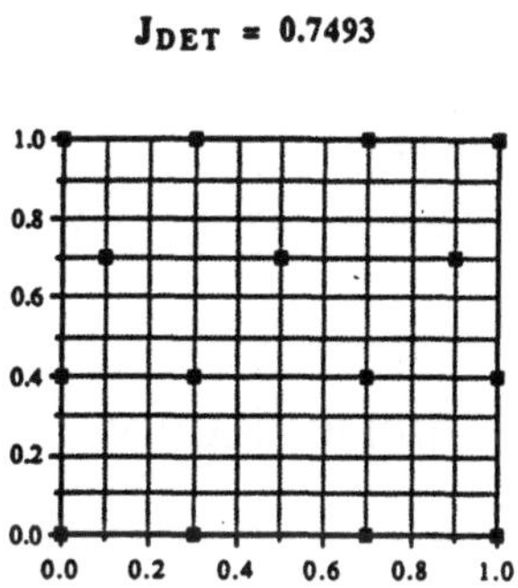

Figure 4.2. Gaussian Covariance Structure, $\sigma^2 = 0.0$.

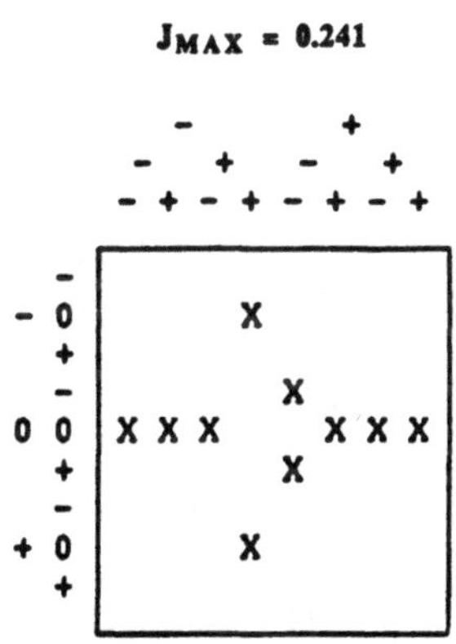

Figure 4.3a. Gaussian Covariance Structure, $\sigma^2 = 0.0$.

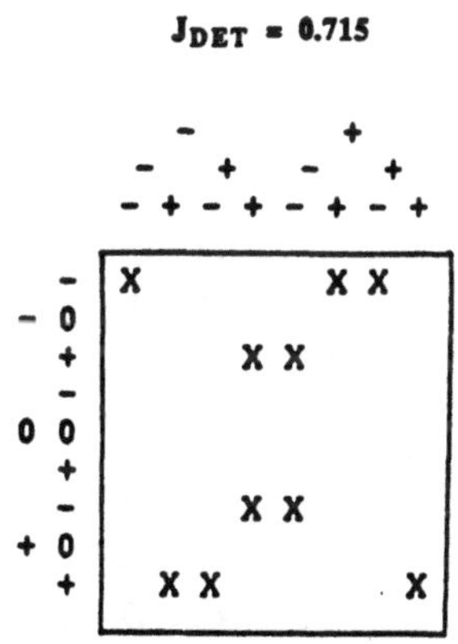

Fig 4.3b. Gaussian Covariance Structure, $\sigma^2 = 0.0$.

ones of earlier sections.

It is surprising and interesting that J_{Det} gives rise to optimal designs which are noticeably different than those of J_{Max}. A number of examples exhibit this characteristic. First consider T, the regular grid of 121 points on the unit square and covariance R_G (see (3.1)). When $k = 4$ it is easy to obtain the optimum designs and as Figure 4.1 shows, there is a great discrepancy. The same holds for $k = 5$ where the optimum designs are obtained by adding one point at the center of the grid to the design for $k = 4$. These results are not the consequence of 'small' k; in Figure 4.2 a similar discrepancy is found when $k = 15$ (compare to Figure 3.1b).

It is apparent that J_{Det}-designs have points closer to the boundary than J_{Max}-designs. This is intuitively explained by the fact that maximizing $\det R_S$ occurs by moving points in S as far apart as possible. In contrast, minimizing J_{Max} suggests placing S so it 'infiltrates' as much of T as possible. When the dimension is large, such as 5 in the case of the $3^2 2^3$ factorial, more boundary points are available and the difference between J_{Det} and J_{Max} is accentuated. In fact, in the factorial experiment (where T is the set of $3^2 2^3$ points and $k = 10$) the J_{Det}-design has no central points, in contrast to the J_{Max}-design which has only central points!

BIBLIOGRAPHY

Bohachevsky, I. O., Johnson, M. E., Stein, M. L. (1986). *Generalized simulated annealing for function optimization.*

Geman, D. and Geman, S. (1984). Stochastic relaxation, Gibbs distributions and the Bayesian restoration of images. *IEEE Transaction PAMI* **6**, 721–741.

Hajek, B. (1985). *Cooling schedules for optimal annealing.* (Manuscript).

Li, K. C. (1984). Robust regression designs when the design space consists of finitely many points. *Ann. Statist.* **12**, 269–282.

Metropolis, N., Rosenbluth, A. W., Rosenbluth, M. N., Teller, A. H., and Teller, E. (1953). Equations of state calculations by fast computing machines. *J. Chem. Phys.* **21**, 1087–1091.

Micchelli, C. A. and Wahba, G. (1979). *Design problems for optimal surface interpolation.* Tech. Rpt. 565, Dept. of Statistics, University of Wisconsin.

Mitchell, T. J. (1974). Computer construction of D-optimal first-order designs. *Technometrics* **16**, 211–220.

Sacks, J. and Ylvisaker, D. (1970). Statistical designs and integral approximation. *Proc. of the 12th Bien. Seminar Canad. Math. Cong.*, 115–136. Montreal.

Sacks, J. and Ylvisaker, D. (1984). Some model robust designs in regression.

Annals of Statistics **12**, 1324–1348.

Sacks, J. and Ylvisaker, D. (1985). Model robust design in regression: Bayes Theory. *Proceedings of the Berkeley Conference in Honor of Jerzy Neyman and Jack Kiefer*, Vol. II (ed. L. Le Cam and R. A. Olshen) Wadsworth, 667–679.

Shewry, M. and Wynn, H. P. (1986). *Maximum entropy sampling.* (Manuscript).

Smale, S. (1985). On the efficiency of algorithms of analysis. *Bull. American Math. Soc.* **13**, 87–122.

Suldin, A. V. (1960). Wiener measure and its applications to approximation methods. *Izv. Vyss. Ucebn. Zaved. Mat.* **6**, 145–158, 165–179.

Wozniakowski, H. (1985). A survey of information-based complexity. *J. of Complexity* **1**, 11–44.

Wynn, H. P. (1970). The sequential generation of D-optimum experimental designs. *Ann. Math. Statist.* **41**, 1655–1664.

Ylvisaker, D. (1975). Designs on random fields. *A Survey on Statistical Design and Linear Models* 593–607. North-Holland, Amsterdam.

Ylvisaker, D. (1985). *Prediction and design.* (Manuscript).

MIX
Papier aus verantwortungsvollen Quellen
Paper from responsible sources
FSC® C105338

If you have any concerns about our products,
you can contact us on
ProductSafety@springernature.com

In case Publisher is established outside the EU,
the EU authorized representative is:
Springer Nature Customer Service Center GmbH
Europaplatz 3, 69115 Heidelberg, Germany

Printed by Libri Plureos GmbH
in Hamburg, Germany